EARTHEN ARCHITECTURE: PAST, PRESENT AND FUTURE

PROCEEDINGS OF THE INTERNATIONAL CONFERENCE ON VERNACULAR HERITAGE, SUSTAINABILITY AND EARTHEN ARCHITECTURE, VALENCIA, SPAIN, 11–13 SEPTEMBER 2014

Earthen Architecture: Past, Present and Future

Editors

C. Mileto, F. Vegas, L. García Soriano & V. Cristini
Universitat Politècnica de València, Spain

CRC Press
Taylor & Francis Group
Boca Raton London New York Leiden

CRC Press is an imprint of the
Taylor & Francis Group, an **informa** business
A BALKEMA BOOK

Cover photo: Adobes drying in the sun in Amantani Island (Peru). Picture by Vegas & Mileto

CRC Press/Balkema is an imprint of the Taylor & Francis Group, an informa business

Typeset by V Publishing Solutions Pvt Ltd., Chennai, India
Printed and bound in Great Britain by CPI Group (UK) Ltd, Croydon, CR0 4YY

Published by: CRC Press/Balkema
P.O. Box 11320, 2301 EH Leiden, The Netherlands
e-mail: Pub.NL@taylorandfrancis.com
www.crcpress.com – www.taylorandfrancis.com

ISBN: 978-1-138-02711-4 (Hbk + CD-ROM)
ISBN: 978-1-315-73687-7 (eBook PDF)

Earthen Architecture: Past, Present and Future – Mileto, Vegas, García Soriano & Cristini (Eds)
© 2015 Taylor & Francis Group, London, ISBN 978-1-138-02711-4

Table of contents

Earthen Architecture: Past, Present and Future – Mileto, Vegas, García Soriano & Cristini (Eds)
© 2015 Taylor & Francis Group, London, ISBN 978-1-138-02711-4

Preface

Sustainability is a word that has monopolised a large number of the scientific debates in a broad range of spheres connected not only with architecture, urban planning and construction, but also with the product market, tourism, culture, etc. Organising a conference on such a broad concept as sustainability, which has unfortunately lost part of its meaning because of repeated, banal use, could look like an attempt to follow the latest fashion and keep up with the times. However, the sustainability proposed as the focal point of the conference is indissolubly linked to vernacular architecture made out of earth or other materials and the lessons this architecture of the past can teach us for the future. The concept of sustainability as it is presented is wide-reaching and covers not only environmental issues but sociocultural and socio-economic questions also. The lessons we can learn from studying vernacular architecture in these three broad spheres are manifold, and can help us not only to further the conservation and retrieval of this architecture already in existence but to rethink new architecture in the light of what we have learned.

In this line of reflection the congress VerSus 2014 | 2nd Mediterra | 2nd ResTapia – International Conference on Vernacular Heritage, Sustainability and Earthen Architecture—was held at the Universitat Politècnica de València on 11th, 12th and 13th September 2014. The main aim of the conference was to discuss and debate the lessons that can be learnt from vernacular architecture to create sustainable architecture today, both for the restoration of traditional buildings and the design and construction of new ones. The conference comprises three important events in a single venue.

The first event is VerSus 2014, which addressed the study of vernacular architecture and the lessons in sustainability it teaches for the future, organised within the frame of the European project: "VerSus: Lessons from Vernacular Heritage to Sustainable Architecture (2012–14)", approved for funding under the European Programme 2000, led by the Escola Superior Gallaecia (Portugal) with the cooperation of the Universitat Politècnica de València (Spain), CRAterre-ENSAG (France), the University of Cagliari (Italy) and the University of Florence (Italy). The congress VerSus 2014 constitutes the closure of this project that represented an important moment of reflection on the themes proposed for discussion with other experts from all over the world. The second event was the second edition of the conference Mediterra – Mediterranean Conference on Earthen Architecture, organised for the first time in 2009 in Cagliari (Italy) by three of the partners of the UNESCO Chair – Earthen Architecture, Building Cultures & Sustainable Development: CRAterre-ENSAG (France), the University of Cagliari (Italy) and the Escola Superior Gallaecia (Portugal). The third event was the conference ResTapia – International Conference on Rammed Earth Conservation, organised for the first time at the Universitat Politècnica de València (Spain) in 2012 with a view to increasing specific knowledge about the restoration of earthen architecture in general and rammed earth architecture in particular.

Given the large number of themes related with these three events, which were satisfactorily brought together under the umbrella of the conference VerSus 2014 | 2nd Mediterra | 2nd ResTapia – International Conference on Vernacular Heritage, Sustainability and Earthen Architecture, on this occasion the international debate addressed five major issues: sustainability concepts in vernacular and contemporary architecture, conservation of urban and rural settlements; documentation and conservation of vernacular architecture; lessons from vernacular heritage for sustainable contemporary architecture; documentation of earthen architecture and proposals for a new architecture in the Mediterranean context (2nd Mediterra Conference); documentation, conservation and proposals of rammed earth and earthen architecture (2nd ResTapia Conference).

The scientific committee was made up of 58 outstanding researchers from 25 different countries from the five continents, specialists in the subjects proposed. All the contributions to the congress, both the abstracts and the final texts, were subjected to a strict peer-review evaluation system by the members of the scientific committee. About 200 papers by 366 authors from 32 countries from the five continents were published, chosen by this process from the over 430 proposals submitted. Apart from the papers, lectures were delivered by two important guest speakers, researchers in the realm of vernacular architecture, José

María Ballester (Director of the Area of Rural Development of the Fundación Botín, Spain) and Marcel Vellinga (Oxford Brookes University, United Kingdom). This constituted an important contribution both to knowledge about vernacular architecture on our planet and the lessons it teaches us and the proposals for the future of architecture at a moment of necessary reflection. All the articles were published in two books. The first of these, *Vernacular Architecture: Towards a Sustainable Future*, contains the texts about the study of vernacular architecture and the lessons it teaches for sustainable architecture, while the second book, *Earthen Architecture: Past, Present and Future*, specifically contains the papers addressing the study of vernacular and historic earthen architecture as a contribution to the sustainable architecture of the future.

The international conference VerSus 2014 received the aegis of: ICOMOS-ISCEAH (International Council on Monuments and Sites – International Scientific Committee on Earthen Architectural Heritage; UNESCO Chair – Earthen Architecture, Building Cultures & Sustainable Development; ICOMOS-CIAV (International Scientific Committee for Vernacular Architecture); PROTERRA – Iberian-American Network on Earthen Architecture and Construction. It also received the institutional support of: IPCE – Instituto del Patrimonio Cultural de España, of the Ministry of Education, Culture & Sport of the Government of Spain; INTBAU-España (International Network for Traditional Building, Architecture & Urbanism – Spain).

The organisation, publication and implementation of the conference were possible thanks to the aid received from the European Union Culture Programme regarding the European Project VerSus (grant nº 2012-2792/001-001 CU7 COOP7), Universitat Politècnica de València, Escuela Técnica Superior de Arquitectura and Instituto de Restauración del Patrimonio of the same university, Conselleria de Infraestructures, Territori i Medi Ambient of the Generalitat Valenciana; companies like ARESPA – Asociación Española de Empresas de Restauración del Patrimonio Histórico, Tarma – Restauración & Patrimonio, Grupo Tragsa, Revista EcoHabitar, IEB-Instituto Español de Baubiologie, KIMIA – Productos y Tecnología para la Rehabilitación; AT Studio; Associazione Nazionale Cittá della Terra Cruda.

Finally we would like to thank all the authors who contributed to the quality, range, diversity and richness of these publications with their articles. We give special thanks to all the partners of the European project "VerSus: Lessons from Vernacular Heritage to Sustainable Architecture" for participating in the conference and their help in spreading the word about it all over the world. We are grateful to the aid of all the members of the advisory committee and the scientific committee for their work throughout the long process of revising the abstracts and papers. And, above all, we thank the organising committee for the complex setting up of the whole conference, the style and language reviewers for their corrections and all the collaborators for their invaluable work in the management and organisation of all the stages of the process

Camilla Mileto, Fernando Vegas, Lidia García Soriano & Valentina Cristini
June 2014

Earthen Architecture: Past, Present and Future – Mileto, Vegas, García Soriano & Cristini (Eds)
© 2015 Taylor & Francis Group, London, ISBN 978-1-138-02711-4

Organization and committees

ORGANIZING COMMITTEE

Camilla Mileto (Chair), *Universitat Politècnica de València, Spain*
Fernando Vegas López-M. (Chair), *Universitat Politècnica de València, Spain*
Mariana Correia, *President of ESG Board of Directors and CI-ESG, Portugal*
Valentina Cristini, *Universitat Politècnica de València, Spain*
Lidia García Soriano, *Instituto de Restauración del Patrimonio—UPV, Spain*
Soledad García Sáez, *Universitat Politècnica de València, Spain*
Paolo Privitera, *Instituto de Restauración del Patrimonio—UPV, Spain*
Maria Diodato, *Instituto de Restauración del Patrimonio—UPV, Spain*
Vincenzina La Spina, *Universidad de Cartagena, Spain*
José Ramón Ruiz Checa, *Universitat Politècnica de València, Spain*

ORGANIZED BY

UPV—*Universitat Politècnica de València, Spain*
IRP—*Instituto de Restauración del Patrimonio, Valencia, Spain*
ESG—*Escola Superior Gallaecia, Portugal*

AEGIS

ICOMOS—ISCEAH—*International Scientific Committee for Earthen Architectural Heritage*
CHAIR UNESCO—*Chair Earthen Architecture, Building Cultures & Sustainable Development*
ICOMOS—CIAV—*International Scientific Committee for VernacularArchitecture*
PROTERRA—*Red Iberoamericana Proterra*

PARTNERSHIP

CRAterre—*Ecole Nationale Supérieur d'Architecture de Grenoble, France*
UNICA—*Universitá degli Studi di Cagliari, Italy*
UNIFI—*Università degli Studi di Firenze, Italy*

INSTITUTIONAL SUPPORT

IPCE—*Instituto del Patrimonio Cultural de España, Ministerio de Educación, Cultura y Deporte, Gobierno de España*
INTBAU—España. *The International Network for Traditional Building, Architecture & Urbanism, Spain*

FUNDED BY

European Union, *Culture Programme. Executive Agency, VerSus Research Project (Grant nº2012-2792/001-001 CU7 COOP7)*
Universitat Politècnica de València, Spain
Escuela Técnica Superior de Arquitectura, UPV, Spain
Instituto de Restauración del Patrimonio, UPV, Spain
Generalitat Valenciana, Spain

ADVISORY COMMITTEE

SCIENTIFIC COMMITTEE

China | Tara Sharma, ICOMOS-ISCEAH, **India** | Thierry Joffroy, CRAterre-ENSAG, **France** | Toshiei Tsukidate, Universidad de Hachinohe, **Japan** | Valentina Cristini, Universitat Politècnica de València, **Spain** | Vicente Más Llorens, Universitat Politècnica de València, **Spain** | Vincenzina La Spina, Universidad de Cartagena, **Spain** | Youcef Chennaoui, École Polytechnique d'Architecture et d'Urbanisme d'Alger, **Algeria** | Yukimasa Yamada, Tokyo Metropolitan University, **Japan** | Zuzana Syrová, National Heritage Institute, **Czech Republic**.

COLLABORATION IN THE PUBLICATION

M. Soledad García Sáez, Maria Diodato, Lidón Castellanos Pla, Davide Covallero, Francisco Javier Gómez Patrocinio, Elena Díaz Rubio, Marta Mestre Sabater.

Earthen Architecture: Past, Present and Future – Mileto, Vegas, García Soriano & Cristini (Eds)
© 2015 Taylor & Francis Group, London, ISBN 978-1-138-02711-4

Conference support

FINANCED BY

With the support of the Culture Programme of the European Union

UNIVERSITAT POLITÈCNICA DE VALÈNCIA

ESCOLA TÈCNICA SUPERIOR D'ARQUITECTURA

IRP INSTITUTO DE RESTAURACIÓN DEL PATRIMONIO
INSTITUTO UNIVERSITARIO DE INVESTIGACIÓN

GENERALITAT VALENCIANA
CONSELLERIA D'INFRAESTRUCTURES, TERRITORI I MEDI AMBIENT

AEGIS

International Council on Monuments and Sites
Conseil International des Monuments et des Sites

UNESCO
Organisation des Nations Unies pour l'éducation, la science et la culture

uniTwin
Chaire UNESCO Architecture de terre

CRAterre ENSAG
école nationale supérieure architecture grenoble

ICOMOS - CIAV
International Committee of Vernacular Architecture
International Council on Monuments and Sites

PRO TERRA

ENDORSER

Leica Geosystems

SPONSORSHIP

TARMA restauración patrimonio

Grupo Tragsa

COLLABORATION

revista EcoHabitar
BIOCONSTRUCCIÓN - PERMACULTURA - TRANSICIÓN

arespa
AR ES PA
ARESPA

IEB Instituto Español de Baubiologie

Kimia
Productos y Tecnología para la rehabilitación

PATestudio

CYRESPA ARQUITECTÓNICO S.L.
Empresa Restauradora

associazione nazionale città della terra cruda

Plenary lectures

Earthen Architecture: Past, Present and Future – Mileto, Vegas, García Soriano & Cristini (Eds)
© 2015 Taylor & Francis Group, London, ISBN 978-1-138-02711-4

Continuity of French rammed earth architecture

H. Guillaud, S. Moriset & D. Gandreau
AE&CC-CRAterre-ENSAG Research Unit, National Superior School of Architecture of Grenoble, Grenoble, France

ABSTRACT: In recent years, more attention has been paid in France to the rammed earth vernacular heritage, considering that it contributes to the cultural and physical identity of a number of territories and represents a substantial economic potential for developing cultural activities. In this approach, the promotion of more eco-responsibility in construction, the valorisation of traditional building cultures and a wider interest on the part of society in natural materials are also taken into account. Thus, with the support of public entities and local governments or through institutions such as regional parks, various policies aimed at revaluing the "pisé" architectural heritage, both rural and urban, have been launched. These include public awareness programmes, professional training courses for architects, engineers, craftsmen and contractors, experimentations on earth renderings and numerous specialized publications. This article briefly presents the French regional rammed earth heritage, focuses on the revival of interest for rammed earth construction with recent architectural achievements and proposes a set of methodological and technical recommendations to revitalize the rammed earth sector in France.

1 INTRODUCTION

Going back two thousand years in time to the Gallo-Roman period, the French rammed earth architectural heritage can undoubtedly be linked to Roman influences that were spread with the colonization of the Empire and the *pax romana*. Archaeological excavations in France provide evidence of the use of mud and wood for the construction of houses, but cannot clearly testify to the building techniques used. Nevertheless, the scientific community globally agrees that wattle and daub walling and straw for thatched roofing were the most commonly used building materials and techniques. More elaborate rural constructions such as large Roman and Gallo-Roman villas that were built in rural and also urban contexts between the 1st and 4th centuries before the beginning of the Middle Ages have shown the use of more massive walling techniques such as cob and unbaked bricks. The *pisé* or rammed earth technique could have been used but there is no evidence of such practices. Evidence of pisé constructions can be found in Provence and in cities such as Carpentras, dating from the 13th century for the earliest examples (De Chazelles and Guyonnet, 2007:109–139). However, it appears that this more massive earthen building technique had progressively been replacing older wattle and daub from the 16th century, as suggested by the famous book *La maison rustique* written by Charles Estienne and Jean Liébault and published in 1564, praising the qualities of the massive earthen construction and referring in particular to its rammed earth walls. In the Dauphiné, this evolution was not only motivated by the will to improve rural living conditions in houses dramatically exposed to fire hazard (wood and thatch), but also by the destruction of the forests to feed local high-consumption industries such as glassmaking or forges (Chancel, 2007:104). We are also aware of the later decisive influence of the French architect and sworn surveyor François Cointeraux (1740–1830)—also recognized as "professor of rural architecture"—and other authors (Goiffon, 1772; Abbé Rozier, 1796; and Rondelet, 1840) in the modern promotion of a new type of *pisé* to improve the rural habitat from the end of the 18th century to the mid-19th century. Thus we can say that the majority of the current French rammed earth heritage observed dates from the end of the 18th and to a great extent, the 19th centuries.

2 THE FRENCH PISÉ REGIONAL HERITAGE

2.1 *The main regional areas*

Several regions in France are blessed with an exceptional rammed earth architectural heritage. This is more specifically the case for the north Dauphiné (between Grenoble and Lyon, Isere department), the Saône and Rhône river valleys (from Tournus to Avignon), the Bresse and Dombes (north-east of Lyon, AIN department), Auvergne (Centre of France) with the territory of Livradois-Forez (around Montbrison) and in the Monts Du Lyonnais (around Feurs), but also, though less known, the south-west region of France in Midi-PyrénéeS, around Toulouse (Klein, 2007:157–170). This heritage displays a large range of building typologies, including farms and barns, fortified houses, wind and water mills, manors and burgess castles, churches and also industrial buildings (textile industry, in the Dauphiné).

2.2 *Two brief case studies of rural pisé heritage*

The French pisé architectural heritage being so vast from geographical, typological and technological points of view, we cannot cover all aspects of it. We present two regional examples that we consider representative: typical farms of the Bresse and north Dauphiné regions where more than half the vernacular heritage is built in *pisé*.

In Bresse, a territory located north-east of Lyon and extended on the north to the boundaries of Burgundy, to Savoy on the south-east and to the Jura on the east, the rural architectural heritage is typical and "born of earth and wood" (Bruno, 2005). Indeed, two types of farms can be observed, both wooden structures filled with daub—and, from more recent times, with burnt bricks also known as *colombages* (half-timbered) and in *pisé*. This is a typical and particularly excellent example of what the local builders, carpenters and masons were able to achieve just using locally harvested materials. Looking at this architecture, it seems that only a few changes have occurred over a long period. This is explained by a long history of geographic isolation of the territory which had been cut off from the main commercial channels (Saône and Rhône valleys) and due to full autonomy of subsistence thanks to the prosperous local culture of cereals (maize, wheat, rye and barley), animal husbandry (cattle and poultry) and pond fishing (south of Bresse, in connection with Dombes where the fishing economy is dominant). All this gave rise to large farms where housing and producing were closely associated under the dependence of old abbeys and lordships. Most of the time the residential part and all annexes of these farms were organized around a large square or rectangular yard with one or two accesses allowing inhabitants and animals to move freely in and out. The Bresse farms are also very typical with their four-sided roofs in canal tiles (that replaced the thatched roofs and locally-produced hollow tiles in the 19th century), with shallow slopes (maximum gradient of 25%). The characteristic feature of these roofs is shown in their famous bona fide and emblematic *sarrasine* (Saracen) chimneys with their square, vaulted or polygonal mitres in burnt bricks. The recently restored *Domaine des Planons*, which now houses the Museum of Bresse, is an excellent example of the building art in wood, daub and pisé (Fig. 1).

Figure 1. *Domaine des Planons*, Bresse region (CRAterre).

Figure 2. Old Barn in Louisias, Charavines, Dauphiné region (CRAterre).

The typical pisé rural house of the north Dauphiné is totally different from the Bresse one but is also very characteristic. We can define its morphology as a cubic shape, with openings on its four façades and caped with a sloping four-sided roof covered with scale tiles. In accordance with local geology, earth is associated with limestone rubble, sandstone for the basements and frames of the doors and windows. More recently, at the beginning of the 20th century, these stones were replaced by concrete. There are very few old barns still covered with thatch, as can be observed in the region near the lake of Paladru (Fig. 2).

2.3 *The urban pisé of the city of LYON*

The pisé was also used for the construction of houses and housing blocks in ancient urban settlements as was previously pointed out in our introduction (Ibid. De Chazelles et Guyonnet 2007). Today several large villages and small market towns in Auvergne (territory of Livradois-Forez), such as MarSac-en-Livradois or Montbrison, but also big cities such as Lyon, provide plentiful evidence of *pisé*. In Lyon, this means of construction was prohibited after extensive flooding of the Saône and Rhône rivers in 1840, which destroyed many pisé houses in the peninsula (between the two rivers and at their junction) and in all areas prone to flooding. A recent global inventory of the existing *pisé* architecture in the city of Lyon *intra muros* and in three communities of Greater Lyon (Alex, 2012) shows that pisé housing is still abundant in the districts of Croix Rousse (Centre of the city), Saint Just, Tassin La Demi Lune and Vaise (Fig. 3).

Most of these houses are one—to three-storey buildings. Very little testimony of François Cointeraux's works still exists in the districts of Croix Rousse and Vaise and some of them are currently under threat of destruction for speculative reasons.

Because over the last fifteen years the vast majority of this regional *pisé*-built heritage has been the victim of bad restoration practices using concrete or hollow concrete blocks with cement plastering, numerous actions in terms of public sensitization and professional training to introduce more compatible materials and practices are being implemented.

Figure 3. Rammed earth buildings in Croix Rousse, Lyon (CRAterre).

3 THE RAMMMED EARTH REVIVAL

In all the regions of France where the rammed-earth architectural heritage is present, the number of restoration and new buildings projects using this technique is increasing.

3.1 *In Rhône-Alpes*

A first step towards the revival of earthen architecture was taken in 1985 with the inauguration, in the new town of l'Isle d'Abeau near Lyon, of the first urban eco-housing district in Europe built from earth: the *Domaine de la terre*. This pioneering achievement of sixty-four social housing units built with earth (*pisé*, compressed blocks and earth and straw) has had a very positive impact on the attitude of building professionals towards earthen architecture in the Rhône-Alpes region. The scale of the operation has raised awareness both in rammed earth property owners, building professionals (craftsmen, companies, contractors, consulting firms, etc.), local authorities and prescribers (politicians, real estate developers, architects and engineers, etc.).

Almost thirty years later, experts gathered at a seminar on the rehabilitation of rammed earth in Rhône-Alpes organized at the Grands Ateliers de l'Isle d'Abeau observed significant progress over the past three decades (Rakotomamonjy, Moriset and Gandreau, 2013). This includes an increasing demand for the rehabilitation of rammed earth buildings, an increasing number of professionals qualified to meet this need, better recognition of the social, economic, environmental and cultural values of rammed earth architecture and the development of more economically affordable advanced techniques. However, they noted that much work remains to be done to revitalize the rammed earth sector in the Rhône-Alpes region and that "the obstacles take several forms: administrative, institutional, technical, financial, organizational and cultural" (ibid., p. 4).

To tackle this situation, they have particularly recommended supporting new successful operations that could serve as references, such as the recent rehabilitation of barns in Isère (one in the village of Vinay by the SCOP Caracol (cooperative and participatory society) and one in the town of Nivolas-Vermelle, by Heliopsis (a specialized contractor in pisé construction) or new constructions such as a house in the town of Corbel in Savoy, designed by architect Thomas Jay and produced by the same SCOP Caracol and also the primary school in Veyrins-Thuellin, in Isère, designed by architects Milena Stefanova and Marielle Bruno, with Vincent Rigassi as *pisé* consultant and conducted by the company Heliopsis for the parts

Figure 4. Primary school of Veyrins-Thuellin, in *Isère* (CRAterre).

Figure 5. European Conservatory of soil samples in Orleans (CRAterre).

built from rammed earth (fig. 4). In 2013 these two last examples won the first national award for earthen architecture (CRAterre-ENSAG, Asterre, Ecologik, 2013).

3.2 *In France*

Beyond the Rhône-Alpes region, the several-times awarded *Domaine de la terre* (1987: UN-Habitat: as an "exemplary pilot operation of international concern"/2007: Region Rhône-Alpes: as a "treasure of sustainable development") appears to have had numerous repercussions on the renewal of rammed earth in France. Projects emerging on a large part of the French territory now show a certain degree of architectural maturity, taking into account the requirements of sustainable development from various aspects: economic, social, cultural—and foremost, environmental.

Progress is also remarkable from a technical implementation point of view, with improved processes that make the work more efficient and financially competitive. As the list of these projects cannot be exhaustive, we refer to the first national award for earthen architecture. Among the twelve winners, in addition to the two projects already referred to for the Rhône-Alpes region (the house in Corbel and the school in Veyrins-Thuellin) six other projects are in rammed earth:

- the rural space for neighbourhood services near Marsac-en-Livradois (first prize);
- the European Conservatory of soil samples in Orleans (fig. 5);
- the centre for the interpretation of archaeological heritage in Alsace;
- Kiethon, the day care space near Rennes for autistic people;
- the sports centre on the Kerlan University campus in Rennes;
- the *Hôtel de la forge* in a village in Auvergne.

At the same time, even if they are less publicized, it is important to recognize the large number of rehabilitation projects for existing rammed earth houses throughout the country.

4 RECOMMENDATIONS FOR SAVING THE RAMMED EARTH HERITAGE

Safeguarding this heritage and revitalizing the rammed earth sector requires the mobilization of many actors. Some entities are already playing this unifying role: Natural Parks, the Councils for Architecture, Town Planning and Environment (CAUEs), the Rhône-Alpes region, the Ministry of the Environment and some municipalities, universities or research centres such as CRAterre. Several actions to promote this heritage, train people and federate stakeholders are also multiplying. But the task is arduous because professionals need to change their working habits and a few negative preconceived ideas about earth should also be wiped out. In the next part we will examine strategic considerations aiming at creating a favourable environment for saving the rammed earth architectural heritage and encouraging the revival of the sector.

4.1 *Inventory*

An inventory of the rammed earth heritage and its outstanding values has already been started, but it still remains insufficient. More effort needs to be made to display the huge importance of this heritage in our region and especially to avoid the systematic destruction we witness in LYON city centre. Earthen architecture appears to be less documented and less known in our territories than other vernacular typologies.

Figure 6. *Grains d'Isère* Festival at its 10th session in 2012. CRAterre students building a cafeteria from rammed earth (CRAterre).

The inventory of skilled craftsmen and competent architects is another necessity. It is not easy for town councils or individuals to find the right professionals to provide reliable advice on the choice of materials or design guidelines on rammed earth buildings. Updated directories of all professionals with their skills (conservation, design or construction) are required.

4.2 *Promotion and awareness*

The current trend towards healthier architecture facilitates the awareness of the assets of earthen architecture. Local governments now have a different view of rammed earth and understand its many benefits. It is now generally admitted that earth is not a poor or an old material, but in actual fact a valuable resource for the future. As from 2013, a national award for earthen architecture is presented annually to promote contemporary earthen architecture in France. This award helps promote new projects and encourages building owners to engage in new rammed earth constructions. To better promote contemporary architectural production, a unique website showing all new projects should be created, since this information is currently dispersed. Site visits during and after construction could also be organized to inspire reluctant developers. It is an established fact that awareness works better when people exchange knowledge and advice *in situ*. We shall also educate people to uphold their local architecture. This should start as early as possible with educational activities in primary schools. Year by year, we observe an increasing amount of festivals, exhibitions, conferences and information sessions, during some of which booklets or leaflets are distributed. This effort should be spread and should also cover those municipalities where rammed earth could be exhibited, even if the municipalities themselves are not seeking this particular information.

One of the key activities in the region is the annual *Grains d'Isère* festival in Villefontaine, which drew 1,300 people to its 11th session in 2013 (Fig. 6). It brings together professionals, students and researchers from all over the world who share their knowledge with the public around multiple activities related with earth architecture. During the festival in 2013, 200 tons of earth were used in different ways; in addition, conferences, activities and exhibitions were held. Having experimented for themselves, participants left the festival convinced of the benefits of using earth in their projects.

In terms of promotion, the exemplary actions of the Livradois-Forez Park should be mentioned. This institution has edited a book explaining the heritage values of rammed earth and encouraging the revival of the sector by showing contemporary examples in the region. One of the most successful activities of the Park is to support municipal councils wishing to organize participatory construction or conservation projects using earth. The interested councils propose projects and mobilize the population while the Park provides professionals to train and supervise the participants. The Park has also produced an exhibition and set up a well-documented website (www.parc-livradois-forez.org). Resources on rammed earth are constantly multiplying: publications, web sites and documentaries, but they are absent from the conventional building sector. Material merchants do not provide information on rammed earth. Straw, hemp or sheep's wool eventually find their way into the Do-It-Yourself shops, but earth does not, and is thus still outside the trade circuits. It is therefore necessary to find a way to make the information on rammed earth available in these circuits, even if earth does not become a commercial product.

4.3 *Training*

Reliable professionals on rammed earth are still too few to maintain the local heritage in good condition. Yet although the skills to restore rammed earth exist, this knowledge is barely accessible and the transmission of skills does not flow smoothly between professionals. Prescribers and users still face difficulties in making the wisest choices to respect the nature of earth materials.

In recent years, vocational training institutions such as AFPA (national association for professional training of adults) have been offering training modules on rammed earth for good practice education of young masons. The French architecture schools also tend to provide a little more space on vernacular building materials, including earth, every year. However we are still far from the idyllic situation wherein all professionals (architects, engineers and builders) receive the minimum knowledge required to deal with rammed earth. This calls for further efforts to develop training and to diffuse educational materials.

4.4 *Support to craftsmen*

Even when they are interested in earth, craftsmen feel reluctant to embark upon rammed earth projects because they do not feel they are competent enough and fear they may put their companies at risk. It is also difficult for them to set aside their professional activities to follow training sessions. Short technical courses enabling them to experiment and upgrade their knowhow on specific issues without upsetting their busy schedules are necessary. Another obstacle to the popularity of rammed earth today is the lack of standardization that deters both craftsmen and insurers. However, the recent publication of professional rules on earth coating and the preparation of several ATEXs (experimental technical advice) facilitate the creation of new projects and illustrate that rammed earth is a technique with a bright future.

REFERENCES

Alex, D. 2012. *Petit Guide des architectures en pisé à Lyon*. Grenoble: CRATerre éditions, 38 p.

Bruno, A. 2005. *En Bresse, construire et habiter, dedans, dehors*. Ed. Musées des pays de l'Ain, Musée de la Bresse, domaine des Planons. 59 p.

Chancel, D. 2007. Pans de bois et murs de terre en Dauphiné. In Guillaud, H., De Chazelles, C.A. et Klein, A. (eds), *Les constructions en terre massive pisé et bauge*. Montpellier: éditions de l'espérou, 328 p., p. 104.

Clair, E., Court J., Moriset, S., Misse, A. 2011. *Rénover et construire en pisé dans le parc naturel régional Livradois-Forez*. Saint-Gervais-sous-Meymont: éditions Maison du Parc Livradois-Forez.

CRAterre-ENSAG, Asterre, Ecologik, 2013. Communiqué de presse: premier prix national des architectures en terre crue. In *Festival Grains d'Isère, du 29 mai au 02 juin 2013*. Grenoble: CRATerre-ENSAG, 2013. p. 26.

De Chazelles, C.A. & Guyonnet, F., 2007. La construction en pisé du Languedoc-Roussillon et de la Provence, du Moyen-Âge à l'époque moderne. In Guillaud, H., De Chazelles, C.A. et Klein, A. (eds), *Les constructions en terre massive pisé et bauge*. Montpellier: éditions de l'espérou, 328 p., p. 109–139.

Klein, A. 2007. La construction en pisé procédant par banchées appareillées en Midi-Pyrénées. Fin XVIII°-début XX°s. Contribution à la reconnaissance des différentes techniques. In Guillaud, H., De Chazelles, C.A. et Klein, A. (eds), *Les constructions en terre massive pisé et bauge*. Montpellier: éditions de l'espérou. p. 157–170.

Rakotomamonjy, B., Moriset, S. et Gandreau, D. 2013. La réhabilitation du pisé en Rhône-Alpes, quelles stratégies? In *Rapport du séminaire du 30 mai 2013 aux Grands Ateliers de l'Isle d'Abeau. Juin 2013*. Grenoble: CRATerre éditions, 113 p.

Lectures

Earthen Architecture: Past, Present and Future – Mileto, Vegas, García Soriano & Cristini (Eds)
© 2015 Taylor & Francis Group, London, ISBN 978-1-138-02711-4

A web map for Italian earthen architecture

M. Achenza
DICAAR—Università degli Studi di Cagliari, Italy

C. Cocco
Associazione Nazionale Città della Terra Cruda, Samassi, Italy

ABSTRACT: Italian National Association of Earth Built Towns, an association counting 38 municipalities of 4 regions, has been founded in 2001 and since then has encouraged, promoted, sustained a huge variety of actions in favor of the diffusion of the knowledge about earthen architecture in Italy. These actions include publications, seminars, conferences, national and international networking, restoration and conservation projects in cooperation with Universities and enterprises. An important investment of energies that helped earthen architecture gain a dignity and an appreciation that was not known just fifteen years ago. The need to make these information available has become a priority in the very last years, encouraging the Association, in cooperation with the University of Cagliari—Faculty of Engineering and Architecture, to reverse into a web site a mapping of all activities related to Earthen architecture in the Country. This includes a list of academies and institutions that are working on the topic, producers and enterprises, professionals and owners. The mapping of Italy is being coordinated together with German (Weimar University) and French (CRATerre) partners in order to start an European data collection on earthen architecture. By now only the German site is fully available, being the Italian and the French almost ready to be put online, as well. This initiative is going to be an important contribution to the dissemination of correct information on the subject and finally giving the real dimension of the significant amount of people and work involved.

1 INTRODUCTION

The Italian National Association of earth built towns has represented in the past ten years an important reality in the context of earthen architecture in Italy. It has encouraged, promoted, and sustained a huge variety of actions in favor of the diffusion of the knowledge about earthen architecture in general and in Italy in particular. These actions include publications, seminars, conferences, national and international networking, restoration and conservation projects in cooperation with Universities and enterprises. A huge amount of work and data collection has been in this way collected though years and the need to make it available to external bodies has become a fundamental need. This is the reason why cooperation between University of Cagliari and the Association was started in order to develop a web tool where the data were organized and diffused to public.

2 AIM OF THE RESEARCH

The research concerns the construction of an information system that provides a nationwide overview of policies, initiatives and actors in earthen building construction.

In countries like Germany, France, New Zealand, Peru, Australia, earth construction is strongly regulated and supported by appropriate policies and actions. In Italy, despite the long-standing debate on the reintroduction of the use of this material in the building process, we often fail to fill the knowledge and regulatory gaps and therefore to remove the constraints that limit its use. Based on these considerations, the research aims to fill a real need of the "Italian earth construction community," formed not only by professionals and entrepreneurs but also by citizens potentially interested in this subject, through a data collection that includes the different experiences directly or indirectly related to the use of earth based materials in Italy.

The creation of an information system together with the offer of a comprehensive and updated knowledge framework, should contribute to disclose information. This process is useful to fully recognize the strategic importance of earth as a building material. It will ensure its fair acknowledgement of value not only in the aspect of recovery and enhancement of the existing heritage, but also as a functional building material and suitable

Table 1. Initials/contacts list.

Initials	Contact
AC	Andrea Cecconi
CR	Chiara Robboni
CC	Chiara Cocco
MLG	Maria Luisa Germanà
GB	Gaia Bollini
IP	Isidoro Parodi
MA	Maddalena Achenza
GC	Gianfranco Conti
SB	Sergio Sabbadini
RC	Rosario Chimirri
RQ	Raimondo Quaresima
LR	Luisa Rovero
UT	Ugo Tonietti
GLP	Gian Luigi Prati
FP	Francesco Poli
GP	Gloria Pasero

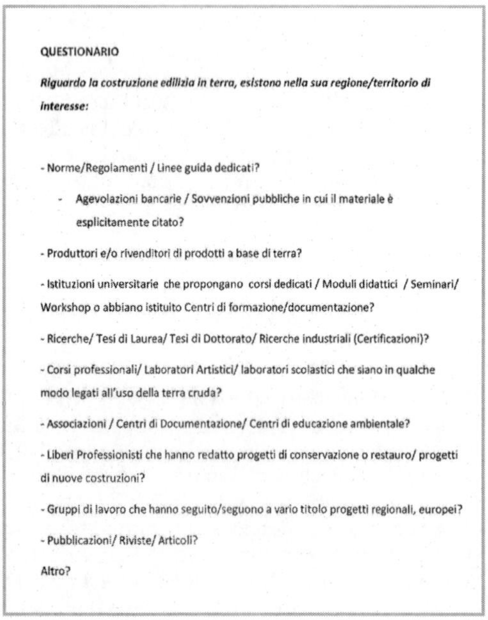
QUESTIONARIO

Riguardo la costruzione edilizia in terra, esistono nella sua regione/territorio di interesse:

- Norme/Regolamenti / Linee guida dedicati?
    - Agevolazioni bancarie / Sovvenzioni pubbliche in cui il materiale è esplicitamente citato?
- Produttori e/o rivenditori di prodotti a base di terra?
- Istituzioni universitarie che propongano corsi dedicati / Moduli didattici / Seminari/ Workshop o abbiano istituito Centri di formazione/documentazione?
- Ricerche/ Tesi di Laurea/ Tesi di Dottorato/ Ricerche industriali (Certificazioni)?
- Corsi professionali/ Laboratori Artistici/ laboratori scolastici che siano in qualche modo legati all'uso della terra cruda?
- Associazioni / Centri di Documentazione/ Centri di educazione ambientale?
- Liberi Professionisti che hanno redatto progetti di conservazione o restauro/ progetti di nuove costruzioni?
- Gruppi di lavoro che hanno seguito/seguono a vario titolo progetti regionali, europei?
- Pubblicazioni/ Riviste/ Articoli?

Altro?

Figure 1. Questionnaire template.

for the most modern and innovative construction techniques. With these targets we have applied a web-GIS application for the divulgation of data and information relating to actors and initiatives that characterize the area of earth-building.

3 PROCESS OF DATA COLLECTION

During the first phase of data collection, experts in the field were contacted, that for a variety of activities (publications, project interventions of restoration or new construction, research laboratories, etc.) have excelled in the Italian scene. After the assembly of the contact list (Table 1), a questionnaire (Fig. 1) was submitted to them by e-mail or through telephone conversation, in which it was asked to indicate which are the main actors and activities in the field of earth building construction in the regions of competence, subdivided by categories.

In order to better focus the research it was decided to limit the data collected to the period January 2009–January 2012. On the basis of the processing of responses to the questionnaire and the data independently acquired through research on the web, the material has been catalogued by associating to each category a brief presentation that defines the contents.

4 ANALYSIS OF THE RESULTS BY RESEARCH CATEGORIES

4.1 *Rules, regulations and guidelines*

The legislation regulates directly or indirectly the aspects of the conservative restoration of earthen buildings, not taking the material for what concern its use in new construction into account. In fact, the Italian Building Norm in force never mentions expressly the material earth. Despite not prohibiting the use of this material *a priori*, it requires however that the bearing structures must be made of materials capable of a minimum mechanical resistance to compression of 5 MP that in the case of handmade earth elements is difficult to acquire, if not with very accurate controls in the production phase of the single elements and in the construction phase.

In addition, both for new buildings and for the renovation of existing ones, in order to meet the energy parameters required by the Decree of the President of the Republic 2 April 2009, n. 59, although the material has very interesting intrinsic performance characteristics, it is necessary to act on the building envelope, destroying as a matter of fact the natural properties of insulation given by earth and putting at risk the health of the masonry itself.

With regard to the regional rules, only the Piedmont and the Abruzzo Regions have legislated specifically on the subject with rules that protect the conservation and promote enhancement of earth construction.

4.2 *Bank subsidies and public grants in which the material is explicitly mentioned*

Although numerous banking facilities are provided for green and environmental sustainable building, none of these refers explicitly to the material earth.

Regarding public grants the regions of Piedmont and Abruzzo with their respective regional laws have instituted contributions to local authorities to undertake a census of the existing buildings and to individuals for restoration works.

4.3 *Producers and retailers of earth based products*

Manufacturers of earth products are almost exclusively located in Sardinia and in Northern Italy. In Sardinia, where the predominant traditional construction technique involved the use of adobe, the production concerns adobe and plaster mixtures. In Northern Italy, instead, enterprises manufacture also other types of products such as air adobe, floor and pisé ready mixes, insulating panels, etc.

In addition to the producers making direct sales, there are companies that deal exclusively with the marketing of earth building products, targeting both domestic and foreign market.

4.4 *University institutions offering dedicated courses, teaching modules, seminars, workshops or established training and documentation centers*

There are several Universities that within their Departments of Architecture and/or Engineering include earthen architecture in their educational offer and their research. Among others, the University of Cagliari, the only Italian Unitwin Partner of the UNESCO Chair on earthen architecture with his one-semester dedicated course, and the University of Florence, that also established a specific module in an integrated course of the master degree program.

However, courses offering teaching modules are becoming more numerous. It is the case of the Polytechnics of Milan and Turin, the School of Architecture of Pescara, Genoa, Palermo and Naples and the University of Calabria.

4.5 *Research, thesis, doctoral dissertation and industrial research (certifications)*

In the years 2009–2012, numerous dissertations on the subject of earth construction have been developed. These works deal with different aspects, such as the restoration and enhancement of existing buildings, the use of the material in green building, the chemical and physical tests on the material, not least the historical and archaeological area.

The research on the topic earth has also been addressed in two doctoral dissertations that have focused on the analysis of degradation and of the structural restoration of the building elements, in Mediterranean sites where earth has been used as a building material since ancient times.

4.6 *Professional courses, artistic workshops or school and workshops that are in some way related to the use of raw earth*

In the three-year analyzed period the training activities on construction techniques and recovery of earthen buildings have been implemented mainly in northern Italy and Sardinia. In the experiences analyzed, the training has been promoted through the fulfillment of appropriate courses coordinated by local authorities, associations of different types, by producers of earthen materials and by individual practitioners. Some courses are aimed at general users without specific skills, others to technical-professional training of operators, public and private. The educational use of the material has also involved the youngest, with the participation of primary and secondary schools, where interactive workshops were organized in order to promote the dissemination of knowledge on the subject.

Ultimately, like earth is used not only as a building material, but also as a raw material for artistic creations, several courses which use earth for aesthetic purposes are dedicated to this second aspect.

4.7 *Associations, documentation centers and environmental education centers*

The National Association "Associazione Città della Terra Cruda" aims to coordinate institutional initiatives and projects for the enhancement of the historical heritage of the municipalities that are part of it, and for the creation of productive activities that can contribute to include earthen building products on the market. The documentation centers and associations who are focused exclusively on earth are limited. However, centers and groups that include it within broader topics, such as environmental education, solidarity in favor of disadvantaged populations, healthy living, sustainable building, and so on, are much more numerous.

4.8 *Professionals who drafted conservation, renovation or new construction projects*

The investigation on liberal professionals who choose earth as a material for construction of traditional and contemporary works, and that deal with the restoration and conservation of built heritage, reveals how their territorial location is influenced mainly by two factors. On the one hand, by the proximity to centers of academic education that into their educational offer include courses related to the construction and restoration of earthen construction, such as the Polytechnics of Turin and Milan and the University of Cagliari; on the other hand, they rely on the presence of areas of greatest concentration of earthen buildings.

4.9 *Working groups that have followed or are following regional or European projects in various capacities*

There are several working groups with the task of carrying out projects of development cooperation, training and research, at regional and international level, supported in various ways by different local authorities, universities and free associations in cooperation with similar institutions abroad.

5 PROCESS OF DATA MANAGEMENT

5.1 *Web mapping tool to collect and manage data*

Web mapping is the process of designing, implementing, generating and delivering maps on the World Wide Web and its product. A web map has to deal with interactivity, usability and multimedia issues, you can interact with it, zoom in and out, pan around, change what you are looking at, etc. The use of web-GIS tool for data management allows their representation on a geographic basis, this make the analysis and interpretation of the information gathered more immediate, accessible and linked to local contexts.

5.2 *Database creation*

The data collected have been subjected to analysis and synthesis and then coded in spatial database and processed in a GIS environment (in particular, have used the following software Microsoft Excel, Microsoft Access and ESRI ArcGIS).

The structure of the data produced is represented here by the modeling language UML (Unified Modeling Language), which is the industrial unified standard language in the in the design and object-oriented programming community.

The Class Diagram (Fig. 2) allows you to describe the database in terms of the types of entities, with their characteristics and possible relationships between these types. Like it can be seen in the scheme (Fig. 3), each class is represented by a rectangle that is divided into two or three sections. In the first section the class name is entered, the attributes are placed in the second and in the third (not relevant in this case study) there are indicated the operations that the objects of the class can accomplish.

5.3 *Geo-referenced data*

Geocoding is the process of converting addresses into geographic coordinates, which you can use to place markers or position the map. The information regarding the various categories of descriptive data of the initiatives and actors in the field of raw earth has been geo-referenced through geocoding services online. The .kml file with the spatial information has been processed in the desktop version of the software ESRI ArcGIS and join with the attributes database. In this way the attributes thematic/descriptive have been associated with geometricospatial ones, which also specify the location in the local contexts.

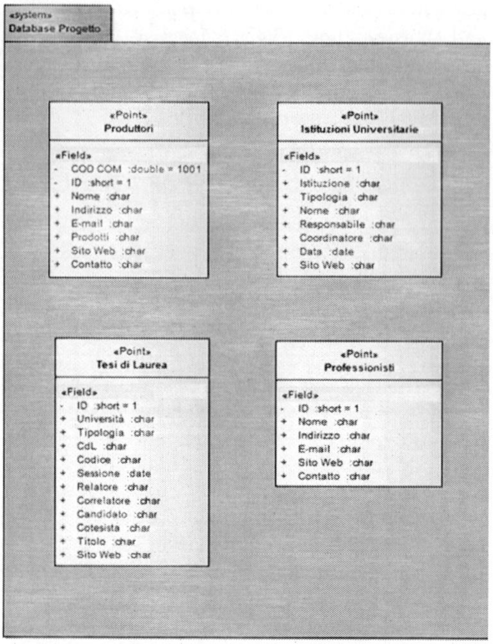

Figure 2. Database UML class diagram.

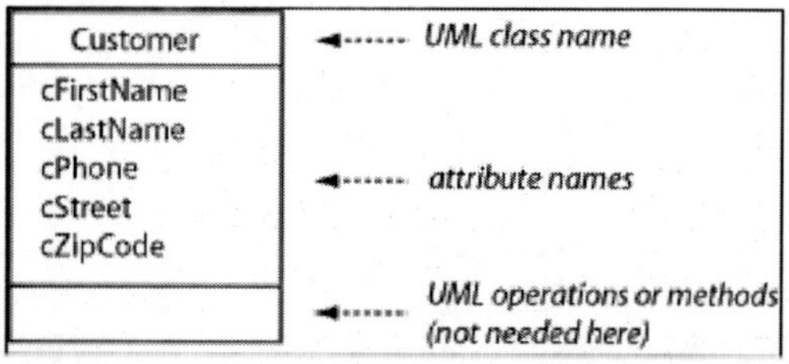

Figure 3. Graphical representation of the fundamental elements in the UML coding.

5.4 *Web Mapping Application*

The basics of spatial data thus created have been published on the web (using the platform Esri ArcGIS Online), where the information system is accessible to the public for consultation. The Web Map Application created (Figure 5) allows users to access data and information about actors and initiatives that characterize the field of raw earth. The application is organized and structured according to a user-friendly interface that facilitates the communication of knowledge to the public, which is also able to interact with the features of the map changing, if desired, also display characteristics.

The model consists of the classes layers of georeferenced data, which are represented by icons on

Figure 4. Pop-up window and text box of a icon.

the map. The pattern scan allows you to display pop-up windows (Fig. 4) attached to each item.

Positioning the cursor over each icon will display the text box that highlights the attributes associated.

The web-GIS application created for the dissemination of data is described here by a use-case diagram (Fig. 7) that, according to the standard UML, allows you to outline the set of features of the system as perceived by the users who interact with the system. The actors that play a role, covered by a set of entities interacting with the system, are two: Content Producer and User.

The main function of the application is to provide access to data and information collected in accordance with a simple and immediate interface that facilitates the communication of the knowledge acquired to the public (Use Case: explore the model). Every user who interacts with the system then has the opportunity to make a number of additional actions that change the map display or allow you to share it.

6 FUTURE IMPLEMENTATIONS

This work aims to represent, as an ever-changing situation, what is the state of the art of the use of earth in building. In the given form it requires by its specific nature a constant and continuous updating. The project in this sense does not end with the research represented by this work, but it will be continued in the future, following in parallel both the evolution of innovations in

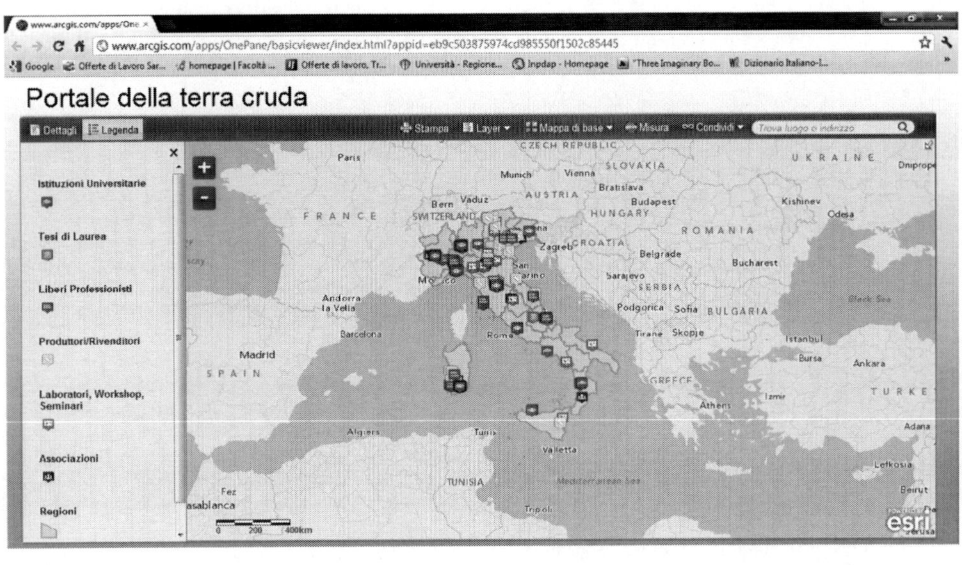

Figure 5. Web Map Application screenshot.

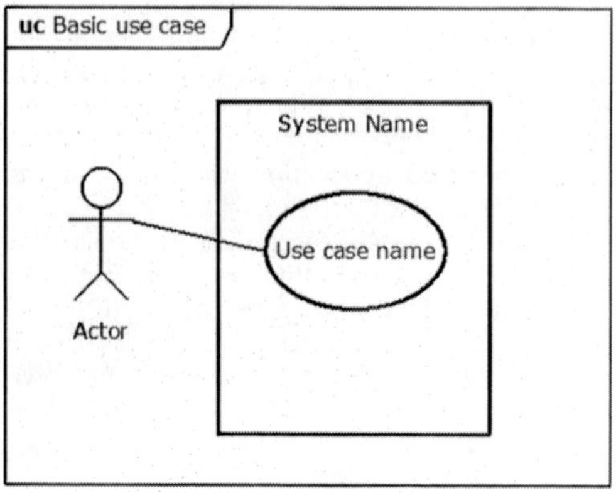

Figure 6. Example of a basic use case diagram.

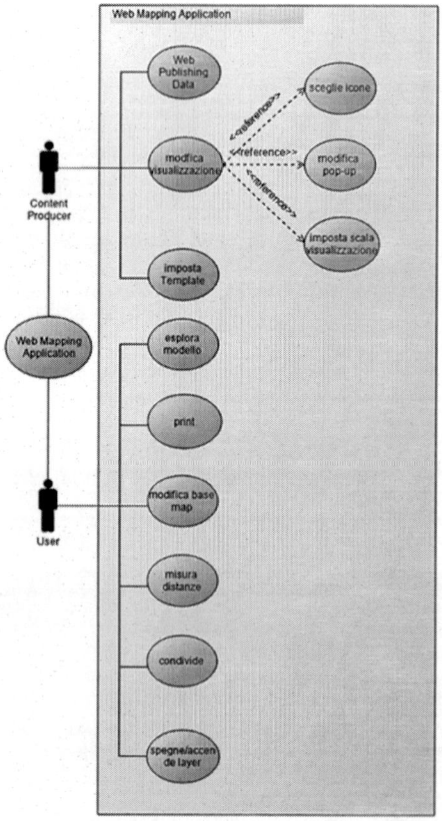

Figure 7. Use case diagram Web Mapping Application.

the field of GIS and the development in earthen architecture.

Additional future implementations considered, include the possibility to initiate active public participation in the construction of the contents of the site: thanks to the interactivity tools of "web 2.0" and the usability of web mapping application, it can become a volunteer data and information provider according to the paradigm of citizens science, a scenario after which citizens as volunteers participate in scientific process together with professionals.

7 CONCLUSIONS

Although the data collected in this research do not represent the totality of experiences in the field of earthen architecture in Italy, the result is an important base on which further implementations can be developed.

The study and the data collected provide a consistent demonstration of how earth as a building material is a deeply rooted and dynamic reality in Italy, showing that the argument of those who assert that the use of earth is an obsolete practice and exclusively linked to an archaic world, is in many ways unfounded or at least inexact.

It is shown, moreover, that the use of web-GIS proves to be functional and responsive to the need to provide access to data and information according to a user-friendly interface that facilitates the communication of knowledge to the public.

The module web-GIS, was finally added to a website—a sort of "portal of earth construction"—full of useful information for visitors, organized and structured with a simple and straightforward interface, that can provide a good level of accessibility.

In addition to the dedicated website, the application that is developed using standard web technologies may be embedded within other websites changing shown characteristics, if desired.

REFERENCES

Baldacci, O. 1952, *La casa rurale in Sardegna*, Firenze: Centro di studi per la geografia etnologica.

Campagna M., Deplano, G., Evaluating Geographic Information usage within Public Administration web sites", ENVIRONMENT AND PLANNING B: PLANNING AND DESIGN, Pion LTD, vol31.1:31–37, 2003.

Campagna M., Iannuzzi Y., Le tecnologie informatiche per la costruzione della conoscenza sul patrimonio insediativo: esperienze in Sardegna. DISEGNARE CON..., vol. speciale, ISSN: 1828–5961, 2012.

Mossa V., Architettura Domestica in Sardegna, Carlo Delfino Editore, Sassari 1982.

Sanna A., Manuali del Recupero dei centri storici della Sardegna. Vol I-V., Dei, Roma 2009.

Sanna U., Achenza M., Il manuale tematico della terra cruda, Dei, Roma 2009.

Earthen Architecture: Past, Present and Future – Mileto, Vegas, García Soriano & Cristini (Eds)
 ISBN 978-1-138-02711-4

The documentation and analysis of the earthen architecture in Saudi Arabia

A.A. Alaidarous
King Saud University, Riyadh, Saudi Arabia
Instituto de Restauración del Patrimonio, UPV, Valencia, Spain

ABSTRACT: This paper will highlight the traditional architecture of Saudi Arabia that seems to be vanishing from the residential construction practices of Saudi Arabia due to the adoption of building materials and techniques of the industrialized nations. The future architects and planners are urged to also pay attention to the historical heritage in order to extract the principles that will help in creating an amalgamation between the modern and the traditional. The practice of unthoughtful adoption of foreign techniques and styles will be discouraged through this paper and the earth construction technology will be introduced to the Saudi building industry.

1 INTRODUCTION

The traditional Saudi building material and techniques were abandoned in the wake of modernization. The dignity and identity represented by the vernacular earthen construction for this specific region was soon forgotten. The present design and construction industry is growing in the absence of the guidance of the master builders and craftsmen, who could have proven to be a treasure of knowledge regarding traditional construction materials and techniques suiting the local conditions. The years and years of experience of these master builders could have been implemented to fulfil the local needs in terms of finance, energy and resources. Unfortunately this bank of invaluable knowledge is fading away.

The vernacular architecture in Saudi Arabia intelligently utilizes the raw earthen materials to provide optimum thermal comfort, natural ventilation and less dependency on energy as opposed to the modern construction that suffers heat losses and over-heating. As Greek architect, Constantine Dioxides said that 4% of the world's buildings were constructed by trained designers, which reveals the dismal picture of the architectural practices that ignores the significance of the "non-pedigree" architecture responsible for the remaining 96% (Agarwal, 1981).

In an interview with one of the traditional builders, Mr Qhtani Mohammad, it was found that he had learned the art of constructing with cob in Najran from his forefathers. This practice had been transferred through the generations over a period of 300 years. The pressure of modernization has eliminated the tradition of training builders over several years under skilled builders and craftsmen. The modern educational system has taken away the time of the younger generation that could have been spent under the guidance and training of their elders. These traditional and expert craftsmen are not revered for their potential by the modern industry that denies them the position that they deserve.

This study is based on a field survey of 2012–2013 related to the building heritage of Saudi Arabia that will focus the issue of modernization replacing vernacular practices. As observed, modern materials, such as reinforced concrete, are being used instead of load-bearing walls of rubble, stone, cob and adobe. Traditional craftsmen who are experts on the local conditions and building materials are being replaced by foreign specialists, who are ill equipped to deal with the local conditions and have no sense about the building traditions of the country. This study is intended to bring this issue under the limelight, although it may not be sufficient to argue the case in its totality.

2 FIELD SURVEY

Students from all over the country come to study at the department of architecture in King Abdul Al-Aziz, Jeddah which has an added advantage of being located in the center of a rich heritage. The field survey was based on the interviews conducted by the author from the students that gave a comprehensive idea about the different types of traditional buildings belonging to the different regions of the country. This data combined with the incomplete government publications on traditional buildings was used to plan the field survey.

Geographical Regions are six regions of Saudi Arabia; the Eastern, Central, Western, South-western, Northern regions, and the large desert of Al-Rub Al-Khali or the Empty Quarter. It is important to relate these regions with the local architectural characteristics and the evolution of their respective traditional building techniques.

The Geographical Najd is clearly indicated along its western, eastern and southern boundaries, but along the north, its boundaries seem to vanish. After the reign of late King Abd Al-Aziz Ibn Saud, the Al-Jawf oasis is considered as the northern section of Najd. Al-Rub Al-Khalil Desert, Al-Hejaz and Al-Hasa will not be covered in this paper, thus leaving two geographical regions: Najd and Asir.

3 SELECTION OF SETTLEMENTS AND BUILDERS

The survey of the various settlements belonging to each region, representing major populations, was done that allowed the author to cover additional cities as well, lying between the vast regions of Saudi Arabia. Local municipalities and museums were contacted to gather the names and addresses of old master builders in each settlement. The researcher was highly appreciative of the cooperation of the architects, engineers and officials of such authorities.

4 INTERVIEWS AND BUILDING SURVEY VISITS

Interviews were the best suited mode of collecting information as it allowed the builders to express their views without a hindrance, while the author took notes. A questionnaire would not have suited this survey as most of the local builders were illiterate. Additional questions were asked by the researcher, when required.

The interview was divided in two sections. The first was related to the personal information about the builders. The second part was related to the building traditions that covered the following points:

1. Building materials types and sources.
2. Construction tools.
3. Construction methods.
4. Elements of the building structure (walls, columns, floor, and ceiling).
5. Finishing and decoration elements.
6. Different features of the earthen architecture.
7. Other various crafts and trades related to vernacular construction.

A photographic survey was conducted of the traditional building, occasionally accompanied by an architect or a master builder of the area. During the visit, the causes behind the deterioration of the building were inspected visually. Samples of the soil and the earth products were also taken from these sites. The survey led to two systems of construction standing out.

5 METHODS AND RESULTS: MUD CONSTRUCTION IN SAUDI ARABIA

Mud was used as the main binding or plastering material in the traditional construction. Najd and the central and eastern parts of Asir specifically show signs of this material being used in construction. The adobe and cob construction were the principal techniques used in the mud-construction of these regions. This study will focus on these techniques.

5.1 *Adobe construction*

The word 'adobe' has been derived from the Arabic word *al-tobe* meaning 'brick'. This word refers to the technique where earth is mixed with water and finely chopped straw and then manually filled into wooden moulds. This technique was used to build the courtyard houses of Najd that would ideally create an internal micro-climatic condition and provide thermal comfort, security, privacy and protection from the external factors (Fig. 1). According to Hassan Fathy, a UN expert in Saudi Arabia, the courtyard houses exemplifies the integration of natural environment and local resources to create thermally regulated socializing spaces (Fig. 2) (Fathy, 1973).

5.1.1 *Design*

The location of the different rooms was marked firstly on site by using gypsum powder or by foot. A 50 cm long stick called *dura* was used by the master builder or *ustad* to check the accuracy of the dimensions. The *ustad* would also specify the location of different columns, in cases of large spanned rooms. Palm tree frond was at times used for longer dimensions instead of the *dura*.

5.1.2 *Traditional building materials and tools*

The materials were limited to those that were locally available owing to the transportation problems present as a result of the vast deserts. Furthermore, traditional construction materials were used without any modern-day influences.

Figure 1. Turaif district adobe building in Al-Diriyah-Najd (Alaidarous).

Figure 2. A Courtyard of Saad place in Al-Diriyah (Alaidarous).

Foundation and columns were constructed with the use of limestone, which is a common, standard technique in Najd; this material was available in abundance. Limestone has a well-defined layer of stone and natural fragmentation properties, which facilitates the extraction and shaping of the material. Limestone quarries were found in the cliffs of Najd or nearby valleys. The stone was shaped and named accordingly. Gypsum and mud mortar was used to bond them.

Adobe is made with soil having high clay content. The soil is dug and brought to the site where it is placed in the form of a circular bed. Water and straw is added. The mixture is left in a moist state for two to three days and then turned over with a spade or a rake and trodden with feet. A sticky mud mixture with an even constituency is obtained as a result. The soil must be plastic enough to be moulded into wooden moulds called *milban*. The mould is lifted to get a mud block. The mould is again placed about 50 mm from the first block and then filled again (fig. 1). The blocks are left to dry in the sun for two days during which these blocks will shrink and gain their strength. Mud mortar was used to bond these blocks which were plastered with mud (Facey, 1999).

Palm tree trunks were used as rafters for roofs. Flooring was also done with palm trees. Palm branches were placed in between the beams and then covered with leaves after which mud was rolled over them. Household items were made from palm tree leaves. Tamarisk tree was an important source of timber. The wood is hard and was used to make household utensils. *Athel* trunks were used as beams, lintels, doors and window panels and frames. Main door frames were made of *athel* with palm tree in-filling paneling (Fig. 3).

Gypsum or *juss* was used in Najd by placing gypsum rocks in kilns for a few days with wood. The timber was burnt for days until it was consumed and the fired material had cooled. This material was the crushed, sieved and stored in racks.

Hand traditional tools were developed locally for construction. These tools were simple as the construction techniques were simplified as well.

5.1.3 *Construction process*

This process was characteristic of Najd, adopted according to local conditions. Settlement pattern was compact with narrow winding streets that acted as cooling wells.

1. Excavation: A continuous trench—with a width of 80–90 cm and 50–120 cm depth—was excavated according to soil conditions.
2. Adobe manufacturing: The adobe blocks manufactured as mentioned above.

Figure 3. Columns, roof and openings elements in Al-Diriyah (Alaidarous).

3. Block layering: A team of workers would work under the *ustad* to layer the blocks in three or four courses in one go. These courses were left to dry before the next set of courses. This process was done until the ceiling level was reached. Scaffolding was used in larger constructions. The adobe walls were made 40 to 70 cm thick at the ground which tapered to 20 cm at the parapet with small openings. The diurnal fluctuations of temperatures were reduced in this way.
4. Columns: Stone columns were used for deep arcades and for large spanned rooms. Roughly cut cylindrical drums were set in gypsum mortar and surmounted by rectangular capitals, *ganaya*, supporting the main *athel* trunks or *sawakeed*. These trunks are positioned on capitals on one side and on abacuses on the other in order to distribute the weight in the adobe walls carrying them (fig. 3). Notably, in this context, an abacus references a ring beam.
5. Floors and roof: The same method of construction was used for walls and floors but with an additional mud layer on them to provide more thickness. The *athel* tree trunk was laid across the space between adobe walls and *sawakeef*. Palm tree fronds were laid across them. Palm tree leaves were laid over the framework and then covered with mud. Roofs were made at different levels according to spaces, which throw shadow on the roof to provide different spaces for sleeping during summer night. A parapet about 60 cm high separates each roof.

Figure 4. Openings and decorative elements on the façade of Al-Diriyah building (Alaidarous).

6. Openings: Few openings are given on the outer facade. The door entrance, windows in the guest room and small windows for ventilation near the ceiling are provided. Larger openings are provided in the interior facades facing the courtyard. The position of openings controlled ventilation and glare (fig. 3).
7. Finishes and decorations: Mud plastering was used for the exterior and interior walls as a finish. The mud was prepared as explained earlier. Exterior facades were left simple, while the interior facades were adorned with relief frieze, crenulations and *sharaf* that a deeply pierced decorative screen at the tops of the external walls or the decorative roof wall (Fig. 3).

5.2 *Cob construction*

This was typical of Asiri where the settlement pattern was tight with independent fort like houses. The cob or *midmak* construction is similar to the adobe in Najd, the only difference being the construction of walls with mud layers instead of mud bricks. This construction was used in castles for defensive purposes. A square plan is followed with 3 to 5 stories tall tower like structures. The mud is prepared and then continuous solid mud layers of about 30 to 45 cm height are built simultaneously without any formwork. When this layer has hardened in a couple of days, the next layer is done until the full height of the wall is obtained. Horizontal bands of (stone) slate creates a series

Figure 5. The cob or *midmak* Tower house at Najran Valley—Asir (Alaidarous).

of parallel parapets that protrude approximately 20–30 cm from the face of the wall and slope outwards slightly to throw away the rain water that would otherwise erode the walls. Structural buttresses are used to support the staircases. The walls taper upwards, thus incorporating a wider base than top (Fig. 5) (Al-Karni, 1990).

6 MODERN METHOD OF HOUSE CONSTRUCTION AND REPLACEMENT PROCESS

The advent of the Portland cement and other various modern building materials has induced a boom in the construction industry, particularly in housing. The increasing demand of cement resulted in cement factories being established in Jeddah in 1959; nonetheless, as of 1977, the industry was still falling short of supplies by approximately 6 million tons, although, at the present time, all factories are producing sufficient cement for local needs (Statistical Year Book, 2001). There was a resultant increase in prices of the building materials, skilled labour and land (Sina, 1989). The unwritten policies of the Real Estate Development Fund (REFD) and municipalities, as well as the early demonstration values of the modern reinforced concrete government buildings, have together boosted the use of reinforced concrete whilst discouraging the use of traditional building materials. The number of traditionally constructed buildings decreased from 1,500 in 1970 to 56 in 1980 in Al-Diriyah owing to a lack of maintenance and the impact of community attention being directed towards modern construction. Importantly, despite its suitability in regard to hot weather conditions, more traditional housing has been downplayed in comparison to the sub-standard modern construction, which is recognised as being less adept to managing the heat, and which is also recognised as offering less strength. REFD's refusal to finance traditional housing projects has played a key role in downplaying traditional construction techniques and materials, with modern architecture becoming more dependent on air-conditioning systems in an effort to deal with the hot weather conditions; this comes as a result of foreign design and the adoption of materials by modern architects with little consideration to its effects. The impractical design has meant more energy consumption as the dwellings were open from all the sides and exposed to the harsh heat. Subsidized energy costs have made the public insensitive about the impracticality of these designs. These designs are representative of the foreign character rather than their own (Salem, 2006). Urban patterns have also been disturbed creating privacy violations that have forced residents to opt for ugly solutions to preserve their privacy.

7 RESULTANT IMAGE

Modern architecture has been widely designed by Western architects or Arabs trained in a Western environment in line with Western requirements. These approaches, owing to being taught in environments that are notably different to where they will be adopted, are recognised as being ill-equipped to deal with the Saudi environment and culture. These buildings have led to costly and inadequate schemes that require foreign material, foreign labour and techniques to be imported adding more costs. Architectural colonialism is now characterised by such practices, with Saudi contemporary architecture recognised as a mix of various cultures' approaches, as shown through the British style of urban planning schemes, Dutch housing projects, German health facilities, American supermarkets and shopping centres, and Italian office compounds, amongst others.

8 EVALUATION OF TRADITIONAL AND MODERN HOUSING SYSTEM

Various locals were interviewed for the survey, based on Dr Mufti's questionnaires, in an effort to garner insight into the general perceptions of the public concerning both traditional and modern architecture, according to cultural, economic, climatic and technological factors. Traditional housing was favoured in comparison to modern housing, as also shown in Mufti's survey of 1981. However, people are now looking for a compromise and good combination of the two architectural practices, combining the suitable features of the traditional with those of the modern in an effort to render structures that are of improved quality and strength, and therefore comparable with modern buildings whilst respecting their environment and culture, as in the case of traditional buildings.

It would be impractical to adopt all the architectural heritage and traditions as the way of living has changed. Nonetheless, valuable lessons, as garnered through Saudi heritage, need to be directed in order to educate future architects so as to ensure that the years and years of experience of these master builders are not lost. This will help to achieve the conservation of this cultural heritage, which currently suffers from a lack of focus and consideration.

9 CONSERVATION RECOMMENDATIONS

Conservation is determined by the future purpose of a building, whether for residential or commercial purposes, such as in the case of museums, heritage tourism, hotels, and so forth. Al-Turaif district in Al-Diriyah, for example, is now recognised as part of a World Heritage Site, with the exception of some plans to rebuild particular areas as open-area museum and hotels. In this vein, maintenance intervention or design intervention methods are possible solutions for historical heritage in the case of earthen architectural conservation in Saudi Arabia. In addition, earth buildings need to be conserved, restored and renovated before they deteriorate and pose risks in their surroundings, as recognised by ICOMOS (Australia). Nonetheless, restoration works should not modify historically significant features of the building. The earthen architecture of Saudi Arabia also needs to be protected from erosion as a result of water penetrating core components. Importantly, restoration will reduce the decaying process and accordingly will eliminate deformities in the building.

10 CONCLUSION

Conservation of traditional architecture is imperative to preserve the Saudi identity and to ensure that the works of our master builders and ancestors are kept intact for the coming generation to revere and learn from. Keeping the decaying state of the heritage buildings, renovation works need to be initiated immediately with the presence of skilled labour and attention to detail. An in-depth analysis of these buildings is necessary in order to devise adequate restoration solutions to respect the traditional earthen construction techniques applied in such instances. The government may have to rethink their ambition of converting these buildings into museums and hotels, as the state of these buildings calls for minimum human intervention.

NOTE

This paper is a part of the PhD research about *Investigate of stabilization procedures for Historical Earthen construction in Saudi Arabia* that done by Ahmed Alaidarous at Instituto de Restauración del Patrimonio—UPV—Valencia, Spain.

REFERENCES

Agarwal. A. Mud, mud: *The Potential of Earth-Based Materials for Third World Housing*. Earthscan, International Institute for Environment and Development, London 1981.

Al-Karni, M. *Study of Traditional Settlement Patterns and Building Forms in Asir Region*, UD, 599, Research in Urban Design—Course Paper, College of Architecture and Planning, King Saud University, Riyadh, Saudi Arabia 1990.

Facey. W, *Back to Earth: Adobe Building in Saudi Arabia*, I B Tauris & Co Ltd. London, England, 1997.

Fathy, Hassan. Constancy, Transportation and Change in the Arab City: From Medina to Metropolis. The Darwin Press, Princeton, USA 1973.

ICOMOS Australia (2013). *The Burra Charter* [Online] Available at: http://australia.icomos.org/publications/charters/accessed 9 March 2014.

Mofti, F.A. *Urban Housing Design in the context of Saudi Arabia cultural and physical conditions*: Potentials and Constraints. PhD. Thesis Rensselar Polytechnic Institute, Troy, New York. 1981.

Qhtani Mohammad, local craftsmen, Najran Municipality, interview, 17 January 2013.

Salem, E. *Building And Energy for Historical of Old Jeddah*. KAU, 2006.

Sina, B. Saeed, S.A. and Al-Hussayen, M.A. An Analytical *Study of the Building Production Systems Recently Introduced in Saudi Arabia*. General Directorate of Research Grants Programs, King Abdul Aziz City For Science and Technology, Riyadh. 1989.

Statistical year book by *Ministry of Trade*, 2001.

Earthen Architecture: Past, Present and Future – Mileto, Vegas, García Soriano & Cristini (Eds)
 ISBN 978-1-138-02711-4

Earth-based architecture linked to charcoal making in north Aragon mountain ranges

Q. Angulo Ibáñez
Universitat Politècnica de València, Valencia, Spain

E. Monesma Moliner
Royal Academy of Fine Arts of San Luis, Zaragoza, Spain

A. Quiterio Calderón
Universitat Politècnica de València, Valencia, Spain

ABSTRACT: The charcoal burning trade, an activity associated with thickly-wooded areas, is on the verge of extinction in the Pre-Pyrenean zone. The northern ranges of Aragón abound with quercus and acorn woods and through countless years the charcoal burners managed to take advantage of their natural resources. Charcoal in northern Aragón was mainly used by ironsmiths, and also for cooking and as fuel for gasification devices. When charcoal began to run short a group of charcoal burners would start looking for a cave or natural shelter in the area close to their work site. If no place fit to withstand the bad weather was found, they would build little huts called charcoal burners' cabins. These buildings had to be adapted to the environment, their shape and size varying according to the number of people they were to house, to the natural characteristics of the terrain and the availability of building materials. The most common types of huts measured about 8 square metres and were 1.6 metres in height. They were very well insulated from the cold and were made with slim tree trunks or some other local material, branches and big lumps of earth, following the same technique as that which was used to build the charcoal pits. They had to be placed close to the charcoal pits in order to monitor the delicate charcoal combustion process.

1 CHARCOAL

From the oldest times charcoal makers have been present in society utilizing natural resources. The mountain ranges of Northern Aragon were thickly wooded with quercus petraea and quercus ilex whose wood was very suitable for making charcoal due to its hot but slow combustion. The best time to harvest this wood was from October to March, when the sap was not rising. If possible, the cutting of wood was done when the moon was on the wane. Once the wood was cut it was left to dry for about four to six weeks.

The post-war years saw a great boom in the charcoal making activity because of the marked shortage of fuel; charcoal was used for cooking, heating, ironworking and even for charcoal-propelled cars. Charcoal making no longer exists as a professional trade.

1.1 *Objectives, techniques and solutions*

This article is centred on the importance that charcoal making had in the Aragon Pre-Pyrenean mountain ranges and its relationship with the vernacular architecture linked to this activity. By doing so we are paying homage to those men and women with weather-beaten skin who lived in harmony with Nature, and who toiled on while enduring the smoke coming from the charcoal pits, and kept the woodland free of dead branches. They managed to adapt themselves to the sites and materials the environment offered them, devising building designs that were simple, practical and functional. On the one hand, they built small constructions called charcoal pits for the manufacture of this fuel and on the other, they built small huts and cabins for habitation. They built the charcoal pits and their huts with logs, branches and soil from their surrounding environment. Unfortunately when the charcoal makers disappeared these uniquely local constructions also went with them; constructions that got their materials from the environment and eventually returned them to it, being recycled when they were no longer necessary. Nowadays there just remain the old black circles of burnt earth within some forests and the accounts of old charcoal makers that take us back to their ancient tasks and constructions in Northern Aragon.

Before any testimony disappears, we must do our best to analyze and describe it in order to let people know of these simple constructions which could be put into practice today as do-it-yourself, low-cost, fast-to-build and durable dwellings, as some of them stood for over two years. In this way we can learn from their know-how and building solutions so that their skills will not fall into oblivion. They can be preserved and integrated into current improvements in sustainable architecture.

The methodology used in this article is wide and complex due to the scarcity both of written documents and extant constructions. Some old charcoal pits have been located by looking for areas which met the criteria necessary for the location of a pit, including area size, and the type and quality of wood. Interviews with knowledgeable people have been carried out and photographs were taken documenting the re-enactment of the building process step by step. Bearing in mind the character of this article a simple line of work has been adopted: brief descriptions and explanations of steps to give the reader a good idea of the processes involved.

Figure 1. "Enramillado". Setting the timber horizontally in a big charcoal pit (E. Monesma).

Figure 2. (Left) Cover of leaves and branches (E. Monesma).
Figure 3. (Right) Setting alight the charcoal pit (E. Monesma).

1.2 *The charcoal pit*

The process of charcoal making would start with the acquisition of a woodland area. If it was a local council property, it was divided into plots and auctioned off. In the case of a single private person a sale agreement was soon reached. The price depended on the potential of the woodland area, that is the quality of the timber and the number of stages required in the charcoal making process.

The task was divided into two stages:

- The first (called "media fábrica" in the charcoal makers' terminology) involved cutting the timber into pieces with an axe within some specified measurements (0.8 to 1 metres in length and 0.03 to 0.18 metres in diameter). Then preparing the area of the pit ("ejecutar la plaza"). This was done by using a hoe and making a circle where charcoal pit would be built. Next came the carrying of the timber from the wood to the location of the new pit. The size of pits varied greatly, some pits could hold up to 40 tonnes. A pit of about 6.5 metres in diameter for example, could yield around 1 tonne of charcoal.

In the second stage ("fábrica entera") the charcoal maker's tasks were:

- Setting the timber horizontally ("enramillar") with the thicker ends towards the centre of the circle. When the first layer reached the middle, a gap was left in the centre itself where a clean dry log of about 4 metres in length was placed with the thick end upwards. Often the pits would have a surface of flat stones to allow the fire to better spread horizontally and avoid getting defective charcoal ("tizas").

Once the timber was properly laid, the whole pit was covered with leaves and little branches of pine and boxwood.

Next, the pit was covered with a layer of mud thick enough so that it couldn't seep into the structure. This outer layer of mud was about 20 cm thick and the earth used for this purpose was of a consistency similar to sand and devoid of rocks. A wood ladder was built "in situ" to help with the last coating and eventually to reach and protect the pit hole ("cabera") in the centre.

- The charcoal pit was set alight from the pit hole so that the fire would spread downwards. Through the pit hole, the fire was fed when necessary, with kindle and small timber.
- This controlled burning took into account such factors as the type and placement of timber, the position of ducts and gaps within the pit and placement of the layer of floor slabs.
- After checking that the pit hole fire was going strong this upper hole was covered and several new holes about 8 to 10 cm in diameter were made on the sides of pit that were protected from the wind. These new holes encouraged the fire to move downwards.

Twice a day the mud lid covering the top hole had to be checked and compacted to avoid having gaps within the charcoal pit. Special care was taken to prevent any type of holes and cracks forming on the mud coating which would result in the whole pit burning to the ground due to the rapid combustion of the timber boosted by the sudden inrush of air.

- Packing the charcoal. When the colour of the smoke escaping from the side vents changed to blue it meant that at that level the charring was complete. The vents were plugged off, new lower ones were opened and the process repeated. Following this the pit was given sufficient time to cool off. By using a hoe and a rake the charcoal was extracted from the pit and its quality checked: A shiny surface, long lumps or even whole logs and a clinking sound while being handled all pointed towards a good product.

2 PRE-INDUSTRIAL CHARCOAL MAKING AND ITS INFLUENCE ON ARCHITECTURE

If near a charcoal pit there existed some cave or natural shelter, it was used for as long as the nearby wood was being harvested for charcoal. However, if none was available in the vicinity the charcoal makers erected their own dwellings to protect themselves from the harsh weather. Using similar techniques to those of the charcoal pits the shape and size of these habitations varied according to the number of people to dwell in them, the characteristics of the terrain and the availability of building materials. Determined by these factors the charcoal makers' cabins and huts came into being.

2.1 *The charcoal maker's hut*

The life of the charcoal makers and their families steadily moved on through the mountains where they carried out their work. This nomadic situation required them to solve their need for shelter with the materials nature offered them, so earth and timber were the basic elements used to make a charcoal makers hut. The main elements were quercus petraea trunks and branches, the most knotted and twisted ones were chosen because they were best suited to making the framework of the cabin. They were cut using using a hatchet and no attention was paid to the moon phases that have such important role in wood preservation.

The shape and size of a hut was related to the number of people who were to inhabit it. The most common ones measured 4 × 6 metres and 1.6 metres in height. Their placement took into account the prevailing winds of the Pre-Pyrenean ranges in that these affected the orientation of the entrance. Huts were built taking advantage of low-lying patches of land so that the interior of the dwelling would be more comfortable. After marking out the desired measurements for the cabin a small excavation was made in the chosen place.

After then cleaning the site, the framework was solidly fixed to the ground and a long log placed on the ridge thus locking all lateral parts together. Great attention was paid to these steps because they were fundamental to the solidity of the whole structure.

To limit heat loss from the inside of the cabin, the door was just big enough to allow the entry of the inhabitants. The chimney was opposite the entrance so as to make sure the fire would draw well. While making the structure of the hut some portions of branches had to be used to reinforce the more twisted vertical timbers. In this way the structure was straightened and stabilized.

A cylindrical beehive was used as a mould to make the chimney. The lower part of it, the cooking area, was made using big stone slabs that also prevented the wooden framework from catching fire.

Figure 4. Setting out a charcoal maker's hut (E. Monesma).

Figure 5. Main structure of a charcoal maker's hut (E. Monesma).

Figure 6. Elements to cover a charcoal maker's hut (E. Monesma).

Figure 7. Covering with earth. "Chasquear y terrar" (E. Monesma).

Boxwood branches were used as external lining to cover the whole hut except the door and the chimney. The earth that was excavated when the site was initially being prepared was used to cover this boxwood lining. Obviously, this process was started at ground level and went upward at a slow and careful pace. This earth layer ensured a comfortable temperature throughout the year.

Round rocks found nearby were stuck on by a thick layer of mud and placed randomly to make up the chimneystack. The charcoal burner took every effort to build himself a strong hut, for the harvest time of a wood patch ranged from six months to two years, and this shelter would have to last him the full period. The special shape of the hut: two half cones plus two slanted rectangular sides facilitated the mud coating. A different design would not have kept the mud coating in place as it would have slid down the structure over time.

The length of the huts varied according to the number of occupants. Many were made to lodge eight or ten charcoal makers who made up a work crew for a specific wooded area. After the earth coating process the hut was ready to be lived in.

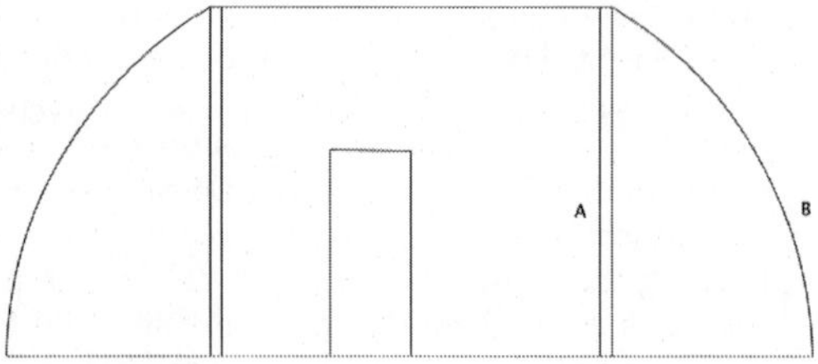

Figure 8. Elevation of a Charcoal maker's hut* (Q. Angulo).

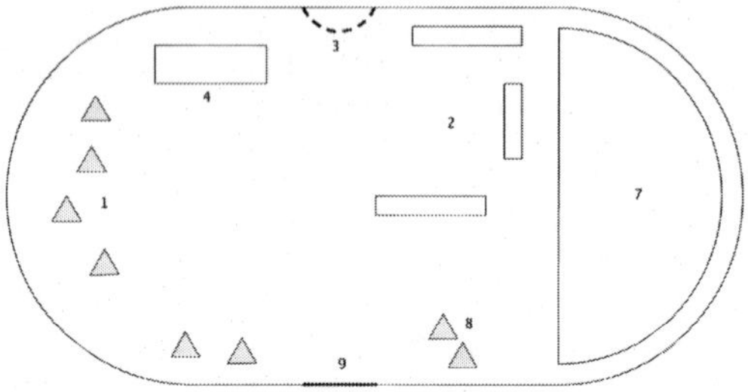

Figure 9. Distribution of a charcoal maker's hut I* (Q. Angulo).

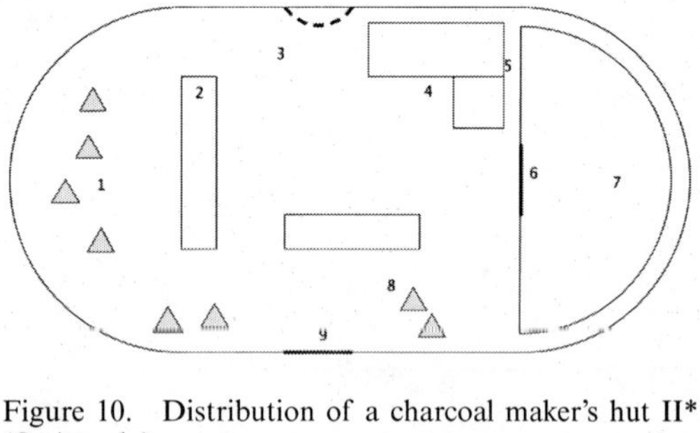

Figure 10. Distribution of a charcoal maker's hut II* (Q. Angulo).

Just a few items accounted for all the comforts the charcoal makers could afford. The fireplace was made with several stone slabs; and the particular slab that would draw the smoke to the chimneystack was supported by some wood beams protected against the fire by two roof tiles. Bunks were made using flat pieces of local timber and flat stones as legs. A bed-frame was made up of small boxwood branches spanning the timber frame and then several layers of sacks on top were the mattress.

When the hut was inhabited by a married couple and their children a dividing wall was raised in one of the round ends. The wall was made of thickly set intertwining twigs of quercus petraea or boxwood. For greater solidity, the dividing wall was partially stuck to the hut side walls.

The outer door was made to measure with a bundle of boxwood branches that fitted tightly in the small entrance opening. Tiny eaves made of stone prevented the rainwater from leaking in.

This small habitation covered all the functions of a regular dwelling. One side was occupied by the couple's bedroom, the centre by the kitchen with its fireplace and on the opposite side there was a bed for the children. The rest of the space was taken up by the tools, personal items, cases and food. The inhabitants had to do their best to keep their valuables out of the reach of rodents.

When the work on the woodland area was over, the huts were pulled down and their wooden framework used in the manufacture of more charcoal. This has been the main factor that has prevented the minutest vestige of this type of structure from surviving to the present-day. Only the memory of the ancient charcoal makers and their willingness to leave their testimony for the future can make us understand the hardship of life up in the sierras for many families who transformed quercus petraea timber into charcoal.

Table 1. Description.

Elements	Description
A Height	1,6 m or minimum to stand up
B Cambered shape	Facilitate the earth covering
1 Personal items	Place for storing the personal items
2 Seats	Seats around fire
3 Chimney	Cooking and heating
4 Kitchen	
5 Interior wall	Separate spaces
6 Interior door	Separate spaces
7 Room	Parents' bedroom
8 Tools	Place for storing work tools
9 Door	Access

* Description of the main elements of a charcoal maker's hut.

Figure 11. A charcoal maker's hut (E. Monesma).

2.2 *The charcoal maker's cabin*

It accomplished the same functions as the hut but its measurements were smaller as there was just room for a fireplace and a bed, as two people commonly used them.

It was built with thin logs, boxwood branches and (this is the main difference with huts) with "tascas", lumps of earth covered with grass on one side, excavated on site. The building processes are very similar but cabins are easier to build as they are smaller. Charcoal makers' cabins are two roofs/ walls constructions, the front one where the entrance is located and the back one where the chimney stack is fixed and two sides that have an isosceles triangle shape.

Over the main logs that shape the lateral wall the ridge log is placed and next the rest of the logs that make up the framework of the cabin. Fully-leaved boxwood branches are entwined over the logs. Over this rustic trellis-work the "tascas" are placed in an orderly fashion, like the scales of a fish and so go to build the walls of the cabin. The slanted position of the logs gave the whole wooden structure ample strength to support this thick layer of earth.

In these constructions, close attention was paid to the quality of the materials and to the careful adjustments required to ensure the solidity of the cabin as hey were intended to be lived in for long periods. Any small opening in the layer of "tascas" had to be closed off to prevent the rain from getting in. This covering technique is a repetitive one: it consists of several processes in which the same actions are repeated until the process is finished: The placing of logs for the framework, making the boxwood trellis-work and finally covering the wall with "tascas" until the top of roof is reached. Special care was taken when covering the ridge of the hut with "tascas". To make the rainwater flow at both sides and then drip down the sides of the cabin, the tascas were laid across the ridge log.

Figure 12. The structure of a charcoal maker's cabin (E. Monesma).

Figure 13. Earth blocks and chimney (E. Monesma).

Figure 14. A charcoal maker's cabin (E. Monesma).

The chimneystack was placed in the gap that had been made during the laying of the rows of side "tascas", while adjusting and levelling these rows for a perfect fit. Next, the chimney itself is coated with mud to give the cabin a homogeneous finishing and also to seal where the chimney and the wall/roof joined. The cabin had to have a certain level of comfort so that its inhabitants could stay warm in the winter and withstand cold, rainy days. With some stone slabs held together with mud, they built a little fireplace that would heat the interior and make their sleeping on the floor or on improvised cot more bearable. The access door is made in the gap previously made during the hut construction. It faces the charcoal pit so charcoal makers could monitor the charring process from inside the hut. For the door they made a framework and over it a trellis-work of green and fully-leaved boxwood twigs resulted in a waterproof element that would allow them to keep the interior of the cabin warm.

3 FINAL CONSIDERATIONS

After the analysis carried out on these vernacular constructions linked to the manufacture of charcoal in the Northern Aragon area, some important facts stand out:

- These buildings were built by their inhabitants.. The charcoal burners built their own shelters with the same techniques that they used when preparing the charcoal pit.
- Wood and earth are the materials used for the construction of the buildings. They were sourced locally, from the environment where charcoal burners work. These materials were also the ones they used in the charcoal burning process.
- The functionality of these buildings: They were designed for charcoal burners to rest in. If necessary they also included a small area for food preparation, storage, etc.
- Durability: these constructions lasted as long as the charcoal burners required them for. This was usually a short period. When work was over, the constructions could be demolished and the materials used in charcoal manufacture or they were simply returned to nature.
- Construction speed: due to the materials being plentiful and available nearby, construction time was a matter of hours.
- Easy construction process: these constructions were built following a know-how that was passed on from parents to children, regarding a logic of structure and materials for what was basically a small framework of thin logs and branches, coated with earth and vegetation.
- Adaptability to users' necessities: the buildings were adapted depending on the number of inhabitants and their relationships. One charcoal burner or a whole crew, single unmarried people or families. Based on this the shelter was adapted to satisfy both the required space to sleep and other possible needs (food preparation room, storage, etc.).
- Dimensions and characteristics varied according to the users' needs: from a one-person cabin (2 × 1 metres) to a hut for crews or families (4 × 6 m).
- Location: due to the characteristics of their constructions, their materials and their environment (woodland), it was important to choose a location within the vicinity of the charcoal pit. Correct aspect and lie were important in order to protect the dwellings from water run-off and cold winds.
- The orientation of the building was made in regard to the location of the charcoal pit. Since charcoal burners constantly had to keep an eye on their charcoal pit, they faced the building towards it. In this way, they were even able to watch it almost as they slept without getting up.
- Thermal insulation: Due to their wood framework which was coated with earth and vegetation, these buildings did not allow much heat loss from within, thereby maintaining a comfortable temperature for the inhabitants.

- These kind of constructions are linked to the utilization of charcoal pits. They can be considered as guilds' shelters.
- Knowledge of the construction of these dwellings was passed orally between parents and children or between members of the same profession.

The rapid implementation, ease of construction, the use of natural materials from the surrounding environment, energy saving characteristics and environmental friendliness make these buildings a simple and practical solution that would be worth considering nowadays for fast and temporary solutions.

REFERENCES

Arbués Possat, J. 2000. *El carbón vegetal en las altas Cinco Villas.*

Bernal Fandos, B. Ontejas, 1997. *Oficios. Los Carboneros* nº 9.

Davasse, B; Galop, D. Le charbon de bois et le pollen: elements pour approche de l'evolution historique du pausage forestier dans les Pyrenees ariegoises. *Écologie et biogéographie alphine.* La Thuile, 2–6 Septembre 1990.

Film: Armendáriz, Montxo. 1984. *Tasio.* Producida por: Elías Querejeta P.C.

Film: García Calvo, A. Comedia de Aristófanes (425 a.J.). Los carboneros. Canal Manu Scriptus.

López Soler, A.; Durán, M.E. Coals from de Teruel Mining District, northeast Spain. Institute of Earth Sciences "Jaume Almera", CSIC. *Applied Geochemistry* 01/1992; DOI:10.1016/0883-2927(92)90070-J.

Lozano Martínez, J. *Los chozos de los pastores y carboneros de Guadalaviar.* CECAL, 2009.

Monesma Moliner, E. Huesca,1993. Carbón vegetal. *Temas de antropología aragaonesa nº 4.* Dianet.

Rivas González, F.A. Etnografía y léxico de la elaboración del carbón vegetal en Cerveruela. *Centro del libro de Aragón nº 12,* Abril de 2011.

Roma Riu, J. *Los oficios y trabajos desaparecidos, celebrados como fiestas nuevas.* Zaragoza, Guara editorial 1980.

Earthen Architecture: Past, Present and Future – Mileto, Vegas, García Soriano & Cristini (Eds)
© 2015 Taylor & Francis Group, London, ISBN 978-1-138-02711-4

A pentagonal block home

G. Barbeta
University of Girona, Girona, Spain

E. Navarrete
Xarxa Ecoarquitectura, Sabadell, Spain

ABSTRACT: Earth construction applied to a bioconstructive and bioclimatic detached house in Sant Esteve de Palautordera, Barcelona. The property is located at the foot of the Montseny massif and is integrated within a fully natural environment. With a pentagonal floor plan, its five facades respond to the needs of the site, property, compositional criteria, and harmonisation. The Compressed Earth Block (CEB), also known as a Bioterre, is a structural material used in walls and arches due to its high thermal inertia and low environmental impact. Solid 30 cm blocks are placed on the south façade to maximise its thermal accumulation. The remaining façades have a 10-cm-thick wood fibre insulation layer, sandwiched between two CEB layers in order to achieve a "passive house" performance.

1 THE CONCEPT

The main objective of the project was to build an earth quality house at the foot of the Montseny Natural Park, approximately fifty kilometres to the north of Barcelona. As such, it had to be both respectful to the natural environment and facilitate the improvement of the health and welfare of the clients, a family with two small children. Applying the principles of bioconstruction as a starting point, it was conceived as a living organism capable of collecting its own rainwater, obtaining its energy from the sun and being rooted to the earth, spreading its roots through the plot and kitchen garden.

The idea of a pentagonal flower took shape, matching the drawings sketched by one of the children, which meant that the central section of the house had to open up to let the sun in.

A key aspect of the initial design stage was the involvement of the whole family. As such, the interaction and fluidity provided by the clay scale model, as shown in the illustration below, was crucial in terms of understanding and appropriating the project as a coherent concept, starting out as clay and ending up as clay. This project has been conceived and executed with care and love towards the dwelling and those who will inhabit it. With such an organic design the walls had to be built in this pentagonal layout, while at the same time curling into corners and extending along buttresses.

The half-buried garage is attached to the north-east façade and forms a ramp leading from the kitchen garden to the rooms on the first floor. With this in mind, Compressed Earth Block (CEB) was chosen as the core material due to the ease it offers in terms of achieving the forms of the design and in order to integrate very thick insulation to increase energy efficiency. The dwelling is situated approximately 300 metres above sea level and its location right next to the Montseny massif means that winter conditions are more severe than in the coastal area. At this point we had to deal with the handicap of having good views of the mountain on the north side.

2 FOUNDATIONS

The dwelling is constructed on very tough quaternary ground composed of round stones, sands and clays. This obstacle is overcome in a straightforward manner with a rigid strip footing measuring 50 centimetres wide and a minimum reinforcing mesh at the base. This reinforcing mesh is composed of bamboo 20 mm diameter treated with borax. The base of the CEB retaining wall is reinforced in a similar manner. Bamboo was chosen taking into account its low environmental impact and the geomagnetic fluctuations generated by steel. The rising wall, which remains under the ground up to the inner ground layer, is composed of 30 cm concrete blocks, filled in turn with concrete. A Tyvek polypropylene membrane was inserted at this point to protect against capillary action. The protection against the capillary action and splashing that occurs on the external skirting of the walls was reinforced by using mixed skirting

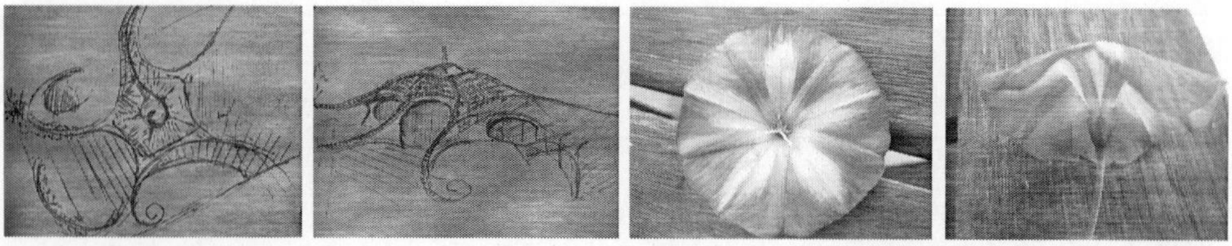

Figure 1. Early sketches, showing a natural relationship with flowers (G. Barbeta, E. Navarrete).

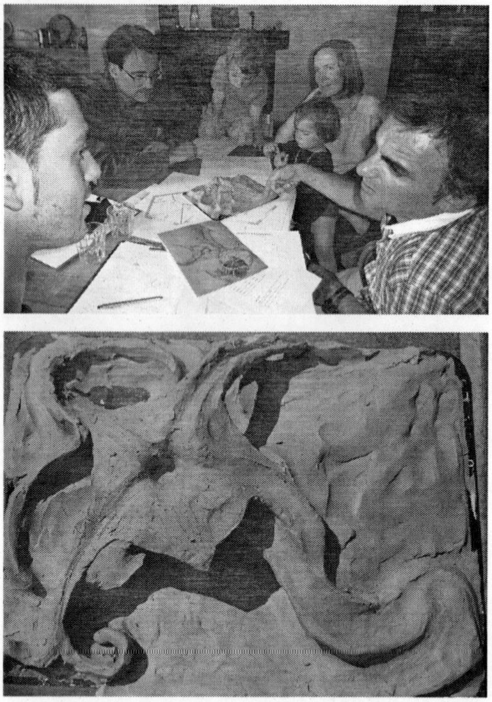

Figure 2. Developing the clay scale model together with the family (G. Barbeta, E. Navarrete).

Figure 3. Laying the first stone (G. Barbeta, E. Navarrete).

Figure 4. Placing of bamboo mesh on blinding concrete. (G. Barbeta, E. Navarrete).

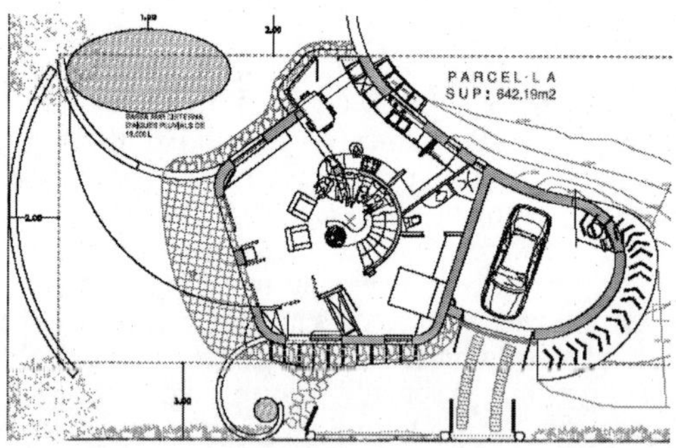

Figure 5. Dwelling floor plan (G. Barbeta, E. Navarrete).

Figure 6. Wall skirting executed through double inner layer of CEB and reused stone from the excavation works (G. Barbeta, E. Navarrete).

partly composed of reused stone from the excavation works. At the base of the exterior wall, the outer layer of CEB has been replaced by a 60 cm high stone layer. This protects the base of the CEB wall and prevents the outer stucco from reaching the ground, thereby preventing problems of efflorescence and maintenance over the years.

3 WALLS

The composition and characteristics of the walls respond to climatic and structural needs, the aim being to achieve the bioclimatic and efficient functioning of the building envelope. The technology used to achieve this purpose is the CEB. The chosen material is earth and the goal is to bring this material into line with other standard construction techniques. With this in mind, it is essential to ensure a comparable minimum resistance, the monitoring of the architectural performance that is obtained and durability against damage by water and, above all, ice. On the basis of the extensive experimentation and methods outlined in the doctoral thesis of Gabriel Barbeta (http://www.tdx.cat/TDX-1105102-161519/index_cs.html), we can state that today it is feasible to obtain high quality, standardised material made out of earth. The CEB is the technological option most likely to fulfil these goals on the basis of the following premises:

1. It is technologically comparable with traditional ceramic brick in terms of size, structural disposition and working method.
2. The skill level of workers does not need to be very high; any builder can easily execute a project with this material.
3. It can easily be manufactured on an industrial scale, producing a more standardised and controllable material.
4. It also enables the use of hydraulic compression machinery, ensuring better compaction and resistance.
5. Given the small size of the individual units, problems related to the drying shrinkage or expansion of clays are overcome and the correct execution of the wall is not affected.
6. The CEB manufacturing plant is located within an acceptable transport radius of 85 km, representing an energy impact increase of just 10%.

The structural walls of the southeast and southwest facades of the pentagon consist of 30 cm solid CEBs in order to maximise thermal inertia and the sun exposure of these façades. Meanwhile, in the case of the north and northwest walls the concepts of inertia and insulation are combined through a 15-cm-thick inner layer of CEB, a 3-cm-thick cavity, a 10-cm-thick layer of wood fibre (Pavatex) and a 9.5-cm-thick outer layer of CEB. The two CEB layers are joined by Murfor type galvanised steel connectors. The design and calculation of the walls were made from a wide of thirty centimeters and a point or height of 2.7 meters, although among them were small dimensional variations. The deck loads transmitted were of 1 KN/ml for compressed blocks, although his maximum compressive strength is $\sigma c = 7 - 10$ N/mm^2. At the same time we considered the variables of wind loads, earthquake and snow and follow the mandatory requirements established by the Technical Code for Construction in Spain (CTE-SE-DB-AE). The design values were 0.4 KN/m^2 for wind and 0.5 KN/m^2 for snow because the altitude is below the 300 meters above

Figure 7. Close-up of north façade (G. Barbeta, E. Navarrete).

Figure 8. Close-up of retaining wall detail with bamboo and Murfor reinforcements (G. Barbeta, E. Navarrete).

Figure 9. View of bare CEBs of the garage wall (G. Barbeta, E. Navarrete).

Figure 10. CEB radiant wall (G. Barbeta, E. Navarrete).

Figure 11. Close-up of Trombe wall and dark lime stucco (G. Barbeta, E. Navarrete).

Figure 12. Close-up of the join between the wall and the woodwork (G. Barbeta, E. Navarrete).

the sea level. Transiting use loads were not taken into consideration because the deck was considered not passable. The calculation of the seismic force resulting from the own loads in relation to the height was made on the basis of Spanish rule NCSE-02.

The garage is half buried on the north façade, creating a ramp to the first floor. These retaining walls are composed of 30-cm-thick CEBs with a 10-cm-thick external layer of insulation cork, plus the corresponding grounding sheets (waterproofing layer, geotextile filtering layer and polyethylene bubble insulation for drainage). In order to withstand buckling, these walls are reinforced with vertical bamboo rods placed 40 cm apart and Murfor type horizontal reinforcements in order to withstand the horizontal thrust of the soil.

For the laying of the CEBs, non-hydraulic lime mortar and Portland cement (1/1/7) was used with 1.5 cm struck mortar joints, ready to be coated. Only the CEBs of the interior garage walls were left visible, which is why the mortar incorporates an earth fraction.

4 CEB AS BIOCLIMATIC ELEMENT

Radiant wall technology is used to heat the rooms on the first floor by means of embedding polyethylene tubes in the inner earth cladding. This makes it possible to use the inner CEB layer and the cladding itself as an inertial system for the accumulation of solar energy through a system of six solar collectors placed on the roof.

In terms of passive uptake, a series of Trombe walls have been incorporated underneath the windowsills of the walls on the first floor of the southeast and southwest façades. They have been executed using 10-cm-thick CEBs with an exterior covering of dark coloured lime stucco internally with the same earth render as the rest of the house. The aim of the colour is to increase the solar absorption coefficient.

5 WOODWORK AND JOINERY

The woodwork is oak and chestnut, prioritising local woods as a basic concept of sustainability. The frames are placed mid-wall and the sub-frame is hidden by the outer layer, creating a step in the style of traditional Catalan farmhouses. The inner vertices are chamfered, creating spills to improve light penetration and to expand the opening radius of doors and windows. For this purpose the CEB is cut with a radial saw.

6 THE STRUCTURE

The floor slab and roof structure consists of chestnut beams with a 14 × 20 cm cross-section, with a reciprocal roof frame. The beams slot numerically into their respective sockets. This structural method, typical of ancient Hopi buildings and Japanese temples, enables large spans to be covered with small elements, with each beam resting on the last. This makes it possible to have an interior open floor plan and a central skylight with a 2.6 metre diameter. The loads are distributed very uniformly, primarily in the corners, which are reinforced with buttresses. The ground floor slab consists of unidirectional wooden beams with ceramic infill.

Figure 13. Reciprocal frame roof, central section with skylight (G. Barbeta, E. Navarrete).

Figure 14. Reciprocal frame in the study with interior earth cladding (G. Barbeta, E. Navarrete).

The garage roof has the same structural system and has a landscaped flat roof, to generate the exterior access ramp. As such, part of the garage building is buried to enhance integration and improve efficiency in withstanding the cold northern winds.

The support system of the wooden beams at the head of the wall consists of installing perimeter tie beams in two phases. In the first phase an 8-cm-high tie beam is put in place with two 8 mm diameter rods and connecting rods in standby mode. In this way the beams are repositioned and, once in place, the second phase is executed, placing tie beams between beams and inserting self-tapping screws in the beams and intermediate rods. The beam thus becomes embedded, achieving the required reduction of the bending moment and the working section.

7 RENDERS

The exterior rendering consists of three layers of non-hydraulic lime mortar (Mortedec brand): a 2-cm-thick render; a 1.5-cm-thick plaster coat and a 3-mm-thick finish coat. This render protects the CEB from weathering but retains the breathability properties of the wall. In our case, special emphasis was placed on the sand colour in order to achieve a perfect finish with no need for subsequent painting. A reinforcement treatment is applied to the exterior in the form of a dilution of potassium

Figure 15. Photo of the exterior (G. Barbeta, E. Navarrete).

silicate and siloxane-based water repellent protection. Plastic mesh has been installed in the areas where two different materials are joined together in order to deal with differential shrinkage and avoid cracks.

Earth renders were executed inside the dwelling in natural tones. These coatings increase internal inertia, working in conjunction with the technology of the radiant wall. They also help to regulate atmospheric humidity due to the hygroscopic properties of clays, not to mention the therapeutic properties of mud. The render was executed in two layers working with local clay, supplied by the Argiles Colades company. The first layer contains a percentage of crushed straw to minimise retraction, while the second finishing layer offers different colour options. This first layer was executed with a wet spraying machine to speed up the work.

8 FACILITIES

The design of the house includes other concepts of bioclimatic and bioconstructive architecture: green pergolas for sun protection in the south-est facade, roof garden with native species on top the garage, recovery systems for rain and waist waters, cross ventilation systems with a big skylight in the center of the deck, biomass boiler, four solar panels for heating sanitary water, underfloor heating. A superior air quality is assured by the absence of pollutants normally used in carpets, furniture, coats and varnishes and minimizing the presence of metals and steel structures.

9 CONCLUSIONS

In conclusion, thanks to this new project we can determine the technical and economic viability of organic bioconstructive and bioclimatic projects using CEBs. The size of the construction, along with its acoustic, thermal and structural performance enables us to design passive dwellings that are well suited to our climate, where thermal inertia is fundamental in terms of summer comfort, an aspect that is overlooked in most current projects that are based only on wood or straw.

REFERENCES

AEN/CTN 41/SC 10 *"Edificación con tierra cruda". Norma 41410 CEB. Edit. AENOR. Madrid 2005.*

Barbeta i Solà, G. 2007. L'Arquitectura de terra. Barnola (ed.), *Ecoarquitectura sostenible amb terra estabilitzada*, Guissona, España.

Barbeta i Solà, G. 2008. Conciencia arquitectónica englobando al ser humano y la energía. COAAT (ed.). *La Punxa A-A63.* España.

Barbeta i Solà, G. & Navarrete, E. & Palau, P. 2012. UPV-Balkema Taylor (eds), *Escuela municipal en Santa Eulalia de Ronçana. Congress proceedings Restapia.* Valencia, Spain.

Kahn, L. 1994. Tursen-Hermann Blume (ed.), *Cobijo.* Madrid.

Minke, G. 2010. Ecohabitar (ed.). *Manual de construcción en Tierra.*

Popovic Larsen, O. 2008. Elsevier (ed.). *Reciprocal frame Architecture. ISBN: 978-0-7506-8263-3. Oxford, UK.*

Wright, D. 1983. Gustavo Gili, editorial revisada (ed.). *Arquitectura solar natural: un texto pasivo.* México.

Earthen Architecture: Past, Present and Future – Mileto, Vegas, García Soriano & Cristini (Eds)
© 2015 Taylor & Francis Group, London, ISBN 978-1-138-02711-4

Thermal improvement of rammed earth buildings by the inclusion of natural cork

G. Barbeta Solà
Universitat de Girona, Girona, Spain

F.X. Massó Ros
Earth Construction Universitat de Girona, Girona, Spain

ABSTRACT: Within the Spanish territorial situation, the walls of rammed earth provide a great thermal inertia and an average insulation that depends on the type of the soil, density and thickness. This is a basic principle for a bioclimatic design, as its energy balance in the southern orientation is more convenient. Thus, when we study in detail the colder sides of a building and we try to optimize efficiency by designing a zero energy building, or almost completely passive, we miss a higher thermal insulation rate in the walls facing north. This work demonstrates, in all cases, a significant improvement in thermal conductivity in a percentage up to 60% in the most optimal compositions. This is accomplished by the contribution of the natural cork shavings with different degrees of fineness. The relationship between optimal stabilization and size and dosage of particles, is carefully analyzed.

1 INTRODUCTION

Thermal inadequacy in rammed earth buildings has, in some cases, become evident when trying to comply with energy saving requirements governing the Basic HE Document of the Spanish Technical Building Code (CTE DB HE in Spanish). In designs where the walls are primarily made of rammed earth and cladding is not included, the insulation indexes are typically in the order of 0.6 at 1.2 W/mK (Bauluz & Bárcena, 1991) for standard thickness between 30 and 60 cm, depending also on the density obtained, amount of clay and soil type.

Therefore, when comparing the standardized conductivity indexes of the core materials of conventional construction and those made of earth, the rammed earth type remain in a middle position.

In order to comply with the maximum thermal transmittance U, which according to climatic areas is 0.74 at 1.22 in W/m^2K for walls and interior partitions of the thermal envelope, the walls would have to be almost one meter thick.

Therefore, the inclusion of internal or lightening insulators in the normal composition of rammed earth walls is a practice that, having been tested in different studies, generates significant improvements in this area, even if this requires special monitoring of the loss of other attributes, such as compressive strength.

material/reference	density	conductivity λ
rammed earth, sun-dried brick, Compressed Earth Block (CEB) *(Eduardo Torroja Construction Sciences Institute, CEPCO & AICIA,2010)*	1770 - 2000 kg/m³	1,10 W/mk
rammed earth *(Bauluz & Bárcena, 1991)*	1400 kg/m³	0,60 W/mk
	1600 kg/m³	0,80 W/mk
	1800 kg/m³	1,00 W/mk
	1900 kg/m³	1,20 W/mk
	2000 kg/m³	1,60 W/mk
CEB *(Dominguez, 1998)*	1700 kg/m³	0,81 W/mk
CEB recycled C UdG BTC UNE 92 *(Alvarez Vazquez & Potrony Serret, 2001)*	1960 kg/m³	0,41 W/mk
CEB recycled 1 UdG BTC UNE 92 *(Alvarez Vazquez & Potrony Serret, 2001)*	1710 kg/m³	0,54 W/mk
sun-dried brick *(Dominguez, 1998)*	1200 kg/m³	0,46 W/mk
BTC Bioterre *(Bioterre technical specifications)*	1790 kg/m³	0,48 W/mk
Cannabrick *(Cannabirck technical spefification)*	1171 kg/m³	0,19 W/mk
reinforced concrete	2300 - 2500 kg/m³	2,3 W/mk
mass concrete in stiu	2000 - 2300 kg/m³	1,65 W/mk
baked clay for masonry units *(Eduardo Torroja Construction Sciences Institute, CEPCO & AICIA,2010)*	1700 - 2000 kg/m³	0,59 / 0,74 W/mk

Figure 1. Comparative values of rammed earth insulation (G. Barbeta & F. X. Massó).

2 BACKGROUND

2.1 *The use of black cork shavings*

In 1994 the author used cork as a stabilizer and additive to improve the grading curve and the thermal quality. The first tests were conducted with

"Licorela" type earth and shale with additions of 30% per volume of arlite, a type of lightweight aggregate derived from expanded clay based on thermal treatment and autoclaving. Thereafter, this work experience continued with two actual rammed earth building projects in Barcelona. Crushed black cork was used from the remains of the cork panel manufacturing process by pressure and temperature. Particle size obtained was in the order of 12/25.

2.2 *Arlite or expanded clay*

Arlite or ripiolite or expanded clay or leca is a lightweight ceramic aggregate.

The S-low modular building system consisting of a wood frame and rammed earth, developed in Barcelona by GICITED (Gonzalez Sanchez, 2013), uses walls stabilized on a central strip 10 cm thick with 12.2% of arlite in volume. Thus an average dry density of 2.07 g/cm^3 is obtained equivalent to proctor compaction of 73.75%, while insulation is improved by 25%.

Thermal conductivity	
Natural rammed earth cladding	1,110 W/mk
Arlite rammed earth—10mm	1,329 W/mk
Arlite rammed earth—8mm	0,830 W/mk

2.3 *Sirewall® System (Stabilized Insulated Rammed Earth)*

Sirewall, registered by a Canadian company, is a structural wall system 45.72 to 60.96 cm thick, which utilises compacted and stabilized earth reinforced with bars, with a 10.16 cm rigid insulation concealed in the centre of the wall. These built walls comprise a complete wall system with exterior and interior finish, similar to sandstone.

3 HYPOTHESIS FOR IMPROVEMENT

3.1 *Type of soil utilized*

The earth employed in this case was extracted from a country property in the surrounding area of the town of Santa Coloma de Farners, in the province of Girona, Catalonia. At the time of extraction there was abundant evidence of the validity of the material for an earth-based building project. The town's traditional architecture has many examples of rammed earth constructions and in addition, various brickyard were known in the area, in disuse today, which had their own quarry at the same spot. The visual and organoleptic analysis of the property carried out in-situ, along with the study of the geological maps of the area provided by the Geological Institute of Catalonia (IGC in Spanish), revealed soil composed of arkosic sandstone, with clay from the Pliocene era, granitoid metasediments of the Paleozoic era, including shale and gneiss and quartz gravels. The DRX has also found minority components such as Albite and Sanidine. The presence of clays is not evident in the conventional powder diffractogram. In the preparation of oriented particles, the diffraction peaks of the Illite and Kaolinite can be observed. Their presence is seen in magnitudes of 5%, which is quite consistent with the behaviour observed in sediment measurements.

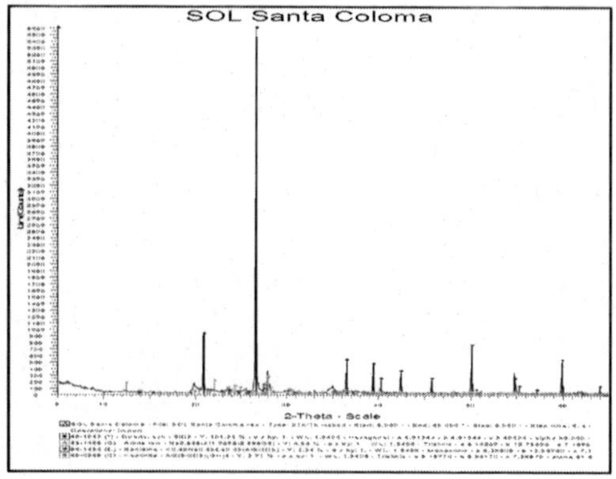

Figure 2. DRX. Quartz and 5% Kaolinite and Illite is noticeable (G. Barbeta & F. X. Massó).

3.2 *Physical and chemical analysis*

Particle size distribution of the particles is unbalanced according to the standards established for the execution of the rammed earth technique. The particle size distribution is comprised of, in a much higher percentage, of <2 mm particles that contrast with a very poor content of medium and large particles (fine gravel, gravel and stones). According to the USDA ternary graph classification, it is clay loam soil.

The plasticity index of the particles <0.32 mm (fine sands, silts and clays) has an approximate value of 12.7%.

The percentage of calcium carbonate is very low or non-existent, according to the qualitative determination of reaction generated by the attack with hydrochloric acid.

The presence of organic matter has been quantified slightly below 1%.

Linear shrinkage value is 3.5%, this factor will be reduced through physical and chemical stability, otherwise it would certainly create pathologies in work implementation where a normally long wall is at issue.

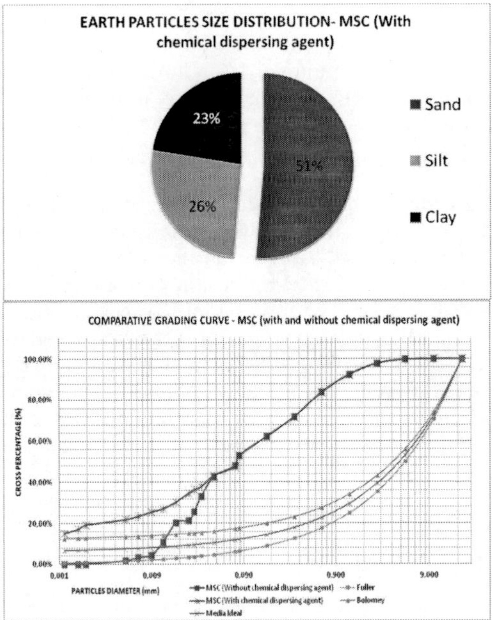

Figure 3. Grading curve graphics (G. Barbeta & F. X. Massó).

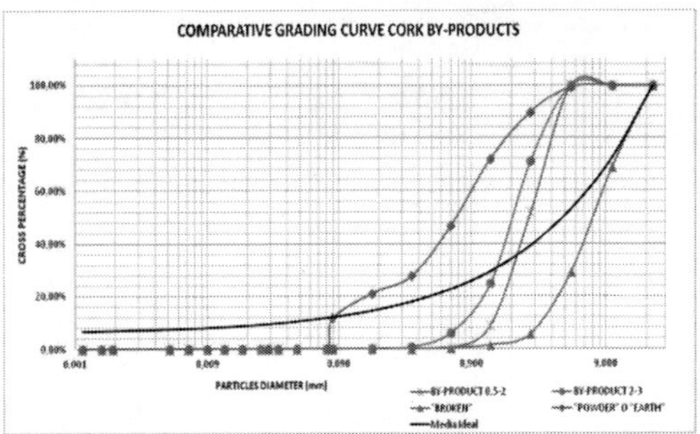

Figure 4. Particle size of the various cork by-products utilized (G. Barbeta & F. X. Massó).

3.3 *Cork characteristics as rammed earth physical stabilizer*

Natural cork is a raw material typical of the Mediterranean region, quite abundant in the Iberian Peninsula and much of the province of Girona in general. It is a natural, sustainable material which by its use fosters forest management in the region.

For this project, four types of natural cork shavings were used (by-product 0,5–2, by-product 2–3, *Broken* and *Powder* or *Earth*), all of these were obtained primarily as a by-product from cork processing for the wine industry.

There are few natural materials in the world with such extraordinary innate characteristics such as cork.

Its two great properties are its mechanical capacity of elastic memory capacity and its porosity. These, among other properties, make it an excellent thermal and acoustic insulation material with a very good fireproof behaviour, the ability to float and high resistance to friction as well.

4 LABORATORY TESTS

4.1 *Dosages*

Twenty-five different compositions were carried out with varying percentages and types of cork shavings, production water, cementitious stabilizers and using different methods of dynamic and static compression.

The inclusion of different percentages of natural cork shavings is directly proportional to the reduction in the percentage of local soil. The maximum percentage of cork shavings used in any of the types never exceeds 35%. This figure was set at the beginning of the study, based on information taken from other similar research and experiences, in relation to the significant loss of resistance caused by excessive lightening.

The cement product dosage in all the compositions which include natural cork shavings has always been maintained at 8%, with the aim of minimising the loss of compressive strength caused by lightened cement.

The static compression method, which was employed using a manual hydraulic press that makes it possible to always apply homogeneous levels of force in a small-scale simulation of the technique used for Compressed Earth Blocks (CEB), has been used to validate the static compression method used to produce the tested samples.

Figure 5. Rammed earth sample with "broken" cork shavings (G. Barbeta & F. X. Massó).

4.2 *Curing and strengthening process*

A curing or strengthening process of over 28 days was established for all the compositions tested under compression.

4.3 *Densities*

The main objective of including cork shavings is getting a material with improved insolating properties. In addition, it weight is significantly reduced. In order to determine the absolute density of the tested compositions, all the samples were measured and weighed and the average level of density of each composition was determined.

4.4 *Volume stability—linear shrinkage*

The local soil not subjected to any stabilising processes had a very high linear shrinkage level (3.5%), which, due to the construction method, rammed earth, compromised its usage. It was therefore considered, on a theoretical level, that applying physical and chemical stabilizers in varying percentages would result in significant improvements in this respect. A linear shrinkage test was carried out for each of the sample compositions (Barbeta Solà, 2002). According to the indications from the same source, an acceptable limit for this property was set at <20 mm/m or 1%, a limit which the compositions in question had to meet to be accepted.

composition nomenclature	natural cork shavings type	natural cork shavings percentage	chemical stabilizer percentage	water manufacturing percentage	linear shrinkage percentage
R1	-	-	-	20,00%	3,50%
R2	-	-	8% CP	20,00%	1,50%
R2.1	-	-	8% CP	12,70%	0,50%
R3	By-product 0,5-2	30,00%	8% CP	20,00%	1,00%
R4	By-product 0,5-2	20,00%	8% CP	20.00%	1,00%
R5	By-product 0,5-2	15,00%	8% CP	20.00%	1,08%
R6	"Broken"	30,00%	8% CP	20,00%	0,83%
R7	"Broken"	20,00%	8% CP	20,00%	1,17%
R8	"Broken"	15,00%	8% CP	20,00%	1,17%
R9	"Powder" o "Earth"	30,00%	8% CP	20,00%	1,08%
R10	"Powder" o "Earth"	20,00%	8% CP	20,00%	1,08%
R11	"Powder" o "Earth"	15,00%	8% CP	20,00%	1,50%
R12	By-product 2-3	30,00%	8% CP	20.00%	1,16%
R13	By-product 2-3	20,00%	8% CP	20,00%	1,25%
R15	By-p. 0,5-2 + By-p. 2-3	30%*	8% CP	20,00%	0,83%
R16	By-p. 0,5-2 + By-p. 2-3	20%**	8% CP	20,00%	0,92%
R17	By-p. 0,5-2 + "Broken"	35%***	8% CP	20,00%	0,66%
R18	By-p. 0,5-2 + "Broken"	30%****	8% CP	20,00%	0,83%
R19	By-p. 0,5-2 + By-p. 2-3 + "Broken"	35%*****	8% CP	20,00%	0,83%
R20	By-p. 0,5-2 + By-p. 2-3 + "Broken"	30%******	8% CP	20,00%	0,67%

CP. Portland Cement CEM I

*By-product 0,5-2 (15%) + By-product 2-3 (15%)

** By-product 0,5-2 (10%) + By-product 2-3 (10%)

*** By-product 0,5-2 (20%) + "Broken" (15%)

****By-product 0,5-2 (15%) + "Broken" (15%)

*****By-product 0,5-2 (12,5%) + By-product 2-3 (12,5%) + "Broken" (10%)

****** By-product 0,5-2 (10%) + By-product 2-3 (10%) + "Broken" (10%)

Rejected

Accepted

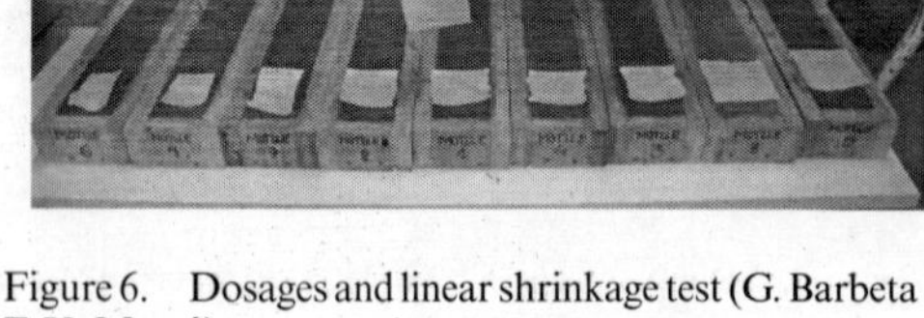

Figure 6. Dosages and linear shrinkage test (G. Barbeta & F. X. Massó).

4.5 *Breakage under compression*

In order to demonstrate and prove that thermal improvements do not compromise the rammed earth resistance capacity, all the sample compositions were subjected to the compression test. This was carried out with six samples of each composition, following the parameters set out in standards UNE 41410 and UNE 772-1. The acceptance or rejection value was set at 3–4 N/mm^2. The dynamic compression system used (Rn compositions) has been validated, as the samples with identical compositions which were subjected to one or another compression method gave very similar results.

4.6 *Thermal test*

This test was carried out using a calibrated hot box that had been made beforehand (Sancho Porcell, 1997) and according to the scale for a standard hot

composition nomenclature	water manufacturing percentage	compaction methods	resistance capacity - standarized levels
R1	20,00%	dynamic	5,4 N/mm^2
R2	20,00%	dynamic	6,29N/mm^2
R2.1	12,70%	dynamic	8,60 N/mm^2
R3	20,00%	dynamic	3,91 N/mm^2
R4	20,00%	dynamic	4,56 N/mm^2
R5	20,00%	dynamic	3,65 N/mm^2
R6	20,00%	dynamic	2,00 N/mm^2
R7	20,00%	dynamic	2,71 N/mm^2
R8	20,00%	dynamic	3,03 N/mm^2
R9	20,00%	dynamic	3,06 N/mm^2
R10	20,00%	dynamic	4,08 N/mm^2
R11	20,00%	dynamic	4,77 N/mm^2
R12	20,00%	dynamic	3,96 N/mm^2
R13	20,00%	dynamic	4,00 N/mm^2
R15	20,00%	dynamic	4,78 N/mm^2
R16	20,00%	dynamic	5,29 N/mm^2
R17	20,00%	dynamic	2,61 N/mm^2
R18	20,00%	dynamic	3,03 N/mm^2
R19	20,00%	dynamic	3,98 N/mm^2
R20	20,00%	dynamic	4,10 N/mm^2
BTC T	12,70%	static	6,65 N/mm^2
BTC C1	12,70%	static	7,09 N/mm^2
BTC Sn	20,00%	static	5,76 N/mm^2

Rejected

Accepted

Figure 7. Levels obtained under compression (G. Barbeta & F. X. Massó).

Figure 8. Thermal chamber used (G. Barbeta & F. X. Massó).

box. In order to determine the thermal conduction coefficient of the Sant Feliu compressed earth blocks, a standard test was carried out using standard UNE 92 204:1995 and the tests performed by the University of Saskatchewan (Hutcheon & WH, 1949) as a reference. In order to identify the thermal conduction coefficient of the material being studied, the flow of heat transmitted from one hot side to a cold side was determined by measuring air temperatures and the surfaces on both sides of the sample, as well as its measurements (thickness and area). The aim is for the flow of heat from one chamber to be in a steady state.

With the aim of complying with the Basic Document of the Spanish Technical Building Code and the basic principles for a passive design with bioclimatic criteria, calculations were made of the necessary wall widths in relation to the thermal conduction coefficient obtained and the maximum transmittance levels according to the code, with the required level being obtained in half of

composition nomenclature	density	conductivity λ	e wall within maximum transmittances*				
			zone A**	zone B**	zone C**	zone D**	zone E**
R1	2080 kg/m^3	1,35 W/mk	0,88 m	1,03 m	1,19 m	1,34 m	1,59 m
R2	1830 kg/m^3	0,82 W/mk	0,53 m	0,63 m	0,72 m	0,81 m	0,97 m
R2.1	1990 kg/m^3	1,27 W/mk	0,83 m	0,97 m	1,12 m	1,26 m	1,50 m
R3	1540 kg/m^3	0,45 W/mk	0,29 m	0,34 m	0,40 m	0,45 m	0,53 m
R4	1650 kg/m^3	0,65 W/mk	0,42 m	0,50 m	0,57 m	0,65 m	0,77 m
R5	1720 kg/m^3	0,64 W/mk	0,42 m	0,49 m	0,56 m	0,64 m	0,76 m
R6	1410 kg/m^3	0,52 W/mk	0,34 m	0,40 m	0,46 m	0,52 m	0,61 m
R7	1540 kg/m^3	0,60 W/mk	0,39 m	0,46 m	0,53 m	0,60 m	0,71 m
R8	1590 kg/m^3	0,48 W/mk	0,31 m	0,37 m	0,42 m	0,48 m	0,57 m
R9	1630 kg/m^3	0,57 W/mk	0,37 m	0,44 m	0,50 m	0,57 m	0,67 m
R10	1720 kg/m^3	0,50 W/mk	0,32 m	0,38 m	0,44 m	0,50 m	0,59 m
R11	1740 kg/m^3	0,59 W/mk	0,38 m	0,45 m	0,52 m	0,59 m	0,70 m
R12	1650 kg/m^3	0,56 W/mk	0,36 m	0,43 m	0,49 m	0,56 m	0,66 m
R13	1690 kg/m^3	0,68 W/mk	0,44 m	0,52 m	0,60 m	0,68 m	0,80 m
R15	1620 kg/m^3	0,74 W/mk	0,48 m	0,57 m	0,65 m	0,73 m	0,87 m
R16	1680 kg/m^3	0,86 W/mk	0,56 m	0,66 m	0,76 m	0,85 m	1,02 m
R17	1460 kg/m^3	0,64 W/mk	0,42 m	0,49 m	0,56 m	0,64 m	0,76 m
R18	1490 kg/m^3	0,71 W/mk	0,46 m	0,54 m	0,63 m	0,70 m	0,84 m
R19	1490 kg/m^3	0,56 W/mk	0,36 m	0,43 m	0,49 m	0,56 m	0,66 m
R20	1520kg/m^3	0,60 W/mk	0,39 m	0,46 m	0,53 m	0,60 m	0,71 m

*CTE DB HE maximum transmittances, equation calculation: e minimum = (1/U-(Rsi – Rse)*λ)
Coefficient Rsi and Rse for e calculation; Rsi=0,13 and Rse=0,04.
**Zones maximum transmittances: Zone A = 1,22 W/m2k / Zone B = 1,07 W/m2k / Zone C = 0,95 W/m2k / Zone D = 0,86 W/m2k / Zone E = 0,74 W/m2k.
Acceptance wall: e ≤0,60m

Rejected
Accepted

Figure 9. Table of thermal conductivity levels obtained and, in grey, cases in which the requirements of the Spanish Technical Code are met with wall widths of ≥0,60 m.

the cases, as can be seen in the table on the following page.

5 CONCLUSIONS

Portland cements is the factor which, in this case, results in the greatest reduction in linear shrinkage, that is of 50%. The inclusion of different types, sizes and percentages of cork, instead of only one, results in slight improvements in volume stability, making it possible to comply with the acceptance limit of ≥1% in all cases. The stabilization applied has proved to be effective in improving volume stability, compressive strength and has substantially improved thermal conductivity. Using a homogeneous mixture in the entire rammed earth matter is a very attractive option, as improvement methods in different layers excessively complicate production, by making compression and pouring of the material difficult. This fact has demonstrated that, despite having soil with low potential, by applying stabilizers with a solid foundation of prior tests, very significant improvements are achieved. Based on the results outlined above, building a rammed earth wall in the future with natural cork shavings throughout its substance, which complies with thermal standards and high thermal requirements, is a viable option, with a view to more efficient and sustainable construction.

REFERENCES

AEN/CTN 41. 2011. *Standard UNE EN 772-1 Test methods for masonry units—Part 1 Determination of compressive strength.* Madrid: AENOR.

AEN/CTN 41 Construction/SC 10. 2005. *Standard UNE 41410 Compressed Earth Blocs for walls and partitions. Definitions, specifications and test mehods.* Madrid: AENOR.

AEN/CTN 92 Thermal insulation ANDIMAT [Spanish National Association of Manufacturers of Insulating Materials]. 1997. *Standard UNE EN ISO 8990. Determination of steady-state thermal transmission properties. Calibrated and guarded hot box.* Madrid: AENOR.

Álvarez Vázquez, J.M. & Potrony Serret, E. 2001. *P.F.C, Anàlisi d'envelliment, i millora del bloc de runa comprimida. [Analysis of the ageing and improvement of the compressed earth block],* Girona: University of Girona.

Barbeta Solà, G. 2002. *Doctorate thesis. Mejora de la tierra estabilizada en el desarrollo de una arquitectura sostenible hacia el siglo XXI. [Improvement of stabilised soil in the development of sustainable architecture towards the 21st Century.]* Barcelona: Higher Technical School of Architecture.

Bauluz, G. & Bárcena, P. 1991. *Bases para el diseño y construcción con tapial. [Basic rules for designing and building with rammed earth.]* Madrid: Spanish Ministry of Public Works and Urbanism.

Domínguez. M. 1998. *Propiedades térmicas de los adobes. [Thermal properties of adobe] CSIC. Architecture of soil.* Madrid: Spanish Ministry of Public Works and Transport.

Eduardo Torroja Construction Sciences Institute. 2009. *Documento Basico HE del Codigo Técnico de la Edificación (CTE DB-HE) [Basic Document HE, Spanish Technical Building Code.]* Madrid: Boletín Oficial del Estado.

Eduardo Torroja Construction Sciences Institute, with the collaboration of CEPCO and AICIA. 2010. *Codigo Técnico de la Edificación; Catalogo de Elementos Constructivos [Technical Building Code; catalogue of construction material.]* Madrid: Boletín Oficial del Estado.

González Sánchez, B. 2013. *P.F.G Informe de las propiedades térmicas y mecánicas del sistema modular S-Low de construcción con estructura de madera y envolvente de tapia. [Report on the thermal and mechanical properties of the S-Low modular construction method with a wood structure and rammed earth cladding.]* Barcelona: Higher Technical School of Architecture.

Hutcheon, N.B and WH. Ball. 1949. *Thermal conductivity of Rammed Earth.* University of Saskatchewan: C. Engineering.

Sancho Porcell, F. 1997. *P.F.C. Mur de terra estabilitzada aïllant. [Stabilised insulated rammed earth wall.]* Girona: University of Girona.

Walker, P., Keable, R., Martin, J. & Maniatidis, V. 2005 *Rammed Earth: design and construction guidelines.* Bracknell: BRE Bookshop.

Earthen Architecture: Past, Present and Future – Mileto, Vegas, García Soriano & Cristini (Eds)
© 2015 Taylor & Francis Group, London, ISBN 978-1-138-02711-4

36 *Thankas* to promote the conservation of the Ladakhi earthen heritage

M. Bertagnin & D. De Antoni
University of Udine, Udine, Italy

A. Skedzuhn-Safir
Brandenburg University of Technology Cottbus—Senftenberg, Cottbus, Germany

M. Oeter
University of Bamberg, Bamberg, Germany

H. Vets
Freelance Architect, Berchem, Belgium

ABSTRACT: Today it is widely accepted that the conservation of the architectural heritage as an essential ingredient for future local development can be ensured through the active involvement of communities. Therefore it is important to heighten the awareness of the significance of heritage conservation within the communities. The paper discusses the experience of an exhibition, as part of a strategy to increase the interest in heritage conservation, and to foster community driven heritage conservation in Ladakh, India. The exhibition is designed within the framework of a cooperation between the Achi Association, and students of the School of Architecture at the University of Udine. The Achi Association is an NGO based in Europe which has been working in Ladakh since 1999 on earthen architectural heritage endangered through climate change related effects such as excessive rainfall, floods, or landslides, and an increased pressure of modernization on living standards and people's expectations.

1 INTRODUCTION

1.1 *Activities of the Achi Association*

The Achi Association is an organization which has been working in Ladakh, in the north of India, since 1999. The main goal of the team is to preserve the Ladakhi earthen architecture, paintings and sculptures, belonging to the Buddhist religious tradition.

Ladakh is also called "Little Tibet" as it shares with Tibet many cultural expressions including building typologies and materials, architectural ornaments or intangible heritage such as ritual dances and songs typical of the Buddhist belief. The villages where the association is mostly active are Kanji, Wanla and Skurbuchen, situated in the Himalayan mountains.

In these places many campaigns were carried out, providing numerous positive results: conservation of Buddhist temples, chörtens and wall paintings. All of these interventions were implemented with the help of the local population. An objective of the Achi Association is in fact to improve the awareness of communities concerning their heritage, striving to involve them in the process of site preservation.

1.2 *Idea of the exhibition*

The Ladakhi earthen architecture heritage is suffering an increasing process of decay due to factors like climate change and its related effects such as excessive rainfall, floods, and land slides. However, also the pressure due to modernization on the expectation on living standards and commodities have caused a shift in the perception of traditional building culture.

For this reason the Achi Association is continuously setting up new strategies in order to improve the impact of heritage conservation, and creating new tools to integrate heritage conservation into everyday life.

The paper presents the design of an exhibition designed in the framework of a cooperation between the Achi Association and the School of Architecture at the University of Udine which was implemented in the course "Sustainable Architecture" as part of the master's programme.

The exhibition presents the territory of Ladakh and its main characteristics tied to the tangible as well as intangible heritage, the economic and social aspects and some of the results of the interventions by the Achi Association carried out during the last 15 years (Bertagnin & De Antoni 2011).

2 THE EXHIBITION

Several years ago the Achi Association was planning an exhibition on the preservation of earthen architectural heritage in Ladakh as part of a three-year youth training programme supported by the Getty Foundation in Los Angeles. Aim of the exhibition was to raise awareness within the Ladakhi visitors to the exhibition but also to sensitize the young Ladakhi trainees who were preparing the material for the exhibition together with the trainers. Due to unfortunate circumstances and difficulties on organizational level the exhibition could not completely be prepared and presented.

The idea to sensitize the people living in Ladakh for the cultural built heritage however remained a significant aim of the Achi Association. Therefore, the idea to prepare an exhibition in Europe would alleviate organization and preparation of material while still maintaining the main intention, namely to raise awareness for conservation related challenges in Ladakh and to ideally stimulate discussion within communities in Ladakh. Seeing your own cultural identity explained and profiled in an exhibition, might encourage self-reflection and self-esteem which could enhance awareness for the tangible and intangible heritage.

2.1 *Design of the exhibition*

Today is widely accepted that the future of the architectural heritage can be ensured through the active involvement of communities. The participatory approach is in fact an essential ingredient for future local development. Therefore it is important to improve the community awareness concerning the preservation of heritage as the strategic goal for a sustainable future development.

The students, assisted by their professor and some of the Achi Association experts were asked to design an exhibition reflecting the traditional way of Buddhist visual communication. Different options had to be identified to find the most appropriate communication tool. It was decided to use a traditional method of communication, the so-called *thanka,* or *thangka.* The *thanka* is a scroll painting with traditional images painted on cotton or silk. It is mounted on a fabric over which a protective cover is placed, usually made of silk. As it is a portable painting roll it can be taken along during journeys characteristic of a semi-nomadic population like that of Ladakh.

The *thanka* is basically used as an important teaching and meditation tool, often depicting a *mandala,* deities or myths linked with them, scenes from the life of Buddha, or historical events connected to specific lamas (Tucci 1949, Kreijer 2001). Therefore the *thankas* as a communication tool linked to meditation were considered appropriate for communicating the contents of the exhibition-as it resembles the traditional form of visual communication to which many people in Ladakh are accustomed.

The 36 *thankas* for the exhibition are also designed to recall the traditional craftsmanship. All the *thankas* in fact are made in a hybrid process consisting of both artisanal know-how, the sewing of coloured cloths, and modern reproduction techniques on a cotton support.

As the *thankas* are easy to transport not only across sea but also within Ladakh, the setting up of the exhibition can be carried out without much sophisticated equipment, within a short time frame and at relatively low cost. This in fact will enhance the possibility to present the mobile exhibition in many villages in Ladakh which contributes to or initiates a mobilisation process aiming to strengthen the cultural identity.

2.2 *Topics of the exhibition*

In the first part of the exhibition social and economic characteristics in Ladakh are presented.

In many Himalayan valleys, the practise of polyandry has not been completely abandoned. Due to social changes the new generations do not longer want to adhere to these traditions which are expected to disappear within a few decades.

The role of the women is still very much valued and important. They are highly involved in taking decisions concerning family matters as well as the working life regarding the agricultural and pastoral activities that are the base of the economy in Ladakh. The cultivation of the fields is concentrated in the four summer months, in a direct relationship with the land resources, to ensure a self-subsistence, necessary to cope with the winter period in particular. Some *thankas* present the breeding of yaks and donkeys, the harvesting and thrashing of barley.

The exhibition also explains the traditional food in Ladakh which is based most of all on vegetable products and barley flour. Yak's milk is used to produce cheese and butter, which occasionally is added to tea.

Other arguments of the *thankas* concern traditional garments (Fig. 1) which are embroidered waistcoats, robes in yak wool, capes in brocade

Figure 1. *Thanka* no. 8 represents the garments belonging to the Ladakhi tradition (Credits: M. Bertagnin).

and goatskin, woven scarves and shawls, silk and felts hats, embroidered pants, fine felt sashes with woven in beads, and in particular *gonchhas* made of velvet. In contrast to the king and the nobility, who had the ability of buying silk, velvet and brocade, the villagers produce their garments of locally woven yak and sheep wool.

Some panels show pictures of the musical instruments employed in Ladakh. The ritual music played in the monasteries is quite similar to Tibetan music where holy chants in Tibetan or Sanskrit are considered to be an integral part of the religion. The chants resemble rather a prayer than a song.

One of the aims of the Achi Association is to involve the local population in its activities. Thanks to the support of the Getty Foundation, in 2008 the so-called "Pilot Youth Training Program" was launched. The project's objectives were to assist local communities in capacity building for the maintenance of built architectural heritage and to introduce methods for preventive conservation through raising awareness. It was developed for young Ladakhi people, monks as well as lay people, who were expected to take responsibility for Ladakh's cultural heritage in the future.

Several of the images pictures employed for the exhibition show different youth group members being trained by the Achi Association members in the preservation of the traditional architecture, sculptures and wall paintings.

The exhibition presents the Ladakhi architectural typologies, which are divided into three groups: buildings used as housing, buildings used for worship (temples and monasteries) and public buildings (schools and local meeting places). 15 buildings are represented in scale mobile as well as folding models realized with the Japanese technique called *kirigami*, which is a variety of *origami* (Figs. 2–3).

In order to understand the main characteristics of these constructions, we present the building materials (earth, wood, and so forth) and their implementation which follows, most of the time, the traditional building technology.

The exhibition shows the work the University of Udine in collaboration with Achi Association carried out in order to prepare the "Kanji Conservation Manual" (Bertagnin et al. 2010).

Some panels present the contents of this future publication (Fig. 4). The manual is designed to be used as a guide for any maintenance and conservation work within the present Kanji urban fabric and in other similar sites like Wanla and Skurbuchen. The conservation manual could also help the appropriate design and construction of new housing in these types of villages.

Figure 2. Building mobile model realized with the *kirigami* technique (Credits: M. Bertagnin).

Figure 3. Wanla temple in a *kirigami* model (Credits: M. Bertagnin).

Figure 4. *Thanka* no. 27 presents the content of the future publication "Kanji Conservation Manual" (Credits: M. Bertagnin).

Figure 5. *Thanka* no. 30 explains the conservation of the sculptures carried out in the Tsuglag-Khang temple in Kanji (Credits: M. Bertagnin).

In other *thankas* of the exhibition pictures and drawings are shown concerning the architectural traditional details such as wooden capitals and roofs of the buildings in Kanji and some of the interventions that should be carried out in order to maintain these traditional building elements.

Some *thankas* are dedicated to the conservation of wall paintings and sculptures and present the work of the Achi Association in the Tsuglag-Khang temple in Kanji and in the Avalokiteshvara temple in Wanla (Figs. 5–6). The conservation work addressed the damage that was mainly caused by irregular maintenance and improper building preservation works.

During the past years, on and off-site investigation regarding paint layer and plaster were carried out with the aim to obtain additional information on materials and work techniques as well as the cause for characteristic degradation patterns.

The exhibition in fact shows the study and results of the (chemical) analyses carried out on the paint and plaster layers of the interior walls in the temples in Kanji and Wanla.

To understand the cultural landscape of Ladakh, a part of the exhibition focuses on the basic ingredients of this indigenous and autarchic society.

The harsh living conditions in this remote barren landscape and extreme climate have influenced the social and economic life in a very respectful and energy-saving way to ensure self-sufficiency. Villages were built as dense homogenous clusters of two or three storied buildings, thus reassuring natural protection and social cohesion.

Figure 6. *Thanka* no. 22 shows the study and results of the (chemical) analyses carried out on the paint and plaster layers of the interior walls in the temples in Kanji and Wanla (Credits: M. Bertagnin).

Buddhist religion is fully embedded in daily life and religious structures are omnipresent in the surrounding landscape. While temples and monasteries were constructed mostly above the villages, ancient ritual paths dotted with (gateway) *chörten*, prayer wheels and *mani* walls allow a spiritual approach and circumambulation or *skora* of this sacred area. Only local resources were used as building materials, mainly stones, earth and wood.

The traditional buildings have masonry foundations and the walls are built with adobe bricks called *pakbu*. Some older buildings, for instance castles or towers are constructed out of rammed earth walls. Roofs are flat and made of poplar tree trunks and willow sticks, padded with dry grass and covered with a layer of mud, a mixture of a specific with clay, locally called *markalak,* small stones, straw, animal hair and water.

The historic settlements are severely challenged by the changing Ladakhi lifestyle and by climate change (UNESCO 2007). Indigenous cultures are very sensitive and fragile to westernization and modernization, certainly when it arrives with an incredible speed and 'aggression', and seems to offer sudden comfort and higher living standards. Ancient social structures are getting disrupted resulting in different living patterns and hence a downturn in traditional customs, values and all aspects of life.

The necessary maintenance of monuments and buildings can no longer be secured through traditional social structures (Bertagnin & De Antoni 2012).

The traditional building practices with local materials are getting lost and replaced by a new form of architecture and architectural language that requires imported materials such as cement or metal and is executed mostly by non-native workers.

This is exacerbated by several other factors, among them the sharp increase in outward economic migration, mainly among the male adult segment, and the growing proliferation of an uncontrolled form of tourism.

In addition, the marked increase in summer precipitation in recent years, make the negative impact caused by the absence of traditional maintenance measures even more dramatic.

One of the aims of Achi Association is to raise awareness and strengthen pride and identity among Ladakhis for their culture and heritage. One way to achieve this is to actively involve the communities and local craftsmen into the various processes and decisions of on-going or future architectural conservation work. Several heritage campaigns of the above mentioned Pilot Youth Training Program, the conservation of the Avalokiteshvara temple in Wanla during the last 10 years and the conservation project of the Kagan Chörten in Wanla in 2011, managed to raise the interest in heritage preservation among the villagers in Wanla, also due to fact that local craftsmen, volunteers and school children were actively involved in the projects. Cooperation and expertise was asked for future conservation/preservation projects and to build local capacity for the revival of historic building techniques and preventive conservation methods.

3 EXPECTED OUTCOMES

A man travels the world over in search of what he needs and returns home to find it

(Moore, 1916).

Devising an exhibition is generally perceived with a specific audience in mind. The content will have to be adapted to the context and the audience, with regard to age group, educational and cultural background, gender and so forth. The underlying question is thus always: Which "message" can be delivered through which methods to whom? The focus, however, is on the so-called "receiver" of a message. Everything is focused on the audience and the expected outcome the information transfer may have: how the receiver may re-act to the new or differently presented information and how his cognitive capabilities can be stimulated?

Preparing an exhibition on traditional and vernacular heritage in a remote area of the Himalayas by a European student group bears several traps. To stimulate the interest of all these different views, the posters are presented in a form similar to *thankas*, which is common to the people in Ladakh, but also interesting for non-Ladakhis. The research is conducted and the texts are written by the students who never have been to Ladakh. The idea is to produce material interesting and easy enough to read for people who never have travelled to the Himalayas or who never before have been interested in cultural heritage in general and the vernacular in particular.

But what does the transfer of information do with the senders, in this case the architecture students? Will the preparation and presentation of information have an effect and added value for them? The student's research on Ladakh's culture and above all building culture involves a much deeper dealing with and at least an attempt to understand the foreign culture than the audience of the exhibition will probably experience. Additionally, the wish to comprehend other cultures and their way of life will ultimately also reflect on how the own culture is perceived. This makes a project of preparing an exhibition with students a very stimulating teaching tool (Fig. 7).

This is the pivotal point, in which the understanding of the vernacular architecture of Ladakh is expected to influence the attitude of the students towards the vernacular architecture in general. In fact, the sensitization for the importance of traditional vernacular heritage is double fold: the receiver in Ladakh ideally would be made aware

Figure 7. Students of the course "Sustainable Architecture", academic year 2013–2014 (Credits: M. Bertagnin).

of the challenges of earthen built heritage, but the same would be also true for the sender in Italy regarding the built traditional heritage in front of the own (vernacular) door step.

4 CONCLUSIONS

The design of an exhibition that can provide a panoptic picture of the experience of an association such as the Achi Association that has dedicated its activities to the conservation of Ladakhi earthen architecture and related art, can certainly help to improve the awareness of the involved communities on the importance of the preservation of their tangible and intangible heritage. The awareness for cultural resources on the part of the local community is, in fact, a crucial factor for the participation in conserving built heritage, cultural landscapes, and elements of intangible heritage.

An exhibition can support the community engagement in the valorisation of local heritage as well as the recognition of the values of each community. This is crucial in defining what should be conserved so that heritage has a continued relevance in society, especially in small rural communities. The mobile exhibition conceived by young people may be the appropriate tool that can be adopted for knowledge transfer ranging from architectural heritage, oral traditions, education and awareness programmes.

An exhibition conceived by young students for young people in remote communities has certainly the merit of being fresh and original as the craft production of *thankas* and *kirigami* models have shown. The exhibition can provide a satisfying result in terms of communication, especially among the main stakeholders, the youth of the community of Ladakh who are the main addressees of the exhibition.

REFERENCES

Bertagnin, M. et al. 2010. Kanji: the Conservation Manual. *In Terra em Seminário 2010*: 48–52.

Bertagnin, M. & De Antoni, D. 2011. A Preventive Maintenance Work Site At Kanji And Observations On The Degradation Of Earthen Architecture. In: Rostislav Drochytka and Stepan Bohus (eds) *Building Materials and Building Technology to Preserve the Built Heritage*. Vol. 2, Part 1: 85–94. Munich: WTA Publications.

Bertagnin, M. & De Antoni, D. 2012. Earthen Architecture After Disaster: Initial Risk-Reduction Measures in Kanji Village for Conserving Non-Engineered Rural Ladakhi Architecture. In: *Proceedings of Terra 2012—XIth International Conference on the Study and Conservation of Earthen Architectural Heritage:* 22–27. Lima, Peru.

Kreijer, H. E. 2001. *Tibetan Paintings: The Jucker Collection*. Boston: Shambhala Publications.

Moore, G. 1916. *The Brook Kerith: A Syrian Story*, ch. 11, p. 123.

Tucci, G. 1949. *Tibetan Painted Scrolls*. 3 vol., Rome.

UNESCO 2007. Climate Change and World Heritage. Report on predicting and managing the impacts of climate change on World Heritage and Strategy to assist States Parties to implement appropriate management responses. In *World Heritage Reports* no. 22, Paris: UNESCO World Heritage Centre.

Earthen Architecture: Past, Present and Future – Mileto, Vegas, García Soriano & Cristini (Eds)
© 2015 Taylor & Francis Group, London, ISBN 978-1-138-02711-4

The first Italian *Rammed earth restoration guidelines*

G. Bollini
Architect, Cassola VI, Italy

I. Parodi
Department of Territorial Management and Urban Planning, Novi Ligure Town Council, Novi Ligure, Italy

ABSTRACT: Rammed earth is one of the earth construction techniques in Italy. Nevertheless the approach proposed in the *Guidelines* can be applied to other earthen-architecture technologies too. Nowadays part of the rammed earth heritage is lost due to neglect and to the idea that a new house, made of "performing" contemporary materials, is synonymous with better living and architectural quality. But something is changing: the growing attention of the Public Administration and the ever more pressing environmental problems demand a more sustainable approach to the building process. The Guidelines evolved on the basis of these thoughts. The starting point is the belief that keeping this heritage alive means keeping it inhabited. To make it possible it is necessary to reintroduce earthen heritage into the normal real estate cycle. This step is not automatic since in the last decades, both the knowledge and the technical know-how were lost, due to the interruption in memory. The Guidelines goal is contributing to fill this gap. This paper illustrates the project that led to the development of the *Rammed earth restoration guidelines*.

1 INTRODUCTION

(Properly) speaking about earthen buildings means, first and foremost, dispelling some deeply-rooted clichés that (improperly) associate such material with a state of poverty and uncertainty (economic, social and cultural). They are of course one of its dimensions, but they are not the distinctive trait of raw earth. Where it is used, it is but a resource like any other.

An earthen built heritage is typical of the architectures and landscapes of a multitude of countries, including Italy.

It tells our story, the social and economic changes experienced by our regions, and the economy that used to support them. Its current, often decayed or forsaken state does not mean that the material performs poorly; it means instead that the national rural model has evolved and the rural world has experienced a crisis, where people had to move to the big cities and have often abandoned such buildings (Bollini 2008).

Differing according to the building options available in each area, they are scattered all along our peninsula as specimens of a high architectural value, smart technology and "know how". Of course, a lot of these have been lost due to neglect, poor upkeep, and the belief nowadays that a new home built with the "innovative" and better "performing" materials of the contemporary age meant a higher architectural quality and better living. But something is changing: people's sensitivity, the public authorities' attention, not least the increasingly-pressing environmental issues, which call for a more rational, sustainable approach to construction.

Earthen buildings, the legacy of that old culture that knew how to perfectly match materials, climate and functional requirements, translating them into "appropriate" building techniques and accurate architectural and typological choices, now find a reason for redemption in this sense as well (Bollini 2008).

2 THE PROJECT

2.1 *The starting point*

Opting for recovery and reuse does not only mean protecting an ancient building method. It can conversely add great value and boost a very sustainable development of an area (Bollini 2008).

Over time, the heritage has grown much poorer due to poor maintenance and wrongful projects.

Such reflections and considerations have led the Municipality of Novi Ligure (Piedmont, Province of Alessandria, the Italian symbol of rammed earth), supported by Associazione Nazionale Città della Terra Cruda (www.terracruda.org) and the Piedmont Region, to have a working tool that could actually respond to the technical-cultural requirements and protect a heritage of proven tangible and intangible values.

The outcome of all this is the book "Terra battuta: tecnica costruttiva e recupero. Linee guida per le procedure di intervento", published by Edicomedizioni in December 2013, the birth and contents of which are presented hereunder (hereinafter referred to as *Rammed Earth Restoration Guidelines*) (Fig. 1).

The starting point is the belief that keeping such heritage alive means keeping it inhabited; it is the everyday life that take place in it that permits it to survive and be passed on to the next generations. But it must be contemporary everyday life, not the slowly dying life of old owners that are still bound to a place out of affection or social status.

To do this, such heritage must be brought back into the normal economic cycle of the property market.

However, such a transition is not so natural, since expertise in the material (earth) and the know-how (design and technical) have been lost over the decades, due to a gap of memory and cultural oblivion (Bollini 2009).

Figure 1. The cover of the handbook.

The goal of a handbook about proper restoration projects is to help fill these gaps; in other words, facing the technical criticalities of recovering an earthen heritage as a solution to the cultural problem that tends to dismiss it.

2.2 *The rationale*

Before going into the contents of the guidelines, we need to comprehend their purpose so as to avoid any misunderstanding.

Over the last thirty years, most of the dimensions of earthen building have been explored (technology, performance, cultural and social), and the results have filled hundreds of pages, videos and now, even, Internet websites.

This paper does not mean to add up to what has already been produced thus far, despite its potential to further explore such issues; its purpose is to act as technical support to policies aiming to protect and recover the built earthen heritage and its landscape, in the sense of coming possessing again an identity, and (why not?) as a propeller of the economy of an area.

The technological dimension explored herein is that of rammed earth (*pisé*, *tapia* or *terra battuta*), with more or less implicit local repercussions on the lower Piedmont region. In this respect, this paper is the umpteenth block in the local promotional choices and policies aiming to add new value to the built-up earthen heritage that have been going on in such regions since 1997. Despite this, they may be understood as taking inspiration, in terms of applicability, to the cob technology, since it too produces a monolithic raised construction that is typical of the Marche and Abruzzi regions; anyway, the message, considerations and approach herein proposed may also apply to other contexts and other forms of raw earth technology as well. The focus is actually the material, its specific traits, criticalities and potentials.

This paper will not try to understand what is or what should be the most consistent and correct cultural approach to recovery or renovation. In this sense, all that has been written and said about this subject so far, regardless of the material that is recovered or renovated, what is written in the European Landscape Convention and, last but not least, the designing sensitivity and skills that are inherent in our profession, are all true.

The focus is instead on the proper technical management of earth as a material, in the attempt to make the professional world, labour and clients trust it again.

From this perspective, the approach to recovery should not be hypocritical and show no emotional or nostalgic attitudes.

The text has been prompted by a few simple ideas that are key to a proper approach to an existing earthen construction:

- any work must be limited to as little as needed to make a building safe and stable;
- any legitimate aspiration to reuse must be sympathetic with the material and consistency of the building and add new resources rather than remove or waste the existing ones;
- the rationale behind the structural and thermal-hygrometric behaviour, as well as the balance that any building has achieved over time, must be understood and accepted.

Translating such beliefs into a work-tool that is fit for the real state of things and compatible with the current understanding of such constructions has resulted in a publication that tries to give directions about how to recover rammed earth constructions (or parts of them), while also trying to help users, professionals and managers make informed choices (Bollini 2013).

Figure 2. Typical wooden connection to tie the walls at each floor (G. Bollini).

Figure 3. The pavilion vaults made of fired bricks (G. Bollini).

3 THE GUIDELINES AND THEIR CONTENTS

Before we started to draw them up, we asked ourselves how they should be, for them to actually become a useful, effective tool, properly poised between valuable academic research and ordinary professional practice; a book developed by "professionals for professionals", halfway between a handbook and an operational procedure.

They may be consulted at different levels, depending on the need, from a full reading to taking just some specific aspects and issues, without it affecting the general understanding.

The book consists of three macro-sections. The first one (Basic References) provides basic notions to understand what earth means as a building material and therefore what a rammed earth building means: how it is made, what e its distinctive traits and structural rationale are, with some typological notes about the local context (Figs. 2–4).

The second one (Damage and Decay) helps find the most serious criticalities that such buildings may exhibit for natural or anthropic reasons, in the attempt to get the most correct insight (Figs. 5–7).

The third section (Solutions and Projects) deals with potential solutions and the proper approach to specific projects (opening a window, redoing the plaster, etc.). Such section was deliberately based on the technical-economic feasibility of a project, above all in the light of what the market of building components and materials has to offer (Figs. 8–12).

A (rammed) earth building responds well to those works that are as simple as possible and are sympathetic with its material and building technology. Lots of materials that are currently available on the market are not always fit for earthen buildings, as they are too rigid, waterproof, un-breathing or unable to adhere to the existing constructions. It is important to be familiar with a construction, with earth (as a building material) as well as with

Figure 4. Typical wooden roof with tiles (I. Parodi).

Figure 5. Lintel not clamped enough for the wide opening and subsidence of the parapet (right side) (G. Bollini).

Figure 6. Plaster detachment due to rising damp caused by erroneous design of lower external coating and paving as well (G. Bollini).

the available products, so as to be able to understand which ones may be compatible and which ones are not. There is no doubt that an "earth on earth" work would be the most desirable one and certainly the most rewarding one, but we think we

Figure 7. One of the numerous designs of the handbook, reworked by Isidoro Parodi. Abacus of the main degradation:

1. Erosion due to rain
2. Run-off due to water infiltration (by the roof)
3. Erosion due to rising damp
4. Efflorescence due to rising damp
5. Plastering detachment
6. Cracking due to bad manufacturing
7. Vertical fracture for erroneous loading distribution
8. Vertical damage due to a concentrated load.

Figure 8. Undo-redo technique to reset a lesion using fired bricks and lime mortar (G. Bollini).

should be realistic and we should admit that the survival of earthen buildings is closely dependent on the option to make them talk with the contemporary building world. Far from being a list of

Figure 9. The (re)connection of a new rammed earth wall to the existing one using wooden ties (G. Bollini).

Figure 10. Placement of new beams with underlying element for load distribution (in this case fired bricks) (G. Bollini).

Figure 11. The open of a new window or door using two wooden lintel: one inside and one outside, then connected together as a one by steel bars (G. Bollini).

Figure 12. The waterproof layer (the grey one; for about 20 cm from the ground) and the bed of gravel and sand for external paving (G. Bollini).

Figure 13. An example of restored rammed earth building (Novi Ligure, AL, Italy) (G. Bollini).

solutions, this section tries to provide a number of proper approaches.

Methodologically too, a choice has been made: in sympathy with earth, a combination of a proper working technique and the contemporary building world. Despite this, in any area in which true working consistency would opt for more traditional solutions, focused on using raw earth and the distinctive technology of an ancient expertise, they have been described, leaving it to the professionals to choose what way to go, in their ordinary design and building practice.

The book ends with an Addendums section, providing a number of further considerations to enrich and complete the notions provided, without making the other sections uselessly dull. The same applies to the "boxes" that have been scattered throughout the text.

A topic that has been addressed, but that has certainly not been solved yet, is the one about the regulations (at least in Italy) for structural adjustments (first and foremost, anti-seismic) and energy reduction.

While, when working on the traditional heritage, it may be handled without too many problems or uncertainties, with an earthen building things get more complicated, and quite a lot. In this sense, the subject has been critically dealt with by illustrating the current (regulatory) constraints and potential insights or procedures, even if they are often experimental. Raw earth is not one of the building materials covered by the Italian technical codes. What is provided by these pages cannot go beyond a critical analysis of the problem, which may only be responded too by political choices that are beyond the scope and specificity of such Guidelines.

4 NEXT STEPS AND FUTURE AIMS

The book was officially presented to the public; soon the validation process by the Superintendence of Architectural Heritage of the Piedmont Region will start.

The intention of the Administration of Novi Ligure is to extrapolate from the manual the official guidelines to be attached with the Rules of Implementation of the local Master Plan to promote (and to define) the recovery of the rammed earth architectural heritage.

The planning of specific training courses will follow (soon); the aim is to organize them in collaboration with the Professional Associations and the local School Building. The *Rammed earth restoration guidelines* will be the starting point and the common working tool as well.

In parallel, the handbook will be a useful reference and support tool to do outreach activities of citizens (monthly articles in the local press, events and thematic days etc.).

Certainly the work configures as an interesting point of arrival as well as a useful systematization of study and experiences, but it is mainly a starting point for the future policies of revaluation and proper recovery of the earthen heritage.

REFERENCES

Barozzi, P. 1975. Le Trunere della Frascheta nella piana di Alessandria. In *L'Universo* n. 3. IGM.

Bertagnin, M. 1992. *Il pisè e la regola. Manualistica settecentesca per l'architettura in terra*. Roma: Edilstampa.

Bollini, G. 2008. La conservazione e la valorizzazione del patrimonio in terra cruda: una scelta strategica. In P. Lombardi (ed), *Riuso edilizio e rigenerazione urbana*. Torino: Celid.

Bollini, G. 2009. Conservazione e tutela: lo scoglio della normativa italiana. In E. Galdieri (ed), *Scritti sulla terra*. Saonara (PD): Il Prato: 65–71.

Bollini, G. 2013. *Terra battuta: tecnica costruttiva e recupero*. Monfalcone (GO): Edicom Edizioni: 11–18.

Coppa Patrini, A. 1935. Costruzioni edilizie di terra battuta nel territorio della Frascheta (Alessandria). In C. Contessa (ed.), *Annuario Scolastico. Il Regio Liceo-Ginnasio G. Plana in Alessandria. Cronaca quinquennale 1929/34*. Alessandria: Tip. Colombani & C.

Fontaine, L. & Anger, R. 2009, *Bâtir en terre. Du grain de sable à l'architecture*. Paris: Éditions Belin.

Galbusera, G. & Panzeri, A. 2010, *Prestazioni estive degli edifici*. Milano: TEP srl.

Galbusera, G. (ed) 2011. *Guida alla nuova Legge 10*. Milano: TEP srl.

Graz, C. & Nourissier, G. et alii, 2008, *Terra incognita. Preserving European Earthen Architecture, vol. 2*. Lisbon: Argumentum.

Gulli, R. 2002. *Il recupero edilizio in ambito sismico*. Monfalcone (GO): Edicom Edizioni.

Jeannet, J. & Pignal, B. et al, 1997, *Le pisé. Patrimoine, restauration, technique d'avenir*: Nonette: CRÉER.

Mattone, R. 2010. *Il paesaggio delle case di terra cruda*. Savigliano (CN): L'Artistica Editrice.

Norme tecniche per le costruzioni e circolare esplicativa. DM Infrastrutture 14 gennaio 2008, Circolare 2 febbraio 2009 n. 617/C.S.LL.PP. Roma: Dei Tipografia del Genio Civile (2011).

Röhlen, U. & Ziegert, C. 2013, *Construire en terre crue. Construction—Rènovation—Finitions*. Paris: Le Moniteur.

Sanna, A. & Azteni, C. (eds) 2009, *I manuali del recupero dei centri storici della Sardegna. Architettura in terra cruda*. Roma: DEI Tipografia del Genio Civile.

Earthen Architecture: Past, Present and Future – Mileto, Vegas, García Soriano & Cristini (Eds)
© 2015 Taylor & Francis Group, London, ISBN 978-1-138-02711-4

Jute fabric to improve the mechanical properties of rammed earth constructions

S. Briccoli Bati, M. Fagone, F. Loccarini & G. Ranocchiai
Department of Civil and Environmental Engineering, University of Florence, Florence, Italy

ABSTRACT: Despite the awareness that earth cannot provide the same mechanical performances of other building materials, it should be asserted that this natural material is suitable for a large part of the construction types. With the ultimate aim to improve the carrying capacity of the earthen constructions, especially in the seismic areas, an experimental campaign was carried out on specimens and structural elements made with rammed earth and reinforced with natural and biodegradable fibre fabric. For these experimental studies we chose to use the jute fabric. The results obtained showed that the strengthening method improve the earthen specimens strength in order to achieve a standard of quality of the earth material. In fact fabric has the immediate consequence of limiting shrinkage and increasing mechanical properties, stability and durability. Acting with appropriate corrections and devices, rammed earth buildings can be constructed even in areas subject to seismic risk, guaranteeing acceptable safety margin.

1 INTRODUCTION

In the current context of sustainable development, the renewed interest in raw earth as a building material derives from several qualities such as local availability and recyclability, suitability to satisfy the need of low cost housing, self-build practice and preservation of natural resources (Pacheco-Torgal, F. & Jalali, S. 2012). These features make earth material and the corresponding construction techniques extremely competitive if compared to the conventional ones. Since the physical-mechanical characteristics of earth material vary from one site to another the questions are: which construction technique is best to use? Which type of reinforcement is more suitable? In order to recover the existing buildings and to propose raw earth as a building material also for the new constructions, it is essential to develop a strengthening system able to attribute at earth material equal dignity as other building materials. As raw earth is a no tension material, it is necessary to combine it with a material able to compensate for its deficiencies. From these reflections came the idea to produce a composite consisting of raw earth and natural fiber fabric. Rammed earth material is ideally composed of sand, clay and gravel. These are mixed with their optimum moisture content and compacted inside temporary formworks. The earth composition varies greatly and always contains clay but should not include any organic components. Clay acts as the binder between the grains, a mixture of silt, sand and gravel up to a few centimeters diameter. Compaction is undertaken on material prepared to its optimum moisture. In several layers of earth. The earth mixture is poured loose in layers into a timber or metal formworks, and is then rammed with a rammer (manual or pneumatic). The procedure is repeated until completion of the element. For traditional rammed earth construction, referred to as "rammed earth" or "non-stabilized rammed earth" the only binder is clay (Bui et al. 2014b). Small quantities of other binders can be added, such as cement, hydraulic or calcium lime and gypsum. In this case we speak of "stabilized rammed earth" (Isik. 2011). The main advantage of stabilization is the increase in durability and mechanical performance (Briccoli Bati & Rovero. 2001). However, stabilization usually increases the construction cost and/or environmental impact. Rammed earth material and a compatible reinforcement system are the focus of scientific research for two main reasons. Firstly, the heritage of rammed-earth buildings in the world, especially in the seismic areas is still important. Secondly, the use of non-stabilized rammed earth in new constructions is possible in several countries, particularly in the context of sustainable development. The maintenance of this heritage needs scientific knowledge on the material to assess appropriate renovations.

Recent severe earthquakes like those of Bam (Iran 2001), Pisco (Peru 2007) and Maule (Chile 2010) showed the vulnerability of earthen structures that, in spite of their "green" appealing characteristics, reveal serious structural fragilities (Bui et al. 2011a). The seismic vulnerability combined with high expressive and environmental potential of this material justifies the research of efficient strengthening methods. Most of researches carried out in the field of the earthen constructions are related to architectural and historical aspects and to thermal and hygrothermal performances, while the structural and/or mechanical problems have been disregarded. Some researchers developed various techniques to characterize soil and to identify the clay minerals composition. Recently other studies focused on manufacturing process optimization, compaction energy, environmental comfort, durability and mechanical behaviour of earthen constructions (Morel & Pkla. 2002). Since 2005 the Civil Engineering Department of Aveiro, University, in Portugal, has been developing several scientific studies on the behaviour of adobe structures and their constituent materials. A research group from Catholic University of Peru (PUCP) is playing an important role on the acquisition of knowledge on earthen constructions behaviour and on the development of reinforcement solutions against earthquakes (Blondet et al. 2011). Several reinforcement solutions for seismic retrofit of existing adobe constructions have been studied with enveloping geosynthetic, plastic or metallic meshes. The use of vertical canes and horizontal ropes was compared and assessed (Figueiredo et al. 2013).

This paper deals with the possibility to improve the bearing capacity of rammed earth elements strengthened with jute fabric. Recently, due to need for environmental protection, the demand of natural biodegradable and eco-friendly fibres is rising worldwide in several fields (Gowda. 1999). Jute is a natural biodegradable fibres with advantages such as high tensile strength and low cost and then could be used to replace the synthetic fibres as a reinforcement. The experimental tests were carried out on rammed earth materials, but the analysis presented can be extended to other earthen materials such as adobe and cob.

2 MATERIALS

The materials employed for the realization of the specimens described are: earth material and a jute fabric chosen for reinforcement. Earth was taken in Musciano (PI) to a depth sufficient to exclude organic substances and various kinds of impurities (50 cm). The reinforcement employed is made of jute fabric in warp and weft. The fabric is made exclusively from natural jute fibres, obtained from the stems of the plants of the genus Corchorus, which belongs to the family Malvaceae. Jute fibre contains three main categories of chemical compounds namely cellulose (58~63%), hemicellulose (20~24%) and lignin (12~15%), and some other small quantities of constituents like fats, pectin, aqueous extract, et al. These fibres have a colour ranging from white to brown, are rough, tough and highly hygroscopic (Wang et al. 2008).The processes followed for the preparation and curing of the specimens are reported and the testing methodology adopted is presented.

2.1 *Test for characterization of the soil*

Characterization tests were realized to classify the soil; tests included diffractometric analysis of fine soil amount, particle-size distribution analysis and determination of the Atterberg limits. The mineralogical composition was determined by X-ray diffractometry carried out at the Florence CNR "Istituto per la Conservazione e la Valorizzazione dei Beni Culturali" (ICVBC).

Mineralogical analysis showed that the raw earth of Musciano is relatively lean, that is suitable for the pisè technique. The objective of the particle-size distribution determinations is to classify, according to fixed classes of size, the particles that constitute the sample in order to determine the percentage distribution according to size. This analysis were conducted by dry method and showed that the soil of Musciano is predominantly large-grained. The bulk density is 2.28 g/cm^3. The consistency limits (or Atterberg limits) are the water content that determines the transition of earth from solid to liquid and plastic state. The determination of con-

Table 1. Main mineralogical composition (CNR Florence, 2012).

Main mineralogical composition	
Quartz	40%
Feldspar	20%
Clay minerals	40%

Table 2. Clay minerals composition (CNR Florence, 2012).

Clay minerals composition	
Kaolinite	20%
Illite	35%
Illite—smectite	25%
Vermiculite	20%

Table 3. Geotechnical parameters of Musciano soil.

Bulk density (g/cm^3)	Ip (%)	Ic	Il
2.28	11.03	1.44	–0.44

sistency limits was carried out following ASTM D4318 standard. The indices that allowed to classify the soil were determined. The table 3 shows the values of bulk density, plasticity index Ip, consistency index Ic and liquidity index Il.

According to the consistency indices this soil is classified as a semi-solid, in fact it is located on the Casagrande plasticity chart between the low plasticity organic clays and inorganic silts of low plasticity. Finally, the linear shrinkage was determined both on raw earth and on specimens stabilized with gypsum powder; These tests were carried out because the matrix of the reinforcement has to be realized with earth and 15% of gypsum, in order to obtain better workability without run the risk of cracking. The linear shrinkage values confirmed that the addition of 15% by weight of gypsum powder (biocompatible stabilizer) greatly reduces the earth shrinkage.

2.2 *Mechanical tests on rammed earth specimens*

The mechanical responses of rammed earth were analyzed with mechanical tests on specimens realized ad hoc. In the absence of Italian regulations concerning the tests on earth building material we referred to the procedures used for natural and artificial stone. Earth specimens of different size and composition were made to be subjected to uniaxial compression test, three points bending test and indirect tensile test.

The values of the mechanical parameters determined on cubic specimens of raw earth and 15% by weight of water are reported in Table 5: compressive strength σc, Young's modulus E, kinematic ductility μc, determined as the ratio between the displacement corresponding to the peak load and the linear displacement corresponding at the peak load; available kinematic ductility μcd, calculated as the ratio of the displacement measured at the ultimate load (conventionally established as two thirds of the peak load) and the linear displacement corresponding to the peak load. The results obtained from the indirect tensile tests provide a tensile strength equal to 0.36 MPa.

Table 4. Average linear shrinkage.

Specimens	Initial length (cm)	Final length (cm)	Shrinkage (%)
Earth	14	12.68	9.37
Earth+15% gypsum	14	13.66	2.38

Table 5. Average values of the mechanical parameters determined by uniaxial compression test.

	E (Mpa)	σ_c (MPa)	μc	μcd
A.V.	294.32	3.76	1.21	1.27
S.D.P.	8.70	0.17	0.09	0.06
C.V.%	2.96	4.53	7.48	4.64

2.3 *Reinforcement characterization*

Reinforcement is characterized by two phases, that is jute fabric and matrix of earth and 15% of gypsum to limit the shrinkage during the curing. In what follows, the tensile tests performed on strips and single wires for the mechanical characterization of the jute fabric are shown.

2.3.1 *Tensile tests on jute fabric strips and jute wire*

Peak load and stiffness were determined on jute strips of about 23 cm in length, natural brown colour and a weight of 155 g/m^2. For the realization of specimens, we cut ten strips of 50 cm of length, in the direction of the warp, and 7 cm in the direction of the weft. The strips were wet and then let them dry, to reproduce the reinforcement application condition. The specimens were glued overlapping the ends for about 3 cm, with mastic polychloroprene and were tested with a device able to proceed in displacement control.

The tensile strength and stiffness values to the strips were determined. The table 6 shows the average value, the population standard deviation and the coefficient of variation of the tensile strength and stiffness of real strips and of conventional strips i.e hypothetical strips each containing 10 wires.

Table 6. Mechanical properties of jute fabric.

	Wires n.	Peak load (strip) (N)	Stiffness (strip) (N/mm × 10^{-5})	Peak load (10 wires) (N)	Stiffness (10 wires) (N/mm × 10^{-5})
A.V.	12.3	238.20	5.29	193.07	4.31
S.D.P.	0.6403	46.72	1.01	34.67	0.86
C.V.%	5.21	0.20	0.19	0.18	0.20

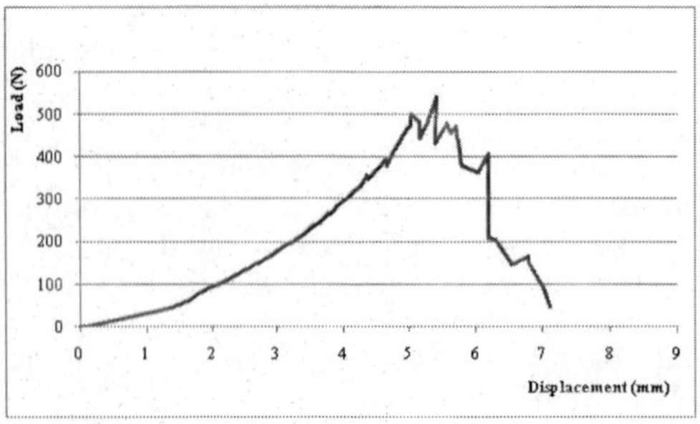

Figure 1. Load-displacement diagram of a jute strip (Graph by authors).

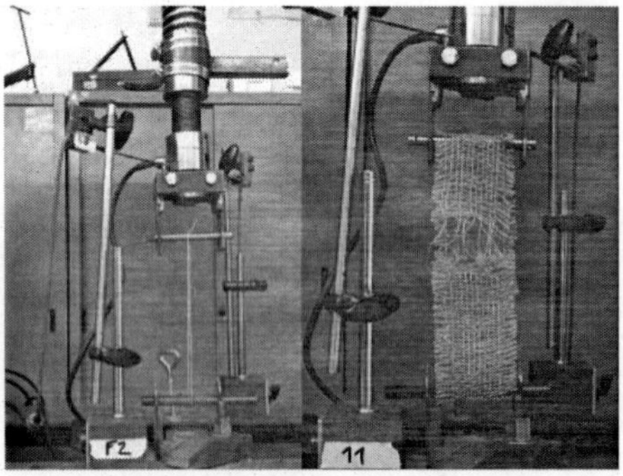

Figure 2. Tensile test on jute wire and jute strip (Photograph by authors).

Table 7. Mechanical properties of jute wires.

	Maximum load (N)	Stiffness (N/μm)
A.V.	27.66	0.006489
S.D.P.	7.28	0.001525
C.V.%	26.34	23.50

3 EXPERIMENTAL ANALYSIS ON REINFORCED SPECIMENS

Three-point bending tests and uniaxial compression tests on structural elements were carried out to evaluate the strengthening capacity of composite material (jute fabric and matrix of earth +15% of gypsum). In the first case the specimens were constituted of separate elements connected to each other with a strip of jute fabric, while for uniaxial compression tests specimens were made with three rammed earth blocks superimposed and intercalated with layers of jute fabric in the matrix of earth, 15% of gypsum powder and 30% of water. In particular, with the three-point bending tests the adhesion properties of jute fabric to the rammed earth were investigated. The increase of strength and ductility produced by the application of the reinforcement were verified.

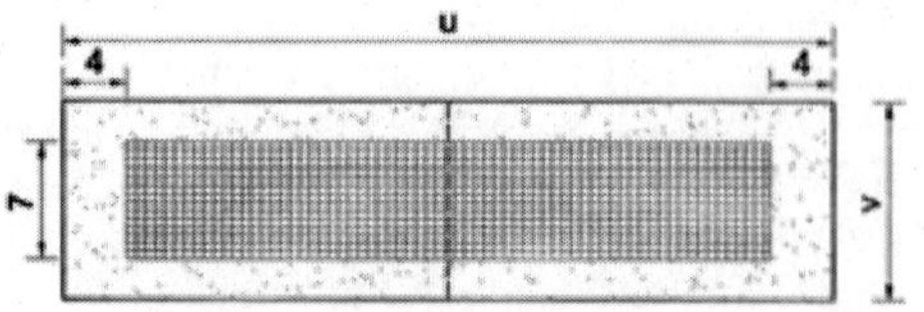

Figure 3. Reinforcement strip (Graph by authors).

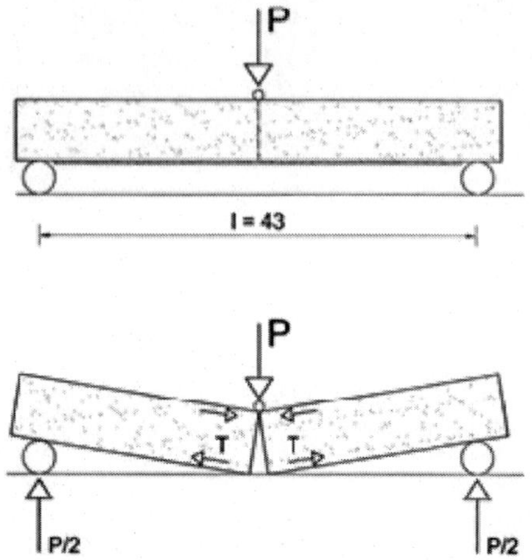

Figure 4. Three point bending test (Graph by authors).

Table 8. Mechanic parameters determined by three point bending test.

	Tensile force first series (N)	Tensile force second series (N)
A.V.	520,81	430,35
S.D.P.	51,49	41,86
C.V.%	9,89	9,73

3.1 *Bending tests on prismatic specimens*

Six prismatic specimens of about 48 × 12 × 6 cm were made of two rammed earth blocks each of about 24 × 12 × 6. Three of these specimens were simply pushed together (second series) and the other three were connected by joints made of earth and gypsum (first series). In both cases a strip of jute fabric of 40 × 7 cm was applied on one of the widest surfaces of the six samples as to produce a beam with reinforcement on the bottom. The reinforcement was placed 4 cm away from end edges and 2.5 cm away from the longitudinal edges of the specimens.

3.2 *Compression tests on structural elements*

Specimens for compression tests were manufactured using three rammed earth blocks 12 × 6 × 24 cm for each specimen. As for the tests above mentioned the material taken from Musciano was mixed with 15% water by weight and subsequently it was rammed manually with a rammer into successive layers of about 1 cm, inside the timber. For the present study six specimens were prepared out, three of which walled with joints of earth, 15% gypsum and 30% water. The other three, on the other hand, were made by affixing a reinforcement fabric of raw jute in the joints among the blocks. The portion of fabric, of 12 × 24 cm, was applied moistened using the same matrix. The fabric was laid with the wires in the warp direction along the largest dimension of the surface of the support. The samples were subjected to uniaxial compression tests at a load rate of 2000 kg/min. The test is considered ended when the loading force reaches approximately 1/3 of the peak load. The average value of peak load registered for the unreinforced specimens was 3765.57 daN, while that registered for reinforced specimens was 6577.94 daN. Table 9 and table 10 show the average value, the population standard deviation and the coefficient of variation of the compressive strength, stiffness,

Figure 5. A specimen after three point bending test (Photograph by authors).

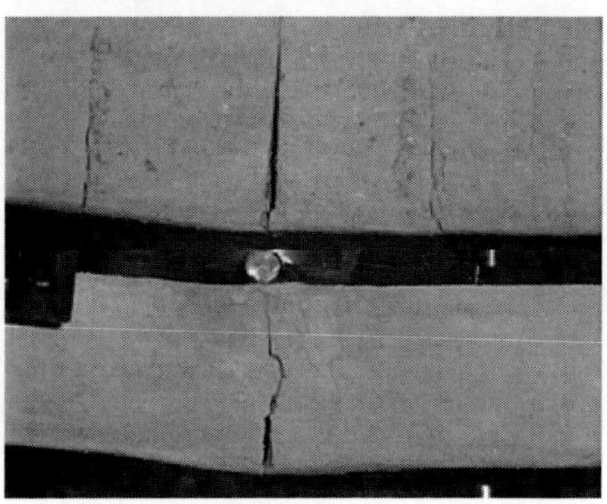

Figure 6. Fracture paths for bending test (Photograph by authors).

Table 9. Average values of the mechanical parameters determined by uniaxial compression test on unreinforced specimen.

	K (N/μm)	E (N/mm²)	σ (N/mm²)	μ_c	μ_{cd}
V.M	10.01	27.46	1.28	1.64	1.41
D.S.P	0.18	0.49	0.31	0.52	–
C.V. (%)	1.77	1.77	23.81	31.9	–

Table 10. Average values of the mechanical parameters determined by uniaxial compression test on reinforced specimen.

	K (N/μm)	E (N/mm²)	σ (N/mm²)	μ_c	μ_{cd}
V.M	12.61	34.60	2.24	1.18	1.74
D.S.P	6.95	19.06	0.22	0.07	0.46
C.V. (%)	55.11	55.09	9.82	5.93	26.22

Figure 7. Unreinforced specimen for uniaxial compression test (Photograph by authors).

Young's modulus, kinematical ductility and available kinematical ductility.

4 CONCLUSIONS

In this paper we describe the results of an experimental analysis developed to check the effectiveness of jute fabric as a natural reinforcement for raw earth elements. Fabric provides wires oriented both in longitudinal and transverse directions.

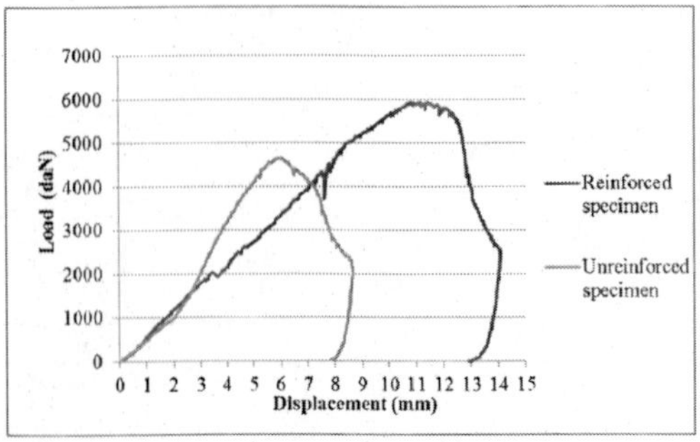

Figure 8. Load-displacement diagram of two specimens respectively unreinforced and reinforced.

Figure 9. Reinforced specimen for uniaxial compression test (Photograph by authors).

Tests for the characterization of the reinforcement material showed that an n-wire strip has an average tensile strength lower than n times the average tensile strength of the single wire. This is certainly due to the heterogeneity of the jute wire and to the test modes used to determine the respective tensile strengths. The experimental tests performed on earth reinforced elements proved that the jute fabric strips laid on the earthen support (block) subjected to three point bending tests show tensile strength higher than the average tensile strength of the jute. Debonding phenomena do not take place as normally happens to other reinforcement systems.

Based on the experimental investigation reported in the paper the following conclusions can be drawn:

- The earth block (bricks) connected together by jute fabric via the above mentioned matrix, exhibits its flexural strength due to the strengthening.
- The elements constituted of three blocks superimposed interposing jute fabric strips exhibits comparatively higher compressive strength of about 39%.
- A layer of jute fabric provides a very good adhesion between the earth blocks. This kind of reinforcement is able to improve the mechanical characteristics of the earth material so to make it more efficient as a building material; depending on specific requirements, it can be laid on any shape and size of structural elements.
- Finally, the presence of jute fabric reinforcement provides ductility resources to the structures thus enhancing their earthquake resistance.

REFERENCES

Blondet, M., Vargas, J., Tarque, N., Iwaki, C. 2011. Seismic resistant earthen Construction: the contemporary experience at the Pontificia Universidad Catolica del Peru. *Informes De La Construccion 63: 41–5.*

Briccoli Bati, S. & Rovero, L. 2001. Natural additives for improving the mechanical properties and durability of adobe building material. *Materials Engineering, 12: 413–425.*

Bui, Q.B., Hans, S., Morel, J.C., Do, A.P. 2011a. First exploratory study on dynamic characteristics of rammed earth buildings. *Engineering Structures 33: 3690–3695.*

Bui, Q.B., Morel, J.C., Hans, S., Walker, P. 2014b. Effect of moisture content on the mechanical characteristics of rammed earth. *Construction and Building Materials 54: 163–169.*

Figueiredo, A., Varum, H., Costa, A., Silveira, D., Oliveira, C. 2013. Seismic retrofitting solution of an adobe masonry wall. *Materials and Structures 46: 203–219.*

Gowda, T.M. 1999. Some mechanical properties of untreated jute fabric reinforced polyester composites. *Composites Part a-Applied Science and Manufacturing 30: 277–284*

Isik, B. 2011. Conformity of Gypsum Stabilized Earth-Alker Construction with ‘Disaster Code97’ in Turkey. *International Journal of Civil & Environmental Engineering IJCEE-IJENS, 11.*

Morel., J.C. & Pkla, A. 2002. A model to measure compressive strength of compressed earth blocks with the ‘3 points bending test’. *Construction and Building Materials 16: 303–310.*

Pacheco-Torgal, F. & Jalali, S. 2012. Earth construction: Lessons from the past for future eco-efficient construction. *Construction and Building Materials 29: 512–519.*

Wang, W., Cai, Z., Yu, J. 2008. Study on the Chemical Modification Process of Jute Fiber. *Journal of Engineered Fibers and Fabrics 3: 1–11.*

Earthen Architecture: Past, Present and Future – Mileto, Vegas, García Soriano & Cristini (Eds)
© 2015 Taylor & Francis Group, London, ISBN 978-1-138-02711-4

Practices and demands in retrofitting of vernacular rammed earth houses

A. Buzo & H.J. Cho
DSA Architectures de Terre, CRAterre, Grenoble, France

L. Genis & G. Paccoud
AE&CC-CRAterre-ENSAG Research Unit, National Superior School of Architecture of Grenoble, Grenoble, France

ABSTRACT: Numerous vernacular rammed earth buildings are today submitted to energetic regulations, especially thermal norms, and contemporary comfort requirements. In such a changing context, the ability to preserve and maintain this rammed earth heritage is questioned. Based on field studies results, the paper exposes the current practices and stakes concerning rammed earth heritage retrofitting in Dombes area, north of Lyon, France. It shows how cultures of building and of inhabiting (household profiles, retrofitting practices and processes, information gathering) can be related to energy efficiency, so as to better preserve heritage quality and to address heritage retrofitting on a wider scale.

1 INTRODUCTION

The ability to preserve and maintain rammed earth heritage is questioned in Rhône-Alpes region, although hundreds of thousand rammed earth buildings from the 19th and 20th centuries form an important part of the region's cultural heritage (Doat *et. al.* 1985, Le Tiec and Paccoud 2006). If the thermal properties of rammed earth are recognized by its inhabitants in the first place, the evolution of requirements in building's energy efficiency leads this heritage to energetic retrofitting. Examples of collapse and damages show the potential danger of bad practices in retrofitting rammed earth. However, the knowledge and know-how of the specificities involved by working with earth material remain distant to many. A study localized in a Rhône-Alpes community of cities permitted to explore the existing practices to understand the pressure put on thermal performances. Current issues in ancient building's retrofitting will be covered in a first part, followed by the objectives and methodology for the study. The results will then be presented and discussed.

2 CURRENT ISSUES IN ANCIENT BUILDINGS' RETROFITTING

2.1 *Energetic approach in retrofitting*

Considering it represents more than one third of existing buildings, ancient houses retrofitting is an essential issue at the European level. Energy retrofitting of rammed earth houses has been intensively studied during the last decade (Allinson & Hall 2010, Faria *et al* 2012, Beckett & Ciancio 2013). However, it often undergoes technical and social issues that go beyond the lone standards compliance (Debizet 2010). Assessment procedures of energetic performance are hardly adapted with the complexity of ancient building (Meeddm 2010). Ancient rammed earth houses are moreover often inhabited and refurbished by private individuals. Before considering energetic performance, the buildings are thus already submitted to an imperative of comfort for simple questions of inhabitability (Berdoulay & Soubeyran 2002). They are eventually bearing "heritage" values, anchored in history, culture and identity, characteristic of the evolution of rural areas in the last decades (Chevallier 2000).

Although the current regulation in France requires minimal insulation when retrofitting existing buildings (Decree n° 2007–363 19 march 2007), those ancient buildings with massive walls remain exempted as their hygrothermal behavior wasn't known well enough when the law was adopted. Even so, it is a frequent concern for inhabitants and owners of old buildings: recurrent national and local awareness campaigns are stressing the need of insulating homes to save energy. These campaigns also promote global approach for the insulation process, when financial backing programs allow both global and step by step implementations.

2.2 *Social practices of ancient rammed earth houses*

This uncertain situation often leads house owners to undertake themselves refurbishment works (Charlier 2012). Some actors (inhabitants, craftsmen, professionals, architects), who experiment and put up rammed earth buildings in everyday life, are advocating the right to live in link with exterior climate, in opposition with ambient standardization (Amsler 2012). However, ancient buildings sensitivity may render the interventions ineffective, if not dangerous, if some rules are not respected (Le Tiec & Paccoud 2006, Levy 2010). Studies concerning energy retrofitting emphasize the need to apprehend context diversity, so as to better target retrofitting situations and question the implementation strategies (Prebat & Puca 2013). In parallel, the literature in human sciences focus mainly on questions related to energy consumption in housing, in a quantitative perspective brought by socio-economic studies (Lutzenhizer 1993) and sociology of innovation (Zelem 2010). The interest to focus on the expression of actors practices during the renovation process is however recognized (Crosbie 2006, Subrémon 2010).

2.3 *Energy retrofitting in Dombes area*

These theoretical issues can be questioned when applied at the local scale. Indeed, the numerous old rammed earth buildings in Dombes area, north of Lyon are no exception to the rule, and are today submitted to energetic regulations, especially thermal norms, and contemporary comfort requirements. The issue was put forward by a community of cities in that area when its call for contributions was answered by a local association interested in local heritage. Their proposal was entitled "preserving and promoting the values of our built heritage considering the new building norms and our contemporary needs, especially those linked to energy savings". The study that is presented here is a part of the response to this call, and was led by the Association Saint Guignefort and CRAterre-ENSAG Laboratory, with the help of DSA-Terre students. The aim of this paper is to present the study realized from that demand, and how it helped us showing to which extend energy retrofitting does imply more than technical answers. In this respect, a better understanding of the inhabitants and their retrofitting practices, as well as the reasons driving them in the retrofitting process are essential.

3 OBJECTIVES AND METHODOLOGY

3.1 *Communauté de communes Chalaronne-Centre, in between ancient heritage and reconfiguration for rural spaces*

The field study has been realized during the last months of 2013 in a French rural district offering a wealth of dispersed rammed earth buildings: *Communauté de communes Chalaronne-centre*, in Dombes area (Fig. 1).

Dombes is a wet zone, where old marshes were turned into a complex of ponds centuries ago. The originality of this area is based on its land tenure, formed of large domains rooted on fish farming and hunting. In the area, housing used to be structured around dispersed farms built on small hills, wherever possible far from flood prone areas and damps. The soil is fine with almost no stone, sometimes clayish but mainly silty. Earth is one of the unique building materials available. It produces a fine yellowish rammed earth. Most of the time, the walls are supported by one meter basement made of pebbles that were harvested in the fields by the builders. Those stones were most of the time laid in "opus spicatum", "ear of wheat" or "fish bone", sometimes separated by horizontal brick layers (carons). Often inhabited by farmers, those houses are today being re-appropriated by new owners, urban workers who invest in this rural area (Manceron 2006).

The area presents one of the highest energy consumption in Rhône-Alpes region. The last studies raise the point that parts of this consumption can be explained by the high proportion of ancient houses whose owners do not have the capacity for thermal renovation (Meyer, 2012).

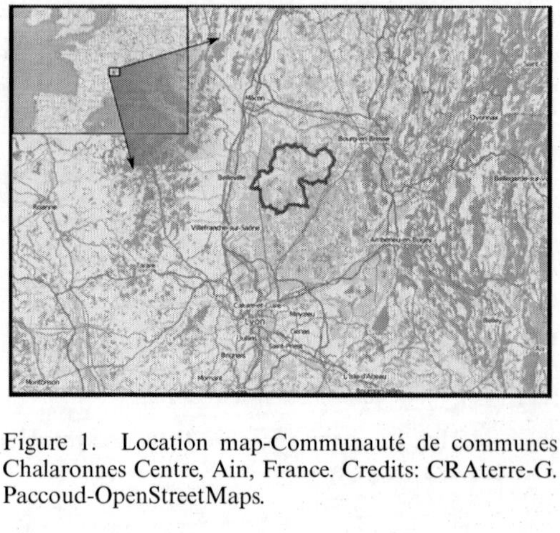

Figure 1. Location map-Communauté de communes Chalaronnes Centre, Ain, France. Credits: CRAterre-G. Paccoud-OpenStreetMaps.

3.2 *Survey process*

The methodology used to collect data was based on two main qualitative tools according to the sociotechnical approach:

- Socio-anthropological semi-directive interviews, life and retrofitting project accounts, socio-demographic data, way of life, insulation, consumption, comfort level, heritage appropriation;
- Technical surveys of surroundings and buildings, history of retrofitting projects, energy use estimates.

A qualitative sample was built to be representative of both the buildings of the area and the period and methods of renovation. 23 owners and buildings were surveyed.

3.3 *Aims of the study*

The aim of this data collection was to describe the interventions made on these buildings, especially in terms of energetic retrofitting: typologies, materials, intervention types and schedules, upkeep condition and renovation level, etc. A special attention was paid to a detailed understanding of the reasons of their interventions, their consequences in terms of energy consumption and comfort, but also in technical terms and inhabitants' representations.

4 RESULTS

The analysis of the qualitative sample showed at first glance the interest to focus on a wider scale on building cultures when considering rammed earth heritage retrofitting current practices and stakes.

Indeed, the comparison of the consumption per square meter of each case studied shows that the highest energy consumption is 2.5 higher than the lowest. Most of the cases are above the national mean, although half are very close to it. The relevant examples have homogeneous characteristics. It seems thus difficult to understand this data without entering deeper into the local cultures of building and inhabiting ancient houses.

4.1 *Evolutions of lifestyles and space management*

The demographic restructuring of rural areas (Chevallier 2000) are well represented in the sample. The evolution from familial farm owned houses and land towards a new model of rural life impacts the evolution of houses appropriation. Rural Dombes houses, conceived to shelter big families and communities (typically family and farm workers), are today inhabited by much smaller families. Most common profiles count two to three people, average-aged couples whose children have left home.

Moreover, the use of the buildings as permanent habitat for non-farmer residents fosters the differentiation in spaces management (Fig. 2). The search for big volumes, spaces and comfort inside the house lead to the spread of two levels spaces distribution, and to an increase of inhabited space at the cost of intermediate spaces. Those were previously associated to the thermal envelope such as, for example, an attic full of straw and grain. In the same direction we observed a tendency to create and enlarge openings to provide wider and brighter spaces.

The changes in the inhabitants' profiles seem then to be linked to an evolution of the social representation of rammed earth houses and farms. Inhabitants' housing trajectory show that rammed earth houses are often associated with atmosphere representation, distinctive of the research of an "ancient" farm.

It should then be "isolated", "calm", offer "spaces" and "volumes". The owners develop a reflective thinking on the project management, so as to rebuild the ancient house according to the spatial representation it has on "tradition" or "territory", (Ruegg 2011).

The increase of spaces and the decrease of inhabitants lead to heat up bigger and less occupied spaces. The tendency observed in the survey (Fig. 3) is that the energy use per inhabitant increases with the size of the heated space per inhabitant, even if it is clear that other factors contribute strongly to the overall performance. How then to reconcile the aspiration of space and volume with the expectations of energetic efficiency?

Some architectural options can help mitigating thermal requirements such as a conception using buffer spaces between outside and heated spaces, in a form of renewal of vernacular principles. As the volumes required increase, the space available

Figure 2. One of the rural Dombes houses from the sample. Credits: CRAterre—H.J. Cho.

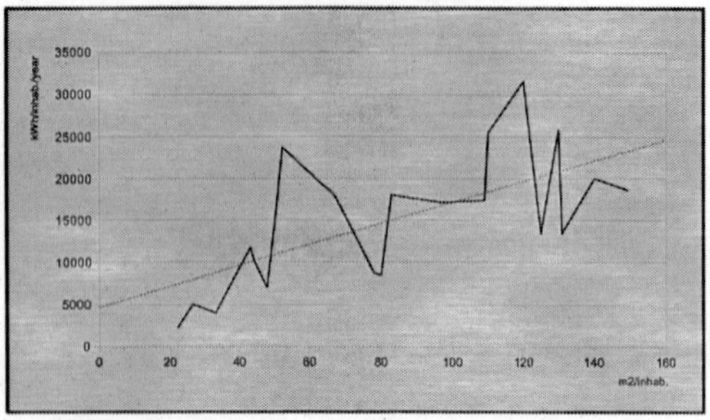

Figure 3. Energy consumption and floor surface per inhabitant. Credits: CRAterre—G. Paccoud.

to design with these buffer spaces is reduced: they are converted into living and heated spaces. Apart from a revision of the space and volumes requirements, which seems to be the easiest solution, technical solutions are promoted since the 80', especially the insulation of the building envelope.

4.2 *Evolution of the technical systems linked with an increased energetic performance*

4.2.1 *Insulation process*

Focusing on the insulation process showed that people who realized a full insulation of their house at once (floor, walls and roofs) were consuming less energy than those who insulated gradually.

The difference is mainly caused by an uncompleted insulation (e.g. only the roof was insulated). However, the overall thermal performance can be reduced by insulation differentials and thermal bridges between gradually insulated zones. Iterative modifications of the inhabited building usually leads to complex configurations of walls with connections between different structural and insulating materials, prone to thermal bridges especially when no global thermal project was designed.

The choice between global and gradual insulation design and implementation has to be placed within the process of getting into an existing building to be better understood. The survey showed that the priority can be to focus on the living spaces first (rooms, bathroom, kitchen), before even considering insulation. It might be difficult then to conceive and realize a global insulation, as both conception process and budget are constrained by the works already done. Global insulation of rural buildings can indeed be a significant investment, especially for such big ones as farms, and it seems it's seldom the chosen path.

Furthermore, the current regulation in France applying to such housing, provides minimum performance requirements for elements and not for the global project, thus encouraging this step by step approach. Financial support programs for energetic performance improvement also refer to elements, but increased amounts are possible for packages of works, which tends to promote a global approach. As budget is often too tight for a full insulation, parts only will be insulated: either entire elements (roof, wall) as required by the standards, or simply rooms when outside of the standards' range. However, this regulation doesn't apply to rammed earth walls as their hygrothermal behavior and the associated risks are still a research topic (Dglan 2013, Heitz 2014, Miget 2012).

Technical recommendations for rammed earth insulation, although existing, remain very limited and advice provided can also be to avoid insulation. The lack of a global insulation is therefore not only a matter of limited money or too big project for one's capacity, but also of limited access to information or choice not to insulate the walls. Other interventions such as roof insulation, openings and heating systems upgrade and installation of ventilation systems are referred to when it comes to reduce energy consumption and improve inner comfort and design a global thermal project.

4.2.2 *Material used*

Some walls were nonetheless declared to be insulated in the surveyed buildings. The main technical solution observed in this respect is drywall. This technical solution promoted since the 80' is still the reference system today for internal insulation. The plasterboard can be complemented with internal insulation but many times is left with a simple still air space between plasterboard and wall. The advantages of the solution as identified by the inhabitants interviewed are esthetic: it straightens up the walls and is referred to with words such as "straight", "clean" or "achieved".

However, both cases are hardly compliant with the hygrometric behavior of rammed earth. Still air will hinder moisture transfers within the wall and lead to water condensation and mould growth inside the cavity, sometimes towards the inner wall. Insulation materials commonly found present a limited lifespan, especially with earthen walls: moisture, water, rodents and insects damages glasswool, the most common insulating materials used. Polystyrene and other foams, also used as insulation materials for these drywall systems, are not allowing the continuity of the moisture transfers within walls and can lead to a dangerous accumulation of moisture in the rammed earth. Such insulation practices consider ancient buildings the same way as conventional construction and can have harmful consequences on the building, ultimately the collapse of the considered wall 5 to 10 years after the change in hygrometric mode caused by the works (Heitz 2014).

4.2.3 *Energy source*

The focus on energy use is also prevalent when considering energetic performance of the buildings. It was then interesting to notice that the specificity of the earth material was barely taken into account when choosing the type of heating system: radiant or convection heating is not differentiated. The dominant factor remains the energy cost. It brings us to consider the motivation for energy saving in buildings, which seems more linked with an imperative to save money than towards an ideal of "environmentally friendly energy saving" (Nemoz, 2011).

For example, the choice of heating fuel used to be prevalent because of its lower price, with relatively low cost boilers. Today, with the increase in the price of heating fuel, the trend is to use the heating fuel boilers as an extra heating system. Thus, we can notice that, apart from the ones who buy their wood (they generally have average energy consumption levels), some inhabitants do have a free access to that resource. They're generally farmers, and have among the highest energy consumption level in this study. These farmers own their house and surrounding lands where they work and get wood. As long as the access to wood is almost free, they don't need to improve their energy efficiency and their heating system to save money.

This data show that, of course, energy performance is correlated with the technical works done on the buildings (in terms of space, insulation, or choice of material) and the choice of energy source. However, is energetic performance only linked to the technical capacities of the building? Are the potential savings relying on technical choices only?

5 QUESTIONS RAISED BY THE REFURBISHMENT PROCESS AND NEEDS

"What to expect from the inhabitants we are, taken between ecological good will and comfort expectations, between heating problems and savings worry, between "household reasons" and technical directions?". Subremon (2011) raises here the dilemmas that concern any individual when involved today in a renovation process. The interest for the energetic issues leads to question the whole renovation process: energy issues cannot be dealt with alone, neither at the beginning nor at the end of the process.

5.1 *The dilemmas of "good practices"*

The aspects mentioned above (changing heating systems, focusing on complete house insulation) are well known in terms of technical features. They are covered in most of the sensitization materials produced and prescribed by norms, rules and good practices as "good ways" to save energy. It gets however problematic when considering the choices that any household has to make concerning energy retrofitting, and even more challenging when considering the specific case of ancient rammed earth buildings.

Hence, implementation of "good practices" in energy retrofitting often represents a substantial financial investment. The stakeholders appear to be more and more informed, even sensitized to the use of materials considered as "healthy" and "natural", but also to the functioning of rammed earth buildings in terms of "hygrometry" or "breathability". A common stance among the persons encountered was that it was "important to let the wall breathe". However, this does not always imply a change in practices which are linked to other stakes and causes, such as money. The point in a renovation is often to decide where to invest: it implies reaching compromises between one's issues and wills. The project process implies then to keep the capacity to reassess, to change ones' mind when encountering a problematic issue. Instead of pressing into compromise between insulation materials, architectural aspects can also be taken into account, looking for solutions in space modularity or in staged implementation of a global project.

5.2 *Act in an uncertain world*

One more issue eventually occurs when facing the lack of widespread technical and normative reference concerning the thermal performance of rammed earth, and the potential for insulation. Homeowners have to find their way in the middle of different opinions, taking advice where they can find it to finally find what they consider being the "good information". In that uncertain world (Callon 2001), where even the professionals do not agree on the good practices, the social capital is important, and will determine the solutions that enter in one's landscape of possibilities. It leads to strategies of self-information whose results can often be difficult to measure.

Data research often starts with preliminary information gathering in a close relation circle (family, friends, local professionals...) to spread afterwards to specialized organizations. This advice-taking process aims at establishing a first diagnosis, to know towards which practices to turn. However, the information collected often lacks details, and makes it difficult to find accurate solutions for one's specific situation. Independent diagnosis can answer that need but it is costly. On the other hand it can be difficult to get specific information from specialized and official structures that provide general directions and are in a tricky situation when it comes to recommend artisans or other professionals. The lack of a specific information desk dedicated to rammed earth retrofitting is often pointed out.

6 CONCLUSION

The complex stances that came out from the field study brought us to face multiple facets of the buildings and inhabitant's trajectories. A complex and interwoven fabric of specific characteristics comes out, hardly compatible with the idea of uniformity conveyed by norms. It shows how spaces appropriation, heating and insulation systems or project processes are elements to be considered to propose adapted solutions for the ancient buildings and those who live daily in it.

Ensuring an outstanding thermal performance with the currently encouraged technical solutions is not obvious. It implies a global conception of the retrofitting project, limited modifications during the process and a global implementation. The study realized didn't come across such process fully implemented, and examples of external insulation aren't common. This supports the idea that these conditions are rather unusual in retrofitting projects, and possibly contrary to the motivations of old buildings' owners who undertake a retrofitting project. Furthermore, global implementation remains expensive enough to drive people away even if they are willing to consider it.

Eventually, it seems essential to think about how to better consider energy efficiency of the old buildings, including economic, environmental and social aspects and not only technical. Some proposals could be:

- Sensitize on the impacts of renovation on the building structure. It could use new indicators for energy efficiency to promote self-responsibility on energy consumptions, g.e. kwh/inhab/year instead of kwh/sqm/year
- Produce reference documentation about intervention on earthen buildings
- Make specialized advice affordable/available, which implies a political will as well as dedicated organization(s).

Such advice should realize the importance of a global thermal conception for the projects, integrating phasing of works and potential for evolution of the spaces. Its implementation calls upon architectural competences to complement today's engineer approaches.

REFERENCES

Ademe. 2013. Rénovation énergétique. *Habitat naturel,* 50. Montgeron: Armada Concepts.

Allinson, D. & Hall, M. 2010. Hygrothermal analysis of a stabilised rammed earth test building in the UK. *Energy and Buildings*, *42*(6): 845–852.

Beckett, C. & Ciancio, D. 2012. A review of the contribution of thermal mass to thermal comfort in rammed earth structures. *2nd International Conference on Sustainable Built Environment*. Sri Lanka.

Berdoulay, V. & Soubeyran, O. 2002. *L'écologie urbaine et l'urbanisme, aux fondements des enjeux actuels*. Paris: La Découverte.

Callon, M., Lascoumes, P. & Barthe, Y. 2001. *Agir dans un monde incertain, essai sur la démocratie technique.* Paris: Ed. du Seuil. Charlier, D. 2012. *Les déterminants des investissements économiseurs d'énergie dans le secteur résidentiel en France*. Grenoble: Presses de l'Université de Grenoble.

Chevallier, D. 2000. *Vives campagnes: le patrimoine rural, projet de société*. Paris: Éd. Autrement.

Crosbie, T. 2006. Household energy studies: the gap between theory and method. *Energy & Environment*, *17* (5): 735–753.

Devron, J.-F. 2012. Energétique du patrimoine, des matériaux innovants pour la rénovation. *D'architectures, le magazine professionnel de la création architectura*, 207: 33–64.

Dglan (Direction générale de l'Aménagement, logement et nature) (ed.), CETE Est, LRA, LMDC & MPF. 2013. *HYGROBA. Fonctionnement Hygrométrique du bâti ancien*. Paris: Ministère de l'écologie, du développement durable et de l'environnement.

Doat, P. & Centre de recherche et d'application pour la construction en terre. 1985. *Construire en terre*. Paris: Éditions Alternative.

Faria, P., Silva, V., Pereira, C. & Rocha, M. 2012. The monitoring of rammed earth experimental walls and characterization of rammed earth samples. *Rammed Earth Conservation*: 91–97.

Heitz, P. 2014. Architecture en terre - La pathologie humide du pisé. *Revue Qualité Construction*, 143: 62–66.

Le Tiec, J.-M. & Paccoud, G. 2006. *Pisé H2O. De l'eau et des grains pour un renouveau du pisé en Rhône-Alpes*. Villefontaine: CRAterre Editions.

Lévy, P. 2010. *La rénovation écologique: principes fondamentaux* (Vols. 1-1). Mens, France: Terre vivante.

Lutzenhizer, L. 1993. "Social and Behavioral Aspects of Energy use". *Annual Review of Energy and the Environment*, 18 (1): 247–289.

Manceron, V., 2006. Le pays de Dombes et ses mises en image: jeu sur les frontières et quête de reconnaissance. *Ruralia. Sciences sociales et mondes ruraux contemporains*. http://ruralia.revues.org/1261.

Meeddm & Maisons paysannes de France (dir.) 2010. *ATHEBA, Amélioration Thermique du bâti ancien*. Paris: Ministère de l'écologie, du développement durable et de l'environnement.

Miget, S. 2012. Isoler un mur en pisé, une hérésie? *Le moniteur.* 23/05/2012. http://www.lemoniteur.fr/183-recherche-developpement/article/actualite/17623527-isoler-un-mur-en-pise-une-heresie.

Prebat & Puca (dir.). 2013. *Réduire la précarité énergétique*. Colloque organisé dans le cadre des «rencontres du PREBAT», le 23 mai 2013. Ecole Nationale Supérieure d'Architecture Paris-Belleville.

Ruegg, F., 2011. *La maison paysanne: histoire d'un mythe*, Collection Archigraphy. Poche. Infolio, Gollion, Suisse.

Subrémon, H .(dir.). 2010. *État de la littérature anthropologique sur la consommation d'énergie domestique en particulier de chauffage*. Charenton-le-Pont: CNRS, Laboratoire Architecture urbanisme sociétés.

Zélem, M.-C. 2010. *Politiques de maitrise de la demande d'énergie et résistances au changement: une approche socio-anthropologique* (Vols. 1-1). Paris: L'Harmattan.

Earthen Architecture: Past, Present and Future – Mileto, Vegas, García Soriano & Cristini (Eds)
© 2015 Taylor & Francis Group, London, ISBN 978-1-138-02711-4

Latest trends in rammed earth restoration in western Andalucía

J. Canivell & A. Graciani
Seville University, Seville, Spain

ABSTRACT: Over the last decade, the significant increase in interventions undertaken in western Andalusia in rammed earth constructions is the result of a better understanding of its features and behaviour, and of the interest shown by those administrations responsible for the enhancement of defensive architecture. Although studies have become increasingly scientific and interdisciplinary, constructive analyses are not always carried out, which would otherwise ensure the compatibility of the intervention with the pre-existing rammed earth constructions. A critical analysis of the latest trends and techniques, based on contributions, innovations, strengths and weaknesses of the most significant interventions carried out on buildings with the greatest architectural and social impact, offers to the professionals in the field, not only an approach to this heritage, but also potential guidelines for action.

1 INTRODUCTION

The growing interest in earthen construction, which has become evident worldwide in recent decades, has led to an increase by the Spanish government in their enhancement of this heritage and a proliferation of scientific studies in this regard. Western Andalusia, with all its provinces (Huelva, Cadiz, Seville and Cordoba), has not remained immune to this process and offers a wide-ranging heritage of constructions built with rammed earth, in which Islamic fortifications, built in the 12th and 13th centuries, are of note.

In this context, three factors have contributed towards a new vision with respect to interventions into rammed earth architectures. On the one hand, the drafting of certain Management Plans, Special Protection Plans, Catalogues of historic sites and archaeological charts have contributed towards the appreciation of this heritage; moreover the implementation of the Andalusian Plan of Defensive Architecture of 2005 (Plan Andaluz de Arquitectura Defensiva (PADA)) of the Junta de Andalucía, which affects over 2000 fortifications, has contributed towards the proliferation of actions on rammed earth buildings, since the majority of military buildings were constructed with this technique. Finally, advances in the theoretical know-how that have built up over time and the progressive specialization of certain restoration firms, which, after several decades, have developed their own increasingly effective intervention techniques, have enabled solutions of a more coherent nature that are also respectful towards pre-existing rammed earth construction.

The present study focuses on interventions in different rammed earth constructions that have been registered during the last decade in western Andalusia. Their analysis has enabled us to assess their principal technical innovations, performance criteria, and results obtained and therefore, to identify strengths, deficiencies and opportunities in the restoration of rammed earth construction.

2 EVALUATION CRITERIA

The assessment of the measures, criteria and applied techniques, in terms of the circumstances and specific requirements of each case analysed constitutes the essence of this work as well as the consideration of their scientific rigour, their physical response and their innovations.

A scientifically rigorous performance implies that the intervention criteria clearly responds to an in-depth prior knowledge of the historical construction technique, of the materials of the rammed earth and of the heritage values of the building as well as to a detailed assessment of the building.

The physical response of each intervention has been evaluated on the basis of the physical results of the corrective or preventive measures applied, since only the passage of time will enable an estimation of the extent to which the techniques used were appropriate and the degradation risks were controlled. The degree of innovation of each solution, in terms of restoration techniques and intervention criteria, has been considered. The conclusions obtained of the assessment of these three assumptions allow us to provide an overview of current trends in restoration.

3 RESTORATION WORK

3.1 *Castle of Aroche (2007–2009), Huelva*

Aroche Castle (Huelva) was built in the 12th and 13th centuries in rammed earth; its complex later history contributed towards the fragmentation of its wall facings at different historical moments. The superposition of many different examples of rammed earth construction on the original wall has resulted in a complex building in which the patina of time has tempered.

The analyzed restoration is the second campaign, directed by Pedro Rodriguez and Maria Luisa Marín, and executed by Freyssinet, S.A., was preceded by rigorous preliminary studies to clarify the constructive processes that the castle experienced; these studies include materials testing and wall-face readings based on the stratigraphic method that, among other things, helped establish the upper level of the stretch of wall, the crenellations, and the module of the rammed earth courses.

The aim of the intervention was to value the original Moorish fortress by maintaining its structural characteristics, but without obviating the historical overlays.

After removing some contemporary structures, the primitive rammed earth structures underwent a conservation process rammed earth. To this end, restoration was carried out on one face (conforming to the traditional technique) of those sections of walls that had lost a considerable percentage of mass. In areas in which the battlements or upper courses had disappeared, the walls were recreated up to their upper level by reconstructing both elements; these new volumes, which are anchored to the base of the stretch of wall with internal metal pins, were executed with poured concrete of hydraulic lime that was lightened with expanded clay aggregate.

So that similarity with the original rammed earth construction would be maintained, great care was taken with the textures, the original dimensions of the formwork were respected and precise dimensions were used (Rodríguez & Marin 2003), and the usual decorative mortar of the Almohad military architecture were reproduced (Márquez & Gurriarán 2008).

In order not to distort the overall image of the castle and to avoid the prominence of any particular historical period, contemporary stretches of rammed earth that were found in good condition were maintained under reinforced lime mortar, as well as those that presented decorative mortar or courses of masonry and brick. Currently, the intervention exhibits a good physical response, since the necessary corrective measures have been taken to repair failures, together with other preventive measures to protect the walls; in particular the establishment of zones of protection in the surroundings and the elimination of degraded urban areas.

The innovation of this intervention, beyond the restoration techniques applied, lies in the focus of intervention criterion. Given both the heterogeneity of the rammed earth work of the castle and the understanding that the building should not be interpreted as a sum of interventions of different historical phases and should be understood as a whole unit, the architect chose not to give preference to a particular historical period, and based the intervention on the principle of maximum conservation. Therefore, in order to obtain a homogeneous but precise view mortar joints were applied in the different interventions (brick, masonry or stones), chromatically coordinated with the tonality of the Islamic rammed earth wall.

3.2 *City wall of Niebla (2010–2011), Huelva*

The city wall of Niebla, dating from the 12th and 13th centuries, has been the subject of several campaigns of intervention since the 1970s (Guarner 1991), and constitutes an essential reference in the restoration of historical rammed earth walls in Spain. The latest, conducted in 2010 by the architect Manuel Vicente López (2009) and the company Joaquín Pérez Díez, SL, was focused on the restoration of the section between towers 10 and 11 and of tower 11 itself, corresponding to an emergency intervention on a sector already intervened in 1974 and 1983. Following the demolition of the damaged areas, these were then restored by means of new rammed earth on one face; although the restoration affected almost the entire length of wall, traces of the original rammed earth were left in some areas. Aware of the mistakes committed in previous restorations, the upper area was protected by a thicker slab and the bottom with a base of masonry. In order to prevent the lack of adherence of restored sections of the wall, lime slurry was injected into the interface, using metal anchors to sew the consolidated areas, and U-shaped metal rods in the restored areas. In an effort to maintain the constructive features of the rammed earth, the original thicknesses were scrupulously respected as were the heights and sequence of the Almohad rammed earth courses.

The physical response of this intervention is appropriate, since there is no new damage; the consolidation techniques and preventive measures at the bottom, at the top and on the surface of the walls have enabled the risk of degradation to be reduced. However, the use of metal anchors, of little compatibility with the rammed earth, raises doubts about the durability of these repairs.

Although there are no technical innovations, the measures employed are more compatible than earlier restorations. However, some formal criteria imply certain developments. Thus, formal reinterpretation of the putlog holes as white patches made with lime mortar, in addition to maintaining an evident parallelism with the decorative lime mortar from the Almohad period, offers a clear reading of the facing in accordance with the scale and with the constructive features of this type of fortification (Fig. 1).

3.3 *Castle of Alcalá de Guadaíra (2003–2010), Sevilla*

The Castle of Alcalá de Guadaíra was originally erected in the 13th century, built with monolithic rammed earth, except for some towers that were implemented with stone-column reinforcements. Until the 17th century, when it fell into disuse and underwent progressive deterioration, the Castle has been subjected to numerous renovations and expansions. The first scheduled interventions took place from the late 20th century when a series of works for its conservation, enhancement and adaptation to tourist circuits were developed. These interventions were directed by Montserrat Díaz-Recasens and Antonio Martín. Those of the City walls of la Villa (2003), executed by Joaquín Pérez Díez, S. L., were focused on the recuperation of the lost outline, integration into the landscape, and improvement of the public use. Still dealing with walls of rammed earth, the new outline were rebuilt by means of a 3-layer intervention, consisting of a double facing of rubble limestone masonry and a cyclopean core of mortar. In order to improve the connection between the castle and the city, after demolishing the worthless contemporary elements, several roads were built and traversed the reconstructed sections of wall.

The next major work took place between 2009 and 2010 in the Sima and Silos Courtyards, these being carried out by Freyssinet, S.A. and Díaz Cubero, S.A. Under the same assumptions, the walls underwent interventions to accommodate visitors' tours, although, given the good condition of the walls, no far-reaching corrective measures were implemented. The cracks were filled with fluid lime mortar; superficial erosion was recovered with lime mortars simulating horizontal joints and putlog holes. In some parapets of the walkway, it was decided to reconstruct the volume with lime concrete poured into wooden formwork. The route for the visitors' tour was paved with coloured precast concrete slabs and stainless steel railings were incorporated which rendered the intervention reversible. Finally, in order to achieve colour uniformity of the wall sections, a thick lime wash was applied.

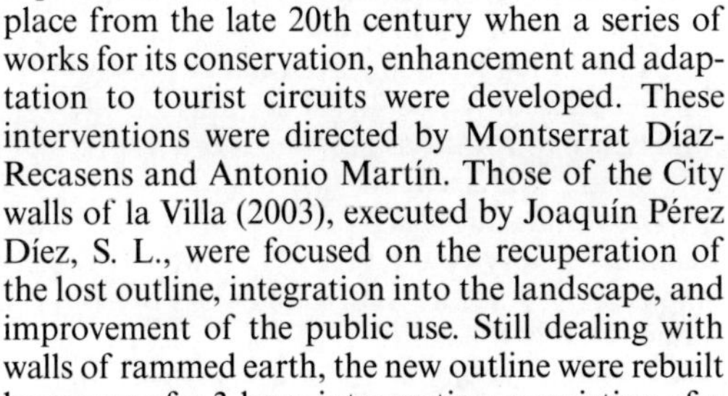

Figure 1. Restoration of one side of a rammed earth wall. City wall of Niebla (Canivell & Graciani).

The intervention of 2003 is more innovative since the techniques applied fail to follow the traditional pattern in the rammed earth. Although the solution of masonry already existed, it was different from the original; the outline of the castle has been restored and integrated into the landscape, but neither the material nor the construction technique reproduces the original techniques.

In general, the technical criteria applied were simple and involved no major innovations. Although a more contemporary language was applied for these interventions (for example, slabs of concrete in the entrances and in the walkway), the image of the castle has not been harmed, and remains recognizable in the local landscape.

3.4 *Torre del Oro (2004–2005), Seville*

This Almohad tower lies on a stone plinth and is built of rammed earth with stone-column reinforcements in each corner.

In 2004 and 2005, the architects María Caballos and Cristina Borrero led an intervention based on the utmost respect of the existing construction (Caballos & Borrero 2004) which was executed by Clar Restauración. This very measured and controlled intervention was based on preliminary studies on construction materials focused on the layers of mortar that, unlike the rammed earth, were not found in good condition. Cement mortars were eliminated, and those mortars remaining were consolidated: with injections of ethyl silicate in those of lime and in the rammed earth fissures; and with injections of mortar in those which had lost their adherence.

Thanks to its good state of preservation, the rammed earth wall was consolidated only superficially. In order not to distort the colour of the tower, a meticulous study was carried out to determine the composition of the aggregates of the new coatings.

After almost ten years, judging by the state of the rammed earth walls, corrective techniques seem to have successfully responded through the elimination of lesions and risk control, especially when taking into consideration that the *Torre del Oro* is located on a river bank in an highly polluted environment and that it is a building in constant use.

3.5 *Hacienda de los Quintos (2008), Dos Hermanas, Seville*

The Hacienda de los Quintos maintains only some sections of wall of the original Almohad castle, integrated into a modern country house, and also some isolated remains found in very poor condition. The most significant construction is the rectangular tower of monolithic rammed earth. Modern rammed earth construction also exists from the 20th century, which corresponds to work combined with brick courses and stone-column reinforcements.

The latest intervention (2008) consisted of the restoration and adaptation of the complex for administrative use. Directed by Mª Dolores Prados, José Ramírez and Antonio J. Pérez, it was executed in a joint venture between Ferrovial and Agroman.

The Almohad rammed earth walls of the tower, being in good condition, were plastered with lime mortar. By contrast, the remains of the sections of wall integrated into the country house were in very poor condition, for which they were partially patched with bricks and lime mortar.

The most modern construction was also treated. In the monolithic walls, structural reinforcements of steel mesh were applied; the compound rammed earth walls were consolidated with lime mortar, thereby left as evidence of traditional construction techniques.

In general, the interventions in the medieval rammed earth wall were technically correct despite not following the construction criteria. Thus, although patches of brick have historically been used for repairs, the facing generated barely resembles a military rammed earth wall. However, the plastic value awarded to the consolidated compound rammed earth wall should be borne in mind since it is exposed within the site and thus recognized as having special value.

3.5.1 *City walls of Seville (2007)*

The last section of intervention in the city walls of Seville corresponds to that between the Church of San Hermenegildo and the first opening. The works were directed by José García-Tapial and Gonzalo Sánchez and executed in 2007 by Sanor, S. A.

The entire site is made mostly of monolithic rammed earth. The sections of wall, although superficially damaged, maintained their structural integrity, and therefore the corrective measures were insubstantial. The rammed earth blocks that presented considerable losses of plaster were restored with rammed earth on one side; although the traditional technique (wood formwork with Almohad modules) was followed, due to the complicated implementation, a filling of poured lime concrete was used; its composition and colour was controlled so that it would not contrast excessively with the rest of the wall section. Finally, an ethyl silicate consolidant was applied to the entire surface.

The interventions were also oriented towards prevention. The passageway between the main and the outer wall was emptied and the waters were channelled to restore the original level; doing so, the uncontrolled accumulation of earth that would have caused future damages to the sections of wall was prevented.

This intervention is aimed towards integrated conservation and also employs materially compatible traditional techniques. Unlike the restorations of the 1980s, only essential alterations are intended, and hence restorations are even more respectful of the heritage values of the city wall.

The results have been positive, even when maintenance is very low and the city wall is directly exposed to urban environmental or physical attacks. Only the solution of the direct contact of the concrete paving of the sidewalks with the wall could bring problems.

3.6 *North façade of the Cathedral of Seville (2013)*

The north façade of the Naranjos courtyard of the Cathedral of Seville was repaired in 2013 as a consequence of major irregularities that arose in its wall coverings.

The third restoration phase was directed by Alfonso Jiménez Martín and Juan Luis Barón and executed by Joaquín Díez Pérez, S.L. The main objective was to renovate the façade of Alemanes Street, by eliminating irrelevant wall coverings and exposing the underlying stonework. Four buttresses divide the façade into five panels, consisting of brickwork and several remaining courses of rammed earth in the central section. It was decided to leave the building structure in view, and to clean the brick and consolidate the rammed earth. The original rammed earth that was in better condition was maintained, although the most affected rammed earth blocks were restored on one side by a similar procedure to that used on the city wall of Niebla. For the recreation of the putlog holes and putlogs, which had always remained hidden, square wooden rods were used with a brick at the top. Finally, surfaces of brick and rammed earth were consolidated with a sprayed coat of ethyl silicate.

Although the physical response of the intervention is not assessable due to the short time elapsed, the criteria for intervention lead to interesting news. Specifically, the exposed brick and rammed earth, it is evident that a different formal reading has been followed, which, despite remaining a risky criterion, denotes an in-depth knowledge of the rammed earth and its techniques. Thus, for didactic or merely aesthetic considerations, an effort has been made to render the putlog system, the sequence of rammed earth courses and even the traces of the rammed earth formwork patently clear. This is transcendental in rammed earth restoration, since this element, largely considered as an example of poor architecture, is used in such an outstanding example of heritage and therefore is awarded with special value (Fig. 2).

3.7 *Murallas del Alcázar Viejo (2000–2005), Córdoba*

The Walls of the Old Alcázar of Córdoba were restored under the direction of Antonio Castro Escobar and Juan Jiménez Povedano and by the company Exisa Construcciones, S.A.

This 14th-century Christian work forms part of an enclosure which surrounded the old fortress. It comprises two end walls with six rectangular towers of monolithic rammed earth, except for the second tower, which presented the remains of rammed earth reinforced with stone inner columns.

The rammed earth was found in very poor condition, affected by intense disintegration, with a significant risk of collapse. Moreover, it was altered by the addition of ceramic and brick or concrete veneers, crudely simulating rammed earth. The intervention first considered an important phase of cleaning and conditioning to remove any foreign element and to recover the original height of the wall. Each end wall was subjected to different treatment; the most degraded side was partially restored by rammed earth on one face of the wall, mainly at its base, in order to stabilize it. On the adjacent side, the one-sided restitution of rammed earth covered the whole elevation, and even recovered the lost dimension of the walkway.

Rammed earth, of 40 to 60 cm thick, was restored, in which it was chosen not to reproduce the putlogs. However, by using wooden formwork, the new facing remained marked by the boards, which followed the module and the sequence of the courses of the original rammed earth. Finally, the high zones were isolated from water seepage by means of a concrete slab.

In this case, the intervention provides no technical innovations nor does it reveal any new criteria. The corrective or preventative measures implemented have been appropriate to address the serious problems presented, thereby better controlling the risk of future damage. The one-sided restitution of the wall was applied in a measured way on the most degraded side, although on the other side, where a lining of concrete was removed, another lining was introduced that was more in accordance with the original technique of the rammed earth.

Intervention criteria have been sustained in the specific analysis of the construction technique. The measures carried out are compatible with the rammed earth on having redesigned the courses of rammed earth, which are executed in a traditional way, and according to the ancient wall.

3.8 *San Romualdo castle (2006–2013), Cádiz*

The Castle was restored by the architect José Carlos Sánchez Romero between 2006 and 2013.

Its sections of wall and towers are made of rammed earth, domes and arches are in brick, stone columns for wall reinforcement are in stone masonry, and veneers in rubble masonry.

After a long phase of archaeological studies, the intervention focused on giving a public use to the building which was badly damaged and degraded after having fallen into disuse. Archaeological studies showed that the building was coated with lime mortar, and therefore the coating remains were analysed to determine a material compatible with the support and with the mortar. After cleaning and conditioning the support of the rammed earth and masonry, the walls were coated with lime mortar similar to the original coatings.

Since no record that any specific conservation treatment had been applied to the rammed earth, we assume that this presented no serious material damage.

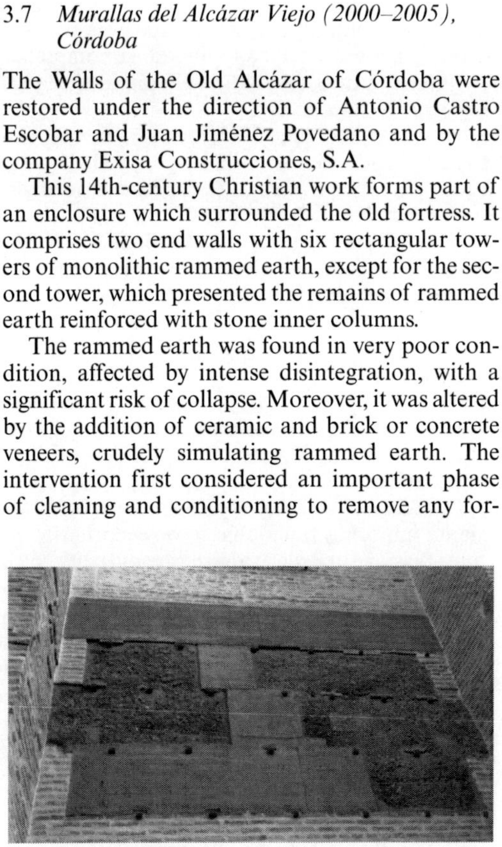

Figure 2. Restored rammed earth walls. North façade of the Cathedral of Seville (Canivell & Graciani).

This intervention follows a criterion of conservation different to those hitherto presented since more importance is given to the continuous coating of the factory, which had virtually disappeared, than to other elements. Although the restored work possesses considerable visual impact, the application of lime mortar ensures its preservation and better maintenance; the current aesthetical contrasts are fading due to weathering.

3.9 *Keep of Setenil de las Bodegas (2011), Cádiz*

The remains of the keep of the ancient castle of Setenil de la Frontera correspond to a fortress of Nasrid origin, which was restored by the architect Fernando Visedo in 2011.

In the absence of cover, the rammed earth wall of the tower was badly eroded, as well as the stonework at the corners was almost disappeared. The intervention was mainly intended to recover the lost volume and indoor spaces in order to accommodate public use. To this end, existing stonework was cleaned and any missing part was replaced.

On the surface of the rammed earth construction, which, despite still being eroded, small surface losses were restored and rammed earth blocks were completed with lime mortar. The courses of brick were maintained or restored, leaving the putlog holes only in the original brick courses. The original volume of the tower was completed with coloured concrete, without joints, putlogs, or putlog holes. The innovation in this intervention, beyond the materials and techniques used, lies in the different interpretation of the rammed earth. On one hand, knowledge of the technique is evident and the interest in applying techniques of consolidation that respect the original construction without masking its values remains clear. On the other hand, the restoration of the work has not been carried out in accordance with the traditional technique, nor with traditional materials, but uses a more contemporary style and leaves clear evidence of replaced elements (Fig. 3).

Figure 3. Restored tower of the Castle of Setenil (Canivell & Graciani).

4 CONCLUSIONS

In the light of the interventions analysed, it can be stated that mostly restoration techniques for rammed earth construction have been properly followed, thanks to the experience accumulated by technical experts, and especially to the expertise of the companies. Materials compatible with earth have been used, adherence and restoration have been improved, and simple solutions have been adopted to protect rammed earth, although only time will provide the ultimate response.

In general, the choice of a criterion of intervention has depended on the state of conservation of many structures and on the necessity to provide a use for the building. In exceptional cases, (Torre del Oro and Seville City wall), when the construction was almost entirely conserved and was not very degraded, it was decided to undertake comprehensive conservation, involving only minimal changes as corrective measures.

On the other hand, for very fragmented construction and when functionality is needed, (which occurs in many cases), the approach involving comprehensive conservation of existing construction and the restoration of lost volumes have been chosen. In this case, there are two lines of intervention: the first includes the execution of new structures with traditional techniques and materials that simulate the original work (Aroche Castle, the Alcázar Viejo de Córdoba, the City Wall of Niebla); the other, resorts to more conceptual techniques, always with aesthetic and materially supported solutions (Castle of Alcalá de Guadaíra, Tower of Setenil).

In other cases, the approach has been more dramatic, imposing functionality or decorum to the conservation; in order to give homogeneity to the construction, without conceding sufficient value to its material reality, the wall is hidden under a new coating, which impedes the understanding of the building (Castle of S. Romualdo and Hacienda de los Quintos).

Selective recovery has constituted an innovative criterion in interventions in rammed earth walls, since in general rammed earth has been relegated to the side-lines. This criterion has been applied in the north façade of the Cathedral of Seville, where, among existing works, rammed earth has been displayed and preserved, and other stratigraphic units considered to be of less value have been eliminated.

NOTE

This paper is a part of the scientific research project entitled *Restoration of rammed earth architecture in the Iberian Peninsula. Criteria, techniques, results and perspectives* (BIA 2010–18921) granted by the Ministry of Science and Innovation under the National Grant Scheme for the year 2010.

REFERENCES

Caballos, M. & Borrero, C. 2004. *Proyecto de Ejecución de limpieza y conservación de los paramentos exteriores de la Torre del Oro*".

Guarner, Ismael. 1991. Muralla de Niebla, Huelva. Paper presented at *Jornadas sobre restauración y conservación de monumentos*: Madrid 24 y 25 de abril de 1989.

López Vicente, M. 2009. *Proyecto Básico y de Ejecución de reparación de las Murallas de Niebla*, Huelva [not published]".

Márquez, S., & Gurriarán, P. 2008. Recursos formales y constructivos en la arquitectura militar almohade de al-Andalus. *Arqueología de la Arquitectura*, Vol. 5: 115–134.

Rodríguez, P. & Marín, M.L. 2003. *Castillo de Aroche. Restauración de los lienzos 1 al 7 de la muralla.*

Earthen Architecture: Past, Present and Future – Mileto, Vegas, García Soriano & Cristini (Eds)
© 2015 Taylor & Francis Group, London, ISBN 978-1-138-02711-4

Albacete fairground: Conservation of a singular rammed earth building

C. Caro Gallego
University of Alicante, Alicante, Spain

ABSTRACT: The building of the fairground of Albacete is a complex which occupies a large area and was built with rammed earth. It is characterized by its vernacular architecture and, with no doubt, by its singular construction creating a fair prototype design of the Enlightenment. These have made it possible to remain standing up for 230 years as a symbol of the people to which its architecture has been serving. During this period, the building has been subject to partial interventions, refurbishments and enlargements that have modified its original shape. However, there is not a study that identifies the different construction areas or a detailed summary of the interventions, nor a study of the current status of the building. This paper gathers up all these reference interventions happened along the years, performs a zoning of the technical characteristics of the fairground in its present situation and describes the actions recently performed.

1 INTERVENTION OVER A RAMMED EARTH MONUMENT

The fairground of Albacete covers around 31,000 m^2, between Benavente and Feria Streets, and Arquitecto Julio Carrillero and De los Toreros Avenues. It is a methodical vernacular building and architecturally unique in its kind (Bonet 1983), with a particular but regular geometry, it is drawn as simple in sketch but has some difficult connection points due to its curvatures. The main access opens up into a rectangular body of 80 × 95 m, broken by two longitudinal pavilions. These are exempt and covered, and include curved ends that convert them into areas differentiating transit spaces. Through the uncovered and longitudinal axis the fairground opens up to two concentric rings roofed with 40 and 55 m radius respectively and of 7 m wide. The complex has a concentric enclosure in its north–south side, with an exterior radius of 82 m, and serves as a containing wall for the squares and open streets. Despite being completely closed to the outside, it gives on the contrary a permeability feeling due to the serenity of its proportions that barely reach one or two floors.

The building of the fairground was declared an Official Asset of Cultural Interest (BIC) in 1991, and categorized as Monument, as a result of the process initiated in 1983 to avoid the variation of the building. Some actions are taken according to its protection level, which are included in first term in the Catalogue of Protected Assets of the General Urban development Plan of Albacete of 1999. The building was granted with a primary protection order that implies the possibility to perform consolidation, refurbishment and conservation works. It is also granted with the possibility to refurbish it to adapt its use or for the improvement and revitalization of obsolete uses. These actions are complemented with the ones referred in the Castilla-La Mancha Historic Heritage Law, which resolves the protection scope of each one of the BIC and attaches them to the report of the Territorial Heritage Commission. However, these measures have not been sufficient to adapt or maintain the building, as an asset of the city of Albacete. It has suffered many refurbishments, partial rebuilding, rehabilitations and conservation works, most of them disorganized. A Development Plan for the fairground should be elaborated with an structured strategy for its use and maintenance (Gutiérrez 2009), and draw the guidelines, avoiding any social, management or lobby matters that can act against the common good of our Heritage (Sanz 2005).

On a lower scale, it is mandatory to gather the building distinctive features before making any intervention, bearing in mind the difficulties attached to the actions taken on historical buildings. The fairground is a regional construction type built with rammed earth. Recently, some conservation and refurbishing works have been performed due to the bad conservation status of the building, fulfilling all the actions accordingly to its protection requirements, and respecting the initial practise. This paper shows the most important interventions made in order to be able to identify the constructive areas of the actual building, making a constructive area plan. It studies specially this last intervention as a part of a more complete and ambitious plan, and it is proposed as a first step to make an integral Plan to regulate the intervention over the building in time as well as in form, just thirty years after being proposed as a Monument.

2 REFERENCE CONSTRUCTION AND INTERVENTIONS

Its construction began on the 5th August 1783, with the project of the local architect Josef Ximénez. The whole complex was designed as a one floor construction, with the exception of the Casa de la Villa, with a first floor. In this first year, 8 one floor walls were built and their foundation was executed with 8 rammed earth sets (Castilla 2005). It is in this way how the square ring, the outer ring and the pillars were configured. The walls formed with seven spans of stone above ground, and over these, two more courses of earth and lime. The pillars were built with stones and mud, with all the corners covered with gypsum and rough cast with lime. At the same time the four access arcades were placed, built with gypsum and stones. Inside the complex and giving its back to the ring, the inner store porches were finished, the slabs made in wood and the area was covered with ceramic roof tiles. 132 commercial stands facing the inner square were set using wood and tiles from the previous posts used for the fair in Los Llanos. In the center of the place a storm-water pit covered with a steel grill was installed, as well as four water supply pits made with stone on the outside of the building, which were executed on the first place to supply the site. The flattening of the whole complex was also performed and the wall of the access street but without a cover, also another drainage point was foreseen in its center. All the works were performed by a team of 72 stonemasons, distributed in groups and led by a foreman, all of them contracted after their proposal by the architect. The carpenters built up to 362 wood stands, the covered ones in the central space, the ones on its back side, the ones close to the access street and the ones leaning on the ring wall. The architect stated to have the wall ready to install the wood slabs on stone columns. The works were postponed in order to use the complex that year, being occupied by the 5th of September of 1783.

Other important constructive works were finished in 1784 by other architect, Antonio Cuesta. This new project included the construction of a green boulevard with elms and poplars, which connected the city with the main door of the fairground (possibly this is the origin of the current boulevard). Despite the fact that supposedly the initial design was modified, his project has not been found, only an etching from 1784, which shows more academic and varying proportions and land services (Sambricio 2011). Antonio Cuesta continued building with local materials through the rammed earth technique, roofs with horizontal structure in wood and ceramic roof tiles, concluding in this way the project that his predecessor and the City Hall had started. The remaining materials from Los Llanos were used to finish the building: wood parts, lime, gypsum, sand and stone. New foundations and rammed earth walls were built in order to double the backside store stands to the street, its wood porch was executed just after. The building is concluded in 1795, with the installation of the four wood doors, decorative steel arch and ashlar fittings.

From this moment, the building suffers continuous decay and looting due to conflicts and epidemic diseases, being adapted with little resources to the new skills and materials of the moment. Thus, between 1813 and 1866 the wood doors were replaced by steel ones in all the accesses, and the stands attached to the backside wall were converted into houses. Within the same refurbishing process, a re-composition of the building was made in different stages: a first intervention was made between 1870 and 1873, where the quarters of the inner ring were rebuilt, the second one made between 1875–1877 when the construction of the inner ring was concluded and a central stand was built. Even though the change in the materials was not significant in the rehabilitation, the constructive requests were very important, which affected positively in the maintenance and survival of the complex: local high quality materials, solid foundations made with masonry, stone ashlar and columns and latticework in steel, all of it made with controlled weight, measures and drawings. In 1883 its centenary was commemorated with only through a few ephemeral architectures. Among other works, a temporary pavilion was planned by the architect of the province Justo Millán y Espinosa, as well as other 20 stands designed by the local architect Juan Peyronnet.

The main investment was used for the refurbishment of the exterior gallery of the building and the inner ring, with columns and lattice works in steel. However, Albacete, the named province capital and city, saw how the construction of the kiosk did not reach the quality level expected. This year the damages were substantial and the kiosk was demolished in 1895 only after 18 years of its construction. The details of the construction process and its specific layout are not clear, but the problems could have been caused by a bad foundation and structure solution. Along with a progressive current of the latest third of the 19th century in Albacete, the City Hall was willing to plan the construction of a new kiosk in the center of the complex. The foun-

Figure 1. Outside view of the building of Albacete fairground (Caro).

dations were executed but the building was not finished. It was in 1912 when a new local architect, Daniel Rubio, designed the kiosk on the main axis and in the center of the fairground. The building is concentric with the circles made by Josef Ximénez, fact that strengthens a better geometrical conception and joins it to the complex.

It was not until the post-war period in Spain when the great enlargement was performed, through an arcaded circle that defines the third ring. The second proposal made by Julio Carrilero Prat and Miguel Ortiz was the awarded in the project tender made in 1944. The concept keeps the vernacular guidelines that tied it with the local constructive tradition, and it maintains the original materials and technique. Its outline under-drawn the illustrated geometry, but it was justified for the fulfillment of the draft tendering specifications. The area of the new fairground changed from 11,000 m^2 to 28,000 m^2, keeping the main door. Adapted to the new dimensions, the building required a new façade. A new proposal was made in 1970, led by the architect Manuel Carrilero, following with the execution of the main door in 1974, replacing the neoclassical door that has remained from the 19th century. This last reference intervention was performed with simple guidelines, as a wink to the traditional and regional architecture.

3 CONSTRUCTIVE AREAS OF THE FAIRGROUND

3.1 *Identification of the constructive areas*

The current fairground is roughly distributed in three constructive areas that show the same global features in their configuration: the entrance door, the pavilion and the rest of the building. They are identified with reinforced concrete, mixed structures and rammed earth buildings. However, it is worthy to know the different interventions made along the time, to be able to recognize what has been modified and what remains in the three areas. The small interventions have not been considered in this paper, as they do not affect the main constructive characteristics established. In order to identify the reference actions, a chronological list has been followed, so that the exposition and recognition of the areas could be clear.

3.2 *Constructive area I*

This space is the oldest of the complex, occupying most of the area of the fairground, so it becomes its constructive definition (Temes 1933). Its technical particularities are its ordinary masonry foundation, walls constructed with rammed earth on both sides; arches, columns and brick partitions stuck with cement mortar; roofs supported on pine wood structure and covered with Arab tiles; walls are covered either with lime or plaster depending on the area, and with floating on the outer side of the fairground façade.

In this great area and according to the age and kind of intervention suffered, four groups can be found, and have been refurbished in small parts during the last year:

1. Intermediate ring wall, including its porch to the inner side. The first partial rehabilitation was made in 1883 by the architect Juan Peyronnet, rebuilt with the same technique the both right and left sides. It was concluded in 1911, repairing 15 stands through the replacement of counters and columns by steel ones. In 1923 the columns were replaced along with some old buttresses, in this case with reinforced concrete.
2. Sanitary facilities on the wrapping walls close to the north entrance. Built as a rectangular space outdoors in 1914, for the same use and as an exhibition room. It was refurbished in 1940, replacing a part of its foundation and masonry walls, floors (supposed to be made with compacted sand), adding a metallic mesh and paving it with hydraulic floor tiles. Other minor interventions as rebuilding the counters with hollow bricks, replacement of columns, false ceilings and general fittings on the roof covering.

   Due to its recent repairing, in the tender of 1944, it was mandatory to include and respect it in the proposal designs, becoming in the justification for the surrounding design that is built, conditioning the new fairground design.
3. Restaurant hall for the City Hall building, the stands in the entrance street extended up to that façade and outer porch of the inner ring, built in 1923. There is no constructive report for these works, and the construction is supposed to be done in the same way than the existing one.

It is known through graphic documentation, the introduction of ceramic slabs and concrete columns which allowed long roof span and created its ground floor pass way and access to the main square with a more permeable inner façade. Some limed ceramic partitions were built, just as the existing ones in the building. In 1984 the upper floor was refurbished, incorporating a chapel and toilets for the stand holders in Talabarteros.

4. Great refurbishment and enlargement of the fairground executed between 1944 and 1946. A new envelope was constructed, with a new livestock area, trade fair, squares and public areas. During this period the auction room was also built (nowadays it is the knife exhibition room, refurbished in 1984). The west door of the complex was modified, just where the poorer houses were previously attached to the wall. Furthermore, this intervention includes the inclusion to the building and indoors rehabilitation of the sanitary facilities area referred in 1914. Another row of modest houses was also joined to the building, and it becomes in the continuation of the wrapping of the entrance street, and was totally rebuilt with the rammed earth technique. The street stands were refurbished for the construction of two new pavilions. Additionally, the mortar of the outer ring was rehabilitated, becoming an open arcade, and expanding it to meet the city hall building.

3.3 *Constructive area II*

It refers to the small building of the bandstand. Built between 1912 and 1913 over the foundations of the previous stand, it became an element differentiated from the rest for its innovative constructive technique. Reinforced concrete slabs of only 15 cm thick made it one of the first public buildings of this type in Albacete, using this material combined with metal beams and columns. It is opposed to the rest of the architectural character, being a modernist octagonal landmark developed on two floors above ground and crowned with a banked blind octagonal dome resting on the flat circular cover. Lightweight, open and spacious located within eight metallic pillars bounded by radial beams that contrast with the curved and vertical elements in the center touching the floor: the staircase, its enclosure and a central support. This ninth pillar does not seem successful, wanting to fill the center of the composition (Gutiérrez 2006) to perform different functions depending on the plan. At the bottom part, it took the form of a reinforced concrete hollow metal cladding being a radial support for the beams and the brick staircase. The second is a circular hollow metal column from where the radial beams start, and on the roof, a metal support of eight brick ribs on which the ceramic dome and lantern are developed.

Vertical and horizontal structures are attached on both floors by tying beams which draw an interior octagon concentric to the façade line and from where the cantilever slabs are supported. The front and underside of the top slab, the radial web beams, the outer surface of the dome and the front bar liquor store, are decorated using the technique of *trencadís*, probably hiring Bausis i Pujol company. Meanwhile, the rest of the building, such as the ornamental cement mortar lining of all 8 pillars and others such as the ceramic brick bar topped with marble, street lamps and cast iron railings were made by Butsems and Fradera until 1913.

Several interventions have been carried out, e.g. 1958 and 1970, being the most important the one held in 1981, where certain rusty pillars were reinforced restoring their ornamental mortar cement finish.

Out of the project, as the platform and as a semi-basement, the small building hides a regular rectangular room, which is accessed from the outside by a linear staircase or by the central one. Its retaining walls were carried out in masonry in areas of both access and ceramic brick of several formats with tiled finishings. Floor made with concrete joists and flat ceramic infills.

Presumably, the existing foundation was made using similar techniques and materials. It has overhead lighting through glass block skylights on the kiosk ground floor and lateral ventilation through a small window to the outside staircase. There is still a piece below this room that is accessed by a trapdoor placed on the storage ground: it is a small disused pit with masonry walls and a timbrel lowered ceramic brick vault.

Perhaps it was that first original rain pit, whom offset location responds the manuscript of 1783 that stated that it had been difficult to run the stone found in the basement. The last intervention is the only comprehensive one, being studied in the following section.

3.4 *Constructive area III*

The main door was built between 1974 and 1975 by replacing the neoclassical one, of which the sides of the stone pilasters remain, embedded in the new perimeter wall. Its architect reflected in the attached report, his intention to maintain the previous character, adapting the existing doors on the place next to the final solution (the so-called Gate of irons). He proposed to use the same materials and techniques, although with great predominance of white cement brick due to their durable qualities and beauty. Finally, the structure was made with

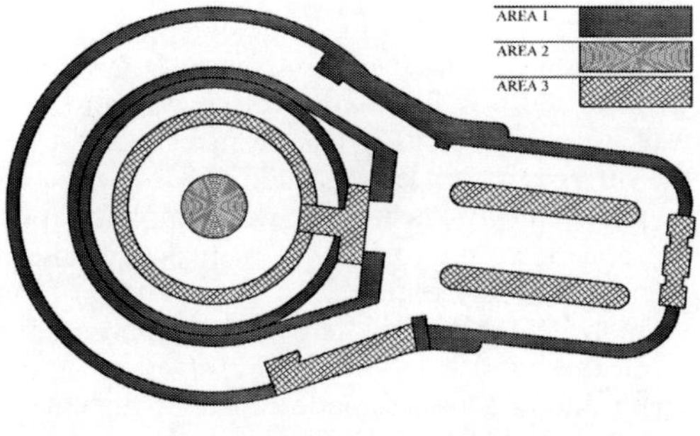

Figure 2. Plan of the fairground constructive areas (Caro).

Figure 3. Fairgrounds roofs review which shows traditional techiques (Caro).

reinforced concrete and columns, beams and slabs aware built with the same brick finishing described. He raised three simple arcades for three entrances (as before) with barrel vault, which height was proportional to its depth, being the central vault the biggest and the more important one. The doors were joined with a pair of open arcades with a gabled roof and finished with ceramic tiles.

It is included among this construction the area of the inner circle, completely demolished and rebuilt in 1983 with reinforced concrete slabs and pillars, and cast iron pillars in the front of the square. The porch was lost and a wooden structure and facing boards were placed in the roofs of the back covered porches, introducing a solution that did not exist in the entire building. Also included in this Area III are the other interventions of the past thirty years: from 1986 with the reconstruction of the Chapel, municipal halls and toilet areas and transformation of certain dependencies and later municipal shelter, school canteen or the adaptation to promote accessibility. The greatest intervention of the last decade was executed in 2010 and includes the renovation of the infrastructure, paving and construction of underground parking below the promenade, leaving aside the Architecture. These last reforms have used contemporary techniques and materials.

4 DETAILED STUDY OF SOLUTIONS OF THE LAST INTERVENTION

In the last year emergency actions on roofs and walls of area I and area II had to be carried out. However, only part of the roofs and support walls of the enclosure of the area I, which have not been upgraded since its construction in 1944 to 1946, were rebuilt. Meanwhile, the conservation works carried out in the whole bandstand are responding to the need for intervention observed during maintenance works in 2010. Previously both interventions were subjected to the Territorial Heritage Committee report.

In the area I the state of roofs and walls of rammed earth walls in an area of 2160 m^2 was revised. It corresponds to the left and right areas on the outer perimeter, barely 210 and 840 m^2 respectively. Also 213 m^2 of the outer roofs of the intermediate circle, at the Talabarteros square were inspected. The reason is the rotting of the timbers of the roofs and deterioration of walls, including deformation, humidity and detachment of suspended ceilings. It is intended to ensure the stability, safety and watertightness of the building complex through the reproduction of the original construction techniques. The works of repair, partial and total replacement of roof panels and their supporting structures are carried out. Once the facilities and suspended ceilings have been dismantled, the uprising of the roof tiles with the subsequent recovery has proceeded, checking the condition of the timber trusses. Their status has been reviewed by the complete dismantling of the trusses that have problems to run the work on the ground, and repositioning them once they are been repaired. The head of the support walls has been cleaned and regularized with a mix of mortar of lime, sand and cement in a height of 60 cm. As a new base of support for the trusses, a small brick partition of 30 cm is done with a rugged perforated ceramic brick, received and leveled with cement mortar. In the roof structure, the damaged woods have been replaced and all existing woods have been tied. A surface treatment has been applied for their conservation.

Instead of the thatched roofs that were supported on straps, fireproof and waterproof 22 mm wooden boards have been placed. The wood has been placed under a 3 cm cement trowelled mortar and covered with recovered tiles. In conclusion in zone I, the suspended ceiling has been replaced in the original way, replacing the detachable one.

In the area II (the bandstand) there have been two key actions: the structure intervention and the stand *trencadís* works. The other elements are also repaired, such as the renovation of the roof and finishing facilities. Both type of beams present oxidation. Those are cleaned by sandblasting and galvanized steel reinforcements are put up on the beams and pillars, protecting them with rust paint. All bolted joints are made with an overlap of 1 m on each side of the beams and on the full height of the I-pillars with two UPN-profiles, metal brackets attach the plates of the foundation and the first level. To be able to access the pillars it has been needed to remove the decorative protection of cement mortar, previously making a latex mold that reproduces the finish again after conservation work has finished. Similar molds for decorative roof elements are performed. To prevent cracking, a fiberglass mesh is added to the mixture and a 6 mm wire mesh. The central pillar, which was burned inside as it was used as a fireplace, has been knocked down on the ground floor due to its poor condition and it has been rebuilt as the staircase slab. The additional materials have been removed throughout the building and part of the original flooring has been recovered without causing vibrations on the slabs, which has been remade with traditional machinery. In the basement, paving has been taken out discovering a soil at 1 m depth, a suspended floor foundation brick wall on stone wall was discovered. It has been regularized and the surfaces have been plastered.

Figure 4. Bandstand first slab before and after the intervention (Caro).

The *trencadís* technique was completely photographed, keeping a copy at actual size on site. It has been only removed and recovered at those points where it is absolutely necessary, proceeding to clean it and a laboratory observation of pigments and materials is made, replacing the missing or irrecoverable parts. After plastering the surface, the tiles are placed following the current drawing with white glue mortar, making all joints and applying a final coat of transparent polyester coating to waterproof them.

5 CONCLUSIONS

A detailed study of the technical and constructive characteristics of the elements which remain today in the fairground of Albacete has not been done. This work should be done, a historical identification and a fieldwork to adapt each construction to the composition that gives it its unique value. At the expense of these actions, works are made with more wills than means. Especially respectful is the action on the bandstand, which has reproduced all damaged parts and original techniques, thanks largely to the work of the designer and the Construction Management. It was not so much the conservation and rehabilitation of the historic decks and walls, for what it should have been demanded the recovery of thatched roofs and rammed earth walls according to the original techniques, being in addition the last walls of this way of building that existed throughout the fairground. Few attitudes and skills have been found over the years such as the team that carried out this intervention, hoping that they can undertake the preservation of the rest of the complex. However, all measures independently will be little while they are not made within a common program: let's propose the intervention as an integral necessity according to its own category through the exposition of its needs and realities.

NOTE

This study is part of the work carried out for the preparation of the Doctoral Thesis: "El Edificio del Recinto Ferial de Albacete: Ciudad y Arquitectura 1910–2010". Ph.D. Student Cristina Caro Gallego, Director of the Doctoral Thesis Professor Elia Gutiérrez Mozo. University of Alicante.

REFERENCES

Bonet Correa, A. 1984. El Edificio del Ferial de Albacete y la Arquitectura de la Ilustración. In Instituto de Estudios Albacetenses, Confederación Española de Centros de Estudios Locales (eds), *I Congreso de Historia de Albacete Vol. III: Edad Moderna* (serie III), Albacete, 8–11 December 1983. Albacete: Instituto de Estudios Albacetenses.

Castilla Pascual, F.J. & Núñez Martí, P. 2005. Estudio para la recuperación de la técnica del tapial en la construcción tradicional de la provincia de Albacete. In S. Huerta (ed), *Actas del Cuarto Congreso Nacional de Historia de la Construcción*, Cádiz, 27–29 January 2005. Madrid: Instituto Juan de Herrera, SEdHC, Arquitectos de Cádiz, COAAT.

Gutiérrez Mozo, E. 2006. *Daniel Rubio Sánchez y su época. Albacete 1910–1920*. Albacete: Museo de la Cuchillería de Albacete.

Gutiérrez Mozo, E. 2009. POM, POT Y POF: Un Plan de Ordenación para la Feria de Albacete, *Cultural Albacete* (III) 14: 105–109.

Sambricio, C. 2011. La arquitectura en el área de Albacete a finales del XVIII. In A. Escario & F. Candel (eds), *El Edificio de la Feria de Albacete 1710–2010*: 13–25. Albacete: Fundación III Centenario.

Sanz Gamo, R. 2005. La provincia de Albacete: una reflexión sobre su patrimonio histórico, *Añil: Cuadernos de Castilla-La Mancha* 28: 55–58.

Temes González de Riancho, V. & Barrios, R. 1933. La construcción del tapial en la provincia de Albacete, *Arquitectura, Revista del Colegio Oficial de Arquitectos de Madrid* XV (175): 297–301.

Earthen Architecture: Past, Present and Future – Mileto, Vegas, García Soriano & Cristini (Eds)
 ISBN 978-1-138-02711-4

Restoration of the Rojano Tower in Lorca (Murcia, Spain)

P.E. Collado Espejo
Higher Technical School of Architecture and Building Engineering, Polytechnic University of Cartagena, Cartagena, Spain

ABSTRACT: In May 2011, Lorca suffered two earthquakes that affected its architectural heritage. The restoration of the Rojano Tower and part of the Islamic defensive wall of the first historic enclosure of Lorca are one of the interventions for the recovery of Lorca heritage. They are rammed earth constructions of the 12th century during the Almohad period, but they were reinforced with masonry in 14th century during the Christian period. The four objectives of the intervention have been: consolidation and restoration of the Tower and sections of the defensive wall; restitution of lost elements in order to improve the comprehension of the layout of the defensive wall; fitting-out of the urban area next to the intervention zone as an urban space for public use and, finally, informative and interpretative actions (didactic panels) that help the visitors to know the medieval urban fabric and the historic context of Lorca.

1 BACKGROUND TO THE INTERVENTION PROJECT

In May 2011, Lorca (Murcia-Spain) suffered two earthquakes that affected its architectural heritage. The actions financed by the Ministry of Development ("1% Cultural" Programme) that were carried out for the recovery of Lorca heritage include the restoration of the Rojano Tower and part of the Islamic defensive wall of the first historic enclosure of this city. These are rammed earth constructions of the 12th Century, Almohad period, which were reinforced with masonry in the 14th century, Christian period.

The restoration has been carried out according to the intervention methodology and criteria established by the "*Special Plan for the Protection and Restoration of the Historic Core of Lorca*" (approved in 2000) and the "*Special Plan for the Lorca Defensive Wall*" (in 2006); consequently, four objectives were established at the project stage: first, the consolidation and restoration of the Tower and adjoining sections of the defensive wall; second, the restitution of missing elements necessary to make it easier to understand the layout of the medieval defensive wall; third, the fitting out of the urban area next to the intervention zone as an urban space for public use, and fourth, informative and interpretative actions (installation of didactic panels) that help the visitors to know the medieval urban fabric, its constituent elements and the historic context of the city of Lorca.

2 HISTORIC AND CONSTRUCTIVE DESCRIPTION OF THE LORCA DEFENSIVE WALL

Lorca has two walled enclosures: the castle walls (upper enclosure), which protected the alcazaba (Citadel), and the medina (City) walls (lower enclosure), where the Rojano Tower is located. Following the Christian Reconquest, walls were not necessary any longer and the city expanded beyond the walled enclosure. From its origins until the end of the 16th Century, the defensive wall of the lower enclosure delimited the Medina and configured the urban fabric that, fortunately, is still preserved and delimits the historic city, especially in the eastern and south-eastern area, where its volumes and sections are currently visible, although intermittently, when proceeding from the Puerta del Pescado (Fish Gate) (where the Castle walls joined the Medina walls) towards the interior of the medieval urban fabric (historic center). As the city began to expand, the lower enclosure wall sections were used to construct adjoining buildings and its materials were even reused for these new buildings. The construction of new buildings adjoining the wall has considerably contributed to its preservation. In those areas where the said buildings were not constructed, the wall and its towers disappeared due to the construction of large historic buildings. Remains of the wall can be found in some of them only, such as the courtyard of the Convent of Las Madres Mercedarias, the School of La Purísima,

Figure 1. Image of the Rojano Tower prior to the intervention (P.E. Collado).

within the old Prison or at the Pósito (granary) or Almudí (grain exchange warehouse). The defensive wall probably runs under the floor of the Ex-Colegiata de San Patricio (Ex-Collegiate Church of Saint Patrick) as well, where two of the main towers of the medieval defensive wall would be located, known as Gil Ricla Tower and Torre de los Santos (Tower of the Saints).

As regards the materials and building system used in the construction of the Lorca defensive wall, it is necessary to point out that the existence of original wall modules or formworks is still evident today. In general, the sections and towers that made up the city walls were constructed according to the traditional technique of rammed earth walls, i.e., made of rammed earth and lime. In some areas, blocks of rough stones were added in order to improve their stability and strength, which reinforced the defensive and protective nature of these elements. Since its very construction, the Rojano Tower was the object of some structural changes of great significance, which increased its volume by means of masonry walls in its upper part. Initially, this Tower was a filled-in section, but in the 18th Century it was partially emptied (a quarter of its total volume) for housing purposes. From 2005 to 2006, the neglect of the Tower caused its upper part to collapse and it lost part of its materials; these damages were further compounded by the two earthquakes that occurred in 2011.

The original layout of the defensive wall presents crooked forms that adapt to the relief of the land, with defensive towers and wall sections or stretches; moreover, the defence of the city was reinforced by means of a ditch or moat that was possibly covered in the 15th Century, upon the expansion of the city beyond the walls. The current remains of the defensive wall are, at some areas, more than ten metres high, and the distance

Figure 2. Rojano Tower and a section of the defensive wall prior to the intervention, right, and after the intervention, left (P.E. Collado).

between the towers is quite variable (there were probably more than twenty-four towers); these towers are separated from each other by a distance ranging from ten to twenty-seven meters. It is possible to observe that, at rugged areas, the length of wall sections is shorter than the length of those located at areas of flat ground next to Guadalentín River, since in the said areas there were no natural elements providing defence. The shortest sections of the wall—less than ten meters long—were used to create the access and exit gates of the Medina, which included Puerta del Pescado, Puerta de la Fuente del Oro (Golden Fountain Gate), Puerta de San Ginés, Puerta de Gil de Ricla, Puerta de la

Figure 3. Rojano Tower prior to the intervention, protected by the provisional upper enclosure made of lacquered steel plate (P.E. Collado).

Figure 4. Interior of the Tower, prior to the intervention and following the removal of rubble and garbage (P.E. Collado).

Judería (Jewish Quarter Gate), Puerta del Postigo (Shutter Gate) or Puerta de los Santos (Gate of the Saints).

3 THE ROJANO TOWER RESTORATION PROJECT

The Rojano Tower, or Fortified Tower T-23 according to the "Special Plan for the Lorca Defensive Wall", is located at Pósito Street and is one of the few surviving towers of the lower walled enclosure of the city and that, furthermore, still retains the original materials used in its construction, rammed earth walls. This Tower is the property of the Town Council; however, prior to the intervention, it was unoccupied and almost in ruins, with a provisional enclosure made of lacquered steel plate that crowned the Tower, as a cover, installed during a provisional intervention carried out by the Town Council some years before to prevent further damages in view of the progressive deterioration of the Tower. Moreover, it was filled with rubble and garbage.

Next to the Tower, there is a stretch of wall—called Section C-27—that is now visible, due to the removal of the adjoining contemporary buildings. The Project suggested the demolition of overlapped walls of the buildings that had disappeared, the restoration of the original rammed earth materials and the intervention in the remaining municipal plot of land to create a free space for public use, where the remains of the defensive wall could be integrated.

The main problem raised by the enhancement of the Tower and the adjoining defensive wall was that the original rammed earth materials had been continuously damaged and repaired since their construction. Damages suffered by the walls were initially repaired by adding rough stones and lime mortars; subsequently, whole wall sections were reinforced by lining them with masonry materials, even with ashlar blocks at their corners. As the walls were integrated into new buildings, they were usually lined with different materials, recently including cement mortars. All of this should be treated and repaired, with the premise to use original construction materials and techniques, to consolidate the original volumes and to reinstate or reconstruct the various rammed earth or masonry elements up to the original levels (according to the thorough historical, archaeological and architectural studies that were conducted). Moreover, the intervention should ensure the reversibility of the works performed. Should these works be totally or partially removed in the future, this may not jeopardise the integrity of the monument; so, when any materials are added, a fiberglass or felt mesh is previously placed to separate the original elements from those added, which makes it easier to understand the intervention and to remove it in the future, where appropriate. Consequently, any repairs should be based on true data obtained from a thorough study on the existing original elements and should also be clearly differentiated from the original work (in order to prevent the so-called "historical forgery"), although they should be integrated both in colour and shape into the whole, so as to prevent them from standing out too much.

Finally, the intervention should take into account the intended use of the Tower and the adjoining defensive wall, and that is why special attention was paid to measures concerning accessibility and safety for future visitors. In this respect, the project envisaged the basic fitting out of the inside of the Rojano Tower, with access to the top roof from the inside, and the treatment—as an open urban space—of the area surrounding the Tower and the adjoining defensive wall, in order to

Figure 5. Stretch of defensive wall, called Section C-27, adjoining the Rojano Tower, during the intervention (P.E. Collado).

Figure 6. Rojano Tower and Wall Section C-27, following the intervention (P.E. Collado).

create a standing area for the restored structures. In these cases, accessibility and safety for visitors were prioritized.

4 RESULTS OF THE INTERVENTION

The first works focused on the general cleaning of the intervention areas, in particular, the inside of the Rojano Tower that contained large quantities of rubble and garbage, which was emptied and disinfected. In the area surrounding the Tower (Sections C-26 and C-27), an excavation was carried out under archaeological supervision, in search of the original levels that were hidden by the filling materials; this excavation properly identified all the original structures located in the intervention area, which were already visible or discovered as a result of the archaeological excavation; all of these works have been detailed and recorded in the corresponding reports. So, at the corner between the Rojano Tower and the Section C-26, following the removal of the existing rubble from the lower part of the said section, a fountain or basin appeared, quite well preserved, of which we had no information, together with the foundations of the defensive wall section, made of rammed earth, which was better preserved than expected; all of these elements were integrated into the architectural ensemble.

As regards the restitution of defensive wall sections and fortified towers, the intervention criteria have varied according to their state of preservation.

As regards the areas with surface damage (specific losses of the external lime coating, with no remarkable internal damage), the intervention involved the surface cleaning, herbicide treatment

Figure 7. Interior of the Rojano Tower following the intervention (P.E. Collado).

and surface restoration with lime mortar and addition of mineral pigments to unify the final aspect. As regards areas with significant damages (collapse of a part of the internal structure of the wall), it was previously necessary to clean their remains, to consolidate and prepare them for the addition of new materials, by means of anchoring devices made of fiberglass mesh; following the preparation of the surface, damages were repaired by using the original construction technique of wall formworks. In the areas where walls had disappeared (upper parts of the Tower and of the adjoining wall sections), the intervention project suggested their restitution by using the traditional rammed earth technique, however, the historical and archaeological research evidenced that during the Christian period they had been replaced with masonry walls, which were particularly visible at the upper part of the Tower, and it was decided to use this kind of wall by first placing a geotextile mesh in order to identify the intervention properly; moreover, natural rough stones were used, the characteristics of which were similar to those of the existing elements in terms of size, texture, colour..., and were fixed with lime mortar with addition of mineral pigments to achieve the uniform chromatic combination of the whole ensemble.

As regards the areas where there were rammed earth or masonry wall fractures, such as cracks, crevices and small collapses (somewhat ancient damages that had been aggravated by earthquakes), the intervention involved the filling, consolidation and repair of cracks, so as to ensure the continuity of structural elements. In the case of the upper finish of the walls, it was decided to use a slightly curve-shaped lime mortar layer for better rainwater discharge. Finally, all surfaces that were exposed to the elements were protected by water-repellents, with the previous addition of mineral pigments to achieve a uniform chromatic combination for the intervention, by differentiating new sections from the original ones and also by preventing any excessive distortions.

Inside the Rojano Tower, after having cleaned and removed any recent additions (such as flooring elements, internal walls, woodworks, obsolete installations, etc.), ancient pavements appeared—quite well preserved—that were consolidated and restored. Likewise, a structure made of wooden beams and lime hollow bricks and a rammed earth staircase appeared, all of them in a state of preservation that enabled their restoration and, thus, they were included in the Project. In order to reduce the visual impact as much as possible, a sliding low-level caisson made of weathering steel has been installed, so as to grant access to the upper part of the Tower.

Figure 8. Information panel, with text and explanatory drawing, about the original construction of the Rojano Tower (P.E. Collado).

Figure 9. Detail of the element made of weathering steel that grants access to the upper part of the Rojano Tower (P.E. Collado).

Finally, the purpose of the intervention was also to adapt the urban area surrounding the intervention zone, in particular, the municipal plot of land corresponding to the Tower and the Wall Section C-27, which has become a standing and reception area for visitors or, in any case, a new space for public use that is fitted out with urban fixtures (protection handrails, wooden benches, litter bins, light projectors for the Tower and wall sections, and signposting for standing areas) and pavement. The intervention included the installation of information panels, prints with texts and explanatory drawings, both for the Rojano Tower and for the two adjoining wall sections (C-26 and C-27) aimed at showing and explaining to the eventual visitors the historic and formal conditions of the Tower and the defensive wall, their relationship with the current urban fabric and their historical context, in order to contribute to the proper interpretation of the urban, social and cultural evolution of the city of Lorca.

5 CONCLUSIONS

The restoration and enhancement of the Rojano Tower and the adjoining medieval wall sections have implied the recovery of an important historical, architectural and cultural referent of Lorca. The intervention has taken into account the traditional construction technique using rammed earth and masonry walls, according to the intervention zones, as well as the use of traditional materials (earth, lime and stone, similar to those existing in the Tower and defensive walls).

The repair and reconstruction of elements damaged by the earthquake and the execution of the new roof (the ancient surveillance roof) of the Tower, together with a proper maintenance throughout time, must ensure the continued existence of these constructions during the decades to come.

The fitting out of the surrounding area and the installation of information panels both at the Tower and at the adjoining wall sections will contribute to the proper interpretation of the urban,

Figure 10. Rojano Tower prior to the intervention, right, and after the intervention, left (P.E. Collado).

social and cultural evolution of the city and to the revitalization of this area of the historic center of Lorca.

Consequently, with the reconstruction of the volume and original image of the Rojano Tower, the city recovered one of its main historical, architectural, social and cultural references, and this contributes to the social, tourist, cultural and economic development of Lorca.

NOTE

- The intervention's technical information is as follows.
- The developer/owner was the Town Council of Lorca.
- The project management was carried out by Rafael Pardo Prefasi (architect), Severino Sánchez Sicilia (architect), Inmaculada González Balibrea (architect) and Pedro E. Collado Espejo (technical architect).
- The contractor in charge of the works was UTE Entorno y Vegetación S.A. – Construcciones y Desarrollos Tudmir S.L.; the foreman, Eduardo Hernández Sánchez; the archaeologists, Juan Gallardo Carrillo and Juan Ángel González Ballesteros.
- The Works Budget achieved the amount of 307,918.49 €.

REFERENCES

Collado Espejo, P.E. González Balibrea, I. Pardo Prefasi, R. & Sánchez Sicilia, S. (2012): *Restauración de diferentes tramos de la muralla de Lorca (Murcia) entre el Porche de San Antonio y la Torre de Rojano.* ALBERCA nº 10, Lorca.

Collado Espejo, P.E. González Balibrea, I. Pardo Prefasi, R. & Sánchez Sicilia, S. (2013): *Restauración de la muralla de Lorca entre el Porche de San Antonio y la Torre de Rojano.* In XXIII Jornadas de Patrimonio Cultural de la Región de Murcia. Murcia.

Font Mezquita, F. & Hidalgo Chulio, P. (2009) *Arquitecturas de tapia.* Colegio Oficial de Aparejadores y Arquitectos Técnicos de Castellón.

Martínez Ríos, C., (2007): *El Plan Especial de Protección de la Muralla de Lorca.* In Memorias de patrimonio 2003–2005. Intervenciones en el patrimonio Cultural de la Región de Murcia n.º 7. Murcia.

Martínez Rodríguez, A. (2004): *Lorca, ciudad amurallada.* ALBERCA n.º 2. Lorca. Muñoz Clares, M., (1999): *El Castillo de Lorca.* CLAVIS n.º 3. Lorca.

Muñoz Clares, M. (2002): *Documentación sobre las torres y puertas de la Muralla de Lorca.(1490–1526).* ALBERCA n.º 1. Lorca.

Vera Boti, A. Pozo Martínez, I. Robles Fernández, A. & Navarro Santacruz, E. (2004): *Plan Especial de Protección de la Muralla de Lorca.* Lorca.

VV. AA. (1999) *Arquitectura de tierra.* Ministerio de Fomento. Madrid.

Earthen Architecture: Past, Present and Future – Mileto, Vegas, García Soriano & Cristini (Eds)

Saint Anthony hermitage: An example of earth rammed wall in a reconquest church

L. Cortés Meseguer, S. Tormo Esteve, R. Marín Sánchez & J. Pardo Conejero
Universitat Politècnica de València, Valencia, Spain

ABSTRACT: Reconquest churches shape is a religious building typology of the Mediterranean gothic architecture. This sort of church is characterized by a quick and easy way to be built, and they are low height, with a rectangular shape with diaphragm arches, gable roof made by wooden beams, backboard and ceramic tiles. The walls are made with native nature materials as regular stone, ashlar stone or in an unusual case as San Antoni de Xàtiva Chapel, made with rammed earth wall. This chapel is located outside of the Valencian town of Xàtiva and, unfortunately, it has a dilapidated condition. This heritage building offers a wide range of building solutions for a rammed-earth wall chapel, and uncommon case of the religious architecture. Even in this building is mixed two Valencian traditional forms of building, *tapia valenciana* and ceramic tile vaults.

1 A NEED FOR RELIGIOUS BUILDINGS

1.1 *From San Antonio de Padua to San Antonio Abad hermitage*

The Saint Anthony hermitage is located outside of the Valencian town of Xàtiva (Spain), in the way to Genovés and close to the Bellus' channel, which flow to Xàtiva's old town.

The first dedication of this hermitage was to San Antonio de Padua, according to the history texts, its foundation dates in 1654 (Cebrián 2007).

In 1790, the Valencian archbishop Fabián y Fuero named San Antonio Abad to this chapel and checked beside it a house for the hermit. This confusion of names is due to popular devotion to both saints, first one an anchorite. But Saint Anthony's order takes care of ill people and let them rest and recover in their buildings, as a hospital of charity, which first one was established in 1095, in Vienne (France). A similar case take place in nearly town called Benigànim, whereas the early Saint Anthony hermitage became to the later convent of Saint Francis.

In 1878 was extended and restored, as we can read in a stone in the main façade. We suppose it only affected to the sacristy, beside the apse.

This hermitage arrived well preserved to the first half of 20th century, but nowadays is closer to disappear than to be restored. It provides a great opportunity to study this kind of chapels and even its constructions, in this case with *tapia valenciana*.

1.2 *The reconquest churches shape*

Despite there are other hermitages in local area, seven in Xàtiva (Lluch 1980), and due to its tumbledown condition, the Saint Anthony hermitage shows how it was constructed and its different phases, offering a great opportunity of knowledge of this kind of chapels and its construction.

The reconquest churches' shape belongs to a religious Mediterranean gothic typology, characterized by a quick, easy way to be built. Low height, rectangular plan, usually with two or three diaphragm arches and closed with walls. There are some reconquest churches with external buttress that are kept inside and presented as pillars. The gable roof is made by wooden beams, backboard and covered with ceramic tiles.

This kind of buildings appeared while Jaume I reconquered Valencia's region and was extended as a catholic symbol in unusual.

Despite the constructors used rammed-earth walls for Saint Anthony hermitage, an unusual technique for these chapels, their size and height were similar to other chapels, especially to the ancient hermitage of present Saint Francis' convent. The width of the nave is around 5'49 m. (6 Valencian fathom) and each frame is between 3'77 m. and 3'85 m. long. Extracting information, it is to appreciate that the ancient constructors didn't mind the size of the formwork, because it's the external masonry wall. The height in the center of the arch is 5 m, and approximately 30 cm below it were the ceramic tile vaults, dilapidated in the 20th century.

Figure 1. An ancient view of Saint Anthony hermitage (J. Gallego. Tivoli collection. Imprenta Marbau.).

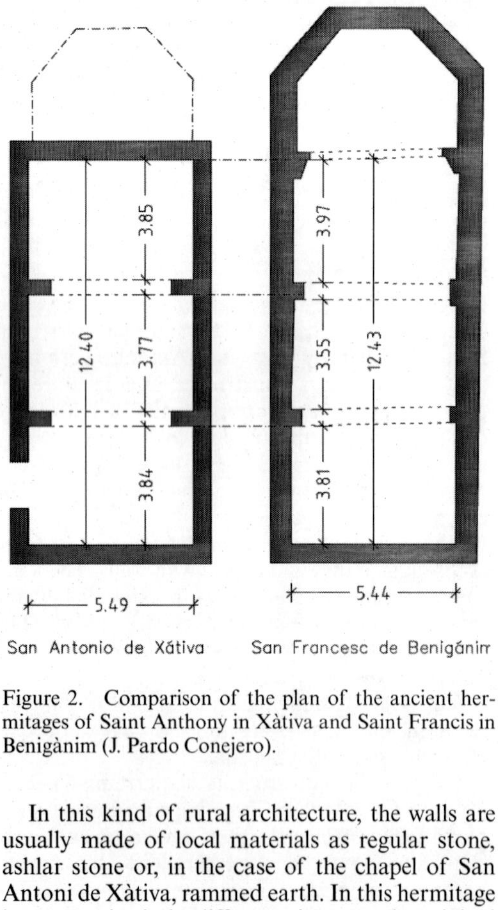

Figure 2. Comparison of the plan of the ancient hermitages of Saint Anthony in Xàtiva and Saint Francis in Benigànim (J. Pardo Conejero).

In this kind of rural architecture, the walls are usually made of local materials as regular stone, ashlar stone or, in the case of the chapel of San Antoni de Xàtiva, rammed earth. In this hermitage is easy to check the difference between the original walls of the ancient hermitage and the posterior ones. The rammed earth wall technique is a mixture of the rural construction tradition.

1.3 *The technique of Valencian rammed-earth wall*

The technique used in the construction of the walls is commonly refered as *tapia valenciana*

Figure 3. External and internal lays of a *tapia valenciana* wall in Saint Anthony's hermitage (Source: L. Cortes Meseguer).

(Valencian rammed-earth). It is a masonry of compacted and confined earth with a reinforcement of brick in the dough which is set in every single layer as a structural reinforcement and that gives its rhythm and characteristic texture, visible from the outside. Its use is dated since the 13th century and the technique was developed until it was necessary the use of drawing tools to be able to line it up and plumb it properly (Font & Hidalgo 2009).

Along the following centuries, this technique became very used in the most important buildings, usually in civilian buildings with defensive needs, or in churches and other constructions that needed a higher strength in their masonries.

In the 17th century, Fray Lorenzo de San Nicolás, in the 39th chapter of the First Part of its treatise "Art and Use of Architecture" had already said that the resistance of its making was good and he also indicated their composition and position in the masonry.

There are many buildings that adopt this type of masonry and are still in a good condition nowadays (Cristini 2008). The dimensions were marked in each case by the treaties of the age or by the books of masonry in which the most interesting constructive aspects were compiled. In many ways, the dimension and width of the walls, the height of the row and the total height were determined by following strict dimensional relations between them and following canonical proportions described in and copied between ages (Galarza 1996).

To summarize, the use of the brick inside the earth wall provides enough dimensional stability, resistance to the surface erosion and mechanical resistance to improve their common behavior. Because of this reason, in the group of the different types of masonry walls, the Valencian is one of the most resistant to the climatic and structural conditions.

Figure 4. Detail of a corner of rammed-earth wall, where the spun separation is about 8 cm., e Valencian fingers (S. Tormo Esteve).

Figure 5. Current inside view (L. Cortes Meseguer).

2 CONSTRUCTING THE CHAPEL

2.1 *Describing the technique*

In the chapel of San Antonio hermitage, the technique of the Valencian wall is in the exterior of the walls while in their interior they show a later facing. The layout of the walls is perfectly clear from outside accentuating boxes of the walls limited by pointers and which are 85 to 95 cm height. The section of the wall is 60 cm, identifying in the damaged and demolished parts, where the layout of the brick was not continuous in the whole thickness, instead two 15 cm external layers have been generated flanking an internal stuffing of compacted earth. The brick that has been used has different dimensions depending on its position in the wall, but most of them are 14–15 cm width and 28–30 cm long, keeping the relationship of 1 long (*soga*) and equal to 2 widths (*tizón*).

In the corners of the building it is possible to appreciate the connection between the superimposed walls; a perpendicular disposition is generated between boxes, so it can be more consistent. It is in this point where it is easier to appreciate the height and the width of the boxes of the wall.

In an organoleptic level the materials which form the wall have a high percentage of clay in the internal part of the wall and a greater proportion and presence of lime in the external parts so they reach a better protection against weather. The aggregates are medium size of rounded shape and are located in a bigger proportion inside, where the smaller ones are find in the outside.

2.2 *Current condition*

The main cause of damage in the building is the lack of roofing which has caused a general damage in the walls and a rising of vegetation. In the rests which are still raised it is easy to see both of the two external layers of Valencian wall and the inside part of the masonry of rammed earth. In all the cases the continuous presence of rain water, the erosion of the wind and the presence of biological material accelerates the deterioration of this sort of wall, so unique in our area.

2.3 *Singular points*

2.3.1 *Foundation*

Water is one of the main enemies of rammed earth walls, as humidity affects to the basement of the wall. In this chapel, foundations are made of materials with different water resistance, but mostly irregular stone and ashlar stone with lime mortar. Its depth has not been checked but is about 50 cm above ground level. This is about 12 cm wider than the wall, and is a platform for the walls and to receive the interior pavement.

2.3.2 *Pillars and arches*

In the case of the construction of the Saint Anthony hermitage, pillars and arches are made as the same structural frame. Its function is just to receive the loads from the beams and to brace the lost ceramic tile vaults that hide the interior view of the roof. Pillars are working as buttress and are made of masonry, using bricks, lime mortar and irregular stones. Arches also are made mixing the same vernacular materials. For the closing walls of the cellar, which its function is to brace against buckling, constructors used *tapia valenciana*. This part was easier and cheaper to do, in contrast to the front door, which is made with ashlars.

The reason for using different material for walls and pillars is both economical and structural. The use of material located close to the building is cheaper than buying and transporting it from far away to the building location. Furthermore, when two materials of different stress are used in the same constructive element, the stiffer one absorbs tractions and elastic deformations.

Figure 6. The basement and staircase of chapel's door (L. Cortes Meseguer).

Figure 7. Detail of top corner (S. Tormo Esteve).

2.3.3 *Top of the wall*

The *tapia valenciana* wall makes three different layers: two external as the formwork and the core is just compacted clay. As the same property of foundation, on top of the *tapia valenciana* wall, ancient constructors also put a layer of rubblework and then the bricks making the cornice.

2.3.4 *Vaults: A formwork with ceramic tile vault*

An unusual kind of formwork appears in this hermitage. There are many different ways to make a vault with bricks: masonry, Byzantine technique and ceramic tile vault. With independence from the technique that we use for the vault, the kidneys should be strong and have been stiffened with dead load.

In the case of this building, the lower part of the vault takes part of the wall, is made by ceramic tile and makes also the internal formwork with a different geometry. That means that, when building the wall, ancient builders made the starting vault and had the initial intention to build the vaults inside the chapel. The bricks are in a diagonal position to transport the loads to the vertical sustaining elements and are another peculiarity of this chapel and a feature of the wisdom of constructors.

Figure 8. Detail of the beginning of the vault and wall (S. Tormo Esteve).

2.3.5 *Windows and doors*

As the usual solution for a void, windows or doors of this chapel are made by an easy solution and use wooden beams. The beams are just logs of trees, 15 cm diameter, and the number of them in a void depends of the width of the wall, but usually is used a pair of beams.

Despite most of the wall is demolished we know that the windows were small, perhaps no more than 60 cm tall and 50 cm wide.

The main door is constructed with a different technique, its exterior layer is made of ashlars, inside there is irregular stone with lime mortar and its width is the same as the rest of the walls.

2.3.6 *Coating*

The external layer, made of ceramic brick, clay, gravel and lime mortar have two functions: the first one, to be the formwork while is introducing the rammed earth wall and, and secondly, it allows a coating. As external surfaces have lime components they are stronger than a clay wall, so it is a perfect surface to receive a coating, offering a durable and a compatible medium. Inside the chapel we can find different layers of mortar and paintings, there are even cornices made of plaster. Outside, as first image of the paper, the hermitage was painted in white, probably with lime painting.

3 THE PROBLEMS OF *TAPIA VALENCIANA* WALL

3.1 *Water*

As we could know, the main enemy of an historic building is humidity and water, becoming the maintenance an obligatory task for preservation. The first phase of degradation occurs when the cover is losing its function; then starts a process

Figure 9. Recent view of the main facade. (J. Pardo Conejero).

of washing mortars that increases in case of earth-rammed walls because they usually are only made of clay.

In the case of *tapia valenciana* wall, as the chapel in study, the water moves inside the wall and separates external layer (formwork) and internal (heart). Due to the lime mortar layer on formwork, interior layer could come down before the external one, like happened in some walls of Saint Anthony's hermitage.

A similar case could happen with humidity; it can destroy the basement of the wall and get it completely demolished, that is why its foundation should be constructed with a water resistant material.

3.2 *Loads*

According to the construction process in reconquest churches' shape, their arches and pillars should be constructed with a resistant material like stone, ceramic bricks or rubblework, neither rammed-earth wall. First of all, because rammed-earth wall couldn't resist tension forces, just compression. We also have to contemplate that a rammed-earth wall arch couldn't resist torsion forces because its fragility, due to the probable eccentric loads in the beams.

So the best way of work for a *tapia valenciana* wall is in compression, its width also depends on its slenderness and it's a bit strange to use this technique for walls thinner than 60 cm.

3.3 *The maintenance works*

Maintenance works are required in all buildings, especially in rammed-earth walls. The most important action is to keep water outside of the wall, so the main maintenance works consists on repairing the roof and conserving the coating. It may other animal pathologies, but it depends on the use and the way to use the heritage building.

4 CONCLUSIONS

The technique of *tapia valenciana* wall was used in Saint Anthony's hermitage because it was easy to build, the material was closer to the building and was a usual technique of building, easy to take and cheaper than others.

Due to the weakness of rammed-earth walls, ancient builders had to take care of the different parts of the building, as foundation, arches, pillars or voids, for example. Even, there is an unusual solution that is a mixture of two Valencian traditional forms of building: *tapia valenciana* and ceramic tile vaults, making the last one the formwork of the wall and serving for arranging the vault, like the Roman "opus caementicium" formwork.

There is not an incompatibility of using *tapia valenciana* wall and continuous coating, mostly lime mortar, plaster or painting.

Regrettably, if there is no maintenance or restoring works, even with its religious use, this little heritage jewel is closer to disappear than to be conserved.

REFERENCES

Cebrián, J.L. 2007. Patrimoni Artístic de La Costera i La Vall d'Albaida en la penombra. In X. Company, V. Pons & J. Aliaga (eds), In *Libro de Estudios Exposició La Llum de les Imatges, Lux Mundi, Xàtiva 2007*: 573–589. Xàtiva: Generalitat Valenciana.

Cobreros Vime, M.A. *La obra de fábrica en los edificios civiles del renacimiento.* Publicación de la Escuela Técnica Superior de Arquitectura de Sevilla.
Cristini, V. 2008. *Estudio de las fábricas de ladrillo en Valencia Analisis mensiocronológico y técnicas de acabado (s. XVII-XVIII)*. Arqueología de la Arquitectura. Madrid/Vitoria.
Font, F. & Hidalgo P. 2009. *Arquitectures de Tàpia*. Castellón: COAAT.
Galarza Tortajada, M. 1996. Actas del primer congreso nacional de Historia de la Construcción. *La tapia Valenciana: una técnica constructiva poco conocida.* Madrid: Instituto Juan de Herrera.
Lluch Garín, L. 1980. *Ermitas y paisajes de Valencia.* Caja de Ahorros de Valencia.
Tormo Esteve, S. García Martínez, V. Cortés Meseguer, L. & Palaia Pérez, L. 2012. Rammed Earth Conservation. In Mileto, Vegas & Cristini (eds). *The Valencian rammed-earth wall (Tapia Valenciana) in the restoration of the Alaquàs Castle*. London: Taylor & Francis Group.

Earthen Architecture: Past, Present and Future – Mileto, Vegas, García Soriano & Cristini (Eds)
© 2015 Taylor & Francis Group, London, ISBN 978-1-138-02711-4

Ceramic connectors & culture connection

V. Cristini & J.R. Ruiz-Checa
Universitat Politécnica de Valéncia, Valencia, Spain

ABSTRACT: The Mediterranean is upheld with millenary constructive synergies, one of them is that of ceramic technology, held as a common base of the many civilizations which have made this sea a cradle of culture throughout the centuries. In this case, we aim to define the precise role of brick, in this geographic environment, as an auxiliary building element implemented in the construction method. In many cases bricks are found hidden, embedded in the stonework or within the cupolas, accomplishing constructive functions, otherwise unbeheld by the naked eye. They are just as important for the Late Roman Empire, as they are for the Byzantine Empire and the High-Medieval Age. Employment of brick, with alternative constructive solutions for local building dilemmas, not only cut down building expenses but also allowed a faster realization. All these interesting applications, which we could name "non-orthodox", emerge among Mediterranean constructive culture.

1 INTRODUCTION

1.1 *Approach*

Bricks count on considerable energy consumption related to their production. However they are really long-lasting and easy maintainable constructive materials. The burning of clay into constructive bricks requires obviously large amounts of fuels and energy, but the results are tangible, as shown by their strength and robustness. Brickwork can last for centuries, considering that finished bricks are "clean" recyclable building unit (Cristini et al. 2013). Bricks are an interesting example of a long lasting material, as History of Architecture can prove.

In this frame the research points out some versatile examples, related with bricks, that the authors have studied along different countries of Mediterranean Basin (Cristini-Ruiz-Checa, 2012), in the frame of the project "Characterization of Valencian Rammed Earth Walls: Documentation, Study and Improved Efficiency". The aim is to define the precise role of the brick, as an auxiliary building element implemented in the construction method.

Particularly, through simplicity and small scale, bricks possess a high generality, which eases integration in historically analyzed constructive cultures, especially the Romans and Byzantines. Brick use in these cultures is not a new topic; but its role, as recyclable historical material, is a fertile research field.

2 VERSATILE USE OF ROMAN BRICKS

Roman bricks undoubtedly record a long saga of foresight and organization trough the Mediterranean Basin. Sun-dried bricks, Vitruvius underlined, should be made only in spring or autumn and, to guarantee complete and uniform drying, they should be made at least two years in advance. The best bricks, like those at Carthage, had been left to dry for five years (Olesone, 2005). Under the Empire, when bricks were visible on the skins of buildings, they were no longer the structural material, in spite of their firing process. They were simply a protective layer covering a strong structure of concrete. Like marble, used in Roman buildings and mainly in slabs for facing, bricks are more "cosmetic" materials than really structural resources (Boorstin, 1992). They cover a solid core of concrete. At the same time this ceramic skin could be separated from the core and reused, as it occurs in medieval architecture (Figs. 1–2).

Roman bricks were made in several shapes and sizes. Often the bricks were cut into triangles which had their hypotenuse laid out and their apex inserted in the core of concrete, as mentioned before, as a facing layer (Fig. 3). As the decades passed, bricks became smaller and took new shapes, while the thickness of the mortar and its strength increased (Boorstin, 1992).

These changes are due to the fact that in the Roman Empire concrete is the main character of the constructions. Romans discover that *pozzolana*

Figures 1–2. Details about *opus testaceum* and fake stones treatment, Pompei, Italy (Cristini-Ruiz).

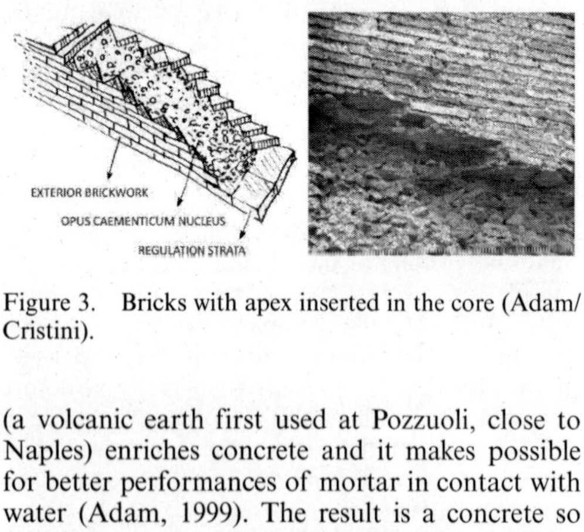

Figure 3. Bricks with apex inserted in the core (Adam/Cristini).

(a volcanic earth first used at Pozzuoli, close to Naples) enriches concrete and it makes possible for better performances of mortar in contact with water (Adam, 1999). The result is a concrete so called *opus caementicium*. This substance, as Vitruvius explained, when mixed with lime and rubble, not only lends strength to buildings of other kinds, but even when piers of it are constructed in the sea, they set solidly under water. Concrete made of *pozzolana* resisted fire as well as water, and would preserve Roman masonries through centuries (Wright, 1924).

The remarkable qualities of perfected concrete with bricks skin would be possible by further trial and error phases, improving the proportions of lime and *pozzolana* and arrange bricks in the best positions. Some detached results are visible, for example, in many *thermae* buildings, where the use of bricks is exploited in walls, in vaults and in hydraulic structures. Bricks, when added in the concrete, gave character, novelty, and grandeur to their works, thanks to coarse-laid brickwork used to face a core of concrete (*opus testaceum,* Adam 1999). Walls of brick required less labor than stonework of the same quality, and could be made of local clay where there was no stone. They perfectly resist humidity and they can shape interior spaces flawlessly. Romans also considered that fired bricks have a good heat and moisture capacity that can help maintain good indoor air quality, above all of the *thermae*. Interesting systems of underfloor heating, used to heat spaces with hot air, called *hypocaustum*, also counts with bricks (figs. 4–5). In this case the suspended floor is raised above the ground by thin bricks pillars, called *pilae* or *suspensurae*, and covers the *hypocaustum* cavity (Adam, 1999).

Another interesting Roman goal is about vaults experimentations. These constructive elements are extended arches and the Romans used them to create large open rooms and high, covered passageways, above all thanks to barrel or groin vaults.

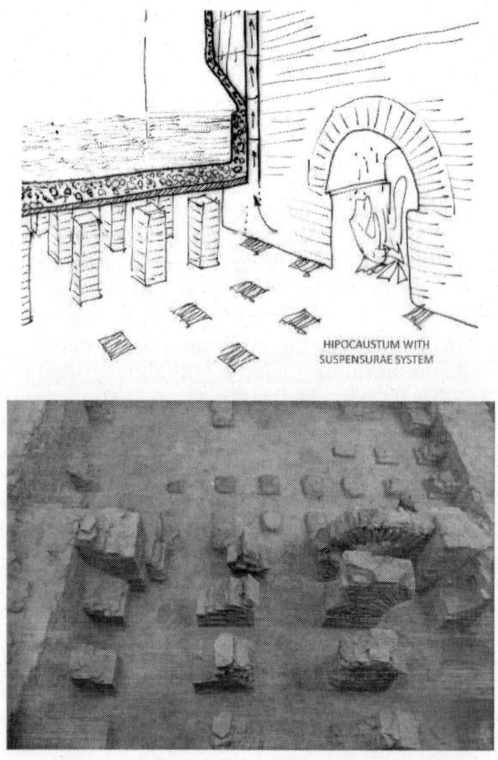

Figures 4–5. Underfloor heating system, Villa La Tejada, Palencia, Spain (Cristini-Ruiz).

The barrel-vault is generally semi-cylindrical in section, and it is used to cover corridors and oblong halls, like the temple *cellas*, or it is bent around a curve, as seen in amphitheater passages. On the other hand the groined vault is formed by the intersection of two barrel vaults, where thrusts are concentrated at all four corners and when several compartments of groined vaulting are placed together over an oblong plan, a double advantage is secured. A lot of apparels of bricks in vaults have been studied and recognized for experts (Humphrey et al., 2008).

Roman brick vaults are constructed in the same manner as common brick arches. The bricks are bonded lengthwise of the vault, with a header course every five or seven courses. Vaults are sometimes built of a combination of brickwork and concrete, using light brick arches and backing up with concrete. They are also built entirely of concrete, the concrete being rammed on the centers in the same way that concrete footing courses are laid (Fig. 6).

According to recent studies (Lancaster, 2005) mortar improved significantly between the late first century B.C. and the second century A.D.

Lighter *caementa* in the upper sections of domes and vaults lessened their potential lateral thrust (fig. 6). Brick linings on the intrados of vaults prevented wooden frames from adhering to the concrete after it had cured (Figs. 7–8).

Dependent on brick industry, these linings were only in use for about a century (between the reigns of Trajan and Caracalla).

The brick industry declined precipitously when political problems ended large-scale building projects in Rome (in the later third century A.D).

Figures 7–8. Brick linings on the intrados of vaults prevented wooden frames from adhering to the concrete after it had cured Villa Adriana, Italy (Cristini).

Figure 6. Bricks on the intrados of vaults, *thermae* of Villa Adriana, Italy (Ruiz).

So due to this factor, a policy of bricks recycling progressively started (Adam, 1999).

3 VERSATILE USE OF BYZANTINE BRICKS

One of the detached legacies of Byzantine architecture is its masonry. At first glaze the Byzantine masonry is very similar to Romans, both used square stones and bricks, fundamentally based on the Roman measures. But in Roman architecture the masonry is used, as said before, "just" for sheet piling, while Byzantine architecture uses bricks for the entire wall. Byzantines probably didn't trust their concrete without *pozzolana* or that former eastern cultures were still influential. But the use of bricks for the entire wall has definitely a radical impact and consequences on the byzantine architecture (Boorstin, 1992).

Byzantine masonry consists above all of *opus mixtum* with alternating layers of bricks and stones (Fig. 10). Decorative masonry connections are also typical, which were possible because of the freedom created by the thick walls and thick mortar joints (Mango, 1978).

There are great varieties of patterns (radial, spiral, zigzag, alternating squares, fishbone and diagonal) as it is proved in a lot of example of this type of architecture, visible above all in Eastern Mediterranean Basin. All these patterns are usually made by recycled and broken bricks (Fig. 11).

Byzantine masonry has thick horizontal joints, almost as thick as the bricks and small vertical joints of just a few millimeters, which gives the wall a rich texture. Because of its popularity the effect is even intensified by the 'hidden layer' method, probably around the tenth century. The method consists of one layer of bricks that is set deepened in the wall and is entirely covered with mortar, so the horizontal joints seemed to be even thicker. This solution (Figs. 12–13) is mentioned as "recessed-brick technique", built with a row of bricks set back from the vertical line of the wall in alternate rows (Ousterhout, 1999).

The presence of masonries with well-marked layers exists all over the Mediterranean countries (Mango, 1978), related with the legacy of post-Roman constructive tradition, featured by masonries with joints the same thickness as the bricks or even thicker (70–90 cm). In fact, there are several techniques from the Orient made with this kind of brick. Some of them combine mixed techniques, using stone or timber, while others just use ceramic elements or applied into rammed earth walls (Ward Perkins, 1954). All these solutions guarantee a reduced quantity of bricks, as an interesting way to save a percentage of raw materials.

Brick grit was also often added to the mortar for its color, but it also worked as catalyst for mortars. This use, inherited by Romans (during the Empire the solution is known as *cocciopesto* technique fig. 14) is quite frequent in all Byzantine architecture.

In this case, the reused bricks act as a waterproof material and raw material to be used for making especially resistant mortars and coatings for pavements (Fig. 15), cisterns and ponds.

It is also interesting to consider that pendentive was made real by Byzantine architecture (Oleson, 2008).

Figure 9. Bricks connectors, Livia's House, Rome, Italy (Cristini).

Figure 10. *Opus mixtum* with alternating layers of bricks and stones, Istanbul, Turkey (Cristini).

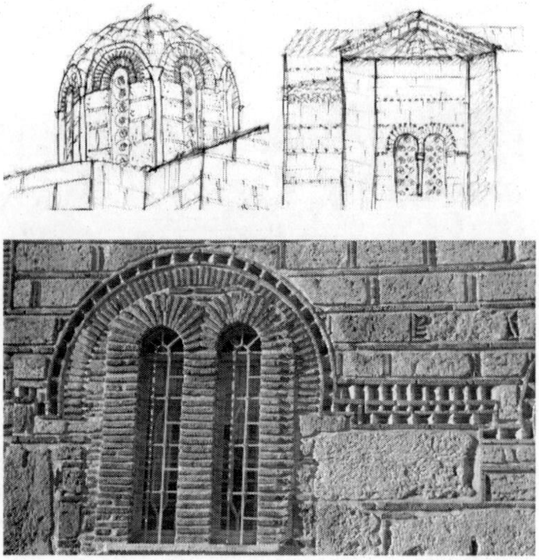

Figure 11. Bricks patterns in Byzantine Greek churches in Agii Apostoli, Agios Thomas, Agia Ekaterini, Athens (Cristini).

Figures 12–13. Recessed brick technique, Old Nicea, Cyprus (Cristini).

Figure 14. Brick grit as catalyst for mortars, Basilica of Maxentius, Rome, Italy (Ruiz).

A pendentive is fundamentally known as a spherical formed triangle that progressively realized the transition between a square plan and a dome (Fig. 16). Setting bricks in parts of circular layers created pendentives.

These parts of circular layers increased in length, but decreased in diameter from the square plan to the circular form of the dome. Undoubtedly the proper use of bricks plays a k-role in the trial and error phases of pendentive constructive experiments.

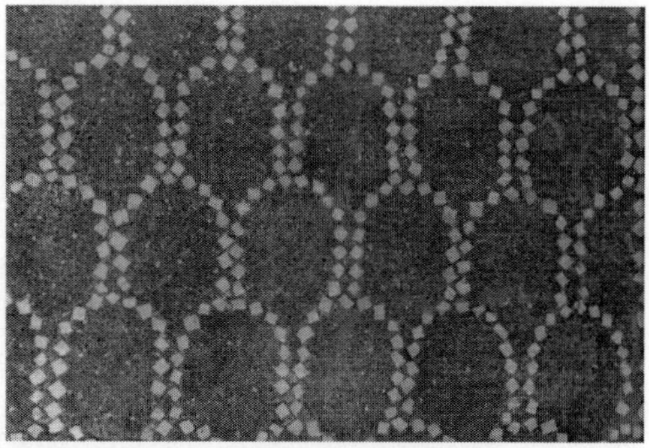

Figure 15. Pavement with brick grit, *opus signinum*, with marble *tesserae*, Ercolano, Italy (Ruiz).

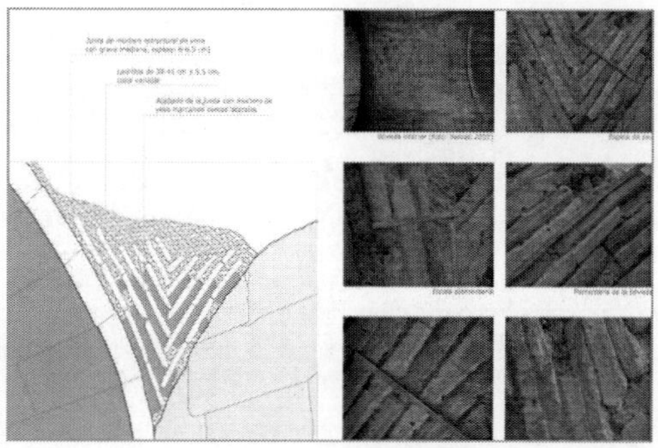

Figure 16. Pendentive in Byzantyne architecture, Pretorium at Zenobya City, Syria (Cristini).

4 CONCLUSIONS

Both Roman both Byzantine historic examples of versatile brick use show interesting concepts about flexibility and reusing capability. Both case-studies appoint that these ancient constructive cultures hold an overall high reusability of bricks, not only due to high scores for reversible connections but for all the criteria regarding the life-cycle analysis of the material. An interesting concept, recently introduced in sustainable architecture studies, called "salvageability" (Nordby et al., 2009), was under lined for centuries by Roman and Byzantine historical use of bricks. It's interesting to consider, in this frame, that historically is feasible to deconstruct an old brick walls, and reuse the material, especially thanks to flexible lime mortar commonly used as in Roman as in Byzantine architecture (Cristini-Ruiz-Checa, 2009). So, indirectly, historical good optimization of bricks and their long-lasting use are obtained due to a proper use of lime mortar, as a binder that is not aggressive for the ceramic connectors. Bricks for centuries, along the Mediterranean Basin, because of this building practice, are used like a recyclable product, and the masonries, cyclically, have been considered as potential "banks" of constructive sources and raw-materials.

ACKNOWLEDGEMENTS

This article would not have been possible without the funding of the research project "Characterization of rammed earth walls with bricks: documentation, study and improved efficiency" (*Caracterización de fábricas tradicionales de "tapia valenciana": documentación, estudio y mejora prestacional* UPV/PAID-06-12/SP20120466/main researcher V. Cristini).

REFERENCES

Adam J.P., 1999, *Roman Building: Materials and Techniques*, Routledge Ed., London, 1999.

Boorstin D.J., 1992, *The Creators*, Random House Ed.,NY, pp. 107–126.

Cristini V., Mileto C., Vegas F., Ruiz-Checa J.R., 2013 "Recycling of bricks in rammed earth walls" in *Vernacular Heritge and Earthen Architecture, Contribution For Sustainable Development,* Correia-Carlos-Rocha Eds. Taylor & Francis Ed., Balkema, pp. 383–388.

Cristini V., Ruiz-Checa JR., 2009, "A historical Spanish traditional masonry technique: some features about tapia valenciana as a reinforced rammed earth wall", in *Canadian Masonry Symposium Proceedings*, Toronto, Canada, CMS Ed., pp. 435–442.

Cristini V., Ruiz Checa JR., 2012, "Core, connectors and crust: Methodological restoration options for reinforced rammed earth wall with bricks" in *Rammed Earth Conservation*, Balkema Ed., London, pp. 297–302.

Humphrey J.W., et al., 2008, *Greek and Roman Technology: A Sourcebook*, Routledge Ed., London.

Lancaster L.C., 2005, *Concrete Vaulted Construction in Imperial Rome: Innovations in Context*; Cambridge Ed., NY.

Mango C., 1978, *Byzantine architecture*, Electa Ed., Milan, 1978, pp.233–56.

Oleson J.P., 2008, *The Oxford Handbook of Engineering and Technology in the Classical World*, Oxford University Press, NY, pp.256–279.

Ousterhout R., 1999, *Master builders of Bizantyum*, Princeton University Press, pp. 157–184.

Sigrid Nordby A., Berge B., Hakonsen F., and Hestnes A.G., 2009, "Salvageability of building materials", *Journal Building Research & Information* (Bri) Vol. 37:1, pp. 55–67.

Ward Perkins J.B., 1954, "Notes on the structure and buildings methods of early byzantine architecture" and "notes of structure" in *The great Palace of the byzantine emperors*, II° report, St. Andrew Ed., pp.56.

Wright G.R.H., 1924, *Ancient Building Technology, Volume 3*: Construction (2 Vols), Brill Ed., Denver, pp.374–383.

Earthen Architecture: Past, Present and Future – Mileto, Vegas, García Soriano & Cristini (Eds)
© 2015 Taylor & Francis Group, London, ISBN 978-1-138-02711-4

A parameterization of brick-reinforced rammed earth in Valencian Region (Spain)

V. Cristini, F. Martella, C. Mileto & F. Vegas
Instituto de Restauración del Patrimonio, Universitat Politècnica de València, Valencia, Spain

ABSTRACT: The study conducted on buildings with strong presence of brick-reinforced rammed earth walls, within the province of Valencia and its surroundings, has permitted the discovery of some details of this constructive technique and has thrown light on their reason for being implemented. Due to this, it has been considered relevant to establish a comparative table of the distinct constructive combinations. The study of the variants of the techniques, in-situ surveys, the bibliography and existent documental sources have helped to reach a first approach regarding the constructive variants found containing the three basic elements constituting the constructive brick-reinforced rammed earth wall essence: nucleus, bricks and lime crust. This first hypothesis of a chrono-typological sequence of the technique, which had previously only been considered unchanged with time, facilitates the understanding of its origin, development and decline; therefore confirming a logical technology which matured along the centuries, adapting itself to specific architectural typologies.

1 INTRODUCTION

Many authors have addressed the subject of the brick-reinforced rammed earth, and rammed earth in a broader sense. However none of the papers nor the articles to date have discussed the variants of rammed earth walls with bricks in detail, nor the varying nature of its constructive characteristics over the centuries. Here lies the novelty of this research.

Starting with a review of existing literature on the subject, we look at notable examples of this construction technique within the Region of Valencia, especially those concentrated in the province of Valencia, Spain (Table 1). Fifty-one buildings of differing types and ages were catalogued for this study. Thanks to this work of comparison, we have been able to put forward an initial theory regarding a possible chrono-typological pattern to this ancient constructive technique.

2 RAMMED EARTH WALLS WITH BRICKS

2.1 *Background and origin*

This technique is one of the most common variants of wall found in the East of the Iberian Peninsula.

The final appearance of a wall made using this technique is clearly identifiable: at first glance one notices a regular rhythm of bricks which appear to arranged in the background compared to the smooth "plastered" facade, as the bricks are slightly recessed inside the wall (Fig.1).

The direct origin of the technique is uncertain, with it being the result of several factors and numerous technological developments linked to the supply of raw materials and varying constructive demands. In reality, the implementation process of the walls is very similar to that of the lime-crusted rammed earth wall. After the completion of a foundation and subsequent base, usually with a stone or brick masonry base of 50–60 cm in height, the formworks are put in place. Until this point, the construction process is similar to that used in other types of rammed earth wall. In the next phase, a layer of lime mortar is poured together next to the forms, followed by a layer of earth (to the same level as the mortar). Finally a brick is laid on top of the layer of mortar and earth against the formwork. After the placement of the second stratum begins (Fig.2). Thus, the brick is embedded inside the wall, except from one side which is placed against the formwork. In addition, the shocks produced during the ramming process of the different layers results in the bricks moving slightly from the surface of the formwork into the depth of the wall, in such a way that some of the lime mortar partially covers the head of the brick (Cristini, 2012).

Table 1. Locations presented in the study, divided by province (authors).

Province	Location
Valencia	Valencia capital
	Xirivella
	Alaquás
	Xativa
	El Puig
	Albalat dels Sorells
	Benifaraig
	Alfara del Patriarca
	Picanya
	Luchente
Castellón	Castellón
	Borriana
	La Villavella
	Mascarell
	Vila Real
	Partida de Fadrell
	Alquería del Niño Perdido
Alicante	Elche
	Cocentaina

Figure 1. Brick-reinforced rammed earth wall; fencing wall of la Cartuja del Puig, Valencia (authors).

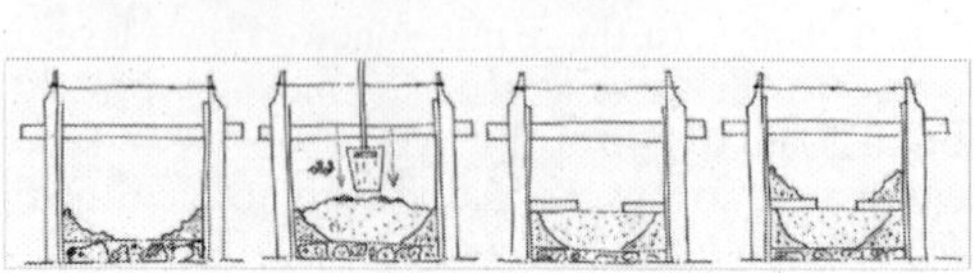

Figure 2. Resume of construction process: rammed earth wall with bricks (authors).

2.2 *Technical details*

A wall made using brick-reinforced rammed earth technique is a double wall: one part stronger and visible, and one interior and weak. What we have is an earth core with a crust composed of bricks and rows of lime mortar (Fig.3).

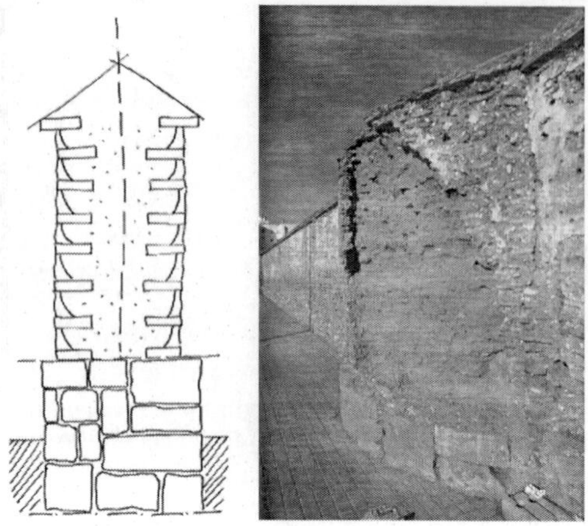

Figure 3. Detail of the fencing of San Miguel de los Reyes, Valencia (authors).

Furthermore, in many cases you can see different types of bricks in the wall itself, of differing colors and dimensions. We must take into account that in the case of many buildings the ceramic components came from other collapsed constructions. The final appearance of many of the facades, if we focus solely on the ceramic elements, is heterogeneous in terms of shapes, colors and textures.

The presence of aluminum silicate, found both in the earth and in the ceramic bricks, gives the structure strength and stability. In fact, this bringing together of earth and bricks is not incidental, but in fact follows a precise spatial arrangement, which gives the structure particular mechanical and physical properties. The presence of the bricks, with their rough and porous properties, prevents horizontal slippage between the layers of earth. In the case of a monolithic earthen wall, the risk of slippage of one layer to that below, caused by a horizontal force such as an earthquake, is much greater than when bricks are inserted. The bricks also act as connectors between the lime crust and the interior earthen filling, and between the layers of rammed earth itself (Cristini, Ruiz Checa, 2009).

3 ANALYSIS AND CASE STUDIES

3.1 *Historical and geographical setting*

The brick-reinforced rammed earth constructive system went through its period of maximum development and employment between the mid-fourteenth century and the second half of the seventeenth century. Indeed, the analyzed case studies are primarily from this historical setting. By the 18th and 19th centuries, the technique was being modified progressively, being replaced by brick or pillars structures. Gradually, the presence of the technique in urban areas is reduced, until its progressive decline and subsequent fall in the eighteenth century.

3.2 *The election of case studies*

The discriminating factors considered for the selection of the case studies were essentially threefold (Fig.4):

1. Reference to documented historical data from literature. A detailed study of existing sources and notable examples of this technique in the Region of Valencia was made.
2. A significant presence of visible rammed earth walls. The selection of buildings where an interesting sample of wall is visible meant that the surface of the rammed earth set to be studied ought to be at a minimum height of 50 cm and with a minimum visible area of elevation of 1 square meter.
3. Dating, trying to identify examples from different origins, covering the indicated chronological range.

After establishing these initial criteria it was necessary to create a template of study: an effective tool to catalogue and subsequently study the selected cases, each of which was distributed according to a specific sequence of items:

- Century of construction
- Previous interventions
- Width of the wall
- Vertical and horizontal distance between the bricks
- Size, colour of the bricks and bonding used
- The composition of the crust and the type of binder used for the crust
- The relationship between the surface area of the bricks and the crust
- Presence of lime plaster in the finishing
- Pathologies present.

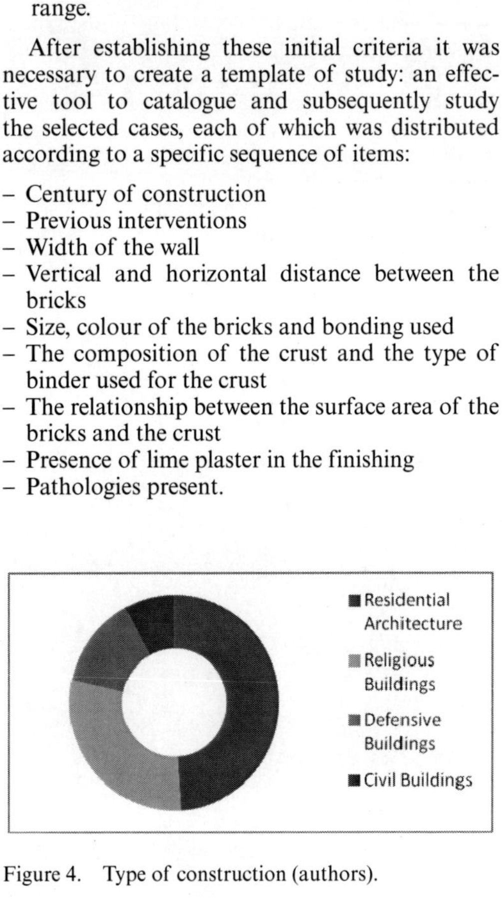

Figure 4. Type of construction (authors).

Thus, every building has a set of data, which can be used in order to compare the fifty-one examples of walls with rammed earth and bricks. By studying and analyzing the selected case studies it soon became clear how some features vary with some regularity and progression, suggesting the existence of macro-families which bring together the constructive characteristics of the different rammed earth walls with bricks analyzed. Using this type of analysis, along with the information seen from the results tables and the comparison of some important buildings, we were able to extrapolate four macro-groups of brick-reinforced rammed earth in Valencia region:

- *Type I*
- *Type II*
- *Type III*
- *Type IV*

This subdivision does not purport to denote marked divisions between one or another variant of type. Rather, it acts more as a guide that can help in the reading and analyzing of some examples of buildings constructed using this technique.

4 THE TYPOLOGIES

According to this basic classification, and following a chronological, technical and constructive logic, four main constructive variations related to certain technological changes could be observed.

The research aims to identify the origin of the technique in the 14th century, beginning with the lime-crusted rammed earth wall and its demise, possibly during the 18th century coinciding with the progressive use of many more bricks in construction. We have wanted to highlight that these typologies are not completely detached from one another; but that each has evolved from the previous variation.

4.1 *Type I*

Chronologically, type I is the first of the identified groups: the analyzed buildings were constructed within a time period that is believed to span the 13th and the 14th centuries. In all the cases studied, this variant of brick-reinforced rammed earth wall is found in defensive buildings (defensive walls or towers).

Coherent with their function, the width of the walls is notable: from 120 cm up to 240 cm. The core of the wall is earthen and, due to its great width, the presence of additional fragments of bricks or stones can be observed. These elements help compacting the interior of the wall, and mean that the main thickness of the wall is not only made of rammed earth. By observing sections of these examples we can see how the lime crust reaches a thickness of between 30 and 35 cm, always with the aim of providing protection. A lime crust of this thickness behaves almost like an additional body to the main rammed earth wall. In the majority of these first examples, the bricks are stretcher laid. In other words, the ceramic components do not join the earthen core with the lime crust. Thus, what you get is the protection of the earthen core from various elements, whether that is from the lime crust or the bricks. The bricks are not arranged in a regular manner, and differing horizontal and vertical distances can be seen between the bricks. This type of arrangement is characterized by rows of bricks which are almost continuous, each very close to the other. On the other hand, the vertical separation between the bricks is vast, reaching up to 25 cm (Martella, 2014) (Fig.5).

4.2 *Type II*

The buildings with variants within this group date roughly from the 14th to the 16th century. The construction technique, unlike that of the previous variant, reveals a broader purpose for the construction, such as religious buildings or examples of residential architecture.

For this reason, according to the building typology, the wall thickness is significantly reduced. The walls in question are between 40 and 60 cm wide. In proportion with the total width of the wall, the thickness of the lime crust also decreases, although it remains relatively substantial (5 to 10 cm thick). As the thickness of the lime reduces, the presence of aggregates in the layer of lime increases. Gravel and pebble of a sufficiently large grain (1–2 cm) are the aggregates of choice. The decision to add these items to the lime mixture is probably due to the improved adhesion with the final lime finish (lime plaster) which they provide. The arrangement of the bricks also changes, moving on from stretcher laid seen in the previous variant, to header arrangement. Without doubt, this arrangement of bricks, along with the narrower width of the lime crust, gives the structure greater stability and strength: the two bodies of the wall are connected in a way that can be likened to a zipper, each part collaborating with the other. The layout of the ceramic components is more homogeneous than in the first group analyzed: in many cases the horizontal and vertical distances between the components comes to the same, ranging from 6 to 15 cm. By decreasing the distance between the laid bricks, compared to the first variant analyzed, the amount of ceramic elements used for the building of the wall increases proportionally, and the amount of externally visible lime decreases (fig.6) (Martella, 2014).

4.3 *Type III*

This solution is typical in household buildings; and outside of urban areas it is especially common in many farmhouses. The approximate dating of the buildings included in this group varies from the 16th century to the 17th century. The width of the wall decreases gradually depending on the type of building, ranging from 55 to 35 cm thick. Buildings such as farmhouses or private homes do not in fact need walls of great thicknesses. Buildings of this kind are made with raw materials found in the local area and at the least possible expense. Other characteristic features of these walls can also be read from an economical perspective and in terms of the speed of construction of the walls: the lime crust has been reduced notably, reaching

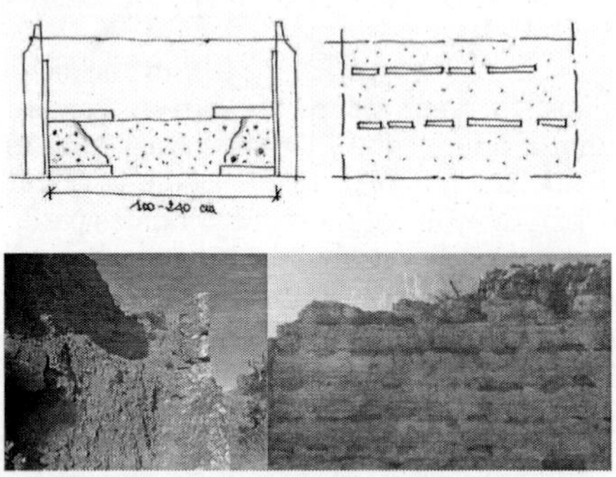

Figure 5. Details of Type 1-wall. Section and elevation of Nules Castle, Castellón (authors).

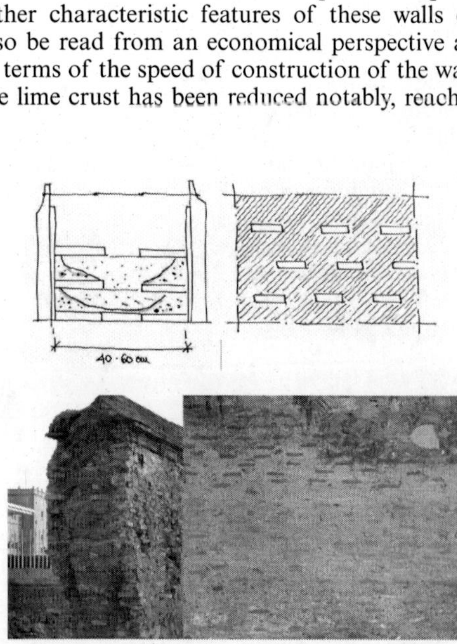

Figure 6. Details of Type 2-wall. Section of the fencing of San Miguel de los Reyes; Elevation Mascarell city-wall, Valencia (authors).

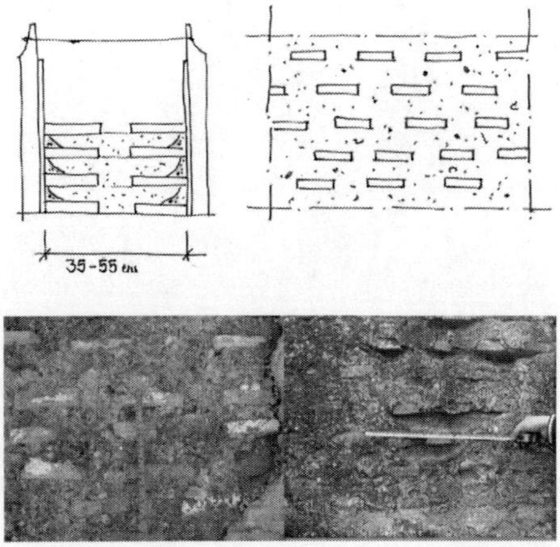

Figure 7. Details of Type III-wall. Detail of the elevation of Convent of San Diego, Alfara Patriarca, Valencia; detail of the elevation of the fencing, Carabona Tower, Burriana (authors).

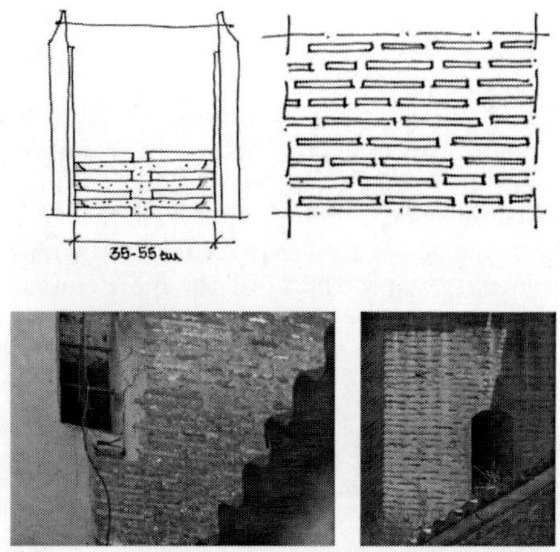

Figure 8. Details of Type IV-wall. Elevation of the Farmhouse of Falco, Godella; Elevation of Saint Ursula Church, Valencia (authors).

widths of just 2 cm, and aggregates of a large size are always included. The reduced width of the wall and the presence of bricks header arranged means that the structure is more resistant and stronger than the other variants. The bricks strengthen the earthen core, fulfilling a role that goes beyond the protection of the core of the wall. This type of solution, with all its features, already begins to resemble brickwork. When we analyze the elevations we see that these walls are characterized by the regular arrangement of bricks. The distance between them, whether that be horizontal or vertical, decreases even further compared to the previous solutions, from 10 to 4 cm. When observing these examples pathologies such as the partial erosion of the lime crust, due to its reduced thickness, can be seen (Fig.7) (Martella, 2014).

4.4 *Type IV*

What we have in this case is a variant which is linked to religious architecture (churches and monasteries) principally from the 17th century.

The width of the walls of these examples remains the same as that of the previous variant, with thicknesses of between 35 and 55 cm. When this technique is used in buttresses, the width increases so that transversal loads can be transferred to the ground. The arrangement of the bricks changes radically compared to the first buildings analyzed: in this case they cover the majority of the outer surface of the wall, in regular and continuous lines. The vertical distance between the bricks, which in previous cases reached up to 25 cm, has now been reduced to just 2 to 4 cm. The final appearance of these walls is similar to that of a brick wall structure, in this case, without the use of lime plaster.

In type IV-walls, the significantly smaller dimensions (compared to defensive walls) mean that the main responsibility for the structure of the building lies with the bricks. In this case the earthen core is almost secondary to the great abundance of ceramic components, which act together as a veritable autonomous structure.

The reduced distance between the bricks further emphasizes this new distribution of roles within the wall. In this case, due to the proximity of the ceramic elements, their function as protective elements intended to bind the wall, is secondary. One can no longer appreciate the "zip effect" which characterized the walls with moderate to thin layers of lime crust.

5 CONCLUSIONS

The study and analysis of all the examples is a task requiring considerable analysis and amalgamation of data, due to the complexity and variety of typologies found. The first investigatory phases were characterized by the search for and study of existing literature on the subject of rammed earth walls, and more specifically, of bricks reinforcement. This was followed by an *in situ* phase of documentation and data collection.

Through the comparison of the different cases it has been possible to establish parameters for this constructive technique, and subsequently the classification of four main macro-families, each of which brings together different yet specific constructive characteristics (Fig.9).

In all the cases, one can note how the brick, combined with lime crust, is the most important element. Its role has changed over the centuries. In the first defensive buildings analyzed, the sole purpose of the bricks, seen to be almost "drowning" in the lime, is the protection of the more vulnerable earthen core.

As the arrangement and bonding of the bricks in the façade changes, so does their function: from a "screen" to an element that provides greater strength to the wall as a whole. In fact, by decreasing the thickness of the lime crust (from 30 cm to 5 cm thick) the bricks connect directly to the earthen core, as such acting as a bond between the outer more resistant layer (the lime crust) and the innermost more fragile layer (the core).

The aim of the research centered on the analysis of a collection of elements which revealed the different uses of rammed earth walls with bricks within the Valencian Community. The aim was for the results to be sufficiently representative so as to permit the identification of the constructive logic, as well as to provide a literary viewpoint which could be used to understand many other cases not covered directly by this research.

In total, we have identified (Fig.10):

- 4 cases of Type I—wall
- 25 cases of Type II—wall
- 20 cases of Type III—wall
- 5 cases of Type IV—wall.

All the data collected and processed provides the following conclusion; how a constructive system shifts and adapts according to the specific requirements of each historical period. None of the essential elements disappears from the ensemble, rather their size and position within the wall is simply changed. The core, connectors and crust each go about playing a different yet decipherable role throughout the lifespan of the constructive technique.

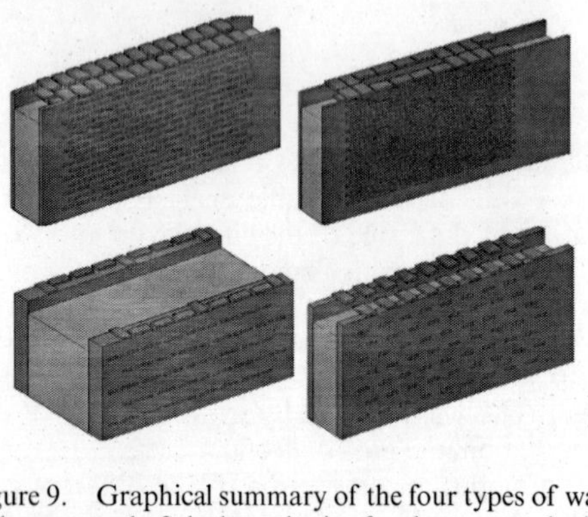

Figure 9. Graphical summary of the four types of walls and a proposal of the hypothesis of a chrono-typological evolution (authors).

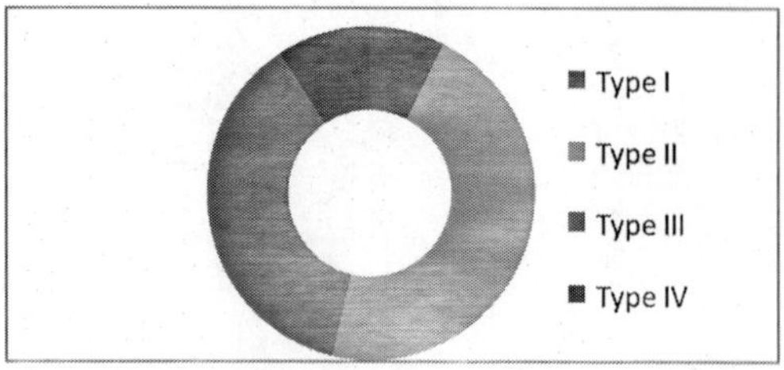

Figure 10. Percentage of different walls studied (authors).

ACKNOWLEDGEMENTS

This article would not have been possible without the funding of the research project "Characterization of rammed earth walls with bricks: documentation, study and improved efficiency" (*Caracterización de fábricas tradicionales de "tapia valenciana": documentación, estudio y mejora prestacional* UPV/ PAID-06-12/ SP20120466/ main researcher V. Cristini).

REFERENCES

Bravo del Fresno, I., Sanchez, J., 2011. *La tapia Valenciana en la ciudad e Guadiz (Granada).* Séptimo Congreso Nacional de Historia de la Construcción, Santiago Octubre 2011.

Cristini, V.; Ruiz Checa, J.R. 2009. *A historial Spanish traditional masonry technique: some feature about Tapia Valenciana as a reinforces rammed earth wall.* XI Canadian Masonry Symposium, Toronto, Ontario, Canada June.

Font, F; Hidalgo, P. 2011. *La tapia en España. Técnicas actuales y ejemplos.* Informes de la Construcción, Vol. 63

Galarza, M., 1996. *La tapia valenciana: Una técnica constructiva poco conocida.* Primer Congreso Nacional de Historia de la Construcción, Madrid Septiembre.

Galarza, M., 2001. *Las técnica constructivas en las construcciones defensivas: la Tapia Valenciana.* IV curso de Cultura Medieval seminario, La fortificación medieval en la Península Iberíca.

Iborra, F. 2012. *About the rediscovery of the Tapia Valenciana.* RESTAPIA 2012 Rammed Earth Conservation, pp. 139–144, Valencia.

Martella, F. 2014. *La tapia valenciana, estudio y caracterización en ejemplos destacados de edificios históricos*, TFM, C. Mileto-V. Cristini, UPV, unpublished, Valencia.

Mileto, C., Vegas, F., Cristini, V. Diodato, M. 2006. *Constructive Features of the historic architecture at Valencia City.* ARCHÉ, Publicación del Instituto Universitario de Restauración del Patrimonio de la UPV, Valencia.

Earthen Architecture: Past, Present and Future – Mileto, Vegas, García Soriano & Cristini (Eds)
 ISBN 978-1-138-02711-4

Transdisciplinarity in rammed earth construction for contemporary practice

M. Dabaieh
Lund University, Lund, Sweden

M. Sakr
Prince's School of Traditional Arts, Egypt

ABSTRACT: For centuries rammed earth has been a method of construction in various parts of the world. It is an economical building technique as earth is an abundant cheap resource. It is durable, safe and a desirable building technique for hot climates. This paper is discussing a workshop activity in building with sustainable rammed earth construction methods in Egypt. It adds to previous research in vernacular rammed earth a transdisciplinary and participatory dimension. Believing that they are assets to promote for rammed earth benefits and applicability in contemporary practice. The central philosophy of this workshop is to enable researchers, practitioners, decision makers and interested laymen from different disciplines to interact in a transdisciplinary action approach to awaken the importance of adopting green affordable building practices. This paper shows how to initiate a low cost and an energy efficient earth buildings and discuss the criteria on which rammed earth structures is assessed.

1 INTRODUCTION AND BACKGROUND

In the world's history of architecture, earth has been the most popular structural building material. However concrete and steel have become the most common and conventional building material in the 20th century due to the modernization movements in building technology. As a result standardized modern architecture has prevailed and has been sweeping away the architectural characters of every region.

Earth behave like no other building material available in the Egyptian market today (Fathy, 1973), in addition it fulfills the current ecological and economic local requirement (Schijns et al., 2003). Earth is an abundant natural resource in the Egyptian desert, it requires few aggregates or additives to improve its properties (Morgan, 2008). Building with earth is an easy yet labor intensive process, a fact that can be considered an advantage in developing countries (UN report, 1986).

The majority of vernacular buildings in Egypt, either in the agricultural or desert areas has been built using adobe bricks (Fathy, 1973). However from our observations there were regions were "*Touf*" technique is used which is laying earth in several courses without formwork, which is similar to the cob technique used in England and France.

Rammed earth is an ancient construction technique that utilizes earth to create thick durable load bearing wall. It is low-cost, with heat-storing capacity and is recyclable (Minke, 2000). It has been developed independently in many parts of the world, and is also commonly known by its French name "Pise" (Keable, 1996). The technique of Rammed earth is simply based on compacting soil between vertical formwork boards, which are then removed leaving a mass soil wall. Usually this technique is used in regions whose soil composition is unsuitable to make sun dried clay bricks (McHenry, 1986).

Rammed earth technique was also used in the arid regions of North Africa and the Middle East, where earth was the only logical building material. The Romans and the Phoenicians transferred the technique of rammed earth to Europe in the early 16th century (Easton, 1996). It is not yet common in Egypt but there are some promising contemporary trials as shown in Figure 1.

This paper is discussing a workshop activity organized in Egypt during February 2014. It is trying to fill in a research gap on wider applicability of rammed earth. It aims at raising the awareness of using such environmental and economic technique for possible contemporary applications. In addition, the main objective is to promote using rammed earth as a sustainable, energy efficient and a low impact building technique. The paper shows how transdisciplinary participatory approach was applied through an action intensive two days

Figure 1. Center for building crafts in Sinai built with rammed earth and stone. Photo credit Nashwa Ibrahim.

workshop training. Where practitioners, decision makers, laymen, members of a local NGO and international initiatives come together to discuss in a practical way the possibilities, challenges and obstacles that hinder implementation and proposing feasible future action plans.

2 PROBLEM DEFINITION, POTENTIALS AND CHALLENGES

Egypt is currently facing a critical problem in terms of energy security. The residential sector accounts for 50.7% of the country's energy consumption (EEHC, 2012). That is for building materials manufacturing, construction and energy needed for running buildings. Moreover, Egypt is suffering from severe housing problems and the government is still unable to provide decent affordable residential units especially for the low-income sector. Which led to the crisis that 50% of the urban housing stock is informal (World Bank report, 2007).

The embodied energy of rammed earth is three time less than those of conventional techniques using fired bricks (Reddy et al., 2014). Also using stabilized rammed earth improves its thermal properties and reduces the energy consumed for heating and cooling (Serrano et al., 2012) which is negligible when compared to conventional methods using concrete and fired brick. However to be able to apply such notion there are still many challenges towards taking further practical steps in the construction field.

Till now there is no national standard or published reference documents in Egypt for earth construction in general or for rammed earth construction in specific. In addition the current regulations and legislations for building permits are considered a main obstacle for any initiative to build using rammed earth.

To create a segment of potential architects and designers who are aware and skilled in building with earth, we have to start by education. Earth construction is not a core curriculum at architecture schools, in some rare cases it is an elective course, but in most cases it is totally missing. When it comes to craftsmanship, there are few vocational trainings on earth construction and skilled masons are rare, and if available their wages are higher than other workers.

There are wide spectrum of good types of soil in Egypt that can be a good potential for the applicability and feasibility of building with rammed earth. However there is no updated precise documentation for the properties of the soil types and the supply chain for raw materials is either not complete or missing. Also the quality control measures for earth structures are not fully established in Egypt or widely recognized and still there are variations between certified tests.

3 METHODOLOGY

The workshop is an experimental research that applied a transdisciplinary participatory action research (PAR) method. Transdisciplinarity was adopted as it suits best the nature of this workshop. The aim of the methodology is to improve this area of concern and to politicize this workshop design process as a base from which power relations can be created and transformed to possible collaboration between researchers and practitioners with national authorities, decision makers and community partners.

It is discussed in literature that transdisciplinarity is an intellectual space where participants from diverse disciplines can express their views and provide input of different types (Somerville & Rapport, 2002). Also transdisciplinarity is a sort of interdisciplinary approach but within a system or structured plan through encouraging participation and involvement of different social members in an action collaborative research plan (Lawrance, 2004). While the participatory component applied in this research is a kind of a social progression. That is aligned with Gramberger's definition to transdisciplinarity as a social process in which a group of individuals decide about the future of a specific issue (Gramberger, 2001).

The authors with two co-instructors were in charge in teaching and monitoring the whole workshop phases. The four instructors are architects with experience in vernacular and traditional earth buildings. They acted as catalysts for the workshop idea, planners for the workshop phases and facilitators for the building process. The transdisciplinary and PAR in this practical training gave the chance to bring scientific research with everyday experience of field practitioners, engineers, decision makers, NGOs and business owners together. The dialogue created between the participants enabled all parties concerned to express ideas about problems, challenges and possible future applications. The workshop process and the project planning after are explained in the below sections.

4 WORSHOP IDEA, HOW AND WHY IT STARTED

A total of 29 participants joined the workshop, including a representative from the Egyptian Housing and Building Research Center (HBRC) who is in charge of green building codes and rating systems. In addition to a representative from MED-ENEC project, which is an EU project for energy efficiency in south Mediterranean countries. Their presence was important as they are in charge of developing guidelines for energy efficiency in buildings. Among the group of participants there were two civil engineers specialist in sustainable structures, two electrical engineers specialist in renewable energy—at the same time they run an NGO-, one product designer and one secondary school student. The rest were architects, six in academia and research, and the remaining are practitioners. There were also three local workers who helped in the building practical phase. The workshop started with an introduction on the site characteristics including local climate, topography, wind direction, and sunlight orientation. Then it was followed by a discussion on how to control energy consumption using energy modeling and simulation programs in designing energy efficient and low-tech rammed earth building. The discussion showed how to achieve preliminary assessment for energy efficiency before construction.

Participants were also introduced to the concept of how to build a low-impact house using practical examples. In addition how to pay attention to the importance of indoor thermal comfort and air quality through cross ventilation, night-flush ventilation and thermal mass. That is to reduce cooling loads and counteract daytime heat gains. Together with how to control daylighting and in the same time reduce glare and heat gain from openings. These factors, together with others, will influence the location and orientation of the building within the boundaries of the site. This phase ended with a theoretical introduction on rammed earth building techniques which was followed by a practical phase for the physical building steps explained below.

5 BUILDING WITH RAMMED EARTH TECHNIQUE: THE PRACTICAL EXPERIENCE

In order to experiment rammed earth as a building technique, practice it, explain it and even promote it, a practical hands on training was held. The following steps were undertaken:

5.1 *Soil identification*

The soil used in rammed earth technique has to be identified to know its components and whether it needs other additives to enhance its characteristics. Accordingly a series of field tests were conducted using a sample from the site taken from a depth of 0.5 meter, to ensure that the surface organic materials are not included.

5.1.1 *The drop test*

A handful of un-sieved soil was taken, moisten and made into a ball, which was held in hand and left to dry for a few minutes before dropping it. Fortunately the soil excavated from the site was suitable as the ball broke into a few lumps.

5.1.2 *The roll test*

Although the drop test applied indicated that the soil dug from the site is suitable, it was important to apply another simple test to double check the suitability of the soil in the building process. So a lump of moist soil was taken and rolled by hand to form a cigar-like roll, which was almost 20 cm in length and 2.5 cm in diameter. The roll was placed on a table and pushed gently over the edge, and the length of the broken part was measured to indicate the amount of clay within the soil. The roll broke almost at half the length, as usually if the broken part is less than 8 cm then this indicates that there is no enough clay and also if the broken part is longer than 12 cm, then this means that there is too much clay.

5.1.3 *The jar test "particle size test"*

In order to know the proportions of different particle sizes of the soil used, the jar test was applied.

Figure 2 & 3. Left: participants testing the soil, the roll test. Right: participants testing the soil, the drop test. Photo credit: M. Dabaieh.

That is to get a preliminary assessment of the ratio of coarse to fine particles in the soil. Two thirds of a bottle was filled by soil taken from the site, and water was added to fill the bottle. The bottle was shaken till all the soil particles were suspended then it was left to settle for a few hours. As the water cleared, we realized the formation of three different layers separated by clearly visible lines. The sand layer settled at the bottom as its particles were heavier, then two layers of silt and clay were formed on top.

5.1.4 *The result*

These simple field tests indicated that the soil of the site contained more sand and gravel than silt and clay, in a rough ratio of 3:1. Although the site soil was suitable to be used in rammed earth, the addition of lime to the soil was suggested as a natural and environment friendly stabilizer that will hold the coarse particles of the gravelly soil together. Accordingly lime was added to the soil in a ratio of 1:9.

5.2 *Soil mixing*

The soil was sieved to ensure that no large size gravel included, and then mixed with lime. After the thorough dry mix process water was added gradually. The mixture had to be turned over while water is sprinkled to ensure that all the particles will be moistened. The amount of water added to the soil is very important, because if the mixture contains little water the soil will not be properly squeezed, and with too much water the soil becomes too wet and water will resist compaction.

The amount of water added depended on a number of factors, the drop test gave an indication of the amount of water needed to achieve plasticity, but the site conditions also had an effect, as water evaporates while mixing in hot weather. Generally speaking water forms 10%–15% of the mixture, and the mixture should look quite dry after adding it.

Figure 4. Participants mixing & preparing the soil for ramming Photo credit: M. Dabaieh.

Figure 5. The formwork when assembled to ram a new layer. Photo credit: M. Dabaieh.

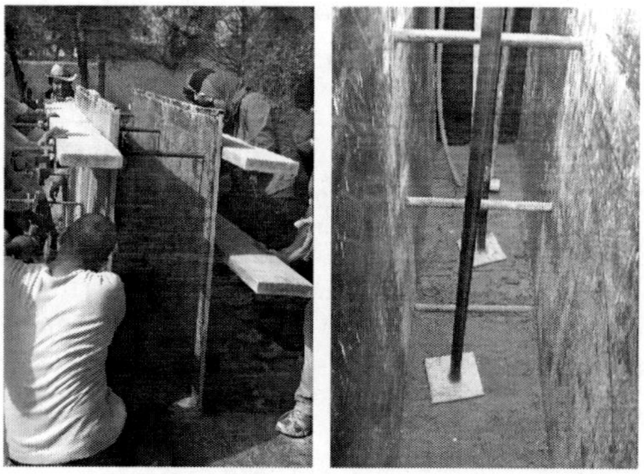

Figure 6 & 7. Left: ramming the first layer. Right: installing the formwork. Photo credit: M. Dabaieh.

5.3 *Formwork*

While the soil was mixed and prepared, another group was assembling the forms. Formwork can range from simple to complicated systems, however in our workshop we decided to use simple, easy to assemble and dismantle and in the same time affordable forms. So these forms were composed of a number of items:

- 16 mm thick plywood boards
- Long metallic bolts to assemble the boards
- Metallic pipes (whose length corresponds to the require width of the wall) to ensure a uniform spacing between the two boards
- Wooden planks that are placed along the length of the plywood boards to keep them in place, and also to be used as scaffolding
- Wooden planks were used as "stays" whenever needed, they were wedged diagonally between the formwork and the ground to keep the formwork in place.

The design of the formwork proved to be practical, as components were easily and quickly assembled to start the next process.

5.4 *Ramming*

The mixed moist soil was poured in the formwork creating a uniform level of almost 15 cm, which after ramming was compressed to 10 cm. As soon as the first layer was rammed properly another was poured to be rammed, and so on.

Metal rammers were used to ram the soil, which were composed of a steel rod with a flat steel plate, the weight of the rammers and the size of the plates differed to suit the users and enable them to ram the corners. A layer was considered to be properly rammed as soon as an echoing sound was heard from the rammers, an indication of the compactness of that layer.

The width of the formwork enabled users to stand inside it and ram, an advantage that ensured that all the corners and the edges were rammed properly.

5.5 *Inserting electricity conduits*

It was important to insert electricity pipes and boxes for electricity sockets in their exact place while ramming. This would save a lot of effort in trying to chisel the walls after drying to place the pipes and then repair them. Placing these pipes proved to the participants that rammed earth is a building technique that is appropriate and suitable to contemporary needs.

5.6 *Dismantling the formwork*

The formwork was left for one day and then the metal bolts were removed slowly and the plywood was slid upwards slightly before being removed from the wall. The process had to be done slowly and gently so that earth does not stick to the formwork and damage the surface. The wall was covered by plastic sheets to protect it from direct sunshine. After one or two days, and depending on the weather conditions, the rammed wall would be dry enough to carry the formwork, and work can be proceeded vertically.

5.7 *Plastering and rendering*

One of the advantages of using smooth formwork is to achieve a smooth fair-faced surfaces. Rammed earth walls do not need plastering at all, it is advisable to sponge the surface with a moist towel immediately after dismantling the formwork. In case there were any small holes in the walls, they can be filled by hand from the same soil mixture. In the workshop we experimented a number of materials to render the walls, such as clay and lime, which are natural materials that are dissolved in water and are easily applied to the walls using brushes or sponges.

6 PARTICIPANTS' REFLECTIONS AND PARTICIPATORY DISCUSSIONS

Although the participants joined the workshop willingly based on their passion and interest to learn more about rammed earth. The majority of them were involved in earth construction in a way or another but the discussion revealed a lot of concerns and fears about using rammed earth. It was crucial to discuss all the challenges and try to propose a road map towards promoting rammed earth construction in Egypt. Some of the participants were still skeptical about safety issues, maintenance cost and construction time in addition to fire and water resistance. Also the land cost and the possibility to extend vertically in case of building in urban areas were among the major concerns.

From the discussion, several questions were raised regarding the applicability of building with such technique. Participants pointed out that building with rammed earth might need more time during the design development phases, something that will be translated to extra cost. But the long term benefits cannot be compared to the long-term running costs and the environmental impacts of our conventional techniques. Which are more expensive to construct and to run in addition result in numerous environmental and health hazards. Their concerns are understood as it is always a major problem that clients pay attention to initial cost but not to the running cost.

Some participants raised the issue of the design of the super structure with the main focus on the compressive strength. Plus water absorption and weather erosion of the earthen walls, including details for visual inspection. That is in addition to the structural stability of the walls. These concerns show a need for and importance of developing tests to improve reliability. Together with performance requirements and tolerances which should be explicitly defined prior to the construction phase.

One important discussion topic was concerning the national building code that should be a guide on soil suitability and moisture content, and sets out requirements for formwork, methods of construction, testing and curing of rammed earth. Also detailed information on material testing is very important. In addition the code should be designed and written in conjunction with all other appropriate building standards. Many countries like Germany, Spain, USA and Zimbabwe have already robust codes and standards. By examining such codes one will find that they are combined experiences presenting the current state-of-the-art in rammed earth construction around the world. The building culture and climate of these countries may differ from the conditions of Egypt but they can still be important basis for the development of national standards and guidelines.

Also the discussion highlighted the material and production parameters and how they have a considerable effect on the quality of rammed earth walls. As optimizing material handling and production can be decisive for the acceptability of the end-product.

Maintenance issues was discussed as being burdensome compared to conventional building methods as absence of regular maintenance can be more damaging in earthen structures than in other building types. Water is a major agent of decay in rammed earth buildings. Maintenance should mainly be for protecting rammed earth from water deterioration.

Concerning the high U value of rammed earth, it was discussed that thick walls which provide high thermal mass could be a common design solution. Through the absorption and release of moisture it regulates the relative air humidity naturally and maintains comfortable surface temperature. Some questions and concerns were on openings, which could be formed easily in various ways, including arched openings, timber lintels, concrete lintels or even steel lintels. When it comes to economics generally building with rammed earth should not be expensive especially when compared to conventional techniques using industrialized building material. The cost ranges from 30 to 40 Euros/m^2 which is even less than 50% of the cost of current conventional techniques for low income housing in Egypt.

Ecological objectives and criteria for the building material and the use of sustainable resources were also of a concern. Rammed earth buildings could be fully recyclable after useful lifetime with low energy production and with nearly zero emissions in the whole building life cycle phases. The conclusion of this discussion revealed that there is a need for low-cost, affordable, energy efficient buildings with minimal carbon and ecological footprints. Building with rammed earth revealed to be a feasible solution.

There are several challenges starting from building standards and legislation till the craftsmanship and implementation. This transdisciplinary approach facilitates taking action steps to move forward towards an action plan. The workshop ended with a design proposal for a pilot project of a model house under the supervision of (HBRC). The participants formed teams for design, building simulation for energy efficiency measurements, material investigation and testing, fund raising and a technical writing team. That will end up with technical recommendations, a manual for rammed earth building best practice in Egypt. Then this technical recommendation could be a draft for a guidelines or standard for earth construction like the German DIN.

7 CONCLUSION

As we are moving towards an era of increasing environmental awareness and with several energy and environmental challenges. The current situation drives responsible researchers and professionals to explore ecological yet economical construction methods. The transdisciplinarily and participatory contribution in this workshop bridged academic research with professional practices together, while involving NGOs and public authorities. The outcome concluded that adopting earth construction techniques should be one of our first natural choices for future building practice. Rammed earth is an optimized construction method with guaranteed product properties for the Egyptian market. Despite all the revealed challenges in this workshop, the demand for earth-based buildings products and services is still developing both on the private and commercial building or construction level. The market is still virgin and there is a big potential to support the local economy and create job opportunities especially in the remotest villages in Egypt. Building using rammed earth can save energy consumption in the building sector and at the same time can offer low cost and self-help housing.

ACKNOWLEDGMENTS

The authors would like to acknowledge Eng. Amr Farouk and Oasis Community Center for hosting this workshop. The co-instructors; Eng. Zeiad Amer & Eng. Ahmed Refay for the effort together with all the workshop participants. Also Axel and Margaret Ax:son Johnson Foundation.

REFERENCES

Easton, D. 1996. *The rammed earth house*. White River Junction, Vt.: Chelsea Green Pub. Co.

EEHC. 2012. Annual report: Egyptian electricity holding company. Cairo.

Fathy, H. 1973. *Architecture for the poor: an experiment in rural Egypt*. Chicago.

Gramberger, M. (red.). 2001. *Citizens as partners: OECD handbook on information, consultation and public participation in policy-making.* Paris: Organization for Economic Co-operation and Development (OECD).

Keable, J. 1996. *Rammed earth structures: a code of practice.* London: Intermediate Technology.

Lawrence, R. 2004. Housing and heath: from interdisciplinary principles to transdisciplinary research and practice. *Futures* 36(4): 487–502.

McHenry, P.G. 1989. *Adobe and rammed earth buildings: design and construction.* Tuscon: Univ. of Arizona Pr.

Minke, G. 2000. *Earth construction handbook: the building material earth in modern architecture.* Southampton: WIT Press.

Morgan, W.N. 2008. *Earth architecture: from ancient to modern.* Gainesville: University Press of Florida.

Schijns, W., Kaper, O.E. & Kila, J. 2008. *Vernacular mudbrick architecture in the Dakhleh Oasis, Egypt, and the design of the Dakhleh Oasis Training and Archaeological Conservation Centre.* Oxford: Oxbow.

Serrano, S., Barreneche, C., Rincón, L., Boer, D., Cabeza, L. F.C. 2012. Stabilized rammed earth incorporating PCM: Optimization and improvement of thermal properties and Life Cycle Assessment, *Energy Procedia* 30:461–470.

United Nations Centre for Human Settlements. 1986. *Earth construction technology. Manual on production of rammed earth, adobe and compressed soil blocks* II: London: Intermediate Technology.

Venkatarama Reddy B.V., Leuzinger G., Sreeram V.S. 2014. Low embodied energy cement stabilised rammed earth building-A case study, *Energy and Buildings* 68:541–546.

World Bank. 2008. *A Framework for Housing Policy Reform in Urban Areas in Egypt: Developing a Well Functioning Housing System and Strengthening the National Housing Program.* Washington, DC. © World Bank. https://openknowledge.worldbank.org/handle/10986/7955. Accessed April 2014.

Earthen Architecture: Past, Present and Future – Mileto, Vegas, García Soriano & Cristini (Eds)
ISBN 978-1-138-02711-4

From refined to popular architecture. Mixed rammed earth walls with adobe reinforcements

M. del Río Muñoz
School of Architecture, University of Valladolid, Valladolid, Spain

F. Jové
School of Architecture, University of Valladolid, Valladolid, Spain
Grupo TIERRA, Research Group of Earth Building Technology, Spain

ABSTRACT: Nowadays, there are still many buildings both vernacular and monumental architecture built with adobe as the main building material. There is a huge range of construction variants, from simple adobe walls to half-timber filled with adobe, and some of them are well documented. However, this is not the case for mixed solutions with adobe. The mixed rammed earth wall between fired brick pilasters and courses was widely documented in the treatises of the 19th century. This system was commonly used in religious and monumental architecture, which used stone or brick to build pilasters that framed the rammed earth blocks of the wall. The transfer of knowledge in certain geographical areas from monumental to vernacular architecture gave rise to a new mixed rammed earth wall, i.e., rammed earth between adobe pilasters and courses. This study focuses on the identification, documentation and characterization of the mixed rammed earth with reinforcements of adobe masonry in vernacular architecture within the geographic area of Castilla y León, Spain. The documentation methodology has brought together traditional and new digital information gathering techniques.

1 INTRODUCTION

One can find endless examples of adobe constructions in Castilla y León, both in vernacular and monumental architecture.

The techniques that have been developed within this area are varied and characteristic: adobe, rammed earth, half-timber filled with adobe, rammed earth walls, mixed rammed earth walls, etc. Earth has been much appreciated by professional builders throughout history. They knew which technique to use for each part of the building.

The system described in this study is the combination of adobe and rammed earth. The technical literature consulted refers to mixed rammed earth walls as the union between masonry, either bricks or similar, and rammed earth blocks. However, adobe is rarely mentioned as potential masonry—the references mainly being found in architecture and building treatises.

Many historians, archeologists, architects and quantity surveyors have done research on the analysis of vernacular building techniques based on earth. Researchers as Alonso Ponga (1994), Jové Sandoval (2011), Sainz Guerra (2012), Font Arellano (2005), García Grinda (2007), etc. have shown a growing interest in the study of construction techniques using earth.

On the other hand, and closely related to the object of study, the mixed rammed earth wall is currently being studied by several researchers—Graciani García and Tabales Rodríguez (2008), Camino Olea (2011), Maldonado Ramos and Vela Cossío (2011), Pavón Maldonado (2012), Font Mezquita (2009), etc.

The aim of this work is to analyze the mixed rammed earth wall built with rammed earth blocks between pilasters and courses of adobe. Although this system has been already named in the past, there are no thorough studies of the matter that achieve a dimensional characterization.

2 TREATISES AND TECHNIQUES

As mentioned before, the architecture and building treatises contain a good description of the rammed earth building technique. These illustrated historical texts, prior to the 20th century, describe how to build the wall, its elements and even the cost, in a way which makes it easy to understand.

We have gone through the treatises from the 18th and 19th centuries and we have found references in each of them to the suitability of this material for the Mediterranean area, due to the climate and the soil composition. Villanueva (1827:8) indicates that "*the use of earth is ancient in Spain; the tales of Pliny, in the 3rd book, chap. 3 of his Natural History, related how, in Spain, there were very ancient earth towers and watchtowers, and even now we can find parts of city walls and castles which have stood the test of time and have survived for centuries in a ruinous state*".

The treatises refer to the mixed technique of rammed earth reinforced with masonry, although not all of them refer to adobe as an example of masonry, but more to brick and stone. For example, Villanueva, refers when talking about the masonry used in the walls' intermediate pilasters: "*This sort of wall is normally built with intermediate pilasters made of stone, brick, rubble or adobe*".

Espinosa (1859) also states as follows:

"Rammed earth walls are built with formworks that are similar to those of concrete walls. As mentioned before, earth is the main material, which is laid in 10 to 15 centimeter layers, and is then tamped down in the formwork; it is sometimes reinforced with masonry, brick or adobe pilasters".

Ger y Lóbez (1898) refers to the system in this way: "*Rammed earth walls are rarely the only component of a wall, unless it is a minor component. They are normally confined between major and minor pilasters or built-in pilasters and reinforcements made of solider materials such as adobe, bricks, flat stones or flagstones*".

3 METHODOLOGY

The first step has been to consult various documentary sources, so as to understand the system and establish a basis for the analysis. Before beginning the fieldwork, a data capture sheet was designed which includes every element of the wall, the materials and the size. This sheet allowed the information to be gathered in a systematic way to avoid the loss of data. The sections of the table included were extracted from a previous analysis of the mixed wall, which resulted in a precise terminology and a global understanding of the system.

The area where the samples are found is the region of Moraña (Moorish land), which includes municipalities in the South of the province of Valladolid, and the Northwest of the province of Segovia—an area rich in Mudejar architecture.

The on-site data gathering process was simple, using visual inspection, the taking of measurements and a series of converging photographs taken with an Olympus Pen LP1 Mini camera. The images were then processed by means of photogrammetric techniques.

Once in the Architectonic Laboratory of Photogrammetry, School of Architecture, University of Valladolid, virtual reconstructions of the 3D models were made to obtain the photoelevations. The Agrisoft Photoscan software was used to process the images and obtain the 3D models and the photoelevations. Later, fuzzy points were removed with Adobe Photoshop and, finally, AutoCAD was used to assign a scale and obtain the necessary size for the analysis.

The last step was the analysis itself to obtain the results. This has been possible thanks to the sketches drawn during the fieldwork, the on-site inspections, the 3D models and the photoelevations.

4 STUDIED SAMPLES

The constructive system under study was found in a total of 18 visited towns. However, only 7 samples have been chosen for a deeper study (Fig. 1).

1. Farm building I in Rapariegos (Segovia). (Fig. 2).
2. Farm building II in Rapariegos (Segovia).
3. Rammed earth wall in Muriel de Zapardiel (Valladolid).
4. Hay barn in Salvador de Zapardiel (Valladolid).
5. Hay barn in Sinlabajos (Ávila).
6. Wall with attached canopy in Sinlabajos (Ávila).
7. House façade consisting of two layers. The inside layer consists of a rammed earth wall between adobe pilasters and courses. On the other hand, the external layer consists of a rammed earth

Figure 1. Map of the fieldwork. The visited towns are highlighted with points. They are located in the surroundings of the area of Moraña, in the north of Ávila, the south of Valladolid and the northwest of Segovia. Map, Junta de Castilla y León, Author: Mónica del Río.

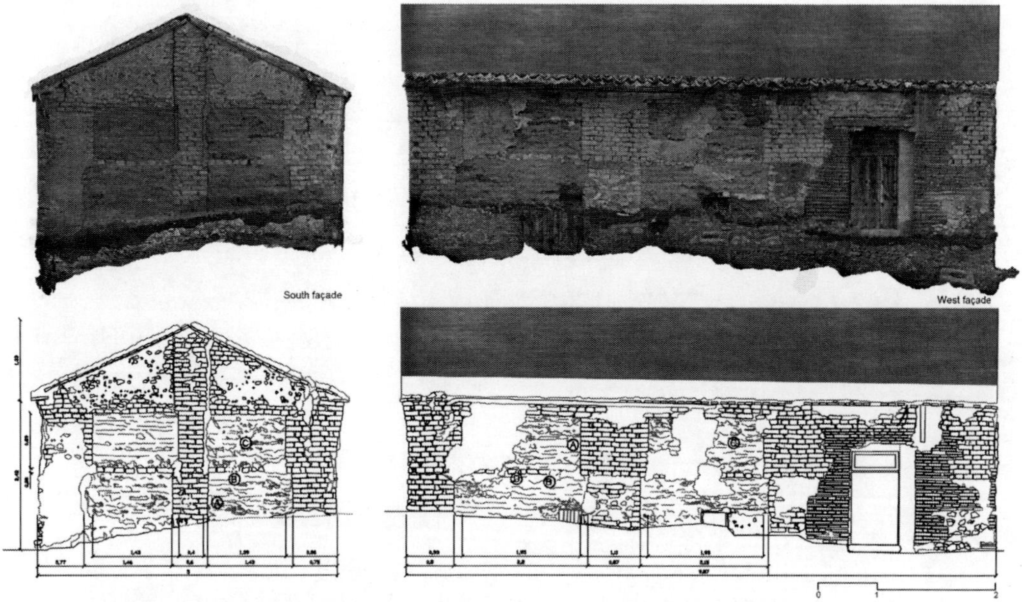

Figure 2. Photoelevations (above) and 2D design of the farm building in Rapariegos (Segovia). A. Adobe pilaster. B. Adobe courses. C. Rammed earth block. D. Putlog holes. Author: Mónica del Río.

wall reinforced with brick pilasters and courses. The aim of this is to make the building look finer. Located in Sinlabajos (Ávila).

5 RESULTS

There are different varieties of mixed rammed earth walls, depending on the masonry—brick, ashlar, rubble and adobe.

The finest masonry, like brick and stone, have been widely used in monumental, industrial, civil and defense architecture. On the contrary, adobe masonry was consigned to vernacular architecture, auxiliary buildings linked to agriculture and livestock, such as hay barns, storehouses and farmyard walls. To a lesser extent, we can also find adobe in houses, in the inner part of the wall (Figs. 3 & 4). Regarding the external wall, the pilasters and courses are made of brick so as to appear finer.

The walls, which are normally around 3.5 m high, and between 0.45 and 0.60 m thick, bear the roof load. The building typology is rammed earth blocks between adobe pilasters and courses. Within this typology, we can find a wide range of morphologies, e.g., adobe-reinforced wall with rammed earth blocks, with plain, jagged, or staggered adobe pilasters etc.

These walls normally have stone masonry, as a form of protection, at the bottom and top of the construction, and a coping or eave in the case of fences, made either of solid bricks or ceramic tiles.

The size of the rammed earth blocks also varies, which means that the wood formwork dimensions were not standardized. The length varies from 1.00 m to more than 2.50 m, and the height ranges from 0.60 m to 1.20 m. The height of the earth layers ranges from approximately 7 cm to 12 cm, which is a very similar thickness. There is no mortar between the earth layers. The additives found from a simple visual exploration were small pebbles, pieces of tile and lime particles.

The reinforcing elements were adobe in the form of vertical pilasters and horizontal courses. The shape of the adobe is not fixed, which means that, in adjacent towns, there was no adobe mold or sample rack with a standard size. On the contrary, each builder had his own mold, or every person who built his own property had his own. The adobe was made of mud, small pebbles and a few plant fibers. The most common, are laid in a stretcher bond (aparejo inglés), although in some reinforced parts they used header bonding.

Finally, the putlog holes for the horizontal timber supporting the formwork side, are normally at the same level as the adobe courses. These putlog holes are sometimes crowned by a specific element such as solid brick.

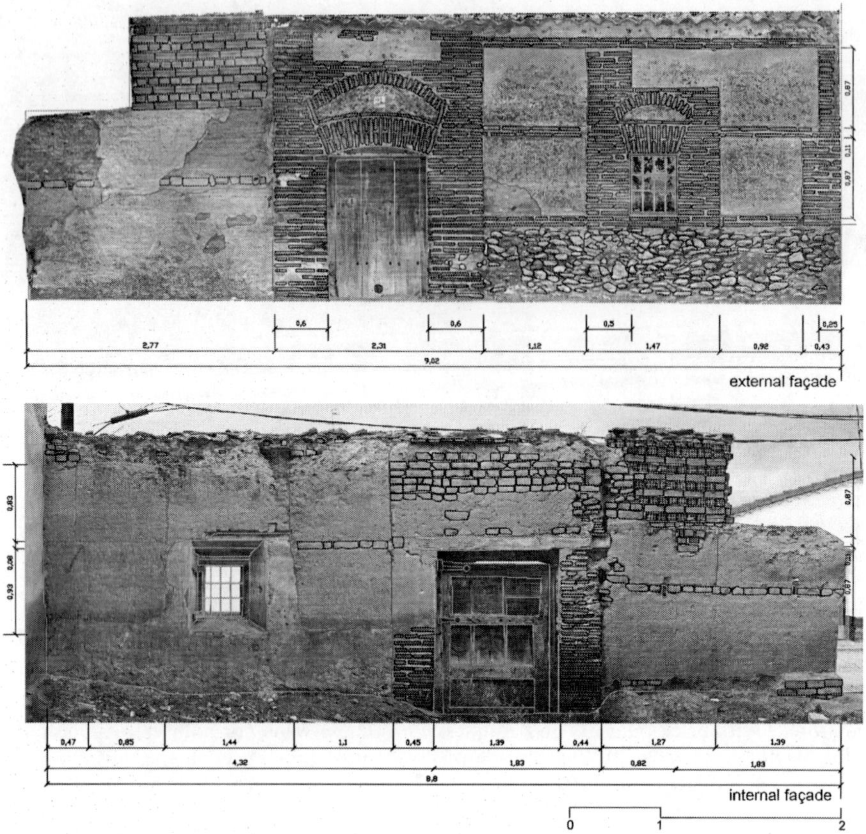

Figure 3. House in Sinlabajos (Ávila). The façade consists of two layers; the inside layer contains adobe masonry in the pilasters and courses and the external layer is a solid brick masonry. Author: Mónica del Río.

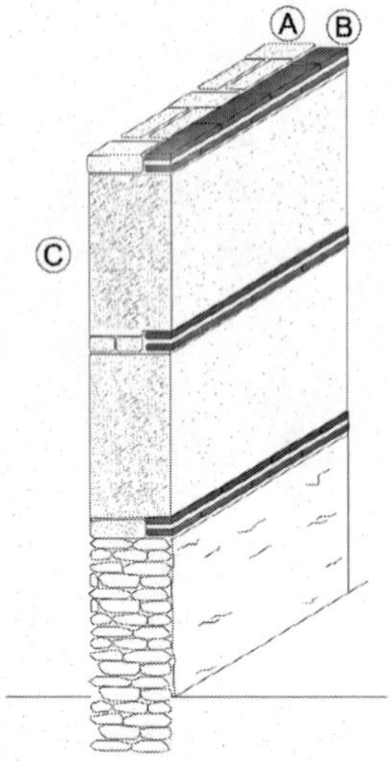

Figure 4. Wall section of a house in Sinlabajos (Ávila). **A.** Adobe course. **B.** Brick course. **C.** Rammed earth block. Author: Mónica del Río.

6 CONCLUSIONS

The fieldwork has proved that this system was mainly used for auxiliary buildings and the internal layers of houses, which means that it had a low cost.

There are very different morphologies within the system, due to the combination of pilasters, brick reinforcements and rammed earth blocks. There is no fixed pattern in the samples.

Adobe did not have the decorative characteristic of bricks, as it was normally covered by a mud or lime mortar, unlike brick pilasters and courses, which were left exposed.

The building procedure was similar to that of brick pilasters, starting with the adobe pilasters and then the rammed earth blocks where layers of earth were poured and tamped.

The system had an economic component: the cost of adobe was much lower than that of bricks, as it did not need to be fired. It is even recorded in the medieval building treatises: Valdés (1870:666)

Table 1. Summary table for mixed rammed earth wall characterization.

		1 Rapariegos	2 Rapariegos	3 Muriel de Zapardiel	4 Salvador de Zapardiel	5 Sinlabajos	6 Sinlabajos	7 Sinlabajos
USE		Farm building I	Farm building II	Rammed earth wall	Hay barn	Hay barn	Wall with canopy	House/wall with two layers
WALL	Type of reinforcement (pilasters and courses)	Pilasters and coursers	Pilasters and coursers	Pilasters and coursers	Pilasters and coursers	Pilasters and coursers	Adobe coursers	Pilasters and coursers
	Height (m)	3.27	3.7	4.1	3.35	3.69	3.28	3.1
	Thickness (m)	0.57	-	-	0.6	-	0.6	0.45
	Socle	Some stone masonry	Stone masonry	Stone masonry	Stone masonry	Rendered masonry	Stone masonry	Masonry
	Cope/Eave	Brick course	Tile	Solid brick coping	Tile eave	Adobe and brick course	Tile	Tile
	Covering	Mud mortar	Mud mortar	Mud mortar	No covering	Covering of the wall	Mud mortar	Inside: mud mortar Outside: lime mortar
RAMMED EARTH BLOCKS	Simple or multiple	Simple	Simple	Multiple	Simple	Simple	Multiple	In: Multiple/Out:Simple
	Size of blocks	1.92/0.88	2.25/1.00	2.13/1.10	2.6/1.08	2.34/0.6	2.0/0.8	Int:1.10/0.85 Ext:1.10/0.85
	Additions	No attachments	Pebbles	Pebbles, tile pieces	Small pebbles/small tile pieces	Pebbles/tile pieces	Round pebbles	Pebbles/Tiles
	Layers of earth (m)	0.1	0.12	0.08 / 0.1	0.07	0.08	0.07	0.1
REINFORCEMENT	Material	Adobe	Stone masonry	Adobe	Adobe	Adobe/Adobe and brick	Adobe	Adobe/brick
	Size (cm)	26.5/14/7	Irregular	32/14/9	30/18/7	28/14/9	27/14/8	Adobe:28/15/10 Brick:29/14/3
	Additives in rammed earth	No additives	Stone assembled with soil	Small pebbles	Small pebbles/plant fibres	Small pebbles	Small pebbles	Small pebbles
	Bonding	English	No bonding	English	English	English	Stretcher	English
	Pilasters shape	Jagged	Plain/ Jagged/Irregular	Irregular/Staggered	Jagged	Jagged	-	Plain
	Pilasters width	0.6/0.4	0.94-0.80	0.7/0.46	0.66/0.88	1.08/0.81/0.54	-	0.45
PUTLOG HOLES	Location	In the reinforcement	In the reinforcement	In the reinforcement/between the formwork blocks of the wall	In the reinforcement	In the reinforcement, covered with stones	In the reinforcement	In the reinforcement
	Distance between them (width/height) m	1.00 / 1.00	1.38 / 1.20	1.57/1.10	1.06/1.08	0.65/0.95	0.6/0.9	1.00/0.92
	Size (width/height) m	0.05/0.08			0.12/0.08	0.09/0.08	0.04/0.08	0.08/0.04
	Ornamental top	No	No	Yes, solid bricks	Adobe	No	No	No
	No. holes/ blocks	6	6	4.,0/6.,0	6	6	8	4.0/6.0

states that 100 adobe bricks cost 0.60 escudos and 100 thick bricks ranged from 1.80 to 2 escudos. Therefore, the whole system was cheaper than having it done with fired bricks.

However, it was not the most popular system when talking about building one's own home, as rammed earth required the use of auxiliary tools, a formwork, putlogs and a tamping rammer. Everyone was able to make adobe and dry it out in the sun, but only those with specific knowledge on the matter were able to actually build with rammed earth.

NOTES

The present document is part of a research project entitled Mixed rammed earth and stone covered earth walls, sponsored by a Research Grant awarded to Mónica del Río Muñoz by the Historical Heritage Foundation of Castilla y León and the company Cabrero Edificaciones, S.A. Jesús San José Alonso, full lecturer at the School of Architecture of Valladolid, is the tutor of the project.

REFERENCES

Alonso Ponga, J.L. 1994. *La arquitectura del barro*. Junta de Castilla y León, Consejería de Cultura y Turismo.

Camino Olea, S.; Labrador, A., Vega, M.. 2011. *La fábrica mixta de tapia y ladrillo, un invariante de la arquitectura monacal de Valladolid, España, hasta el siglo XX. En: Construcción con tierra. Tecnología y arquitectura.* Congresos de arquitectura con tierra en Cuenca de Campos 2010/2011. (on line). Valladolid: Cátedra Juan de Villanueva. Universidad de Valladolid.

Espinosa, P.C. 1859. Manual de construcciones de albañilería. Madrid.

Font, F., Hidalgo, P. 2009. *Arquitecturas de tapia.* Colegio Oficial de Aparejadores y Arquitectos Técnicos de Castellón. Castellón.

Font Arellano, J. 2005. *De Cancho Roano a La Olmeda: Mil años de Contrucción con tierra en la Hispania preislámica.* Actas del Cuarto Congreso Nacional de Historia de la Construcción, Cádiz, 27–29 enero 2005, ed. S. Huerta, Madrid: I. Juan de Herrera, SEdHC, Arquitectos de Cádiz, COAAT Cádiz.

García Grinda, J.L, Gonzalez Fraile, E.M, Nieto González, J. R. 2007. *Patrimonio arquitectónico de Castilla y León. 5. Arquitectura tradicional, arquitectura preindustrial, industrial y del hierro*. Junta de Castilla y León, Valladolid.

Ger y Lobez, F. *Tratado de Construcción civil.* 1898.

Graciani García, A. y Tabales Rodríguez, M.A. 2008. *El tapial en el área sevillana. Avance cronotipológico estructural.* Arqueología de la Arquitectura, 5, enero-diciembre 2008, págs. 15–158. Madrid/Vitoria.

Maldonado, L. y Vela-Cossío, F. 2011. *El patrimonio arquitectónico construido con tierra. Las aportaciones historiográficas y el reconocimiento de sus valores en el contexto de la arquitectura popular española.* Informes de la Construcción. Vol. 63, 523, 71–80, julio-septiembre 2011.

Pavón Maldonado, B. 2012. *Murallas de mampostería, sillarejo y ladrillo en el islam occidental. Los despojos arquitectónicos de la Reconquista.* Inventario y clasificaciones.

Sáinz Guerra, J.L. 2012. *Edificios y conjuntos de la arquitectura popular en Castilla y León* (on-line). Junta de Castilla y León.

Valdés, N. 1870. *Manual del Ingeniero y del arquitecto.*

Villanueva, J. 1984. *Arte de albañilería.* Editora Nacional. Madrid.

Earthen Architecture: Past, Present and Future – Mileto, Vegas, García Soriano & Cristini (Eds)
 ISBN 978-1-138-02711-4

C.A.S.A.—Central American Sustainable Architecture

L. Di Virgilio
AK0, Rome, Italy

Sara Parlato
Dipartimento di Architettura, Università degli Studi RomaTre, Rome, Italy
AK0, Rome, Italy

ABSTRACT: C.A.S.A. is a trade center for the community of Cerro la Granadilla, in Guatemala, built during February 2013. It is a design and construction experience that merges aspects of education, experimental research and international cooperation. Students and young professionals from Europe gained experience with building procedures and detail solutions. On the other hand their presence on site stimulated local craftsmen to (re-)use some traditional techniques like rammed earth or wattle and daub. The cooperative elaboration of technologically, economically and aesthetically appropriate solutions not only provides a useful building, that will help to improve the life of some persons from the local community, but can be replicated in other processes of self-construction for private houses and small public buildings. The manual process of building together enhances the communication between designers, builders and users of the architectural work. The borders between the three categories blur and a spontaneous but intense knowledge flow starts having the building site as an exchange hub.

1 INTRODUCTION

C.A.S.A—Casa dei Mestieri is a trade center established within the community of Cerro la Granadilla, at 50 km from Guatemala City (Fig. 1). In this training school, which has been designed with a strong focus on ecological matters, young persons from the local community will learn the art of making bread and cakes. This shall give them the perspective for independent activities as alternatives to their nowadays' main economy: the production of fireworks. This extremely dangerous business involves children and young people, and is often the only source of income for their families

The complete project includes the construction of a building in a shape of a C that is articulated around a courtyard opened to the village (figs. 2–3).

All the employed materials, such as wood, clay and bamboo, were sourced locally (figs. 4–5).

A wide longitudinal structure defines the main room, a flexible space that can accommodate classes or other training activities. Two kitchen spaces have been designed docked at the main room; they are built in rammed earth.

A succession of wooden frames links the rooms and holds a unique roof that provides shade and protection against rain. The walls are closed with foldable bamboo-screens that modulate the light. According to their position they can frame views on the valley and allow passages towards the inner courtyard (Fig. 6). The rammed earth walls of the kitchens are situated on a stone plinth which protects them from rising damp and from splashing rain. The great thickness of the wall controls the humidity and temperature regime due to its thermal inertia, ensuring a stable and comfortable inner climate (Fig. 7).

On some surfaces plaster protects the walls. They are made with different mixtures for the inner and outer surfaces. Their composition for the internal surface is made of a part of lime, a part of fine sand and for the external ones is made of a part of fine earth, a part of lime and two parts of sand. The use of lime makes them more resistant to wind and rain. Lime was also used within the rammed earth in order to make the edges stronger.

During the installation, the lime is mainly poured to the edges in a thin layer and added to the mixture of earth before ramming. The alternation of layers in which the lime is absent, and of others in which the earth layers have different shades, defines the surface. So it is characterized by a succession of layers of different color that give the walls a recognizable aspect (Fig. 8).

The plaster color is given by the light color of the local soil and the other components. Only the most vulnerable spots of the surface, such as edges and window openings are plastered while in the other parts the rammed earth surface is shown.

Figure 1. General view of C.A.S.A.

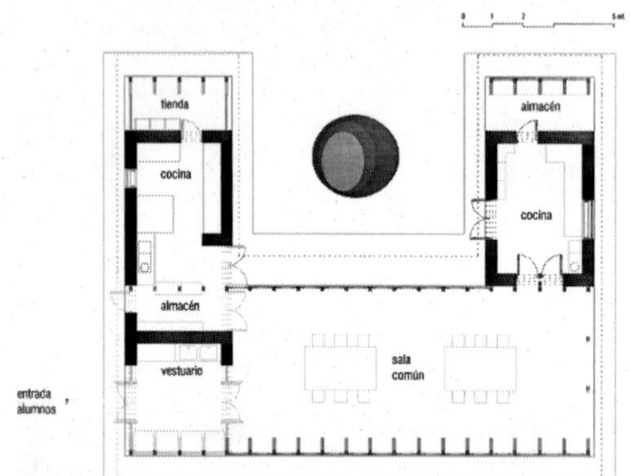

Figure 2. Plan of the project.

Figure 4. All the matirials used were sourced locally.

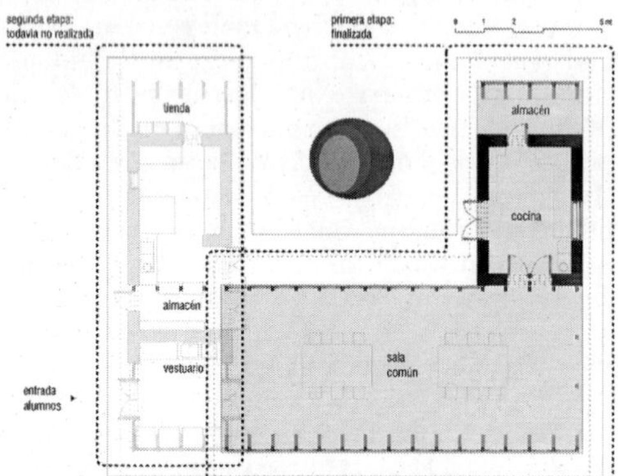

Figure 3. Part of the plan that was actually realized.

Figure 5. Construction of quincha panels.

Figure 6. Windows bellows made of bamboo.

Figure 7. View of the kitchen.

Figure 8. The alternation of lime and earth.

In this way the layered rammed earth patterns get framed like they were paintings.

The same textures and colors can be found in the pavement pattern. A layer of 7 cm of concrete and yellow color was superimposed on a bedding layer also formed of rammed earth.

Figure 9. The sewing laboratory.

Figure 10. The sewing laboratory.

The chromatic effect affects all the environments. In the large training room the yellow surface creates a contrast to the dark wooden spans of the timber structure while in the kitchen it adds a color layer to the gradient of earth tones that characterize the walls and the plastered surfaces.

In a first step of the project, one kitchen and the large wooden training room have been built. Thanks to the versatility of the environments, the functions that take place within them, adapt easily to the needs of the inhabitants. The trade school is in use since last autumn and the local community already introduced educational offers that were not planned at the moment of design, the manufacturing of herbal products and tailoring among them (Figs. 9–10).

2 DESIGNING

In the design phase, the site gave us important hints for our choices conditioning typological, morphological and physical features. These features are all but independent from each other. They influence themselves mutually.

Spatial organization defines the aspect of architecture; at the same time morphology and materials characterize and declare the building's typology. In other cultures spatial hierarchy can totally diverge from ours. Spaces distribution is strictly connected to habits and to social interactions.

In Guatemala, for example, corn plantation is one of the most important activities that guarantee local sustenance, and its production conditions the typology of native architecture. In fact the houses often have a patio where people store corn while drying. This is also the place where tortillas are formed and baked; a welcome opportunity to meet and share experiences. It is, thus, quite easy to understand the importance of the exterior space in the spatial organization of the house. Elements like these, which are deduced from local culture, can help in creating a sense of identity. Several authors, Kevin Lynch among them, demonstrated how important this is for human well-being.

The research of environmental comfort has always induced design choices that can take advantage of natural local resources. In our case, circumscribing the materials to be employed to locally available resources has been taken as a crucial design choice.

We have tried to achieve maximum performance with the local resources maximizing the climate benefit through the spatial organization and choosing the more efficient materials among the available one. In Cerro la Granadilla the ancient knowledge in the field of construction has been abandoned and gradually forgotten in favor of a foreign knowledge that apparently provides people a more dignified social status.

Casa dei Mestieri proposes a design scheme that is very familiar to the community but built with natural materials such as wood and rammed earth. The use of raw earth was rather widespread from the times of Maya people until few decades ago, when concrete was introduced and pretended to be quicker and safer to use.

This led to a gradual loss of knowledge on how to build with earth or other natural materials. Proposing the use of local materials to local builders made it necessary to transmit the techniques especially with respect to detail solutions that can grant durability and performance. Moreover it is important to spread the knowledge on these techniques among the local community in order to assure the maintenance of the building in time.

A crucial design effort has been dedicated to the aesthetics of natural materials in order to communicate their intrinsic dignity, to inspire confidence and to encourage the inhabitants to appreciate their solidity. We wanted to create something new but not something foreign. The challenge of revising tradition in an architecture that aims at combining modern features with the spirit of place is to create confidence and to intrigue at the same time.

The distinctness of the work becomes essential. While respecting the spatial organization of the typical Guatemalan architecture, the Casa dei Mestieri is very recognizable: this helps the initiation of a process of imitation and repetition, which allows the absorption and diffusion of knowledge and building techniques.

Although the project was born from careful studies on the culture of the place, we often had to adapt it to evolving needs. So we reserved the possibility to transform the use of the spaces, as well as the overall shape and the facade details. Involving the local community in these processes of readaptation on site gives confidence for the future life of the building. Other changes and evolutions can be implemented once the European design team left Guatemala and keep the architecture alive.

3 PEOPLE AT WORK

During the whole construction period, which lasted less than three months, approximately 30 people have contributed with different roles to the building site. Each participant had a different kind of training and different work experiences. Both, students and workers, got the opportunity to prove their skills and learn new things. In addition to this large group of volunteers, four local workers have consistently participated in the project (Figs. 11, 12). Involving more community members as volunteers on site, as proposed, proved to be difficult since the work in the field and other primary duties didn't leave time for voluntary work.

The local construction team was responsible for the concrete-foundation and participated during the installation of plumbing system as well as in the roof-construction. Several teams have been organized.

Figure 11. Assembling of wood pieces.

Figure 12. Volunteers at work.

Figure 13. Volunteers and local workers.

Figure 14. Local workers making the pavement.

Figure 15. Window in wood and bamboo.

The distinction of raw earth structures and timber constructions allowed for a certain amount of prefabrication on site, which helped in avoiding spatial interference among the single teams (Fig. 13). One team was responsible for the mentioned prefabrication of timber frames. In order to simplify the wood-construction work, the whole design was based on a single standard section of 15×3 cm to be pole, beam or brace according to its position.

Nevertheless, the pieces had to be cut and assembled before being attached to the foundations (Fig. 14). At the same time another team was concerned with the construction of the rammed earth walls. This phase is naturally more tiring than others and used to occupy a larger number of workers. Other teams cared about the prefabrication of window frames and the preparation of bamboo components: each barrel was cleaned, treated and cut before being inserted into the wooden frames (Fig. 15). The different tasks allowed everybody in the rather irregular working team to find the most suitable job for his own skills and inclinations.

4 CONCLUSIONS

The strength of this project can be found first of all in the effort to consider architecture according to the needs of the inhabitants and taking into consideration the environmental and cultural aspects in which they live. The attention paid to the natural and climatic elements and the socio-cultural dynamics defines the guidelines on which the project is built. Furthermore technological solutions can be reused in tropical areas as well as a functional articulation that can be easily adapted to different needs. At the same time another important aspect was the combination between traditional local aspects and technical skills that has generated a strong project, deeply rooted in the whole community.

Figure 16. The team.

This project was adopted by locals as a gift, so people began immediately to plan educational activities in order to exploit the new building.

Even if C.A.S.A. was realized to accommodate cooking laboratories, also sewing and herbal workshops took place there.

The knowledge exchange between the young volunteer architects from abroad and the local work team turned into a profitable, mutual training. The volunteers, some of them without former experience on building sites, got the opportunity to test themselves in manual labor and organizational tasks. At the same time they acted as motivators for the locals and triggered in them the wish to apply their skills to new construction methods. The fact that they had a lot to teach in terms of operational hints, working techniques and use of tools and the pride that goes along with the sense of feeling useful within the team made them more inclined to deal with apparently precarious materials like raw earth and bamboo. This feeling will hopefully turn into a valuable experience they can spend on other building sites in the region (Fig. 16).

REFERENCES

Houben, H. & Guillaud, H. 1994. *Earth Construction – a comprehensive guide.* Rugby (GB): ITDG Publishing.

Kapfinger, O. & Rauch, M. 2001. *Rammed Earth, Lehm und Architektur, Terra Cruda e Architettura.* Basel/Berlin: Birkhäuser.

Keable, J. 1996. *Rammed earth structures: a code of practice.* Rugby (GB): ITDG Publishing.

Lynch, K.A. 1981. *A theory of good city form.* Cambridge, USA: MITpress.

Minke, G. 2000. *Earth Construction Handbook—The Building Material Earth in Modern Architecture.* Southhampton: WIT Press.

Minke, G. 2001. *Construction manual for earthquake-resistent houses built of earth.* Eschborn: Gate-Basin Gtz.

Minke, G. 2005. *Manual de construcción para viviendas antisísmicas de tierra.* Kassel: Universität Gesamthochschule Kassel, Forschungslabor für Experimentelles Bauen. (http://www.asl.unikassel.de/~feb/wissenswertes/manual_span.pdf).

Minke, G. 2006. *Building with Earth - Design and Technology for a Sustainable Architecture.* Basel/Berlin: Birkhäuser.

Pollak, S. 2007. Emerging Knowledge, Students as Intermediate Agents in Construction and Transformation Processes. In *Transactions, the online journal of the Centre for Education in the Built Environment*. Cardiff: Jebe.

Pollak, S. 2012. *Terra – Ragionamenti e Progetti.* Roma: Aracne.

Earthen Architecture: Past, Present and Future – Mileto, Vegas, García Soriano & Cristini (Eds)
© 2015 Taylor & Francis Group, London, ISBN 978-1-138-02711-4

Stratigraphic analysis of earthen architecture: The mosque of M'Hamid El Ghizlane

O. Domínguez Martínez, A. García Hermida & C. Moreno Adán
Terrachidia Association, Madrid, Spain

ABSTRACT: This paper presents the results of some of the studies developed by the interdisciplinary team of Terrachidia, focused on the earthen architecture of the last oasis in the Drâa Valley (Morocco), M'Hamid, one of the main gates to the Sahara desert. The most monumental building in the M'Hamid Oasis is the ancient mosque of M'Hamid El Ghizlane. This building is the result of an extensive process of alterations and extensions which needed to be studied and understood before proposing any solution for its future conservation. This study was developed in 2013, during January and May documentation campaigns, when the evolution and main historical phases of the mosque could be completely figured out.

1 INTRODUCTION

This paper presents some of the results of the work developed by Terrachidia Association in Southern Morocco, in M'Hamid, the last oasis in the valley of the Drâa River, where the roads end and the caravan routes through the Sahara Desert once started.

Terrachidia is an independent association of interdisciplinary professionals whose aim is to contribute to the preservation of the traditional sustainability of the built environment, mainly through national and international teaching activities. Working for this sustainability is understood as preserving local building traditions alive and worldwide promoting the main values they embody. This work is developed through practical workshops specially focused on earth building techniques (Domínguez Martínez & García Hermida 2013). Through them we study, restore and promote the architectural heritage of this oasis, a fragile environment where building traditions are still alive, but more endangered than ever.

Palm groves are gradually being abandoned, allowing the desert dunes to progressively swallow this manmade landscape (Dłużewsk 2008), while Western architecture and urbanism practices are menacing its traditional balance. In this particular context, and due to its extreme climatic and social conditions, the main lessons which can be learnt from tradition are shown in a particularly clear way, making this place a wonderful school for our teaching purposes.

In these workshops participants learn the traditional building techniques of the oasis directly from local masters, who are hired and encouraged, along with young apprentices, to keep their crafts alive, while contributing to the sustainable development of the region.

But sustainable development in a region whose current economy is mainly based on its touristic resources, thanks to the appeal of the Sahara Desert, also requires enriching the way this tourism is understood, as an attempt to improve its connections to local society and its cultural heritage. In these southern oases, by now, tourism is mostly focused on rallies through the desert and expeditions to Er Chigaga dunes.

Meanwhile, the hospitable culture and the rich architectural heritage of these oases remain completely unknown and little by little disappear without any awareness of their possibilities.

Once we identified this problem, working to solve it became an important factor of our activities there. Extensive and scientific studies of the local heritage are being developed as a way to discover, underline and promote its multiple values. And this is, in fact, one of the aims of this paper.

Before every restoration, we carefully study and document the building utilizing the little resources available. The frequent lack of historical data makes the archeological analysis of its materiality the main source of information about its history and its architectural evolution.

The stratigraphic researches we are conducting, along with the previously nonexistent chronotypological studies of the local architecture, are helping us to cast light on the history of this remote and generally forgotten place and its architecture.

The results presented here are those of the archeological analysis of one of the most complex and interesting buildings in the M'Hamid Oasis: the mosque of M'Hamid El Ghizlane, the main and, according to popular belief, most ancient still inhabited *ksar* in the whole area. In fact, Djinn Jacques-Meunié, the author of the most complete historical study of the settlements of southern Morocco hitherto, links it to the ancient Taragale, which could have lost his previous Berber name after the arabization of the oasis (Jacques-Meunié 1982). Accepting her hypotheses, it was already an important population in the sixteenth century

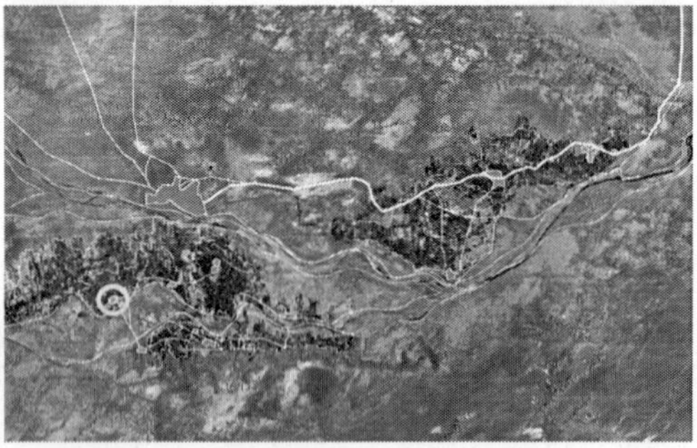

Figure 1. Location of M'Hamid El Ghizlane at the M'Hamid Oasis (Terrachidia).

Figure 2. The mosque of M'Hamid El Ghizlane (Terrachidia).

and could have been mentioned—as *Tart'aggal-* as early as in 1131.

Once more the historical sources about the evolution of the building barely exceed those references about the origin of the current location of the *ksar*. The date of its establishment and its historical vicissitudes are completely unknown. Nevertheless, this mosque is locally acknowledged as the most ancient in the oasis, what the intense and rich history still readable on its walls seems to have confirmed.

2 TRADITIONAL BUILDING TECHNIQUES IN THE M'HAMID OASIS

To understand the importance and singularity of the ancient mosque of M'Hamid El Ghizlane, it is critical first to introduce its architectural context.

Till recent times the oasis population was gathered inside their fortified *ksur*, defensive settlements compact enough to protect their inhabitants not only from the nomadic tribes which threatened them, but also from desert storm and the extreme temperatures of the area.

The architecture of these *ksur* is always made of the same materials, the few ones available here, mainly earth and wood from tamarisks and palms.

Walls are commonly made of rammed earth, using adobe bricks for the top floors, the joints with the horizontal structures and all internal arches and piers. Whenever an arch or a *qubba* are built, due to the scarcity of suitable wood, they are made as corbel arches or corbel vaults. Every building is protected by earth renders, though some singular elements, such as *qubbas*, entrances, and several interior spaces, especially sacred ones, use to be rendered with a mixed lime and earth mortar instead. Earth and lime proportion vary depending on the richness of each construction. Foundations are often built from rammed earth directly on the ground. This method does not become a noteworthy problem in terms of durability, thanks to the low humidity of local soil. Hence, some mosques, palaces and other important buildings lie on richer rammed earth, lime and stone mixture, as it is the case of the M'Hamid El Ghizlane mosque. For the horizontal structures, several layers are placed over palm or tamarisk beams. In the simplest examples earth is directly rammed over weaved palm leaves, so that the structure can stand the compression from the load it carries. Nowadays it is frequent to find another layer made of reeds separating the beams and the palm leaves, and, in roofs, even waterproof plastic coating over the last one. However, reeds are typically from more humid oases, such as the Skoura Oasis (Soriano Alfaro 2006) and traditionally were not used here.

3 LOCAL RELIGIOUS ARCHITECTURE TYPOLOGIES

Among the monumental buildings of the region, usually religious ones were particularly well crafted. In the M'Hamid Oasis two kinds of religious buildings can be found: mosques and *marabouts*, which are mausolea where local civic and spiritual leaders were buried. These public figures are often associated with miraculous powers and venerated as saints.

The architecture of these *marabouts* is always based on central compositions, usually of 3 × 3 modules. The central one is a *qubba* standing over four corbel arches which surround the *marabout* tomb itself. The structure of surrounding spatial modules can be lower secondary *qubbas* or horizontal wooden structures as the ones already described.

On the contrary, the interior spaces of mosques never are centrally arranged, but always built as parallel rows of corbel arches, standing also parallel to the *qibla* wall and allowing these spaces to grow into any direction when any enlargement is needed. The parallel aisles always have horizontal wooden structures, while the *qubba* is only used for the space right in front of the *mihrab*. This *qubba* becomes the main symbol of the mosque from the outside, since minarets are typically not used. In fact, minarets are rare in the ancient mosques of southern Morocco, existing only in some particular places, such as Amezrou or Roda (Mimó 1996; Soriano Alfaro 2006).

It is very interesting to compare the plan of the mosque of M'Hamid El Ghizlane, composed of 7 of these parallel aisles, 9 arches each, to the other remaining ancient mosques of the oasis. The mosque of Ouled Youssef has 4 aisles of 5 arches, in Zoauia, 3 aisles of 4 arches, and the one of Ait Aissa Brahim is the smallest of them all, with only 2 aisles of 3 arches. This is composing a plan of 63 structural modules in M'Hamid El Ghizlane, while the second largest mosque still standing, of those we have studied, located at Ouled Youssef, only has 20 modules.

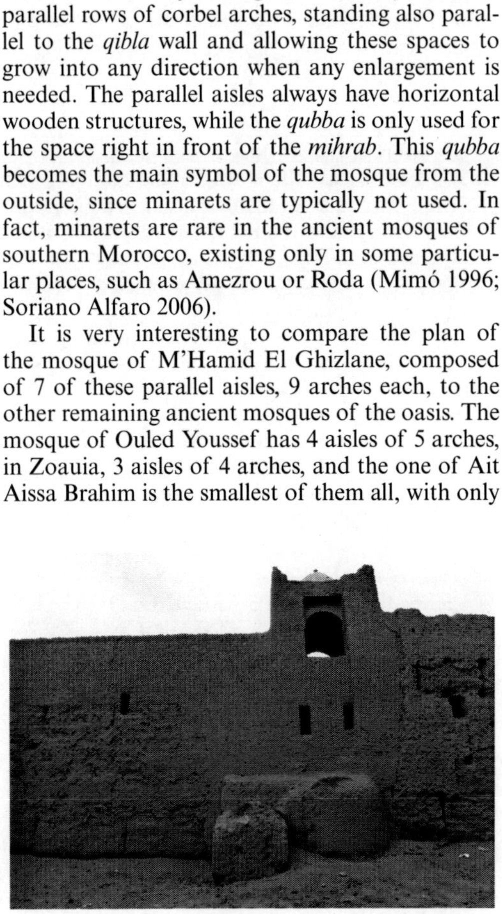

Figure 3. Exterior view of the *qubba* and the *mihrab* of the mosque of M'Hamid El Ghizlane (Terrachidia).

We do not have any historical data about when all these mosques were built, though it is known that both Ait Aissa Brahim and Ouled Youssef were later settlements, probably from the nineteenth century (Mimó 1996), and their current inhabitants know about several transformations since then in their layouts, though they can not specify their range.

In the mosques including a *sahn* or open courtyard, it is placed inside the praying room and not at its side. This same layout is common in most of the mosques of the southern valleys of Morocco (Nijst et al. 1973; Mimó 1996; Soriano Alfaro 2006). This arrangement is probably due to the extreme weather conditions and the subsequent need of ventilation of the main room during the summer.

The mosque complex always includes public baths, a well in the middle of a roofed space and a room to heat the water from the well which is used later in the baths. Some mosques also had a little Quranic school, but nowadays They are found in ruins or lost, as it is the case in M'Hamid El Ghizlane.

The usual urban location of the mosque is opened to the space between the double fortified gates at the entrance of the *ksar*. This is the place where voyagers and caravan traders were hosted in the past, so they could use the mosque without having access to the residential area of the village, but the mosque was protected within the walls. The mosque of M'Hamid El Ghizlane is an exception to this rule, probably due to a later growth of the village to the east, surpassing the former limits of the ancient *ksar* and defining a new and unusually curved perimeter (Landau 1969).

That is not the case of *marabouts*, generally located outside the *ksur* walls, in the surroundings of the villages, in a similar placement to Spanish *ermitas*, and often related to cemeteries.

Nowadays many of these *marabouts* are abandoned and progressively falling apart, while the main threat for the ancient mosques is demolition, as the lack of maintenance and the idealization of concrete structures are often leading to their replacement for newly built ones. These new mosques are designed and built with no respect for local traditions and with a remarkably worst performance facing local climate, what make them almost unusable during the hotter summer days. This permanent threat makes the development and diffusion of these studies all the more urgent.

4 THE MOSQUE OF M'HAMID EL GHIZLANE

The mosque of M'Hamid El Ghizlane is, as it has been said, the biggest and probably the oldest of the M'Hamid Oasis. It was miraculously saved from demolition, even though a new mosque has been built in M'Hamid El Ghizlane too. In this case, instead of the usual replacement, people decided to build it outside the *ksar*.

The 7 mentioned aisles have a width of approximately 4,50–4,80 meters, while the span of each arch is from 4,50 to 6 meters, depending on the area where they are placed. The open *sahn* is roughly in the middle of this room.

Each arch lies on piers which differ from some areas to the others. Their corners and capitals have different shapes which will be analyzed below.

The orientation of the *qibla* wall is almost the canonical one and the *mihrab* is placed close to its center. To the right of the *mihrab* is the *mimbar*, built as a niche and still guarding the ancient mobile wooden structure, and another little niche to keep the sacred texts. Both the *mimbar* and the *mihrab* lay across the wall and protrude from the exterior façade.

The *qubba* in front of the *mihrab* lies on an octagon built over the 4 arches which define this space using tamarisk wood. As all the existing vaults in this oasis, it has an irregular shape, becoming a circle little by little from the octagonal plan. Its outer surface is rendered using lime mortars. Its top, difficult to solve because of being a corbel vault, is instead protected by a piece of green glazed pottery, probably imported from Tamegroute.

The complex of the mosque is entered from its west and south sides, with these two accesses

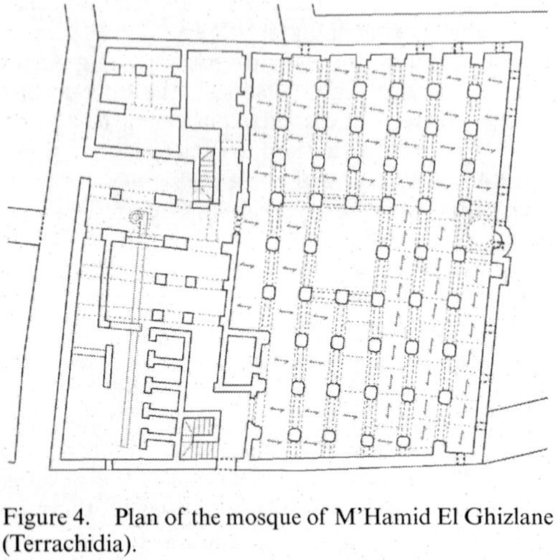

Figure 4. Plan of the mosque of M'Hamid El Ghizlane (Terrachidia).

Figure 5. Stratigraphic units of the northern elevation of the praying room (A. García Hermida).

respectively leading to the praying room through the baths area or directly from the main street of the *ksar*.

The baths occupy the southwest corner of the complex. It also includes the shaded room where the well is open, close to the western entrance, and another room to heat the water, at the northwestern part of the building.

There are two stairs leading to the flat roof, each one close to one of the two entrances. The southern one was also connected to an old Quranic school which was located over the baths of the mosque and whose ruins can still be appreciated today.

Building techniques used in this building are exactly the ones described above as the typical ones in this oasis.

To study the historical evolution of this building, we have used the stratigraphic method, identifying the diverse historical-constructive units which have shape it and the relative chronology between them. This allowed us to get beyond the lack of historical sources and to reach a clear view about how its building process developed through history, though to reach a firm hypothesis about the absolute chronology of its diverse phases is still out of our reach.

Meanwhile, the whole building was analyzed and documented for any future study using ortophotography and making file cards of all its stratigraphic units, including pictures and detailed descriptions of each identified element.

5 RESULTS OF THE ARCHEOLOGICAL ANALYSIS OF ITS HISTORICAL EVOLUTION

It may seem difficult to understand the entire building process of an earthen architecture structure with multiple and frequent maintenance interventions, always using exactly the same traditional building techniques. Nevertheless, this scientific analytic method leaded to surprisingly enlightening conclusions about the succession of the diverse building phases which shaped the current building. We have been able to divide these phases into four main groups of units, each one belonging to a hypothetical different moment of the building process.

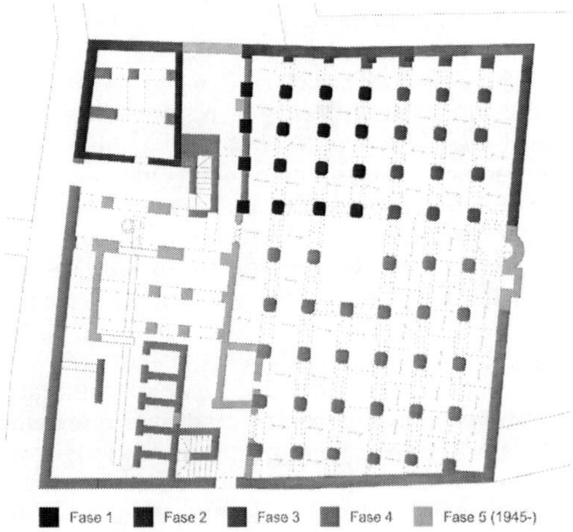

Figure 6. Hypothesis of the main historical building phases of the mosque (A. García Hermida).

Figure 7. Exterior view of the piers attached to the buttresses of the old building (Terrachidia).

The mentioned lack of precise historical data about the building, as it has been explained, did not allow ascribing most of these phases to a particular time period, but material evidences make clear their relative chronology, which is a remarkable outcome itself, providing key additional interest to this heritage for the future development of the region.

The units belonging to the first identified phase define a little mosque which would have been the origin of the wide existing structure. It was a space of 3 × 3 structural modules and was placed in the current northwestern corner of the praying room. The three arches which once formed its façade are now partly hidden and incorporated to the western wall of the praying room. The hypothetic *qibla* wall of this early mosque would have been demolished for its first enlargement, which would have been developed to its eastern and northern sides. The corbel arches of this area stand over square plan piers and the space is roofed using only tamarisk beams.

Stratigraphy shows that its first enlargement implied adding four more aisles to the East with a new *qibla* wall and one more module to the North. This last transformation would have been made absorbing what previously were an open street and also the wall of an ancient building which stood at its other side, fact which we will explain now.

Those piers attached to the current northern wall of the mosque were built to support the roofs of the new aisles and were added when that wall was already standing there. But the most interesting piers are other leaning ones which belong to the original building process of that wall. These last ones undoubtedly were existing buttresses once standing on the street and later reused as piers for the praying room of the mosque. The internal structure of the building they belonged to was demolished to open the new street we may see nowadays to the North of the mosque, probably created to replace the previous one. Some of the new piers were in fact attached to these existing buttresses too, which can now be seen from the outside since the top of the ancient building wall has fallen down. This probably happened because of the demolition of the internal structure of the previous construction, leaving that wall standing free and pressed by the interior arches of the mosque. A small trial excavation on the renders of the interior side of the wall revealed the same link between the new piers and the previous buttresses.

The *qibla* wall of this second identified mosque, according to our hypothesis, would be still standing in its original position, as the walls to the South of it are clearly later additions belonging to the next phase. As there is no trace of its ancient *mihrab* on its outer façade, we can conclude it was a simple niche excavated in the wall, without getting through it, as it is found in other mosques of this oasis, such as the one of Ait Aissa Brahim, where the *mihrab* does not appear in the outer façade. This hypothetical niche would stand now hidden under the lime renders of the interior façade, being an interesting point to study in future campaigns.

The new aisles contained capitals which were carved in the corners and which would later on be awkwardly imitated in the piers of the earliest mosque to make them resemble the new ones The difference between both details can be easily identified in a close view.

Finally, the last and more ambitious enlargement of the mosque was made to its southern side. After its completion, the resulting space would have been more than twice the size of the previously existing one. This could have been due to an enormous growth of the local population, what could relate this transformation to the time following the French occupation and pacification of the whole area, which leaded to an unusually long lasting period of peace and prosperity. Fortunately, we have more than these data to found this hypothesis. The current *mihrab* and the existing *qubba* in front of it were built apparently soon after this wide addition. An inscription on the renders of this work says the qubba was built by Mohamed Ibn Hamed in 1364. This year, translated to our Gregorian calendar, means this last enlargement of the mosque took place in 1945, or at least its new *mihrab* was built by then.

Details about the evolution of the complementary spaces of the complex of the mosque have also been determined. The walls of the room where the water is heated were built before the surrounding spaces were created and they were also already standing when the streets to the North of the complex were modified. Meanwhile, the walls of the baths were attached to an existing perimeter and the space where the well was open was the last to be roofed. The last area was an open space between the bath walls and the northwestern room for some time.

6 SOME FINAL CONCLUSIONS ON THE STRATIGRAPHIC STUDY OF EARTHEN BUILDINGS

As it has been explained, this earthen building was repeatedly modified and restored with the same traditional building techniques which were used in the first constructive phase of its historical evolution. This was always made without intending to create any material difference between the existing elements and the new additions. Nevertheless, studies like the one described here demonstrate that the continuity within local building traditions does not necessary prevent attempts at tracing of the constructive history of these buildings. On the contrary, their stratigraphy can be read, studied and understood, as in any other building, by specialists. This leads to two main key reflections on some of the restoration criteria to be applied in earthen architecture, especially in places like M'Hamid, where fortunately building traditions are still alive.

The first of them is: it is not necessary to abandon traditional techniques when restoring these buildings to avoid historical misunderstandings and escape what in theory of restoration terms is called *falso historico*. Modest but identifiable discontinuities can exist and be analyzed without any need to crudely exaggerate it.

The second conclusion is: instead of renouncing to tradition, we can still use in our restorations the same tools which allow us to recognize the historical evolution of these buildings. They could be used as a set of tools to make our new interventions readable within a cultural continuum and understandable by any interested researcher. The same stratigraphic details we found in the M'Hamid mosque, for instance, could be used in its restoration to ensure the preservation of the rich history we have been able to unveil on their walls without neglecting its architectural integrity and its coherence with the place and the people which still nowadays shape it.

REFERENCES

Dłużewski, M. 2008. Reasons for development of the dune fields in the Coude du Dra region, in Skiba S. & Krzemień K. (eds.): *Contemporary evolution of the natural environment of the region between Antiatlas and Sahara (Morocco)*: 51–63. Kraków: Instytut Geografii i Gospodarki Przestrzennej.

Domínguez Martínez, O. & Garcia Hermida, A. 2013. Learning from Tradition. The Experience of Terrachidia in the South of Morocco, in García Hermida A. (ed.): *Architecture in the Age of Austerity*: 61–64. Madrid: Mairea Libros.

Jacques-Meunié, D. 1982. *Le Maroc Saharien des Origines a 1670*. Paris: Librairie Klincksieck.

Landau, R. 1969. *The Kasbas of southern Morocco*. London: Faber and Faber.

Mimó, R. 1996. *Fortalezas de barro en el sur de Marruecos.* Madrid: Compañía Literaria.

Nijst, A.L.M.T. et al. 1973. *Living on the edge of the Sahara. A study of traditional forms of habitation and types of settlement in Morocco.* The Hague: Government Publishing Office.

Soriano Alfaro, V. 2006. *Arquitectura de tierra en el sur de Marruecos*. Barcelona, Fundación Caja de Arquitectos.

Earthen Architecture: Past, Present and Future – Mileto, Vegas, García Soriano & Cristini (Eds)
© 2015 Taylor & Francis Group, London, ISBN 978-1-138-02711-4

Air lime-earth blended mortars—assessment on fresh state and workability

P. Faria, I. Dias, N. Jamú & V. Silva
Department of Civil Engineering, Faculty of Sciences and Technology, UNIC, NOVA University of Lisbon, Caparica, Portugal

ABSTRACT: Air lime-earth blended mortars have been widely used as masonry, rendering and plastering mortars, mainly for stone or earth masonries. Nevertheless, there is a gap in the scientific knowledge about the characteristics of this type of mortars. Fresh state characteristics of mortars affect workability but also hardened state characteristics. Due to these reasons, an experimental campaign was started with air lime mortars,where lime or fine sand was partially replaced by a clayish earth. Consistency, bulk density, air and water content were determined, compared with water intake of the fresh mortars and with mechanical strength. Simultaneously, workability and shrinkage were evaluated through the applications on substrates—on brick in laboratory conditions and on a rammed earth wall on natural exterior environment. The results show the influence of the earth percentage on air lime-earth blended mortars but particularly that this type of mortars can have very good workability, mechanical resistance similar to pure air lime mortars and fulfillthe requirements for rendering mortars.

1 INTRODUCTION

Earth mortars have been widely used in the past and they still belong to the craftsman knowledge in many regions of the world. They were used as stone or adobe masonry mortars and as plastering mortars to protect the interior surface of walls. Whenever they needed to be applied as rendering mortars, generally an amount of air lime was added to the earth and air lime-earth blended mortars were formulated. Earth-based mortars were applied as renders and plasters on earth-based walls (adobe masonry walls,rammed earth or wattle and daub walls) but also over stone masonry walls.

For some decades the use of earth-based mortars almost stopped in many countries of Europe, mainly substituted by cement mortars. Since the 80's the incompatibility of cement mortars was perceived and lime mortars slowly re-began to be used (Veiga et al. 2010; Arizzi et al. 2012). Only in recent years the advantages of the application of earth-based mortars regained interest. An example of the interest in non chemically stabilized earth mortars are the new Germany standards for earth plaster and earth masonry mortars, showing the need of defining requirements for this type of building products (NABau 2013a,b).

In the past, renders were applied in successive layers, with air lime mortars or with air lime-earth blended mortars, often with lime-wash paint finishing; earth mortars were not normally used due to insufficient resistance to the action of rain.

In Portugal there are still almost entire villages, mainly in the Alentejo region, which are built with earth (Correia 2007). The earthen walls can only be detected when the renders are no longer efficient; that can be due to natural aging deterioration, often accelerated by problems of capillary rise of water.

When degradation appears, the owners often resort to unskilled craftsmanship, who applies cement-based mortar renders and plasters, not compatible with the earth walls.

This type of mortars, instead of effectively contribute for the walls' protection, accelerate its degradation (Gomes & Faria 2011; Gomes et al 2013). In fact the protection and conservation of existing earth walls depend on the application of renders that have to be compatible with the wall materials and the pre-existing mortars.

Beyond the necessary compatibility between materials, another relevant aspect is the sustainability of the construction. The growing concern about this matter leads to an increasing research about earth-based materials. In fact, compared to other building materials, the life cycle energy cost of earth is minor because there is no energy consumption forits production process, a small amount of energy is required for milling and normally the energy for transportationis reduced (sometimes can be dug on site).

Conventional air lime mortars have beenthe subject of numerous studies and therefore there is now a wide range of reference values related to this type of mortars (Faria et al. 2008, Veiga et al. 2010, Arizzi et al. 2012). Nevertheless, most of the available results are related with hardened mortars.

Available results about earth mortars are substantially less frequent compared to air lime mortars. Several authors have been researching topics related with earth mortars (Gomes at al. 2012; Hamard et al. 2013; Faria et al. 2014) but very few results have been focusing on the mortars characteristics in the fresh state (Azeredo et al. 2008). Gomes et al.(2012) have studied the influence of water in the workability of earth mortars to be applied for the repair of rammed earth walls and demonstrated thestrong influence that the water/dry material ratio really have on the workability and shrinkage of these mortars. However, almost no researches have focused onstudying air lime-earth blended mortars (Fernandes et al. 2012). Therefore, and although the large use of air lime-earth mortars for earth or stone wall renders in the past, nowadays there is a gap in the knowledge on air lime-earth rendering and masonry mortars and namely about its characteristics in the fresh state. With the growing of interest on this type of mortars, their characterization has become extremely important.

Motivated by all the previous aspects, the present paper will characterize air lime-based mortars with replacements of lime or fine sand by a clayish earth. The objective is to achieve a method to design mortars that are compatible with rammed earth and other masonries with similar characteristics (like adobe and stone masonry) and that may be used for the new rendering of walls but particularly for the repair or substitution of existing renders and re-pointing of masonry joints.

2 EXPERIMENTAL CAMPAIGN

Two customary air lime mortars were formulated, with volumetric proportions 1:2 and 1:3 (air lime: sand). They are considered reference mortars L2 and L3. The study follows two strands, both with increasing use of clayish earth. Based on the 1:2 proportion, four partial replacements of the binder with earth were made, while based on the 1:3 proportion, four partial replacements of the finer fraction of aggregate with earth were made. The entire process of mixing, preparation of samples and testing was performed in sequential batches.

Tests were performed in all the fresh mortars for the assessment of flow table consistency, fresh bulk density, air content and water retention.

A layer of all the mortars was applied on the surface of ceramic hollow bricks and let to dry in laboratory controlled conditions, while the reference mortars and two other mortars with defined earth percentages were applied as rendering systems on a rammed earth experimental wall.

To be able to compare themortarin terms of mechanical resistances, flexural and compressive strength was also assessed.

2.1 *Materials*

The air lime L used for the mortars was a CL90 supplied by Lusical (Lhoist Group),which has been characterized chemically by XRD analysis elsewhere (Gameiro et al. 2014) and was characterized for loose bulk density (Table 1), based on EN 1097-3 (CEN 1998)—with a free fall of the dry material on a beaker of known volume without any compaction, only eliminating the excess of material.

The sand used for the mortars was composed by a mixture of three washed siliceous graded sands, based on a volumetric proportion of 1:1.5:1.5 (coarse sand CS, medium sand MS and fine sand FS). Each one of the sands and the mixture of sands were characterized in terms of loose bulk density (Table 1) and particle size distribution by dry method, performed according to EN 933-1 (CEN 2000) (Fig. 1).

Table 1. Loose bulk density of air lime, earth and aggregates.

Material	Loose bulk density [kg/dm^3]
Air lime L	0.362
Earth E	1.084
Coarser sand CS	1.412
Medium sand MS	1.405
Finer sand FS	1.388
Mixture of sands	1.463

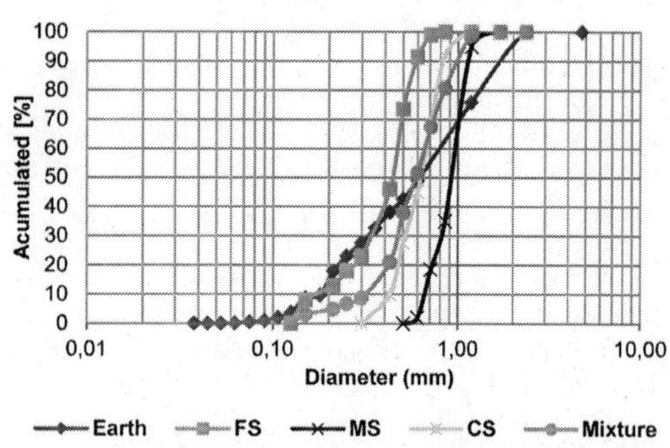

Figure 1. Particle size distribution of each sand, the mixture of sand and the clayish earth by dry methods (authors).

The earth used was supplied by Sorgila (a clay supplier company) and was from Redinha, Pombal, Portugal. It was subjected to crushing in a grinding mill, in order to disaggregate the lumps initially present and then characterized: in terms of particle size analysis, by dry method, also performed according to EN 933-1 (CEN 2000) (Fig. 1); by the loose bulk density, according to EN 1097-3 (CEN 1998) (Table 1). In a previous publication (Faria et al. 2013) the same earth has been also characterized in terms of particle size distribution by wet methodand mineralogical composition by XRD.

2.2 *Mortars and samples*

Based on the 1:2 reference mortar (L2), the following weight percentages of air lime were replaced by earth: 5%, 10%, 25% and 50%. Regarding the 1:3 reference mortar (L3), the following weight percentages of the finer aggregate were replaced by earth: 5%, 10%, 15% and 25%. The ten mortar compositions are presented in Table 2. The designations of the mortars include the reference mortar (L2 or L3), the percentage by weight of clayish earth (a number followed by E), replacing lime (L) or Fine Sand (FS). The composition of the mortars is presented in terms of weight and volumetric proportions, as well as by the Water/Lime (W/L) and the Water/Earth (W/E) weight ratios (Table 2).

The mixing of the mortars was based on EN 1015-2 (CEN 1998) procedures established for mortars but adapted for the particular materials involved. The methodology comprised the following steps: firstly, the dry constituents were weight, manually homogenized and placed in the container of the mechanical mortar mixer; the required amount of water was measured and introduced into the container in the first seconds of mechanical mixing; after 150 seconds, the mixer was stopped and the borders of the mortar were integrated with a spatula; the mixer operated for another 30 seconds.

The amount of water added for each mortar was defined to obtain applicable, workable mortars, with values of flow table consistencywithin the range of 150 ± 10 mm.

Prismatic samples 160 mm × 40 mm × 40 mm for mechanical resistance determination were prepared in metallic moulds. The compaction was mechanical whilethe leveling was manual. Samples were left to dry during 90 days in controlled laboratory conditions with 20 ± 2°C temperature and 65 ± 5% Relative Humidity (RH).

Table 2. Mortars composition and water/lime and water/earth ratios.

	Weight proportion	Volumetric proportion	W/L	W/E
Mortar	L:E:S	L:E:S	[–]	[–]
L2 1:0:7.7	1:0:7.7	1:0:2	0.17	–
L2_5E(L)	1:0.05:8.1	1:0.02:2.1	0.18	35.4
L2_10E(L)	1:0.1:8.6	1:0.04:2.2	0.18	17.7
L2_25E(L)	1:0.3:10.3	1:0.1:2.7	0.18	7.1
L2_50E(L)	1:1:15.5	1:0.3:4	0.18	3.5
L3_25E(FS)	1:1.1:10.5	1:0.4:2.7	0.28	2.6
L3_15E(FS)	1:0.6:11	1:0.2:2.8	0.26	4.0
L3_10E(FS)	1:0.4:11.2	1:0.1:2.9	0.26	6.0
L3_5E(FS)	1:0.2:11.4	1:0.07:2.9	0.26	12.0
L3	1:0:11.6	1:0:3	0.25	–

2.3 *Tests*

The consistency was determined by the flow table test based on EN 1015-3 (CEN 1999), which allows assessing the influence of water (Fig.2, left).

The bulk density of fresh mortars was determined by the quotient of the mass of mortar in the fresh state divided by the volume of the vessel that it occupies. It was determined according to EN 1015-6 (CEN1998) by the shock method—tilting the vessel filled with the mortar about 30 mm on alternate sides, allowing to fall a total of 10 times into the rigid working table.

The air content of each mortar was determined in accordance with EN 1015-7 (CEN1998) and is obtained by filling the air spaces of the mortar with water and its measurement using a specific equipment (Fig.2, right).

The capacity of the mortars to retain water in the fresh state was evaluated by the water retention test that was performed based on the draft standard prEN 1015-8 (CEN1999) but the contact between the fresh mortar with filter paper lasted for 10 minutes.

The workability of the mortars was directly assessed when all the mortars were applied creating a layer 1.5 cm thick on individual hollow ceramic bricks, with a surface area of 29 cm × 19 cm (Fig. 3, left). The shrinkage of the mortars was assessed by visual observation of the mortars' layer on hollow brick while drying in controlled environment (T = 20 ± 2°C and HR = 65 ± 5%).

Figure 2. Table test (left) and air content test (right) (authors).

Figure 3. Mortar layer on hollow ceramic brick (left) and rendered panels on the rammed earth test wall (right). In the middle, the renders with air lime-earth blended mortars (authors).

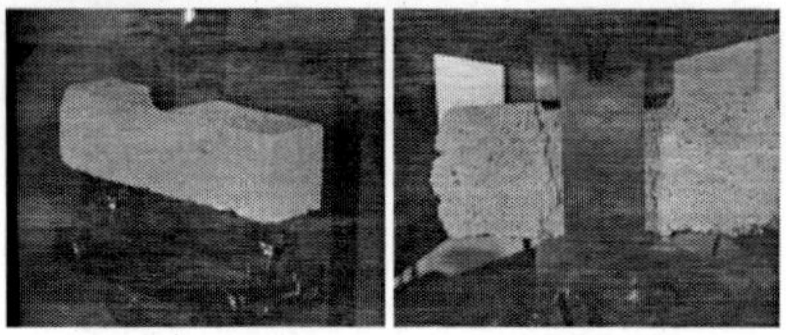

Figure 4. Flexural (left) and compressive (right) strength tests (authors).

The workability and shrinkage were further assessed for four of the mortars that were applied as renders. Each render have 3.04 m^2 of surface and was applied on a rammed earth test wall facing East in exterior natural environment (Fig.3, right). The renders were composed by two layers of the same mortar. The first layer, 0.5–1 cm thick, was mixed with 10% more water than the following and was applied with an irregular surface. The following layer had the same amount of water that was used for the samples production and completed the 2 cm thick render with a smooth surface. The mortars applied asrenders were: both reference mortars, without any added earth (L2 and L3), and two air lime-earthmortars which showed good fresh state characteristics, L3_10E(FS) and L2_25E(L).

As the application of the renders was made in winter time (with a lot of moisture) there was no need of water spraying the walls' surface or the mortar layers after application. Both substrates (the hollow ceramic bricks and the rammed earth) have been characterized elsewhere (Faria et al. 2013).

The prismatic samples were tested for flexural and compressive strength with a Zwick Rowell equipment with a 2 kN load cell for flexural (Fig. 4, left) and 50 kN load cell for compression (Fig. 4, right), based on EN 1015-11 (CEN 1999), after 90 days.

3 RESULTS AND DISCUSSION

The consistency measured with the flow table (average value and standard deviation), the water retention and the water/earth weight ratio of the mortars are shown in Figure 5.

For the mortars based on composition 1:2 (L2…), the replacement of lime with earth causes

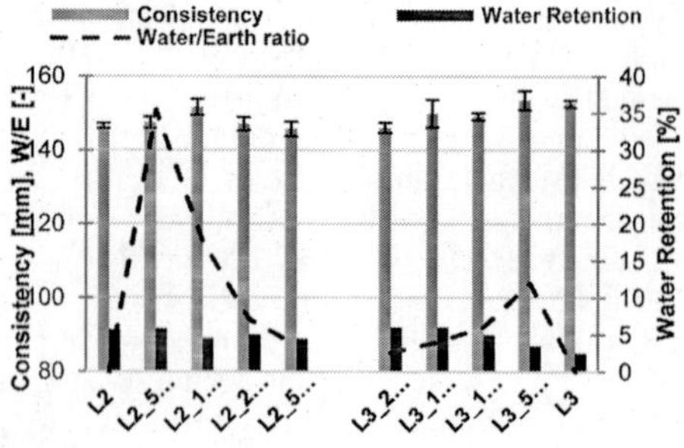

Figure 5. Consistency by flow table test, water retention and water/earth weight ratio of mortars (authors).

an initial increase in consistency, except for the mortars with 25% and 50% of earth—L2_25E(L) and L2_50E(L).

This last mortar registered the lowest value, with 145.8 mm. The mortar L2_10E(L), having 10% of earth, recorded the highest consistency.

Comparing with the water/earth ratio of the mortars it can be remarked that, in general, the consistency decreases with decreasing ratio (with the exception of the mortar with 10% earth—L2_10E(L)). For the mortars with composition 1:3 (L3…) the mortars L3 and L3_5E(FS) are those with the highest values of flow table consistency—even higher than L2_10E(L). During this test it was possible to visually verify these results, since the increase in the percentage of earth significantly decreased the mortars' workability. The mortar L3_25E(FS) (with 25% of earth) had a very low workability. Due to the water/earth ratio of each mortar it is found that, in general, the consistency decreases with decreasing ratio.

The water retention seems to decrease with the percentage of earth in 1:2 mortars, while a clear increase is noticed in the 1:3 mortars, with the replacement of fine sand by earth. The difference in results and trends appear to indicate that the earth has less ability to retain water than lime (mortars 1:2), but obviously has higher water retention capacity than the fine sand (mortars 1:3). Although the maximum values recorded by the two families of mortars were in the same order of magnitude, the lower values were recorded by L3 and L3_5E(FS) mortars.It may indicate that, for these two mortars, curing process should be more careful to avoid a very fast release of water.

The bulk density and the air content of the fresh mortars can be observed in Figure 6. For the mortars based on composition 1:2, where earth partially replaces lime, it is possible to observe that the density tends to increase with the percentage of earth, indicating a greater compactness of the fresh mortar. For the mortars based on composition 1:3, where earth partially replaces fine aggregate, the progress suggests a reduction on density with

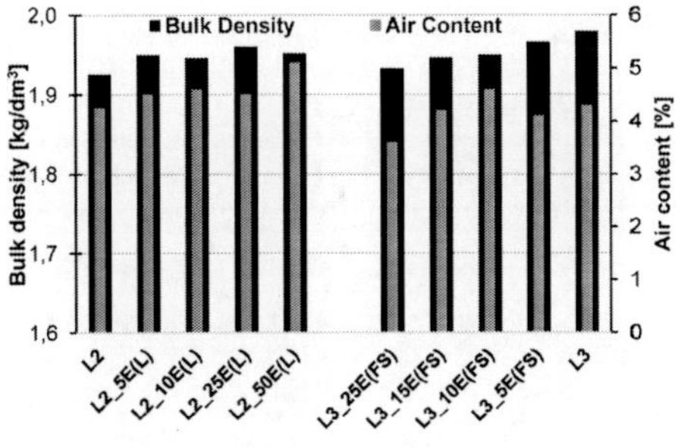

Figure 6. Bulk density and air content of fresh mortars (P. Faria, I. Dias, N. Jamú & V. Silva).

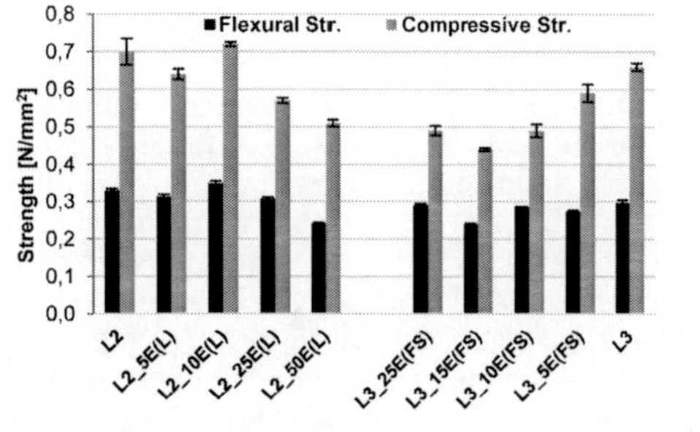

Figure 7. Flexural and compressive strength of mortars after 90 days (P. Faria, I. Dias, N. Jamú & V. Silva).

increasing amount of earth. For that reason, the pure lime mortar with composition 1:3 (L3) have a much higher density than all 1:2 based mortars. Nevertheless, L2_10E(L) and L3_15E(FS) register quite similar bulk density.

In terms of air content it can be noticed a slight tendency for increase with the percentage of earth on 1:2 mortars while for 1:3 mortars an inverse situation is indicated.

Concerning the application of the mortar layers on the brick, no difficulties appear in terms of workability. No shrinkage cracks were observed during drying. In terms of applicability of the mortars asrenders on the rammed earth wall (rendering systems and bigger dimensions compared to the brick layers), the two mortars with earth demonstrated better workability than the reference mortars (without earth) during the application process. Visual observations were conducted during the hardening process of the four renders and there was no shrinkage cracking.

It is also important to notice how the earth affected the color of the rendered sections (Fig. 3, right—two middle sections). This aspect can be particularly interesting because it may eliminate the need for an aesthetic coat of paint.

The flexural and compressive strength of the mortars assessed after 90 days arepresented in Figure 7. It can be noticed that, both for mortars 1:2 and 1:3, the compressive strength decreases with the percentage of earth. However the mortar L2_10E(L) can be competitive compared with the reference mortars (L2 and L3). According to flexural strength, it is also perceptible that mortars with low percentage of earth (again both for L2… and L3… mortars) register high strength compared to the reference mortars, in particular L2_10E(L) and L3_10E(FS). It is interesting to notice that mortars L3_25E(FS) and L2_25E(L) show similar flexural resistances.

Veiga et al. (2010) suggested that the most compatible mortars to be applied on rendering and plastering systems or for re-pointing of historic masonry should register characteristics within defined ranges. Nevertheless these ranges are merely indicative.

For rendering and plastering, the indication is thatmortars should have flexural strength of 0.2–0.7 N/mm² and compressive strength of 0.4–2.5 N/mm²; for re-pointing the ranges recommended are 0.4–0.8 N/mm² and 0.6–3.0 N/mm², respectively for flexural and compressive strength. Comparing the flexural and compressive strength of the formulated and tested air lime-earth blended mortars with the ranges suggested by Veiga et al. (2010), it can be noticed that:

- all the mortars are within the range for rendering and plastering in terms of flexural strength and compressive strength;
- none of the mortars (not even the reference pure air lime mortars) fulfill the recommendation of minimal flexural strength for re-pointing;
- only mortars based on 1:2 formulation with 5% and 10% of earth, based on 1:3 formulation with 5% of earth and the reference mortars fulfill the recommendation of compressive strength for re-pointing mortars.

4 CONCLUSIONS

After the experimental campaign and analyzing the results, some observationscan be taken. The quantity of water needed for the mortars production differs considering the amount of earth (and lime). The water conditions the workability of the mortars, which affects other characteristics, in the fresh and hardened state. Nevertheless, when tested on a small surface (bricks), all the formulated air lime-earth blended mortars seem to present adequate workability to be applied on site. That workability was even better than the one of pure air lime mortars—with 1:2 and 1:3 volumetric proportions—for mortars with 25% of earth replacing lime and with 10% of earth replacing fine sand, when tested for the application of renders on a test wall.

No other fresh state characteristic indicate problems for the tested mortars and drying shrinkage cracks were not detected.

The renders acquired natural pigmentation with the incorporation of earth, which can be beneficial when paint protective systems are not required.

Mechanical characteristics of the mortars after 90 days could be considered similar to pure air lime mortars for most of the air lime-earth blended mortars, particularly in terms of flexural strength. All the blended mortars indicate to be adequate for rendering and plastering historic masonry. Earth-based masonry can be considered to present mechanical and physical characteristics that are often comparable to historic masonry—with low mechanical strength and very porous, water vapor permeable microstructure. For that reason, and also based in traditional practices, it can be assumed that this type of blended mortarsmay also be adequate particularly for rendering earth-based masonry walls.

The use of air lime-earth blended mortars for rendering and plastering existent masonry walls canrepresenta significant eco-efficient improvement. In formulations where earth partially replaces the air lime, mortars can become more sustainable, with low embodied energy. Simultaneously, the knowledge of this traditional type of mortars and renderings can be re-gained, showing a path for more efficient interventions.

ACKNOWLEDGEMENTS

Authors are grateful to the Portuguese Foundation for Science and Technology for supporting project EXPL/ECM-COM/0928/2012, to Lusical–Lhoist and to Sorgila for supplying the air lime and the clayish earth.

REFERENCES

Arizzi, A., Viles, H. & Cultrone, G. 2012. Experimental testing of the durability of lime-based mortars used for rendering historic buildings. *Construction and Building Materials*, 28 (1): 807–818.

Azeredo, G., Morel, J.C. & Lamarque, C.H. 2008. Applicability of rheometers to characterizing earth mortar behavior. Part I: Experimental device and validation. *Materials and Structures*, 41: 1465–1472.

CEN 2012. EN 933–1. Tests for geometrical properties of aggregates. Part 1: Determination of particle size distribution - Sieving method. Brussels.

CEN 1998/2006. EN 1015–2+A1. Methods of test for mortars for masonry. Part 2: Bulk sampling of mortars and preparation of test mortars. Brussels.

CEN 1999, 2004, 2006. EN 1015–3+A1+A2. Methods of test for mortar for masonry.Determination of consistence of fresh mortar (by flow table). Brussels.

CEN1998, 2006. EN 1015–6+A1. Methods of test for mortar for masonry - Part 6: Determination of bulk density of fresh mortar. Brussels.

CEN1998. EN 1015–7. Methods of test for mortar for masonry - Part 7: Determination of air content of fresh mortar. Brusels.

CEN 1999.prEN 1015–8. Methods of test for mortar for masonry – Part 8: Determination of water retentivity of fresh mortar. Brussels.

CEN 1999, 2006. EN 1015–11+A1. Methods of test for mortar for masonry—Part 11: Determination of flexural and compressive strength of hardened mortar. Brussels.

CEN 1998. EN 1097–3. Tests for mechanical and physical properties of aggregates. Part 3: Determination of loose bulk density and voids. Brussels.

Correia, M. 2007. Rammed earth in Alentejo. Lisbon: Argumentum.

Faria, P., Henriques, F. & Rato, V. 2008.Comparative evaluation of aerial lime mortars for architectural conservation. *J. Cultural Heritage* 9 (3): 338–346.

Faria, P., Santos, T. & Silva, V. 2014. Earth-based mortars for masonry plastering. In *9th International Masonry Conference*, 7–9 July, Univ. Minho, Guimarães, Portugal (accepted for publication).

Faria, P., Silva, V., Jamú, N., Dias, I. & Gomes, I. 2013. Evaluation of air lime and clayish earth mortars for earthen wall renders. In M. Correia, G. Carlos, S. Rocha (eds.), *Vernacular Heritage and Earthen Architecture: Contributions for Sustainable Development*: 407–413;CIAV2013, Vila Nova de Cerveira, Portugal, Oct. 2013. London: Taylor & Francis.

Fernandes, M., Mirão, J., Velosa, A. & Candeias, A. 2012. Estabilización de tierra com cal—Ventajes y desventajas. In *Terra 2012*, Lima, Peru (CD).

Gameiro, A., Santos-Silva, A., Faria, P., Grilo, J., Branco, T., Veiga, R. & Velosa, A. 2014. Physical and chemical assessment of air lime-metakaolin mortars: Influence of binder: aggregate ratio. *Cement and Concrete Composites* 45: 264–271.

Gomes, M.I. & Faria, P. 2011. Repair mortars for rammed earth constructions. In V.PFreitas, H.Corvacho, M.Lacasse (eds.), *XII DBMC—12th International Conference on Durability of Building Materials and Components*, Vol.2: 689–696. Porto, Portugal: FEUP.

Gomes, M.I., Gonçalves, T.D. & Faria, P. 2012. Evaluación de la influencia delcontenido de aguaen la trabajabilidaddelmortero de tierra. *Apuntes—Journal of Cultural Heritage Studies*, 25 (2): 258–277.

Gomes, M.I., Gonçalves, T.D. & Faria, P. 2012. Earth-based repair mortars: experimental analysis with different binders and natural fibers. In Mileto, Vegas & Cristini (eds.), *Rammed Earth Conservation*: 661–668; Valencia, June 2012. London: Taylor & Francis.

Gomes, M.I., Gonçalves, T.D. & Faria, P. 2013. The compatibility of earth-based repair mortars with rammed earth substrates. In HMC2013—Historic Mortars Conference, University of West Scotland, Glasgow, Sept 2013 (CD).

Hamard, E., Morel, J.C., Salgado, F., Marcom, A. & Meunier, N. 2013. A procedure to assess the suitability of plaster to protect vernacular earthen architecture. *J. Cultural Heritage*, 14: 109–115.

NABau 2013a. DIN 18947. Earth plasters – Terms and definitions, requirements, test methods(in German). Berlin.

NABau 2013b. DIN 18946. Earth masonry mortar –Terms and definitions, requirements, test methods (in German). Berlin.

Veiga, R., Fragata, A., Velosa, A., Magalhães, A. & Margalha, G. 2010. Lime-based mortars: viability for use as substitution renders in historical buildings. *Architectural Heritage* 4: 177–195.

Earthen Architecture: Past, Present and Future – Mileto, Vegas, García Soriano & Cristini (Eds)
© 2015 Taylor & Francis Group, London, ISBN 978-1-138-02711-4

Earthen construction in Alto Arnoia river valley, Galicia

A. Fernández Palicio
Laboratorio de Tecnoloxías Apropiadas, Vigo, Galicia, Spain

ABSTRACT: Maceda region is a sedimentary plain in the south of Galicia where we still find constructions made with half timbers with earth. Nowadays, there are barely a few constructions remaining which show us an interesting vernacular construction made with earth. Furthermore, the region was an important production center of pottery which, at the end of the 19th century, started to diversify its production to manufacture bricks. This evolution allowed the emergence of a new earthen architecture that was built using traditional constructive techniques keeping traditional shapes but employing new materials coming from the first industrialization, which started by replacing the ancient half timbers with earth and *terróns* walls, and subsequently it generated a new series of building typologies keeping the traditional construction logic.

1 INTRODUCTION

The Maceda depression, located in the south of Galicia at 600 m altitude, is flanked by a series of small mountains between 700 m and 800 m high at Southeast, by a wall between 300 m and 400 m of vertical drop and 1600 m high at North and East and by a series of elevations of up to 800 m at West. It has slopes of the less than 7% (Juana López 1996).

Regarding its geology it is composed by granits of two micas, particularly in the centre and western area. In the eastern area there are also schists and sandstones, specifically granodiorites and e gneisses called *ollo de sapo* (Romero 1985).

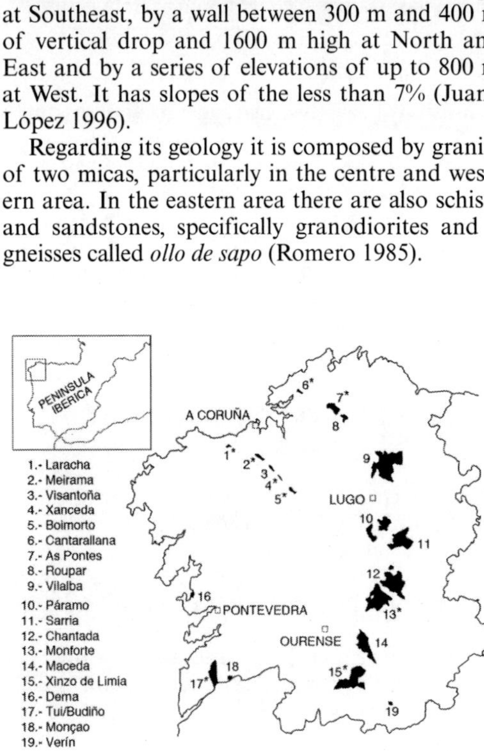

Figure 1. Large sedimentary areas in Galicia. Maceda is number 14. (Barrón, & Santos 1998).

The low basin, over which the main rivers of depression flow: Arnoia, Tioira, Castelo and Maceda; contains sediments of several meters deep of clays, sands and loose stones, sometimes with the topsoil layer covered with pebbles (Sanz 1985).

The climate is oceanic with a continental nuance, characterized by an average annual temperature of 12.2°C with annual average variations between 6.5°C and 19.2°C. The average annual rainfall is 1388 mm, and rainfall is more abundant in winter (37%) and autumn (30%) than in spring (24%). Summer is the most aridity and driest season (8%) and rainfall deficits can rise up to 200 mm. (Martínez & Pérez 1997).

Soils in depression (according to the FAO classification) are mainly litosols, fluvisols and regosols. However there is an abundance of dystric and humic cambisols in the region and, to a lesser extent, brown soils. Seismic risk is measured VI on the EMS-98 strength values scale and 0.04 on seismic acceleration (IGN, 2002).

Figure 2. Clay quarry near Lamelas village (Author).

1.1 *Distribution and structure of settlements*

The traditional population units are located on the western gentle mountainside and at the foot of the eastern slopes. They have a significant number of houses in relation to other Galician's regions (Bouhier 2006); and often they are formed by dwellings row houses forming a street configuration called *rueiro*, particularly in the low areas.

In the last years, the suburban houses have been settled in large fields' areas in any order, primarily in areas surrounding the town of Maceda (1500 inhabitants, the services centre of the region) and the village of Tioira; forming a rur-urban agglomeration typical of the current spatial organization of Galicia and North of Portugal.

The continuous production fields organization near villages is still clear, with well-preserved closures (Bouhier 2006) where we can often distinguish the old enclosed earthen walls yet. The old fallow lands with biannual rest and weak closures have been steadily disappearing and are covered with bushes nowadays.

There are no other settlements or crop areas in the rest of the territory, which was one enormous tapestry for cattle circulation covered with bushes (*ulex* and *calluna* mainly) (Bouhier 2006), it has been now turned to forestlands.

Figura 3. A village forming a street configuration with half timbers with earth. (Author).

Figure 4. Continuous production fields organization with enclosing stone walls close to Vilar village (Author).

2 EARTHEN CONSTRUCTION IN ALTO ARNOIA RIVER VALLEY

2.1 *Need for this research and urgency*

Thus far, stone constructions are the only ones referenced in Galician vernacular architecture monographs without realising that, in many Galician sedimentary areas where stone was scarce, vernacular earthen construction techniques were often the most popular. They were unknown to the majority of the population and even to building professionals. Their use in Galician contemporary architecture is practically zero or even completely rejected due to "earth not being an acceptable building material for Galician climate".

Sustainable contemporary architecture needs to look to earth like appropriate material to the future: healthy, ecological and sustainable. In this way, the strategy to achieve an appropriate construction starts in the dissemination of richness and variety of vernacular architecture.

This research aims to contribute to the dissemination and knowledge of one of the vernacular earthen techniques in Galicia almost disappeared, so that its knowledge may contribute to define applicable principles in sustainable contemporary architecture.

2.2 *Methodology*

This study began two years ago. Its aim was, firstly, to identify the geographic area and extension of the use of this technique and to find out about the characteristics and dimensions of the main constructive systems found. The field work was accompanied with mapping survey of the buildings, with research literature and interviews with old local residents.

3 EARTHEN VERNACULAR TECHNIQUES IN ALTO ARNOIA RIVER VALLEY

Earthen construction in Maceda depression was common in the lowlands close to the main river banks where access to stone material was more difficult. In the rest of the territory, we can only find another earthen construction in the place of Niñodaguia, traditional ceramic area.

The constructions systems that we find here are half timbers with earth as structural system, half timbers with earth as structural closure (both called *pallabarro*) and many remains of former walls of *terróns*. It is worth noticing the old enclosing earthen walls, similar to those from Alta Limia.

3.1 *Terróns, an extinct technique*

Terróns walls were a common constructive system in the neighbouring Alta Limia region (Fernández Palicio 2013) which may have been also usual in Arnoia valley. Even though our research failed to find remains of *terróns* walls or testimonies about their earlier existence, we have found granitic structures similar to those preserved in Alta Limia region which are closed using *terróns*, and that make us suspect that they could be filled in the same way.

Nevertheless, the area where we find this remains is limited to the South of the depression, in a large meadow with similar characteristics and very close to Alta Limia, and includes the settlements of Arnuide, Aldea de Arriba, Aldea de Abaixo, A Eirexa, Cancillós, Calvelo and Xocín. We do not find any others outside of here.

Figure 5. Ancient cow shed in Aldea de Arriba with granitic structure really similar to those with *terróns* walls. The current is higher and was filled with bricks (Author).

3.2 *Half timbers with earth structure systems*

There are many standing constructions in the region with half timbers with earth structure systems, called *pallabarro*, mainly in settlementes around the town of Maceda, in Vixueses and Foncuberta. Our research only found three buildings in region.

These constructions have walls of granite stone with regular masonry in the corners and major parts. The half timbers with earth found in this region are not used for building the whole floor, which is common practice in other regions in Galicia, but only for small parts of the building which also support the roof structure.

Figure 6. Out-of-context auxiliary building in Cancillós, in which there is still preserved the granitic structure of constructions with *terróns* (Author).

The structural wooden frame is made of a weak main skeleton of wooden pillars from the sleeper beam to the roof beam, reinforced half height with other horizontal pieces. On this skeleton the *pallabarro* framework is built. It is made of wooden planks called *fitoiras* which link together with horizontal and diagonal strips (diagonals are more frequent in region). Formerly they were also linked together with willow branches, but we do not find them in this area. The structure is filled with a twisted mud kneaded with straw, pebbles or wood shaving and coated with clay mortar and whitewashed as outer finish.

The structure weakness and the limited woodworking technical development are reflected in the small dimensions of the hollows.

3.3 *Half timbers with earth as not structural closures*

It is the most common system. We find it in all the low areas in region without exception, with many structures still visible despite being a small in

Figure 7. Half timbers with earth with corridor on the first floor. Vixueses village (Author).

Figure 8. Detail of half timbers with earth. Vixueses village (Author).

Figure 9. Main skeleton and half timbers framework (Author).

Figure 10. Facade wall with half timbers with earth (Author).

Figure 11. Facade wall with half timbers with earth (Author).

number those well-preserved. Many of them were substituted for bricks or other materials.

Its construction system is the same described for half timbers with earth structural frameworks. There are not any other different construction systems like *barrotiño* in region. (Fernández Palicio, 2012).

It is always used to close the facade wall that was protected on the first floor's balconies or to close the end of balconies and even to close them from the exterior.

3.4 *Clay mortars*

In many preserved buildings we still find clay mortars used like internal linings. They are mortars applied to improve the thermal inertia, the comfort and the moisture regulation. Their thickness was at least 5 cm but it could be thicker because those which we founded were in very poor condition.

Preliminary analysis of the composition indicates a good proportion of clays, silts and fine sands together with blades of straw that are no more than 4–6 cm.

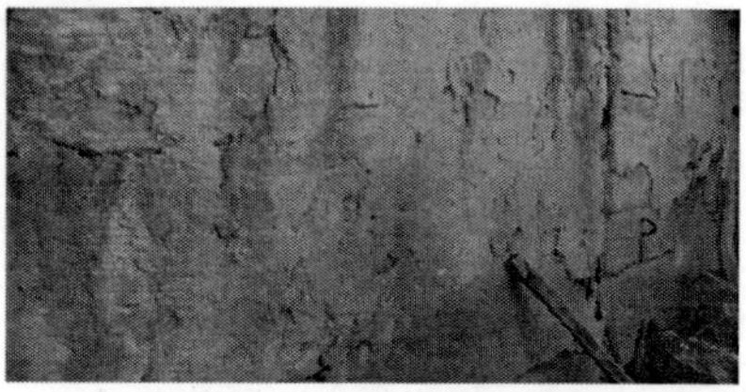

Figure 12. Detail of clay mortar. Vixueses village (Author).

Figure 13. Mortar sample from Vixueses village and preliminary analysis of the composition. (Author).

Figura 14. Enclosed earthen walls close to Francos village, Maceda (Author).

3.5 *Earth used like fencing*

Even though, in many of the low areas in region, fences are made with stone, in another large part of the continuous production fields it is more common to find earthen walls as fields closure, similar to the Lake of Antela, which were still in use in the 70's (Bouhier 2006). Their distribution, according to the found remains in our research, includes the banks of Maceira and Tioira rivers before joining Arnoia river.

4 THE NEW CERAMIC MATERIALS

4.1 *The ceramic workshops transformation and the disappearance of vernacular techniques*

At the end of the XIX century, due to the economy improvement and the increase in trade, in the parish, *freguesías*, of *Niñodeaguia* and *Tioira* the old ceramic workshops incremented their number and increased their pottery production which diversified with tiles and bricks too. In the 30's they began to decline. (García Alén 1983).

The vernacular techniques with raw earth were disappearing being replaced by the new ceramic materials, which were manufactured in great number, first with solid bricks and afterwards with hollow bricks. Thus, we find many examples of buildings where their half timbers with earth, located in the facade wall that was protected on the first floor balconies, were substituted by solid and hollow brick walls; and little by little, they were replaced by stone like structural material in the major walls.

4.2 *The traditional logic of the new material*

These new materials, which came from the first industrialization, allowed the emergence of a new earthen architecture which, keeping traditional shapes and using traditional constructive techniques, preserved the same environment relationship as vernacular construction had or as sustainable construction today asks for: local and close materials with a low footprint (although higher than using raw earth) and that promotes local development.

But this emerging architecture did not have sufficient time for its adaptation due to the rapid arrival of new industrial materials from the 60's for which people were not prepared. The lack of knowledge of their technical properties brought about inadequate criteria to build and a decontextualization of the materials as a consequence. This bad construction rapidly expanded reaching our present days.

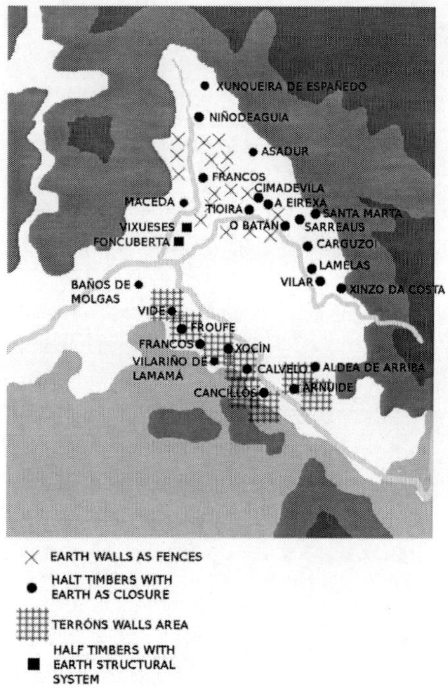

Figura 16. Scheme of Maceda depression, where there are reflected the places where there were found buildings with earth construction systems (Author 2013).

Figure 15. An example of building with the facade wall made with solid brick in Vixueses (Author).

Figure 17. An example of building stone wall where stone was replaced by brick in Calvelo (Author).

Figure 18. Building wall in Tioira where the stone walls were substituted by solid bricks on the left hand and by industrial bricks on the right hand (Author).

Figure 19. Auxiliary building in Parada village built with traditional logic with wood and brick (Author).

Figure 20. Building in Xunqueira de Espadañedo where brick is used with inadequate criterions using a normal brick like face brick being very bad thermally (Author).

5 CONCLUSIONS

This research aims to contribute to the identification and knowledge of the Galician earthen constructive systems almost disappeared in such a way that its knowledge could be used to establish applicable principles in contemporary architecture. The strategy to achieve a sustainable architecture starts by disseminating the richness, variety and evolution of its vernacular heritage.

The disappearance of the half timbers with earth and *terrón* walls have prevented us to know how their technique evolution could have been with current technology. But it is still possible to implement them. It is necessary to develop new earthen techniques which update vernacular techniques to build and restore: Improved half timbers with local earth, clay mortars and local bricks, which could be considered a vernacular, sustainable and ecological material in many regions if they are manufactured promoting local development.

Sustainable contemporary architecture in Alto Arnoia river valley needs to look to earth like appropriate material to the future. It is necessary to spread this technology, making public studies about the social, economic and physical benefits of its use and about its thermal and healthy properties; and promoting pioneering initiatives in earthen construction.

REFERENCES

Barrón, E & Santos, L. 1998. Síntesis Paleobotánica crítica de las cuencas terciarias de Galicia. *Coloquios de Paleontología* 49: 41–53.

Bohuier, A. 2006. *Galicia: Ensaio xeográfico de análise e interpretación dun vello complexo agrario*. Compostela: CAGM.

Caamaño, M. 2003. *As construccións da arquitectura popular: patrimonio etnográfico de Galicia*. A Coruña: Hércules.

Caamaño, M. 2006. *Dun tempo e dun país*. Ames: Laiovento.

Carmona, X & Nadal, X. 2005. *El empeño industrial de Galicia. 250 años de historia (1750–2000)*. A Coruña: PBM.

FAO 2009. *Guía para la descripción de suelos*. Roma

Fernández Palicio, A. 2012. Earthen architecture in Galicia (Spain). Rammed earth in Terra de Lemos. In Rammed Earth Conservation. Taylor & Francis. London.

Fernández Palicio, A. 2013. Constructions with terrons in Alta Limia, Galicia: A lost sustainable architecture. In *Vernacular Heritage and Earthen Architecture*. London: Taylor & Francis.

García Alen, L. 1983. *La alfarería de Galicia*. A Coruña: Fundación Pedro Barrie de la Maza.

Instituto Geológico Nacional 2002. *Mapas de peligrosidad sísmica de España*. Madrid.

Juana López, J. 1996. *O alto Arnoia (o cerne ourensano)*. Vigo: Ir Indo.

Martínez, A & Pérez, A. 1999. *Atlas climático de Galicia*. Xunta de Galicia.

Olmo Sanz, A. 1985. Estudio Geológico Sedimentario de la cuencas terciario-cuaternarias de Monforte de Lemos, Maceda y Quiroga. *Cadernos do laboratorio Xeolóxico da Laxe* 10. A Coruña.

Romero, R & García, C & Pérez, A & Macías F. 1985. Mineralogía de la fracción arcilla de suelos y sedimentos de la cuenca de Maceda. *Cadernos do laboratorio Xeolóxico da Laxe* 9. A Coruña.

Rueda, J & Muezca, J. 2001. Sismicidad, Sismotectónica y peligrosidad sísmica en Galicia. *IGN* 35.

Earthen Architecture: Past, Present and Future – Mileto, Vegas, García Soriano & Cristini (Eds)
 ISBN 978-1-138-02711-4

The vernacular earthen architecture near Turin (Piedmont, Italy)

F. Fratini, S. Rescic & C. Riminesi
CNR-Istituto per la Conservazione e Valorizzazione dei Beni Culturali, Sesto Fiorentino, Italy

M. Mattone
Dipartimento Architettura e Design, Politecnico di Torino, Turin, Italy

L. Rovero & U. Tonietti
Dipartimento di Architettura, Università degli Studi di Firenze, Florence, Italy

ABSTRACT: Piedmont and Sardinia are the regions which have the richest collections of historical and architectural earthen heritage in Italy. The present work is a contribution to the knowledge of the rich earthen heritage that can be found in the Piedmont region and, in particular, in the province of Turin. The research aims therefore to deepen the knowledge of the earthen buildings present in different areas of the province of Turin, examining types, construction techniques and analysing the characteristics and performance of the constitutive materials through in situ tests (rebound test, water absorption test), characterization of the earth of bricks and mortars (XRD, particle size distribution) and identification of possible additives. This type of research will be useful to promote again the use in the Turin area of these construction techniques, evidence of skills and abilities of people to take advantage of the local environmental resources.

1 INTRODUCTION

The earthen construction heritage which is located in Piedmont (Italy) represents a significant testimony of vernacular architecture. It is constituted by buildings which, although primarily designed for utilitarian purposes, are characterized by cultural and architectural values and represent a significant testimony to that society history.

They represent "the fundamental expression of the culture of a community, of its relationship with its territory and, at the same time, the expression of the world's cultural diversity" (ICOMOS Charter 1999).

The acquisition of a growing awareness of the value characterizing certain cultural expressions has led to the inclusion of new categories of heritage to be protected in the Code of Cultural Heritage (Law 42/2004). For examples "rural architectures having historical or ethno-anthropological value since they are testimonies to rural and economic traditions".

The widening of interest does not concern only the physical aspects, but also the cultural implications which stand out from the vernacular landscape to traditional buildings, as well as from tangible to intangible heritage.

They are the result of age-old traditions and a testimony of our rural life, knowledge that ask to be adequately safeguarded.

Unfortunately the survival of earthen architecture «is threatened world-wide by the forces of economic, cultural and architectural homogenization» (ICOMOS Charter 1999), as well as by a lack of recognition of its true historical and cultural value. The preservation of this heritage requires not only its identification, but also a good knowledge of the typologies, construction techniques and materials which characterize it, in order to facilitate the implementation of appropriate restoration and reuse interventions.

2 EARTHEN ARCHITECTURES NEAR TURIN

Piedmont and Sardinia have one of the richest collections of historical and architectural earthen heritage in the whole Italy, whose characteristics change from one area to another.

In the province of Turin there are quite a lot of examples of buildings characterized by the presence of earthen bricks used for the construction of both external and internal walls. Although there are examples of houses and industrial sites, they are mostly rural buildings, spread over the landscape, that are a testimony of the type of settlement whose presence "is documented in a stable form since the late fifteenth and early sixteenth century in all the plain of Piedmont, from Turin, as far as Cuneo and Casale» (Palmucci 1988)

The constructions on which studies and tests were conducted are both rural buildings and provide an interesting example of peasant culture.

The one located in Rivasecca, near Pinerolo, is a nineteenth-century farmhouse and it consists of a single construction that hosts the house, the barn, the hayloft and a shed for the tools.

Definitely more articulate is the complex located in Borgo Cornalese, near Villastellone, a few kilometers away from Turin. The village was founded on the XI century by Hungarians and Bulgarians populations. It still preserves the characteristics of a medieval settlement and hosts many farmhouses aimed at accommodate cows that were used to graze in the neighboring territories.

The farm of san Pietro, which was already represented in the napoleons registers of landed properties, is composed of two different buildings that are developed around two yards. One of these is characterized by the presence of stables built with brick load-bearing pillars and mud-brick walls. Used until the nineties of the twentieth century, it is today almost totally abandoned and presents a state of advanced decay.

3 TYPOLOGIES AND CONSTRUCTIVE TECHNIQUES

Earthen constructions in the province of Turin represent an interesting and important example of vernacular architecture (Figs. 1–3), even if they are less known than the ones of the Fraschetta area in the province of Alessandria.

Essentially there are two typologies which characterize rural buildings: the *cassina*, a rural building complex partly devoted to host housing and partly animal shed and/or tools and product storage (Palmucci 1988) and the *ciabot* or *chiabotto*, a small basic construction, erected with poor materials (Comba 1988).

The *cassina*, which spread from the sixteenth century, consists of the house itself (*domus*), of the stable and the barn (*fenera*). The kitchen, the dining room, the cellar and/or the storage are located on the ground floor. The bedrooms and the barn can be found on the first floor and can be reached by a wooden staircase or, sometimes, with a brick or stone one (Fig. 4).

The *cassina* have a structure made of bricks while the curtain walls are made of earthen bricks (Palmucci 1988). The buildings are generally which mud-brick walls were erected, in order to protect earthen walls from the phenomenon of water capillary rise.

Figure 1. Rural building in Borgo Cornalese, Villastellone (Torino) (credit Manuela Mattone).

Figure 2. Cassina in Rivasecca, Buriasco (Torino) (credit Manuela Mattone).

Figure 3. Cassina in Rivasecca, Buriasco (Torino) (credit Manuela Mattone).

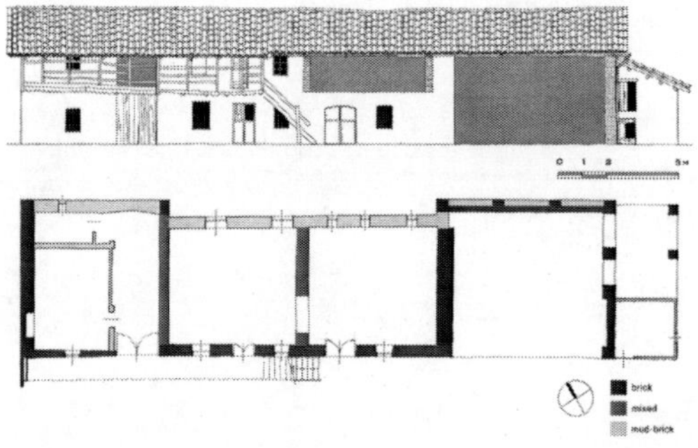

Figure 4. Cassina in Rivasecca, Buriasco (Torino), plan and south front (credit Roberto Mattone).

The horizontal elements are mostly made of wooden beams (placed at a distance of about 80 cm) which support wooden boards or brick vaults. In the more recent buildings wooden beams gave way to H-beams. Often there are also brick barrel vaults poured with a water and gypsum mixture, made without ribs, but using a sand mold (*grasia*), supported by flat plank scaffolding. Horizontal beams (*ciav*), made with wild cherry or black locust, can often be found near the vault in order to ensure a greater stability to the structure (Volterrani 1998).

The openings of such buildings are generally small and have a wooden frame secured to the masonry during the process of construction.

The roofings mainly have wooden structures, whose beams lay on bricks. The roof covering uses tiles and, when there isn't eaves gutter, the overhang of the roof is greater in order to allow adequate removal of rainwater and ensure more protection to vertical surfaces.

4 MATERIALS AND METHODS

The following materials coming from the two earthen building of Rivasecca and Borgo Cornalese villages were analyzed: earthen bricks, earthen bedding mortars and mortar of the plasters.

4.1 *Mineralogical composition*

The mineralogical composition of the earthen bricks, earthen bedding mortar and plasters has been determined through X ray diffraction (X'Pert PRO diffractometer by PANalytical equipped with X'Celerator detector and HighScore software for acquisition and interpretation of data according to the following operative conditions: Cu $K\alpha 1$ = 1,545 Å radiation, 40 KV, 30 mA, $2\theta = 3 - 70°$).

Concerning the analysis of the clay minerals (earthen bricks and earthen bedding mortars), the Philips PW 1729 diffractometer has been utilized according to the following operative conditions: Cu $K\alpha 1$ = 1,545 Å radiation, 40 KV, 20 mA, $2\theta = 3 - 20°$).

4.2 *Petrographical study of the mortars*

A petrographical study of the mortar of the plasters has been carried out on thin sections with optical microscopy under transmitted light (ZEISS Axioscope A1).

4.3 *Grain size distribution*

The grain size distribution of earthen bricks and earthen bedding mortar has been carried out through sieving in order to separate the following fractions: sand (Ø > 63 μm), silt (4 μm < Ø < 63 μm) and clay (Ø < 4 μm).

4.4 *On site test*

4.4.1 *Rebound test*

An analysis by the hammer sclerometer (rebound test) was carried out on a wall of Rivasecca building and on a wall of Borgo Cornalese building, in order to determine the compressive strength of the earthen bricks. In particular, for the Rivasecca building, the rebound test was carried out on the north wall both outside and inside; for Borgo Cornalese building, the test was determined only inside on the north wall.

The determined values are indicative values, because the rebound test is an indirect test, which correlates the surface hardness to compressive strength. The employed hammer sclerometer was the Proceq pendulum hammer type PM (range 0.2–5 N/mm^2), suitable for a wide range of materials with low compressive strength and where small impact is required.

It consists in measuring the rebound R, which is in relationship with the hardness and strength of the material. For each wall, the average rebound was determined and the average compressive strength was determined from the corresponding conversion curves.

In addition, as the earthen bricks exhibited not completely homogeneous surfaces, a statistically significant number of tests was carried out to obtain reliable results. In particular 40 tests for each wall were carried out.

4.4.2 *Water absorption test*

The water absorption tests have been carried out using Karsten pipe (Karsten 1983) and by measuring, for a maximum of 15 minutes and at regular intervals of time (one minute), the rate of water absorbed from the graduated pipe (Fig. 5). The

Figure 5. Karsten water absorption test in Rivasecca, Buriasco (Torino) (credit Manuela Mattone).

interior walls of Rivasecca and Borgo Cornalese and the external wall of Rivasecca were anaysed.

5 RESULTS AND DISCUSSION

5.1 *Earth and mortars characterization*

The results of the XRD and the grain size analysis are reported in the following tables (tables. 1–3).

It seems that the earthen mortars, in both cases (Rivasecca and Borgo Cornalese), have been realized with a leaner earth, starting from the same raw material utilized for bricks.

On the whole in both cases the earths are particularly lean and this is due both to the strong amount of sand + silt and to the type of clay minerals which do not display expandable characteristics.

The comparison with the composition and grain size of other earthen buildings (Fratini 2009; Prati 2012), shows a discordance regarding the granulometry of the earth of the mortar respect to that of the brick.

In fact, for example, both in Alessandria area that in Aleppo region (Syria), the bedding mortars were made with an earth of finer grain size with respect to that of the brick, in particular purifying the same raw material.

With regard to a comparison with the optimal granulometric curve reported by CRATerre (for earthen bricks), it is observed that the granulometry of the samples analyzed (Rivasecca and Borgo Cornalese) falls within the zone of acceptability.

The petrographical study of the mortar of the plasters show that the plaster of Rivasecca is constituted by a scarce binder (Binder/Aggregate 1/3–1/4) with micritic texture made of magnesian air hardening lime (fig. 6).

The aggregate is well mixed, with unimodal grain size (Ø = 400 – 600 μm) and constituted by sub angular fragments of methamorphic rocks (quartzite, quartz micaschists, anphibolites).

The palsters of Borgo Cornalese are constituted by a scarce binder (Binder/Aggregate ~1/3) with micritic texture made of air hardening lime (fig. 7).

Table 1. Principal mineralogic composition.

	Quartz	Feldspars	Anphiboles	Calcite	Clay min + acc. minerals
Rivasecca (earthen brick)	XXX	XX	X	–	XX
Rivasecca (earthen mortar)	XXX	XX	X	–	XX
Borgo Cornalese (earthen mortar)	XXX	XX	–	X	XX
Borgo Cornalese (earthen brick)	XXX	XX	–	X	XX

XXX = abundant; XX = medium; X = scarce.

Table 2. Clay mineral composition*.

	Kaolinite	Illite	Chlorite
Rivasecca (earthen brick)	30	55	15
Rivasecca (earthen mortar)	35	50	15
Borgo Cornalese (earthen mortar)	25	65	10
Borgo Cornalese (earthen brick)	15	75	10

*relative amount.

Table 3. Grain size distribution.

	Sand	Silt	Clay
	Ø > 63 μm	4 μm < Ø < 63 μm	Ø < 4 μm
Rivasecca (earthen brick)	60	22	18
Rivasecca (earthen mortar)	60	27	13
Borgo Cornalese (earthen mortar)	68	26	8
Borgo Cornalese (earthen brick)	52	20	28

Figure 6. Thin section XPL (credit Fabio Fratini).

Figure 7. Thin section XPL (credit Fabio Fratini).

The aggregate is well mixed, with unimodal grain size (Ø = 300 – 500 µm) and constituted by sub rounded fragments of carbonatic rocks and quartz grains.

5.2 *On site test*

5.2.1 *Rebound test*

The following compressive strength were determined: 2.4 MPa (coeff. of variation = 0.09) for Rivasecca earthen bricks and 1.8 MPa (coeff. of variation = 0.14) for Borgo Cornalese earthen bricks.

A comparison with the compressive strength of other earthen bricks carried out by some of the authors in previous tests (Baglioni et al. 2010; Bricoli Bati et al. 2008; Fratini et. al 2012) shows that the values are similar.

5.2.2 *Water absorption test*

As can be seen in the chart in figure 8, the values obtained from the tests show that the walls

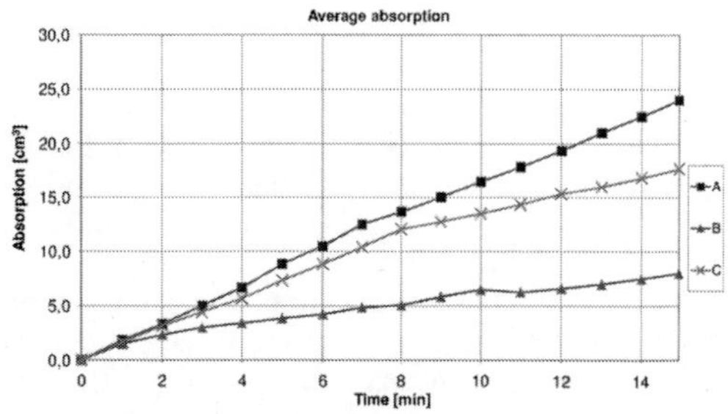

Figure 8. Diagram illustrating the results of water absorption tests by Karsten method. A = Rivasecca interior wall; B = Rivasecca external wall; C = Borgo Cornalese interior wall (credit Manuela Mattone).

protected from rainwater have a higher capacity of water absorption (A = Rivasecca; C = Borgo Cornalese).

This result can probably be due to the presence of a minor quantity of water inside these earthen bricks that are thus characterized by a greater absorption capacity when compared with those used to make the external wall of the buildings (B = Rivasecca).

6 CONCLUSION

Earthen architecture in the province of Turin is an interesting, even if sporadic, testimony of vernacular architecture, expression of the traditional and natural way by which communities used to house themselves.

It is an architecture that has suffered, and will suffer again, transformations and changes in order to give a response to social and environmental constraints.

The appreciation of this fragile and extremely vulnerable architecture and its successful protection requires, first of all, the recognition and appreciation of its value.

Furthermore, since the preservation of historical and cultural values of this vernacular heritage (and not only) cannot consist only of its museification, but requires its continuative use for activities related to daily life and production, it is necessary to involve the community, which could guarantee its continuing use and maintenance.

Although in many cases earthen buildings are perceived as «the painful memory of an atavistic misery» (Musso 2006) that the inhabitants wanted and still want to erase, it is now necessary to rediscover the forgotten values of this heritage.

The detailed study of these architectures, their construction characters and materials is an essential prerequisite to safeguard them and might also contribute to the rediscovery of the earth as a construction material.

The results of compositional, granulometric and mechanical investigations give useful information in order to understand characteristics of the material and technologies used in earthen building construction near Turin.

In both cases (Rivasecca and Borgo Cornalese) the earth is particularly lean and this is due both to the strong amount of sand + silt and to the type of clay minerals which do not display expandable characteristics.

The comparison with other earthen buildings shows a discordance regarding the granulometry of the earth of the bedding mortars respect to that of the bricks; in fact the bedding mortars were usually made with a finer particle size of the earth with respect to that of the bricks, in particular purifying the same raw material.

The granulometry of the earth samples analyzed falls within the zone of acceptability of the Fuller curve reported by CRATerre. The compressive strength of the earthen brick is in agreement with values reported in literature.

Aware of the need to conduct further tests and analyzes aimed at assessing the performance characteristics of the earthen buildings near Turin, it is believed, however, that the research conducted so far can be a first useful step to promote once again the use in the Turin area of these construction techniques, evidence of skills and abilities of people to take advantage of the local environmental resources.

NOTE

To be noted that all the authors should be considered as principal authors.

REFERENCES

Baglioni, E. et al. 2010. The mineralogical and mechanical characteristics of the earth utilised in the earthen buildings sited in the Drâa Valley (Morocco). *Acte de RIPAM 2, 2eme partie*, n° 20, Minbar Al Jamia Ed., pp. 222–231.

Barosso, L. et al. 1993. Constructions en terre crue dans la region turinoise, in *7th International Conference on the Study and Conservation of Earthen Architecture* (Silves, Portogallo, 24–29 october 1993), Lisbona: DGEMN, pp. 43–48.

Briccoli Bati, S. et al. 2011. Considerations on methods to evaluate the compressive strength of earth building materials. *Proceedings of Terra 2008: The 10th International Conference on the Study and Conservation of Earthen Architectural Heritage*, Bamako, 1–6 February 2008, Hong Kong: Getty Pubblications.

Comba, R. 1988. L'insediamento rurale fra medioevo ed età moderna, in Comoli Mandracci, V. *Piemonte*, Bari-Napoli: Laterza.

Drdácký, M. et al 2012. Standardization of peeling tests for assessing the cohesion and consolidation characteristics of historic stone surfaces, *Materials and Structures*, 45 (4), pp. 505–520.

Fratini, F. 2009. The archaeometric analysis of building materials. *In Earthen Domes and Habitats: villages of Northern Syria, an architectural tradition*, shared by East and West,

Fratini, F. et al. 2011. The Earth in the architecture of the Historical Centre of Lamezia Terme (Italy): characterisation for the restoration. *Applied Clay Science*, 53, pp. 509–516.

Fratini, F. et al. 2012. Earthen domes in the Aleppo region (Syria): materials and building techniques. In Congress Proceedings *Domes in the World*, Florence, March 19–23, 2012, G. Tampone-R. Corazzi-E. Mandelli Scientific Editors, Firenze: Nardini Editore.

Karsten, R. 1983. *BauchemiefürStudium und Praxis*, Heidelberg, Germany.

Mecca, S. & Dipasquale, L. (eds.), Pisa-Italy: Edizioni ETS, pp. 313–321.

Houben, H. & Guillaud, H., 1989. *Traité de construction en terre de CRATerre*, Marseille: Editions Parenthèse.

ICOMOS 1999. *Charter on the Built Vernacular Heritage, ratified by the ICOMOS 12th General Assembly*, Mexico, 1999.

Mattone, R. 2010. *Il paesaggio delle case in terra cruda*, Savigliano: L'Artistica edizioni.

Musso, S. 2006. «In viaggio» per vedere, conoscere e progettare, in Bosia, D. *Guida al recupero dell'architettura rurale del G.A.L.*, Bossolasco: Blu Edizioni.

Palmucci Quaglino, L. 1988. Continuità e innovazione nella casa rurale di pianura tra Cinquecento e Ottocento, in Comoli Mandracci, V. *Piemonte*, Bari-Napoli: Laterza.

Prati, G. L. 2012. *Manualetto pratico illustrato dell'architettura in terra cruda, riferito al caso alessandrino*, edited by Municipality of Alessandria, Area Servizi Pianificazione, Sviluppo Territoriale Economico e LL.PP. Direzione Lavori e Opere Pubbliche Servizio Manutenzione Immobili e Impianti.

Volterrani, E. 1998. Terra: una tipologia costruttiva presente anche nel torinese, in Gilibert, A., Mattone, R. (eds.) *Terra: incipit vita nova*, Torino: Politecnico di Torino.

Earthen Architecture: Past, Present and Future – Mileto, Vegas, García Soriano & Cristini (Eds)
© 2015 Taylor & Francis Group, London, ISBN 978-1-138-02711-4

Evaluation of seismic behaviour of rammed earth structures

R. Gallego & I. Arto
Department of Structural Mechanics, University of Granada, Granada, Spain

ABSTRACT: Rammed earth is one of the ancient construction materials and has been present in the history of mankind for thousands of years. At the same time, it is one of the building materials which we have less data about, regarding their mechanical properties and their structural behavior under stress. A large number of houses built with this material are located in seismically active areas so that knowledge of their behavior under seismic stress becomes very important as a way to predict rupture and retrofitting which can be used on them to preserve the lives of their inhabitants. Reinforcement techniques on such structures should take into account the economic and technical capacities of the populations in which they are intended to be implemented, so as to design effective and low-cost retrofit systems.

1 INTRODUCTION

1.1 *An approach to seismic risk*

The rammed earth is present in a large number of countries, making it the structural system used in many homes. Many of the countries in which these structures are found, also present a high level of seismicity, which involves a great risk to the lives of its inhabitants.

Homes built with this material have shown limited ability to withstand the effects of a seismic event, as there is no technology regarding its construction or reinforcement systems, to which we must add above the limited capacity of the material itself.

This fact is evident every time an earthquake ensues where this is the material used as the basis of the structures of most of homes, collapsing a large number of them with the loss of lives.

For these reasons, the study of the seismic behavior of these structures, the knowledge of material constitutive laws and the development of techniques for strengthening or improving the behavior under stresses imposed by an earthquake are fields of work to develop in the future to reduce seismic risk and the cost in human lives in these areas of the planet.

If we compare a distribution map of constructions of rammed earth and adobe with another showing seismic hazard, we can see how the scope of greatest risk is concentrated in areas of the west coast of South America, Anatolian peninsula, inland China, Haiti, New Zealand and neighboring countries of Afghanistan.

There are examples of that type of risk in these areas, such as the case of the earthquakes in Bam (Iran) where its citadel, declared as World Heritage Site, was devastated by an earthquake of magnitude 6.6 Mw of 23 December 2003, with a balance of 27,000 dead, 30,000 wounded and more than 85% of collapsed homes.

1.2 *Analysis of standards*

Unlike other materials and construction systems, earthen structures have limitations in terms of rules and standards to their characterization, design or testing.

Only a few countries have drafted their own rules for this material, often using other types of similar structures in their analysis scheme.

In the United States, there are standardized tests for stabilized adobe construction, compression tests, bending strength, moisture cycles and dryness, erosion and absorption (ASTM D559-44 and D560-44).

New Zealand drew one of the most comprehensive guidelines, the *NZ 4297 Engineering Design of Earth Buildings*, published by the Earth Building Association of New Zealand in 1998.

In the UK *The Rammed Earth House* was published in 2007, and Building Limited Research Establishment Association edited in 2005 the *BRE.EP62 Rammed Earth: Design and Construction Guidelines* and the *BRE.EP80 Earth Masonry: Design and Construction Guidelines* in 2008.

Meanwhile in Australia a manual called *The Australian Earth Building Handbook* was drafted in 2002. In France two guides of design called *L' architecture en Pisé dans le Canton de Boën* and *Techniques de Construction et Restoration* were published in 2005.

In Colombia the *Manual Para la Rehabilitación de Viviendas Construidas en Adobe y Tierra Pisada* by the Colombian Association of Seismic Engineering was published.

In Spain the closest to these is the normative *UNE 41410:2008, Bloques de Tierra Comprimidos para Muros y Tabiques. Definiciones, Especificaciones y Métodos de Estudio* which was elaborated by the Technical Sub-Committee AEN/CTN 41 Edifica SC10.

There are also a number of other standards and recommendations such as the UBC- 1958, REEF-CSTB, some recommendations of the UN and the International Association for Earthquake Engineering and the ones drafted in Peru on stabilized adobe through their Instituto Nacional de Investigación y Normalización de la Vivienda (ININVI) which published the *Norma para la Construcción con Adobe*.

1.3 *Mechanical properties of the material*

Tables 1 and 2 have been elaborated from the data collected by ten different authors. In them, a significant uniformity can be seen in some features (such as density or Poisson's ratio) while a significant deviation from others.

The data obtained depend largely on the type of specimen used, size and slenderness, and are one of the causes of the deviation data.

As it can be seen, the order of magnitude depends a lot on the source consulted.

Analyzed these data and the type of specimen used to obtain them, we conclude that the possible values that characterize the material can be the indicated in Table 3, matching values in addition to the tests conducted by Maniatidis and Walker for cylındrical samples of rammed earth 200 mm high and 100 mm diameter, in which it can be seen that the material has a linear behavior that is about 85% of rupture value, which given the rest within the non-linear behavior.

Table 1. Rammed earth mechanical properties I.

Author	Density kg/m³	Compression MPa	Tensile MPa	Shear Mpa
Hall	2160	0.75–1.46	–	–
Lilley	2170	1.80–2.00	–	–
Maniatidis	1850	3.88	–	–
Jaquin	–	0.60–0.70	–	–
G.Limón	2250	2.45	0.29	–
Novamooz	2000	–	–	–
Aymat	2000	1.50	0.29	–
Bui	2000	–	–	–
Gomes	1900	0.67	0.13	0.08*
Angulo	2000	–	–	0.03

Table 2. Rammed earth mechanical properties II.

Author	E MPa	Poisson	G Mpa	Sample cm
Hall	–	–	–	cube 10.10
Lilley	–	–	–	cube 15.15
Maniatidis	250	–	–	cylind.d = 10 h = 20
Jaquin	60	–	–	samp. 10.10 h = 30
G.Limón	922	0.30	355	cylind.d = 15 h = 30
Novamooz	–	0.33	–	cylind.d = 7 h = 2
Aymat	–	–	–	several authors
Bui	100–500	0.22–0.40	41–180 samp.	40.40 h = 65
Gomes	200	0.35	74	*
Angulo	500	0.20	208	several authors

* NZ 4297 data.

Tabla 3. Rammed Earth mechanical properties. Data for the calculation.

Elastic modulus	E	250 MPa
Yield compression	σ_{c0}	1.85 MPa
Ultimate compression	σ_{cu}	2.00 MPa
Tensile yield strength	σ_{t0}	0.29 MPa
Shear yield strength	σ_{s}	0.22 MPa
Unit.def. elastic compression	ε_{c0}	0.0074
Unit.def. ultimate compression	ε_{cu}	0.0080
Unit.def. tensile	ε_{t0}	0.0012
Density	ρ	2000 kg/m³

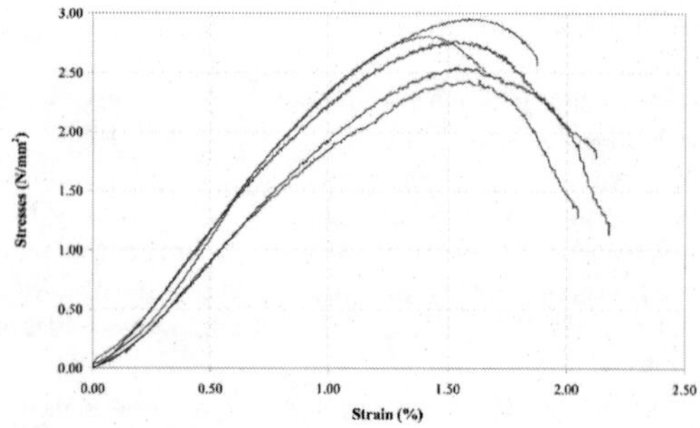

Figure 1. Stress-strain curves for specimens of rammed earth. (Maniatidis & Walker, 2008).

2 DESCRIPTION OF FAILURE MODES

In order to describe different modes of failure, pictures and break diagrams appearing in the *Manual para la rehabilitación de viviendas construidas en adobe y tierra pisada* by the Colombian Association for Earthquake Engineering have been used.

2.1 *Shear failure in the plane*

Failures in the plane are produced by excessive bending or shear efforts, which will depend on the ratio between the length and the height of the wall. For low ratios l/h, with high walls and short length, the flexural failure will occur, while in high l/h relations, with long, low rise walls the shear failure will occur.

This second case, the shear failure, is also associated with walls that receive large horizontal push due to heavy roofs or overload, which are augmented by the distribution of doors and windows openings.

In this failure the characteristic diagonal cracking double X-shaped occurs.

When the size of these cracks becomes important, system degrades quickly, and the strength and stiffness of the walls may collapse due to a decrease in the structure ability to support gravitational loads.

2.2 *Flexural failure*

In this failure the characteristic diagonal cracking double X-shaped occurs.

When the size of these cracks becomes important, system degrades quickly, and the strength and stiffness of the walls may collapse due to a decrease in the structure ability to gravitational loads.

2.3 *Failure by configuration and discontinuity of the structure*

There is a kind of failure, very important in the seismic behavior of structural assembly, due to errors in design or implementation. Among these the following can be cited:

1. Lack of anchorage and wide enough links in the corners of the walls connections that produce cracks and falls in the corners of the structures, concentrating a lot of tension
2. Lack of union in the heads of the beams causes the disarrangement of its elements during seismic shaking.
3. Openings too big or too near to other holes or corners.
4. Lack of connection between the heads of the walls and the floors, or consecutive walls produced by tearing effect of the presence of the heads of beams in the wall.
5. Lack of diaphragm effect in horizontal floor structures, either by poor execution, unable materials or non-existence of horizontal floor.

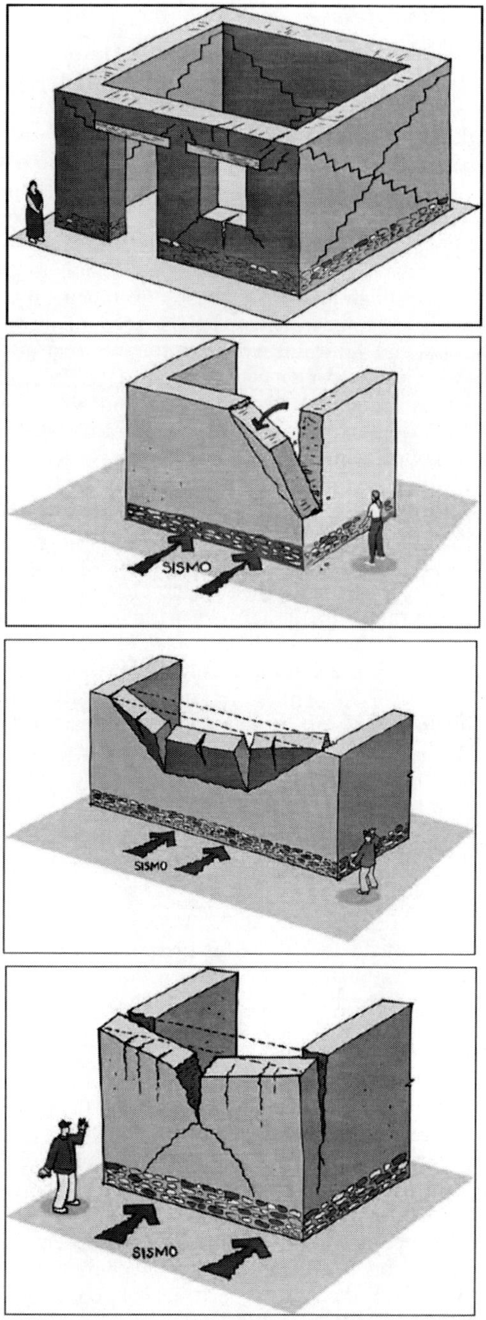

Figure 2. Break schemes. Shear, broken corner and out of plane modes (Gómez, F. Asoc. Col. Ing. Sísmica).

6. Roofs falls into the structure by faults out of plane and bending of bearing-walls, or the presence of heavy cover elements along with their deterioration.

2.4 *Out of plane failure*

This type of failure is very common in structures of rammed earth, even in earthquakes of moderate magnitude. In this case the wall falls outside the plane in which it is in a steady state and explosively causing instability of vertical load resisting system. Within such failures two different mechanisms depending on the geometry and the boundary conditions of the wall occur.

The first with a perpendicular flexion to the plane of the wall with an horizontal cracking in the base, or to an intermediate height, accompanied by cracks and vertical fractures in the wall that make it a non supporting mechanism.

It is a common type of failure in walls very long compared to their thickness and without the cooperation of perpendicular walls to brace them.

The second with a bending also perpendicular to the plane of the wall, but in this case with a vertical fracture in the center of the wall and a diagonal cracking forming the mechanism of fault.

Separation in the corners with the remaining wall facades also occurs due to lack of confinement and link. It is a common fault in short high walls. In some cases and under certain configurations the failure modes can be conjugated with each other simultaneously, so that the result will be a combination of several of them.

3 MODELING THE SEISMIC EFFECT

3.1 *Description of the models*

The synthetic model used is presented in this section to estimate the overall behavior of a possible structure in a seismic event. Thus, the chosen model is a single-storey house between walls in the Vega of Granada, with 7 meters of facade and 10 meters deep.

The height of the walls of the facades is 3 meters and 2 gables. The doors have 2 m of height and a width of 0.80 m. Meanwhile windows have an average height of 1.00 m and a width of 0.80 m.

The perimeter walls, both the facade and the dividing ones are built with 40 cm thick rammed earth, while the interior partitions are constructed with solid brick tile placed on edge.

The roof weight rests equally on the two walls facade, so the lineal load on the façade walls is 6 kN/m which translates to a pressure of 2.4 kN/m^2.

Interactions with other houses or constructive heterogeneities were not taking into account in the calculation. Similarly it is understood that the entire structure rests on the same ground and in a plane.

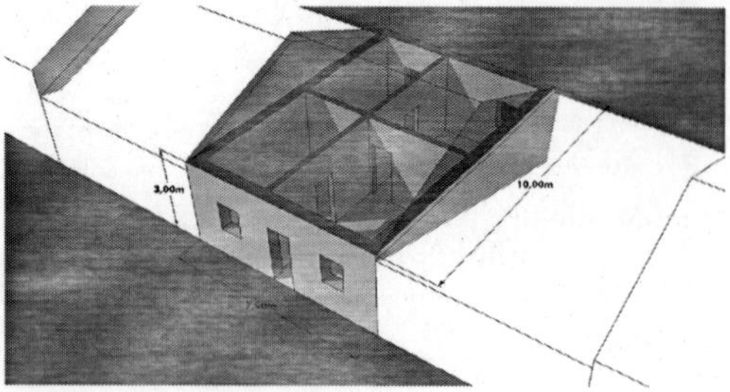

Figure 3. Sketch of synthetic model used. (Arto, I. 2014).

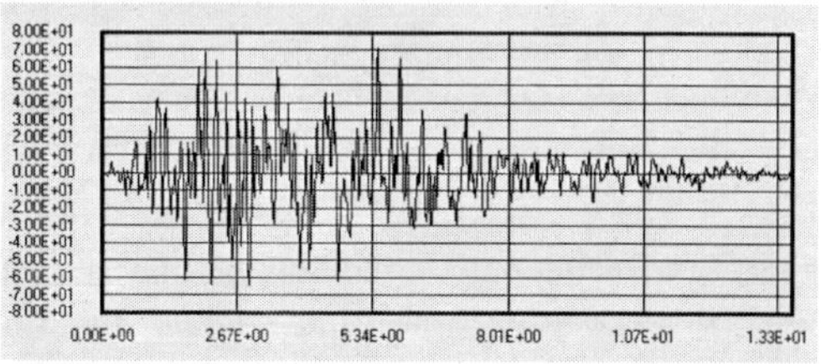

Figure 4. X component of the synthetic earthquake (Arto, I. 2014).

We subject the structure to the effects of an earthquake of magnitude Mw = 6.7 with epicenter at a distance of 30 km, (similar to the earthquake of December 25, 1884 location), using an accelerogram artificially created with the DegtraA4 software (program developed by the Institute of Engineering of the Universidad Nacional de Mexico). The interactions of the soil on the seismic wave during propagation is not taken into account, as well as soil structure under the dwelling. The accelerogram obtained calculation appears in fig.4 with peak values around 70.1 cm/s^2.

After calculating the response spectrum acceleration is observed as the highest value of acceleration is in a strip around the period T = 0.19 s (5.26 Hz) with a value of 217 cm/s^2 or 0.22 g

3.2 *Material constitutive law*

Nonlinearity effects are incorporated into the model through a constitutive law which will govern the behavior of the material by two stress-strain curves, bilinear in the case of compression, and a discharge curve (softening) in tension. Both curves are shown in Figure 5 being the values that delimit those shown in Table 3. Knowing the mechanical properties of the material more accurately is one of the research needs arising from this work. It would be possible to use constitutive laws such as Drucker—Prager with contrasted and acceptable angle of internal friction and cohesion of the material

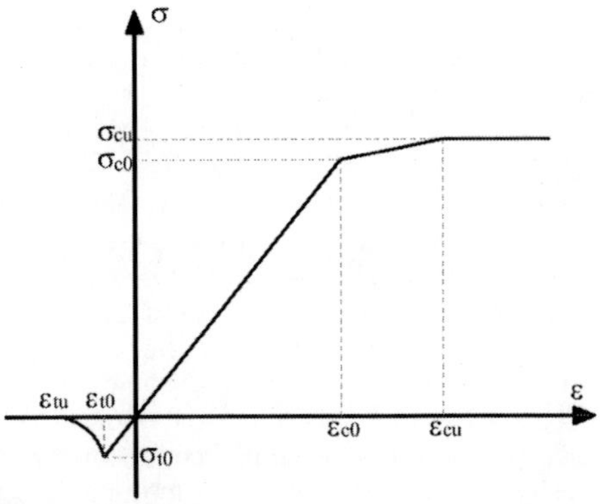

Figure 5. Material constitutive law (Arto, I. 2014).

values, although the authors have been consulted in this work do not mention this values.

Thus incorporating in the model this pattern of damage and failure criteria limiting linear and non-linear behavior we obtain several results that make us understand a little better the behavior of a rammed earth structure.

3.3 *Result of numerical calculations I. Stress*

We used the commercial software of finite element ANSYS Workbench for modeling the structure and get the results that we shall comment below.

We carry out a transient analysis with 20 load steps in which we incorporate the imposed accelerations by the earthquake in the two horizontal axes that we have obtained previously with synthetic acelerogram as well as the value of gravity and overloading the roof.

The stress and strain results obtained are evaluated in three time instants corresponding to 1.46 s, 3.65 s and 5.11 s.

As we can see in Figure 6 to 1.46 seconds the maximum stresses are located at the corners of the holes in the front and in the corners of the meetings of the facade walls and corners. In these areas the values reach 1.65 MPa, causing the first cracks in the material begin to draw diagonal cracks in the corners of the holes moving towards the top of the wall facade and downstream areas. Similarly, in the encounters between the walls that form the structure will begin a cracking that will separate the walls.

At 3.65 seconds the structure has gone through several loading conditions that increase the fissures that had appeared in the previous period.

Stress reached 1.71 MPa in the most sensitive areas, which are again the hollow corners and walls connections. Tensions are spread throughout the structure and two damaged areas appears in the

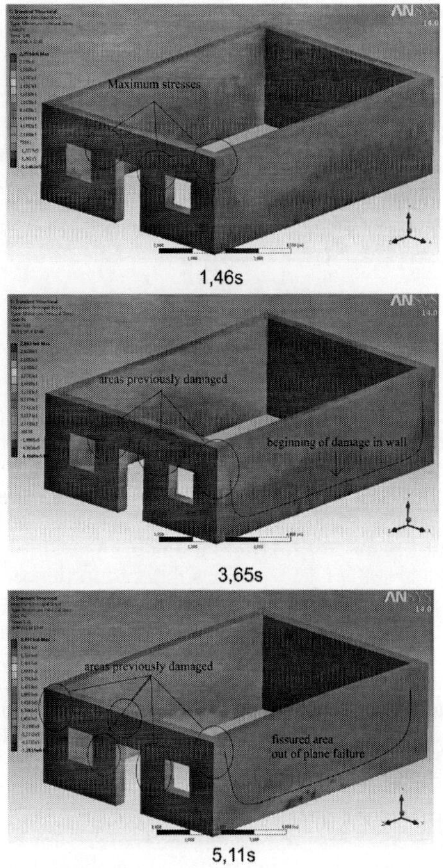

Figure 6. Stresses at the instants 1.46 s, 3.65 s and 5.11 s (Arto, I. 2014).

dividing walls due to the thrust perpendicular to its plane. In these areas the stress reaches 1.46 MPa. In the last instant studied, 5.11 seconds, the entire structure is damaged and the greatest tensions draw paths that make the house becomes an unstable mechanism to cause its collapse.

On the front façade there have been diagonal cracks in the piers and with the similar to shown in Figure 2(1). In addition to the configurations shown the dividing wall as indicated in Figure 2(3).

3.4 *Result of numerical calculations II. Plastics deformations*

Regarding the plastic strain experienced by the structure we see in Figure 7 strains that appear at time 5.11 s, when shaken the structure by one of the highest perpendicular to the planes of the dividing walls accelerations. The plastic strain show a clearly areas where the structure will concentrate

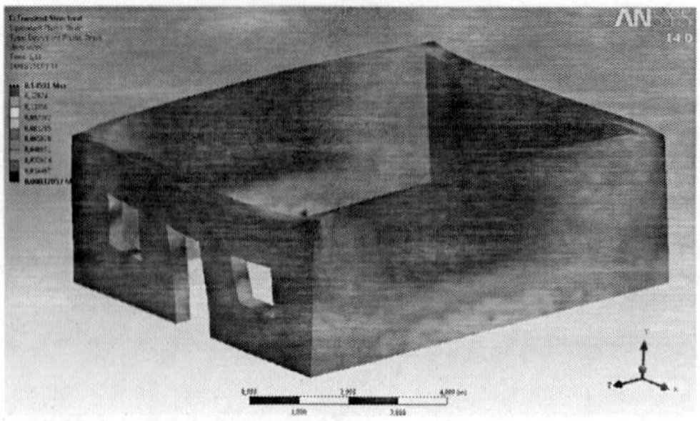

Figure 7. Plastics strains in 5.11 s (Arto, I. 2014).

rupture zones, fitting them with those previously announced.

Furthermore, displacement then reaches 12.9 cm in the heads of the dividing walls that inevitably cause its fall, as indicated in the previous section and Figure 2 (4). At that moment the wall of the facade suffers deformations induced in the central axis around 6 cm.

In this Figure 7 all light colored areas have undergone plastic strain. It can be seen the scope of such strains that encompass the entire facade and a large part of the dividing walls. The dark areas have no plastic but we see that out of plane displacement makes them unstable. The figure has dramatized the movement to have a clearer idea of the situation that has gone the structure.

Interactions produced by internal divisions were not taken into account in calculating the collaboration. In a more complex model could include studying such collaboration and tensions or discharges that occur on the structure.

4 CONCLUSIONS AND WORK LINES

Rammed earth structures located in seismically active areas have a high risk of collapse during a seismic event since its mechanical capabilities are very small and are easily vulnerable. In order to delve into the behavior of these structures in an earthquake it is necessary to be able to create synthetic models that reproduce as realistically as possible the conditions to which these structures are subjected. This synthetic model should incorporate constitutive patterns of damage or laws capable of simulating the stress or strain which will be the actual subject. The dispersion of data between different authors, tied in part to the nature of the material, is another obstacle to be overcome.

By submitting these structures to earthquakes of moderate importance (about 0.2 g, with an intensity around V–VI) we observe how quickly they fail so that it is necessary to implement systems for improving their skills. When developing these systems, it should be taken into account that they are low cost

systems available to the communities in which these structures appear. It is precisely the further development of these reinforcing systems the justification and the future study to be continued by the authors of this article. With the current synthetic model cross diagonal cracks in the pillars and areas close to the holes and corners of the building were obtained. These cracks reproduce the usual breaking schemes so that it can be understood that the model used reproduces the real conditions. Taking to the limit the capacity of the structure we see how finally most of the material of the structure has plastic behavior and failures out of plane added to the initial shear failures. This makes the structure to inevitably collapse.

REFERENCES

Angulo-Ibáñez, Q., Más-Tomás A., Galván-Llopis, V., Santolaria-Montesinos J. 2012. *Traditional braces of earth constructions*: 389–399.

Arto Torres I. 2014. *Evaluación del comportamiento sísmico de estructuras de tapial mediante análisis no lineal.*

Aymat Escalada C. 1993. *Patología y Rehabilitación de Cajones de Tapial.* Consejo General de la Arquitectura Técnica de España. Sumario 54.

Bui, Q., Morel, J., Hans, S. Meunier, N. 2009. *Compression behaviour of non-industrial materials. Materials and Structures* 42(8): 1101–1116.

Gómez F, C.A. *Manual para la rehabilitación de viviendas construidas en adobe y tierra pisada*. Asociación Colombiana de Ingeniería Sísmica.

Gomes, M.I., Lopes, M., De Brito, J. 2011. *Seismic resistance of earth construction in Portugal. Engineering Structures*, 33(3): 932–941.

Gonzales Limón T., Álvarez de Burego M., De las Casas Gómez A. 1.998. *Estudio de los materiales de las fábricas de la torre de Comares de la Alhambra*. CEDEX.

Hall M., Djerbib Y. 2004. *Rammed earth sample production: context, recommendations and consistency*. Constr Build Mater 18: 281–286.

Jaquin P.A., Augarde C.E., Gerrard C.M., 2006. *Analysis of historic rammed earth construction. Structural analysis of historical constructions.* New Delhi.

Lilley DM, Robinson J, 1995. *Ultimate strength of rammed earth walls with opennings.* Proc ICE: Struct Build 110(3): 278–287.

Maniatidis, V., Walker, P. 2008. Structural capacity of rammed earth in compression. In *Journal of Materials in Civil Engineering*, 20(3): 230–238.

Maniatidis V., Walker P., Heath A., Hayward S. 2007. Mechanical and thermal charac. of rammed earth. In *Int. symposium earthen structures. Bangalore:* 205–211.

Nowamooz, H., Chazallon, C. 2011. Finite element modelling of a rammed earth wall. In *Construction and Building Materials*, 25(4): 2112–2121.

Earthen Architecture: Past, Present and Future – Mileto, Vegas, García Soriano & Cristini (Eds)
 ISBN 978-1-138-02711-4

Restoration of monumental rammed earth buildings in Spain in the last decade according to the 1% Cultural Program

L. García Soriano, C. Mileto & F. Vegas López-Manzanares
Instituto de Restauración del Patrimonio, Universistat Politècnica de València, Valencia, Spain

V. La Spina
Universidad Politécnica de Cartagena, Cartagena, Spain

ABSTRACT: This research work proposes an analysis of the interventions carried out on rammed earth architecture in Spain in the last decade according to the 1% Cultural Programme. Since the analysis focuses on interventions paid by public funds, the selected study sample consists exclusively of monumental buildings, while rammed earth constructions of vernacular architecture fall outside the scope of this paper. The research presented here consisted in reviewing the contents of the archives of the Ministry of Development, with a view to providing an initial approximation to the interventions on buildings made of rammed earth in the last ten years (from 2004 until the present day) and funded by the Spanish government. The paper strives to provide a preliminary global analysis that will make it possible to draw certain more specific conclusions about intervention criteria and the construction solutions adopted in the different interventions.

1 INTRODUCTION

1.1 *Aims of the study*

This work is part of a broader research plan addressing the study and analysis of the interventions performed on rammed earth architecture in Spain in the last thirty years, as part of a research programme funded by the Ministry of Science and Innovation "The restoration of rammed earth architecture in the Iberian peninsula. Criteria, techniques, results and perspectives" (ref. BIA 20210-18921; chief researcher: Camilla Mileto).

To carry out this research, the ministries of the Spanish Government connected with interventions on heritage, mainly the Ministry of Culture and the Ministry of Development, were the major source of information. It is important to point out that, since this research focuses on the analysis of interventions paid with public funds, specifically state funds, the buildings studied are monuments, while vernacular rammed earth buildings fall outside the scope of this work.

The specific objective pursued in this particular study was to analyse the interventions carried out on monumental rammed earth architecture in Spain in the last decade according to the 1% Cultural Programme of the Spanish Government.

1.2 *The 1% Cultural Programme*

The 1% Cultural Programme is currently being developed by cooperation between the Ministry of Culture and the Ministry of Development. It began with the 1985 Law of Historic Heritage, which stipulated the obligation of allocating at least 1% of contracts for public works to the conservation or embellishment of the Spanish historic heritage. In this programme, developed in collaboration with the Ministry of Culture, the Ministry of Development generates most of the funds for the 1% Cultural Programme (Sánchez Llorente 2010).

In order to establish criteria and priorities for the interventions to be carried out, an inter-ministerial agreement was reached, setting up a mixed commission between the Ministry of Culture and the Ministry of Development. This cooperation between the two ministries started on 3rd November 1994 and is still in force pursuant to "The 6th collaboration agreement between the Ministry of Development and the Ministry of Culture to act together on the Spanish Historic Heritage according to the 1% Cultural Programme" which was entered into on 15th October 2013 and which increased the sum from 1 to 1.5% for the years to come.

The 1% Cultural Programme, as it has been applied in recent years, is basically put into practice to carry out conservation, restoration and rehabilitation works on buildings belonging to the Spanish historic heritage with an investment even greater than the State Department of Culture possesses for this purpose (Lafuente Batanero 2004).

In this programme, the main applicants for the subsidy are local administrations (town and city councils), although some provincial or autonomic administrations also apply (Junta de Andalucía, Diputación de Castellón or the Government of Aragón, among others). Some applications have also been made by religious foundations and congregations.

1.3 *Research methodology*

In order to carry out the analysis of the interventions on rammed earth architecture in the last ten years funded pursuant to this programme, the main source of information was the archives of the Ministry of Development, which contain the intervention projects that fall under this programme since the mixed commission was set up in 2004 until today. From the general list of interventions performed during this period (with a total of 627 dossiers), we have selected 77 works on buildings originally made with the rammed earth technique as our study sample.

2 APPROXIMATION TO THE STUDY CASES

2.1 *Distribution of the study cases over time*

The intervention projects forming part of our study sample took place between 2004 and 2011. If we compare the total number of interventions carried out each year with the number of interventions selected (those performed on rammed earth buildings), a fairly direct relationship can be established between them (Fig. 1). The group of interventions analysed comes to approximately 12% of the total.

2.2 *Geographic distribution of the study cases*

As regards the geographic distribution of the sample, the 77 dossiers are spread all over Spanish territory, with the autonomous communities of Andalusia (with 25 interventions) and the Valencian Community (with 28 dossiers) heading the list of interventions on rammed earth, followed by the other communities (Murcia, Aragón, Castilla La Mancha, Castilla León, Catalonia and Extremadura) (Fig. 2).

This geographic distribution is not a random occurrence, it responds largely to the territories in the Peninsula where there are more examples

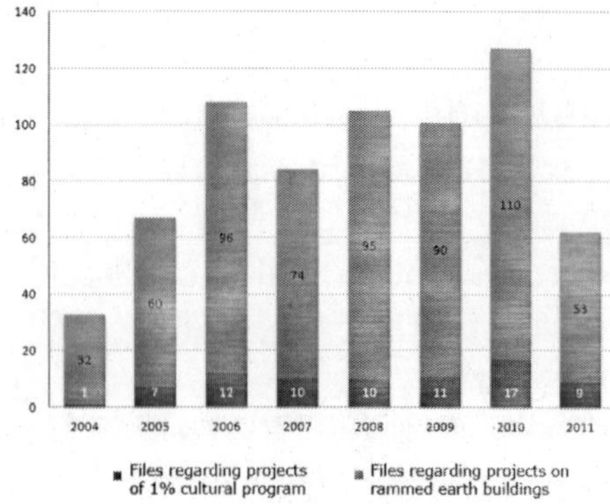

Figure 1. Graph of the temporal distribution of the dossiers (L. García Soriano).

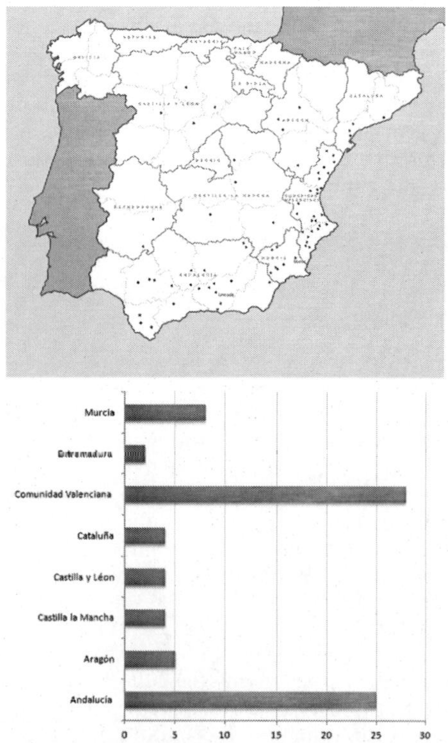

Figure 2. Geographic distribution of the study cases in autonomous communities (L. García Soriano).

of monuments made with this building technique, rammed earth monumental constructions.

As regards the typology of the buildings analysed, the sample is made up almost exclusively of military constructions, with a total of 56 works on castles, 6 on towers and 4 on battlements. This is due partly to the existence of the programme "Urban battlements, castles and other elements of defensive architecture" within the 1% Cultural Programme.

The sample also comprises some examples of interventions on civil and religious architecture attaining to other programmes like "Cathedrals, abbeys, monasteries and convents", but this group is a minority, with a total of 5 interventions on religious buildings (mostly convents) and 6 on civil buildings.

3 ANALYSIS OF THE CASE STUDIES

To carry out the analysis of the case studies, the approach to this work was made from two different fronts: the intervention criteria used as a guideline for each of the works and the constructive techniques and materials used in them.

3.1 *Analysis of the intervention criteria*

It is important to take into account that the intervention criteria do not always appear with this label in the dossiers of the projects analysed. In some cases it is possible to find explicit references to these criteria in the text, but not always, and it is only possible to discover the reasons for applying them by studying the works performed.

To prepare this analysis, the specific intervention criteria were divided into four groups: consolidation, reintegration, reconstruction and demolition. Of these four big groups into which the criteria have been divided, the most common practices are basically the consolidation and reintegration of the fabrics, with some reconstruction works, especially on the copings. An example of this type of intervention is the one carried out on the Castillo de Illora (Granada) in 2011, where the report states that "it is advisable to understand restoration mainly as protection and conservation. This protection and conservation does not exclude some strictly restorative operations that make it possible to save the stylistic, historic, spatial and architectonic image of the monument" (dossier 01-18192-02099-11). Furthermore, as we mentioned above, these consolidation works are usually accompanied by small reconstruction works, in some cases minimal, as in the intervention on the Castillo de Penella (Cocentaina, Alicante) in 2003, which expressed the desire "to maintain the current volume without regularising merlons and without performing permanent rebuilding on any part of the castle itself" (dossier 17-03056-00023-04) (Fig. 3).

The smallest group is made up of interventions driven by the criterion of reconstruction, where the intention is to restore the building to its original state, restoring the walls to their full height, in some cases without rebuilding the coping, like in the works carried out on the Castillo de Calatayud in 2009, whose project states that "the crowning of the parapets will be left flat, without merlons, for despite the fact that the graphic documentation conserved shows that the citadel was castellated, since there were no remains to go by, the walls shall be crowned without merlons" (dossier 02-50067-01964-10), although there have been small interventions that even reconstructed the crowning of the coping. This is the case, for example, of Javier Borobio Sanchiz's intervention on the Castillo de Cadrete, where he says "the final result we wish to obtain with the intervention is to retrieve the original shape of the tower (...) We propose to complete the fallen walls, from the base to the crown, using the same materials as the original. This intervention will include the merlons according to the original model" (dossier 02-50066-01283-06). Another example where reconstruction was contemplated was on the Castillo de Almonecir in 2007, where they speak of "reconstruction works on the rammed earth walls degraded by the effects of the elements on the mortar, with a serious loss of bonding in places and the disappearance of a large section of masonry" (dossier 17-12125-00189-04) (Fig. 4).

Figure 3. Image of the exterior of a wall of the Castillo de Penella in Cocentaina (Alicante, Spain) after the intervention (L. García Soriano).

Figure 4. Image of the exterior of a wall of the Castillo de Almonecir (Castellón, Spain) (L. García Soriano).

As regards the parameters concerning aesthetic issues, it is very common that the cases studied pursue the criterion of distinguishability. The newly added elements (reintegration or reconstruction fabrics) in most cases stand out very noticeably from the rest but seek intentional integration (according to the principle of neutrality) by means of materials, colours or textures. In many cases, chromatic integration of the new structures is sought, but always reinforcing the criterion that they must be distinguishable from the existing parts. For example, in Alba Soler Estrela's intervention on the Castillo de Alcalá de Gallinera in 2006, the dossier states that "the materiality shall be conserved as much as possible, the intervention shall be recognisable, avoiding falsification, keeping a harmonious distance" (dossier 17-03136-01126-05).

In the intervention carried out on the Castillo de Cervera del Maestre in 2009, this criterion is also expressed in the text and in the report it states that "the intervention will use the same elements and the same techniques as the original (...) All the same, it must be respected and historic falsification must be avoided by means of using certain elements of colour, texture, separation, etc. to distinguish the existing parts from the newly built ones" (dossier 17-13044-01718-19).

Nevertheless, despite the fact that distinguishability is usually an objective in the intervention, in nearly all the cases analysed the same materials as the original are used, like, for example, in the intervention performed in 2010 on the Castillo de Miraflores in Piedrabuena (Ciudad Real), in which Javier Navarro Gallego said "in those areas where missing volumes or surfaces must be replaced, reintegration shall always be carried out by using materials identical to the existing ones, safeguarding in any case the identity of the original building" (dossier 08-13063-02068-10).

Regarding the criterion of minimal intervention, in some dossiers they go so far as to say that the intervention on the building should only consist in what is strictly necessary, as it says in the dossier of the works on the Torre de Agua in Segura de la Sierra (Jaén) in 2008, namely "the principle of minimal intervention is of transcendental importance. Any manipulation of the building involves a risk, so only what is strictly necessary must be done, and the natural degradation caused by the passage of time must be assumed" (dossier 01-23081-01879-09). Another example where this criterion is mentioned is the intervention on the Castillo de Alcalá de Guadaira in Seville in 2006, where it states "work must only be carried out on the parts that require it to obtain the objective sought" (dossier 01-41004-01572-08) (Fig. 5). However, this criterion is very subjective and depends on the judgement of each author and, even though the initial intention may be to carry out only indispensable tasks, on many occasions the final result of the intervention is very different from this original intention.

Figure 5. Image of the exterior of the Castillo de Alcalá de Guadaira (Seville, Spain) after the intervention on the walls (L. García Soriano).

It is important to bear in mind that the interventions analysed are contemporary (mainly from the first decade of the 21st century), and the criteria applied to restoration at this moment are very similar. Therefore, the possible variety of interventions and criteria depend basically on the judgement of the author or the personal interpretation made of them in each case.

This reflection and analysis of the criteria is fundamental, because they will dictate the formal choice of intervention concerning the tasks proposed and the constructive techniques and materials used in the new works, the next step in the analysis of this work.

3.2 *Analysis of the constructive techniques proposed in the interventions*

The intervention constructive techniques used in the different dossiers can be assured to be closely linked to at least two factors: the original constructive technique and the author's own intervention criteria. Nevertheless, it is important to bear in mind that these interventions tend to be directed specifically at the pathologies that exist in each case.

In military and defensive architecture, the group to which most of the case studies belong, the walls are usually the main problem and the works are performed directly on them. As we have seen above, the general attempt to make the new intervention fit in with the pre-existing building in pursuit of constructive harmony, chromatic and aesthetic integration, etc. calls for the use of the original building technique, rammed earth, as the intervention method. Therefore, the interventions analysed in this study were nearly all carried out in this way.

In some cases, the objective of using the same constructive technique also requires the use of the original material, as in the case of the 2004 intervention on the Castillo del Piñar, where the report states "the original technique shall be used with wooden planks or formwork and aggregate and lime filling. Parts of the old wall lying at the foot of the castle may be used as aggregates. The size of the caissons, the layout of the pins, the height of the caissons, etc. will depend on the walls that are to be repaired or reconstructed" (dossier 01-18159-01523-07) (Fig. 6).

Another example following this criterion of using the same materials according to the place of the works is the intervention carried out that same year on the Castillo de Nogalte (Puerto Lumbreras, Murcia), where it says "once all the existing structures have been consolidated, a layer of screed will be applied. For this task the traditional building method will be used, with solid lime-crete poured inside the formwork on the oldest part and lime-crusted rammed earth on the extension, just as each of them were built" (dossier 13-30033-00562-04) (Fig. 7).

In some cases, although the general trend was to use the original materials, new materials like

Figure 6. Image of the Castillo del Piñar (Granada, Spain) after the intervention (website of the Ministry of Development).

Figure 7. Image of the Castillo de Nogalte in Puerto Lumbreras (Murcia, Spain) after the intervention (website of the Ministry of Development).

cement were used either out of ignorance or scepticism; on the one hand, the author may be more familiar with this material, but, on the other hand, it is a far cry from the original construction.

We can cite some examples where this method was used, such as the intervention carried out in 2010 on the Castillo de San Juan in Calasparra, where it stated "the rammed earth walls will be rendered in the traditional manner, in a similar way to the existing ones: a paste made with natural soils, stabilised with slaked lime and a small proportion of white cement and a colorant if necessary" (dossier 13-30013-01908-09). In some cases, even though the intention was to use the same technique, it was adapted to today's means to reduce costs, as in the case of the 2006 intervention on the Alcazaba de Reina, where it says "building with those same materials today, handling them as they used to do years ago, is economically impossible" (dossier 01-06110-01440-06).

Another interesting aspect worth mentioning is that, in spite of the fact that the most common intervention method used in the cases analysed is by following the original constructive techniques, the conflictive point open to different interpretations and proposals is the union of the original material and the new one, which will be different according to the constructive detail that each author puts into practice.

Therefore, there are interventions where this combination is enhanced by executing a certain physical layout between the original and new materials, which somehow require partial demolitions in the surfaces of the walls in order to make these unions.

On the other hand, the most frequent proposals opt for improving the adhesion of the new materials to the old ones by increasing the connecting surface with elements like rods, although the most traditional methods use timber stakes to execute these unions. However, in most of the cases analysed they opted for the incorporation of modern materials, mostly fibreglass. An example of these interventions is the work carried out in 2006 on the walls of Marchena, where they proposed that "the reconstruction be carried out on the outside with materials similar to the existing ones by using plank moulding and metal anchors and combining the old bonding with the new manually with filling material and using fibreglass mesh" (dossier 01-41060-01285-05).

Another example is the above mentioned intervention on the Castillo de San Juan in Calasparra, where the proposal was that "the new rendering should be anchored to the naked rammed earth wall by means of fibreglass rods (10 mm) and, if necessary, mesh made of the same material, while attempting to ensure the chromatic integration of the ensemble" (dossier 12-30013-01908-09).

4 CONCLUSIONS

After preparing this general analysis of study cases regarding intervention criteria and constructive techniques, the following conclusions can be drawn.

With regard to the constructive techniques, the fact that traditional methods are commonly used is interesting, for one of the objectives most often pursued is to attain an exterior image similar to the traditional material, but, on the contrary, the unions between the original material and the new are practised in most cases with modern-day materials. Therefore it can be said that, whereas in the direct work on the walls, there is a general tendency to use traditional constructive techniques (rammed earth) with the incorporation in specific spots of new materials, perhaps because they are more familiar and readily available, and, on the other hand, in the implicit belief that they have better technical qualities and enhance the adhesion of the new walls.

As far as the intervention criteria are concerned, reintegration and consolidation are the most common tasks, almost always accompanied by little reconstruction works, mostly in the coping, which, as well as completing the volumes are understood as a proposal that responds to the need to protect it (Fig. 8).

With this overall view, we can say that this first decade of the 21st century was a period during which a large number of interventions were performed on rammed earth buildings as part of the works paid by the 1% Cultural Programme, and these interventions were spread rather homogeneously over the whole Peninsula. Nevertheless, this first examination and analysis has provided general information about a group of interventions and, for that reason, it must be seen as a first step in a broader research programme that is still under way.

Figure 8. Image of the Castillo de Alcalà de Xivert (Castellón, Spain) after the intervention (L. García Soriano).

ACKNOWLEDGEMENTS

In order to carry out this research, we are grateful for the help of Rita Lorite, coordinator of the 1% Cultural Programme, and Laura Collado, Head of the Documentation and Archive Department, which has been fundamental in giving us access to the archives of the Ministry of Development concerning the works funded under this programme, at the Sub-Department General of Architecture and Building.

REFERENCES

Dossier 02-50066-01283-06 of the Ministry of Development Archives. Consolidación y Restauración del Castillo de Cadrete y excavaciones arqueológicas complementarias (Cadrete–Zaragoza).

Dossier 17-03136-01126-05 of the Ministry of Development Archives. Acondicionamiento de accesos y consolidación de muros en el Castillo de Alcalá de Gallinera.

Dossier 08-13063-02068-10 of the Ministry of Development Archives. Intervención en el Castillo de Miraflores para la consolidación de los Muros y puesta en valor del recorrido interior.

Dossier 17-03056-00023-04 of the Ministry of Development Archives. *Restauración del Castillo de Penella*

Dossier 17-12125-00189-04 of the Ministry of Development Archives. *Rehabilitación del Castillo de Almonecir.*

Dossier 17-13044-01718-19 of the Ministry of Development Archives. Consolidación de la torre y muralla norte del Castillo de Cervera del Maestre, Fase III.

Dossier 02-50067-01964-10 of the Ministry of Development Archives. Restauración del Torreón Este y Lienzo de la Muralla contiguo del Castillo Mayor de Calatayud (Calatayud–Zaragoza)

Dossier 01-23081-01879-09 of the Ministry of Development Archives. *Restauración de la Torre del Agua (Segura de la Sierra–Jaén)*

Dossier 01-41004-01572-08 of the Ministry of Development Archives. Restauración del Castillo de Alcalá de Guadaira, Frente Norte de las Alcazabas Occidentales (Alcalá de Guadaíra–Sevilla)

Dossier 01-18159-01523-07 of the Ministry of Development Archives. *Consolidación y Restauración del Castillo (Píñar–Granada).*

Dossier 13-30033-00562-04 of the Ministry of Development Archives. *Rehabilitación del Castillo de Nogalte o de Puerto Lumbreras.*

Dossier 10-06110-01440-06) of the Ministry of Development Archives. Rehabilitación de la Alcazaba Arabe, Fase I: Consolidación del Frente Noroeste, Reina (Badajoz).

Dossier 13-30013-01908-09 of the Ministry of Development Archives. Rehabilitación y Consolidación del Castillo de San Juan, Calasparra (Murcia).

Lafuente Batanero, L. 2004. Las medidas de fomento. Aplicación de la nueva Ley de Mecenazgo en los museos [*Website of the Ministry of Culture: General Office of State Museums*], [Online].

Sánchez Llorente, A., 2010. El 1% Cultural. Una visión práctica. *Revista de Patrimonio Cultural de España* 3 (La economía del patrimonio cultural): 129–142.

Earthen Architecture: Past, Present and Future – Mileto, Vegas, García Soriano & Cristini (Eds)
 ISBN 978-1-138-02711-4

Early modern period adobe in Sicily: Recent finds

M.L. Germanà
Dipartimento di Architettura, Università di Palermo, Italy

ABSTRACT: Earthen architecture in ancient Sicily can be linked to a technological continuity common to the whole Mediterranean basin. The island possesses numerous finds, dating back to periods ranging from pre-history to the Hellenistic-Roman age, during which *adobe* was a widespread technique. The archaeological field had been the only one investigated, also due to the fact that more recent evidence had shown that the use of earth as a principal element in wall-building was unknown in Sicily. Recently, traces of *adobe* have been discovered in the historical centre of Cefalù, on the northern coast of Sicily; these few scraps are to be found in residential buildings made primarily of stone. Albeit modest in quantity, these traces are important for our historical knowledge and to reinforce the hypothesis of contemporary earthen architecture in Sicily.

1 INTRODUCTION

Earth has been detected in buildings dating back to various epochs and in various geographical areas. Earth is a building material that is still pertinent at all latitudes, and has negligible environmental impact; one has only to think of its widespread availability and the consequent reduction in transportation, the extremely inconsequential amount of energy required for its transformation, the elimination of the need for disposal, recycling of waste and dumping. In the last decades great progress has been achieved in the contemporary use of earth in building, though not in Italy, where the material and (perhaps more) non-material values of the links with the contextual reality hinder the use of earth building techniques in ordinary situations. Compared with others Italian Regions, this is more evident in Sicily, where the typical *immediacy* of earth building techniques (deeply rooted in the physical and cultural identity of the productive context) has been lost, because of the lack of flourishing building traditions.

Any hypothesis for bringing earthen architecture into present-day utilization needs to connect to local roots, if we wish to elude dreamland; for this reason, in order to improve historical knowledge it is important to reinforce it.

2 EARTHEN ARCHITECTURE IN ANCIENT SICILY

A National Research Project financed in 2005 by MIUR (Scientific, experimental and tacit knowledge and conservation actions of Earth Architectural Heritage in Southern Italy; Programme Earthen Architectural Heritage in Sicily: knowledge and conservation processes) provided the opportunity to define a less fragmentary framework for earthen architecture in ancient Sicily, a kind of constructed heritage technique that is almost extinct and certainly less well-known than the same building material employed in more exotic contexts (Germanà 2009; Mecca et al. 2011). Before these studies, we only possessed rather scanty and partial knowledge. Sicily was almost entirely missing from the mosaic of earthen architecture in Italy, although there have been numerous occasions for comparing notes and disseminating results, as a consequence of which knowledge about many other regional situations has been consolidated, following research-lines such as: typological and technological investigation; the study of present conditions and particular conditions of deterioration and pathology; analysis of the mechanical behaviour of walls; possible legal repercussions, for the salvaging of existing constructions and the building of new ones (Bertagnin 1999; Scudo & Sabbadini 1997).

A brief text in 1999 outlined the state of the art regarding knowledge of the earthen heritage in Sicily: the well-known 4th cent. BC fortifications of Capo Soprano at Gela; some pani di terra (lit. cakes of earth) cemented together with lime and clay mortar, cropping up occasionally in Cefalù in 17th/18th cent. religious buildings; a few observations regarding pietra e tayu (lit. stone and mud), emerging from the archive studies of Henri Bresc (Mungiguerra 1999). Now that local archaeologists from the Sicilian heritage offices (e.g. Sebastiano Tusa, Rosalba Panvini and Francesca Spatafora) have become involved in research, Sicily is being described as a privileged field of observation, with regard to the ancient employment of earth for building purposes. Numerous finds, dating back to periods ranging from the Mesolithic to the Hellenistic age and representing various building typologies (residential constructions; craftworks areas, fortifications, religious buildings and graves), show the extent to which earthen architecture was widespread in Sicily. In addition, protective interventions carried out in the last fifty years have provided further interest, demonstrating the peculiar problems of conservation of these finds; these are particularly delicate, because of the intrinsic inconsistency of the building material, which is only durable if subjected to maintenance and adequately protected from moisture (Germanà 2011).

In the earliest examples and in subsequent developments in Sicily, earth was used in a mixed technique, along with other raw materials, ranging from stone to vegetal; depending on their on-site availability, these were employed in accordance with a *technological continuity*, common to the whole Mediterranean basin, deeply rooted in the Mesopotamian, Egyptian and Minoan-Mycenaean cultures (Tusa 2011; Spatafora 2011).

When the island opened up its frontiers to non-indigenous migrations, (e.g. Phoenician-Punic and Hellenic), ancient Sicily adopted adobe, a common technique in the Mediterranean basin from the 7th cent. BC onwards (de Chazelles 1995; de Chazelles 2011), of which many more examples, and better-preserved ones, have come down to us.

The best-known example is certainly the fortification of *Capo Soprano* at Gela (a monumental structure—over 350 metres of walling—which was unknown until 1948 because it was covered by a layer of sand over 12 metres deep, and which has also attracted interest as a result of the particular conservation problems). The utilisation of earthen bricks is also evident in the buildings on the same site; for instance, the *Emporio* of *Bosco Littorio*, dated 6th BC, where almost three-metre high adobe walls remain standing (Panvini 2008). Other sites, such as, for example, San Cataldo, Eraclea Minoa (Fig. 1), Solunto, Mothia, Marsala, Kamarina prove that adobe was widely used all over the island until the Hellenistic-Roman period.

The widespread availability of stone in most of the region may have been a crucial factor in Sicily not utilising earth as a principal element in masonry techniques.

However there remains the question as to why there was a clean break with building technique such as adobe, which was also widely used all over the island until the Hellenistic-Roman period. This interrogative gathers strength from an observation from neighbouring Calabria, where the same tradition of earthen bricks in the archaeological field was subsequently carried on until the

Figure 1. Eraclea Minoa (Agrigento). Domus, 2nd/1st century BC (photo by M.L. Germanà, 2013).

last century, and to such an extent that it is still present in numerous relatively recent constructions (Cavalcanti & Chimirri 1999). The scarcity of documented material evidence seems to prove that, during the medieval period and onwards, in Sicily, only in a few cases did walls continue to be built entirely from earth. The term *tabia* has been found in medieval archive sources, indicating walls built by a sort of rammed earth technique (Pezzini 2003). More well-known, and certainly more common in Sicily, were the masonry *pietra e tayo* (lit. stone and mud; rubble masonry) and *a sacco* (rubble masonry used as an infilling between cut stone walls). In these kind of masonry, earth was used not as a principal component, but in an integrating role for the stone elements (Germanà 2011b).

3 RECENT FINDS OF EARTHEN BRICKS FROM THE EARLY MODERN AGE IN SICILY

A document dated 1561 provides information that a potter who usually produced baked bricks, working in a mountain village in Palermo province, was obliged to supply forty thousand earthen bricks along with six thousand unfired bricks with straw (Fatta 2011). This somewhat isolated evidence could prove that earthen bricks were used in Sicily during the medieval and modern age as a low cost technique, and as an alternative, or in addition, to baked bricks or stone elements. This perception is now supported by direct evidence: new studies are providing knowledge about a sort of adobe technique found in buildings of the early modern age in Cefalù, on the northern coast of Sicily.

These advances have been possible thanks to several fortuitous and lucky circumstances and thanks to the collaboration of Fabio Vaccaro (Vaccaro 2012).

The first circumstance was hearing that the data regarding the *pani di terra* (Mungiguerra 1999) had in fact been provided by Pasquale Culotta, the well-known architect (also Dean of the Faculty of Architecture in Palermo), highly active in Cefalù until his sudden death in 2006. Thanks to his daughter Tania, photos were found in Culotta's archive, dated 1995, showing a wall that seems to be made of earthen bricks (Fig. 2). These finds are no longer visible, and it has not been possible to ascertain whether or not they still exist.

The second circumstance occurred in 2010, during works aimed at creating bed & breakfast accommodation in an inconsequential building in the historical centre of Cefalù, when a wall in adobe was found by Mario Caliò and Monica Guercio, architects skilled in bio-architecture (Figs. 3–4). This wall stands in the second, and partly the third, of the three storeys of the building, which is made mainly in stone (Fig. 5). One might surmise that, in the absence of financial means, the adobe wall was built to meet the need for more space.

Figure 2. Inner wall in historical building in Cortile Rosariello, near S. Domenico (17th/18th cent. AD) in Cefalù. Photos dated 1995, found in Archivio Culotta in 2012.

Figure 3. Wall in the second storey of a historical building in via Spinuzza in Cefalù (photo by M. Caliò & M. Guercio, 2010).

Figure 4. Third storey of a historical building in via Spinuzza a Cefalù; on the lower right, the wall in adobe (photo by M. Caliò & M. Guercio, 2010).

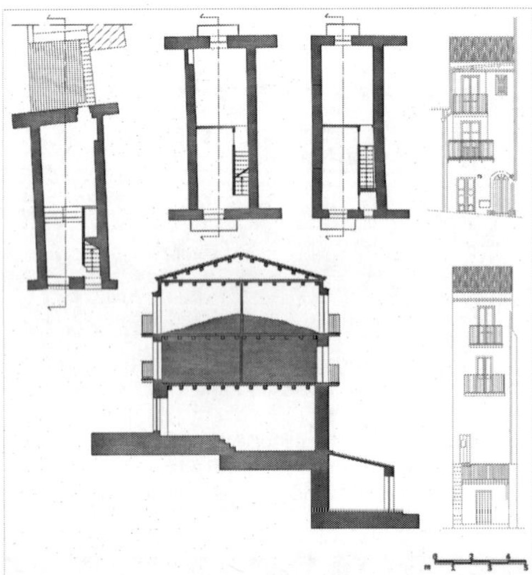

Figure 5. Building in via Spinuzza a Cefalù; in darker grey the wall in adobe in the second storey and in part of the third one (drawings by Fabio Vaccaro, 2012).

Now the adobe wall is covered by plaster, and only in two little recesses has it been left visible. In 2012 a sample was removed (Fig. 6), to examine the components of the mixture forming the bricks (whose dimensions are about 8 × 18 × 40 cm). Preliminary analysis showed that slightly clayey earth was kneaded with sand and with heterogeneous aggregates: straw and terracotta fragments, bone splinters and slivers of wood, pieces of reed or little shells (Fig. 7). Radiocarbon dating carried out by CEDAD (CEntro DAtazione e Diagnostica, University of Salento) on the vegetable fibres contained in the sample, date this find to a period range most probably around 1640 AD.

Figure 6. A piece of the sample of adobe extracted in via Spinuzza a Cefalù (photo by F. Vaccaro, 2012).

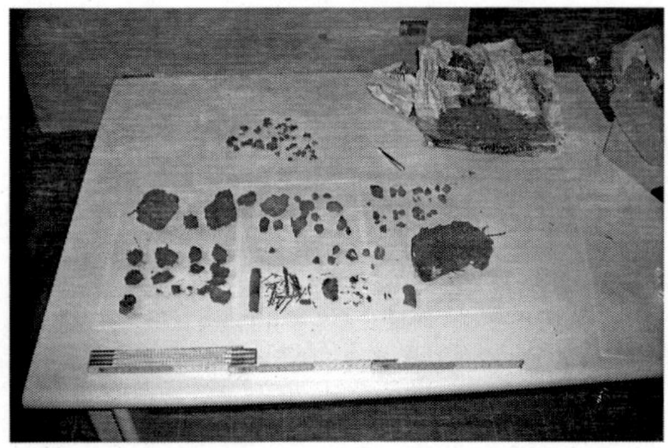

Figure 7. Aggregates found in the sample extracted in via Spinuzza a Cefalù (photo by F. Vaccaro, 2012).

The sample removed from via Spinuzza had a surprise in store: for nearly four centuries, the brick mixture had hidden a small head of a papier-maché Christ crucified (Fig. 8). This unexpected discovery was surprising because of its fortuity and there was consistency with radiocarbon dating. In fact, it is possible to link this find to one of the more evident effects of the Counter-Reformation, i.e. the accent placed on the educational purposes of Christian images and their consequent widespread diffusion among the lower social classes.

The conditions of the fragment and its combining with the brick mixture demonstrate that the Christ crucified had been added before drying. Apart from the underlying significance of salvation in the image (Caiazza 2007), this might suggest a sort of propitiatory custom in construction sites.

In 2013 Fabio Vaccaro found other traces of adobe in Cefalù, whilst browsing around, as was his wont, during works on historical buildings. These, along with the previously found traces, are in the same area of the town, which had flourished in the early modern age (Fig. 9); until now, only

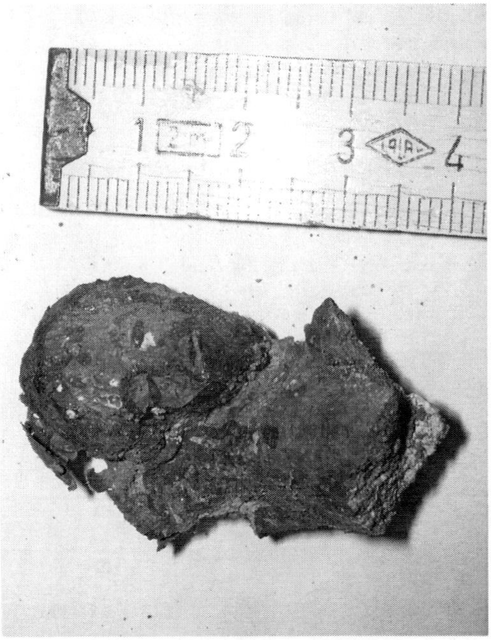

Figure 8. Crucified Christ papier-mache fragment found in the sample from via Spinuzza, Cefalù (photo by F. Vaccaro, 2012).

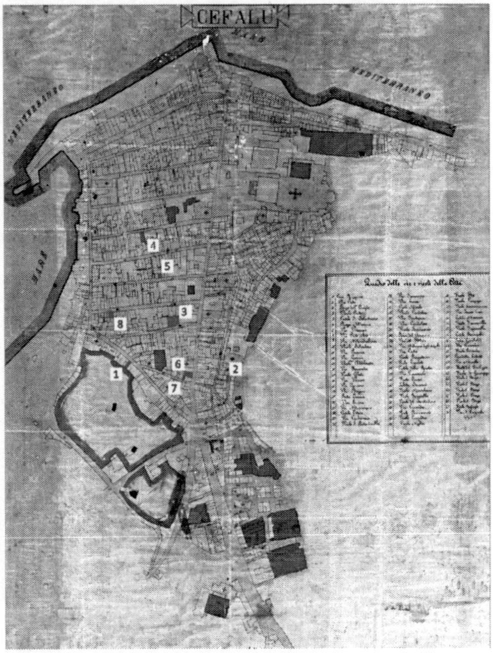

Figure 9. The town of Cefalù in a cadastral map dated 1850/60, reporting the finds of adobe traces.

Figure 10. Filling made of earthen bricks over an arch, in a building in via Botta in Cefalù (photo by F. Vaccaro, 2013).

Figure 11. Empty in a stone-wall, filled-up by earth bricks, in a old building in via Vanni, Cefalù (photo by F. Vaccaro, 2013).

Figure 12. Solunto, Isolato 7 UA XIV. *Casa a peristilio*, I/II sec. BC (photo by M. Schiera, 2006).

little cavity fillers or completion of stone-walls have been found (Figs. 10–11).

4 CONCLUSIONS

By comparing the direct evidence of utilization of adobe in ancient Sicily with the recent finds described here (Figs. 11–12), it has been proved that this building technique (at the same time both remote and current) was used in Sicily for a period of over two millennia.

Albeit modest in amount and in quality, these traces are important for historical knowledge and to reinforce the hypothesis of contemporary earthen architecture in Sicily, thanks to the individuation of the necessary, specific, local roots.

NOTES

The contents of this paper were partially presented at the International Conference "Rencontres d'archéologie et d'histoire sur les constructions en terre crue", held in Pont du Gard (F) on 7–9 December 2012.

REFERENCES

Bertagnin, M. 1999. *Architetture di terra in Italia. Tipologie, tecnologie e culture costruttive*, Monfalcone (GO): Edicom.

Caiazza, G. 2007, E il Verbo si fece… Immagine. Dai simboli cruciformi ai crocifissi tridimensionali. *Bollettino del Gruppo Archeologico Aquileiese* (XVII): 3–17.

Cavalcanti O. & Chimirri R. 1999. *Di Fango, di Paglia… Architettura in terra cruda in Calabria.* Soveria Mannelli (CZ): Rubbettino.

De Chazelles, C.A. 1995. Les origines de la construction en adobe en Extrême-Occident. *Etudes Massaliètes* (4): 49–58.

De Chazelles, C.A. 2011. La construction en brique crue moulée dans les pays de la Méditerranée, du Néolithique à l'époque romaine. In C.A. de Chazelles, A. Klein, N. Pousthomis (eds), *Le cultures constructives de la brique crue*: 153:164. Montpellier: Editions de L'Espérou.

Fatta, G. 2011. Earth in Sicilian wall: notes in progress. In Mecca *et al.* (eds): 291–299.

Germanà, M.L. 2009. Terra cruda nel patrimonio architettonico siciliano: conoscenza e conservazione. In M. Achenza, M. Correia, H. Guillaud (eds), *Mediterra 2009, I Conferenza Mediterranea sull'architettura in terra cruda*: 151–158. Monfalcone: Edicom.

Germanà, M.L. 2011a. Earth in ancient Sicilian architecture. In Mecca *et al.* (eds): 166–188.

Germanà, M.L. 2011b. *A still to be explored heritage: stone and earthen architecture in Sicily*. In Mecca *et al.* (eds): 280–290.

Mecca S., *et al.* (eds) 2011. *Earth/Lands. Earthen Architecture of Southern Italy.* Pisa: ETS.

Mungiguerra, C. 1999. Sicilia: dalle mura di Gela alla 'pietra e tayu'. In Bertagnin: 247–249.

Panvini, R. 2008. Strutture in mattoni crudi dell'antica Gela. In M.L. Germana & R. Panvini (eds). *La terra cruda nelle costruzioni: dalle testimonianze archeologiche all'ar-chitettura sostenibile*. Palermo: Nuova Ipsa: 87–98.

Pezzini, E. 2003. Alcuni dati sull'uso della terra nell'architettura medievale a Palermo: fonti documentarie e testimonianze materiali. In Fiorillo R. & Peduto P., (eds), 2003, *Atti III Congresso Naz. di Archeologia medievale*, Salerno 2–3/X/2003: 624–628.

Scudo G. & Sabbadini S. (eds) 1997. *Le regioni dell'architettura in terra in Italia*. Rimini: Maggioli.

Spatafora, F. *et al.* 2011. The use of earth in central-western Sicily: attestations and documentary evidence. In Mecca *et al.* (eds): 201–225.

Tusa, S., 2011. Origins and initial developments of Sicilian earthen architecture in the Mediterranean context. In Mecca *et al.* (eds): 195–200.

Vaccaro, F. (2012), *Cefalù in* adobe*: testimonianze in terra cruda* report for the final exam of first degree course in *Restauro, Recupero e Riqualificazione dell'Architettura* at University of Palermo, supervisor M.L. Germanà.

Earthen Architecture: Past, Present and Future – Mileto, Vegas, García Soriano & Cristini (Eds)
© 2015 Taylor & Francis Group, London, ISBN 978-1-138-02711-4

The *Ksar* Sidi Bou Abdellah in Mdagra Oasis, Morocco

T. Gil Piqueras & P. Rodríguez-Navarro
Universitat Politècnica de València, Valencia, Spain

ABSTRACT: In the southeastern Moroccan pre-Saharan valleys we see how earthen cities (ksar sing., ksour pl.) represent a culture, the Berber culture, which has influenced, through historical events and its social patterns, the development of cities built with earth, following a unique and characteristic urban design. In this writing we will study the Sidi Bou Abdellah ksar, one of the most important ones from the sixteenth century onwards, in the so called caravan route, an archetypal earthen city of the Mdagra oasis, located in the Er Rachidia province in southeast Morocco. From a historical viewpoint, we will start studying its shape and layout, aspects related to its establishment within the territory, its external morphology, internal pattern and facilities, aiming to determine the factors that give this city its Islamic-Berber character.

1 INTRODUCTION

Within the Tafilalt region, one of the most politically historic and significant areas in southern Morocco, we find the Mdagra oasis, one of the largest concentrations of *ksour* in the region, forming a wide and comprehensive group of earthen cities, which to date have remained beyond any study or research and beyond any biased references made about them. These *ksour*, considered being the oldest across Morocco, stand out due to the purity of their architecture and the simplicity of their forms, blending into a living landscape, relating closely to daily life and the environmental conditions.

The Mdagra oasis is located off the upper Ziz River, belonging in its entirety to the province of Er Rachidia (الرشيدية) in the Meknès-Tafilalt region. With a length of approximately 24 km. the Mdagra oasis is situated at the southern edge of the High Atlas, bordering the Khenkh oasis to the north and the Rteb oasis to the south. Its course is almost perfectly straight and runs with a vertical drop of 110 m from the northwest to the southeast, generating around it a fertile area of around 29 km^2 that is characterized for having a great area of palm trees and for its many olive trees, which in turn means that its population participates actively in the trade of dates and oil.

Nowadays even though we cannot state with absolute certainty when the *ksar* is initially set up as a social and urban structure within the Mdagra oasis, we do know that already in the eighth century this area started to become relevant within the business landscape across north Africa, due to being located within the trade route linking Fez with Sijilmassa. The commercial boom led to the development of new communication routes inside the country, specifically through the Tizi n'Talghoumt passage to the north of the Mdagra Oasis, and the lifestyle of the population living around the fertile irrigated areas by large rivers becoming sedentary.

This is how the first cities emerge, cities in which earth is the main material, the result of skilled labor coming from the south. In a rural setting and with a nomadic population, the *Maâlem* have been the ones who have managed and accomplished the construction of *ksour* along this whole region. Some groups that stand out are the Tafilalt *Maâlems,* considered great masters builders, who are credited with building the *ksour* in the Guir, Ziz, Rheris and Drâa valleys (Laoust 1932).

In the case of the Upper Ziz, the first reference to its people dates from the time when the Ait Idrassen confederation of Berber tribes was founded (Couvreur 1968), arising from the sedentary population's need for protection from to the

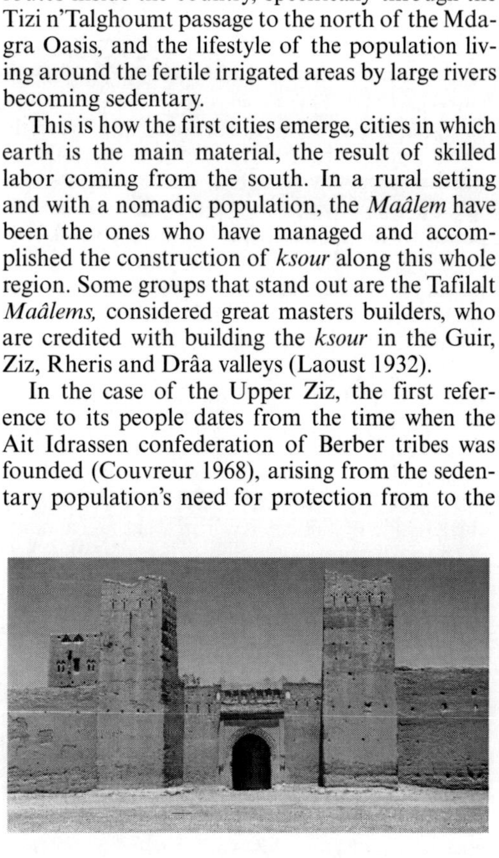

Figure 1. Entrance flanked by towers. *Ksar* Sidi Bou Abdellah (P. Rodríguez-Navarro, 2013).

constant attacks carried out by the nomadic population and the successive invasions from the Arabs coming from the south and east of the continent.

2 *KSAR* DESCRIPTION

The term *ksar* has always been controversial among researchers (Laoust 1934), mainly due to the many transcriptions made of this name into different languages (Arabic, Berber, French or Spanish) and the variety of meanings this term acquires according to the geographic area in which we find ourselves. However, the term *ksar* implies a closed and fortified space where sedentary people seek refuge and protection from external attacks, generally coming from the nomadic population. Usually, several family groups live within the ksar, among which there can be kinship or not, while there is also a range of community services available.

The Sidi Bou Abdellah *ksar*, with its 44,500 m^2, is possibly one of the oldest and most historically important among those in the Ziz valley. Its ground floor, excessively large compared to the other *ksour* in the oasis, has a square shape and is surrounded by a wall which includes a total of 11 tall towers among which there are two placed flanking the entrance door. This door is topped by a pointed arch that leads to two other doors situated in different walls, which in turn lead to two clearly distinctive areas: the social one, where daily activities take place and where the square, the *hammam*, the access to the mosque, the *ksar*'s most important buildings and the access to the housing neighborhoods are located; and the economic and trading area, where the *almácera*, the warehouses and the stables of the *ksar* can be found.

Apart from the photograph taken by Henry Terrasse in 1938 (Terrase 1938), the only three references to this city are given by Charles de Foucauld, Denise Jacques-Meuniè and Roger Mimó. Between 1883 and 1884 Foucauld (Foucauld 1888) went down the High Atlas trail, situating this *ksar* on the left side of the Ziz River in what he called the Metrara district, estimating a population of 300 rifles at that time, considering the number of resident men old enough to fight.

In 1982 Jacques-Meuniè (Jacques-Meuniè 1982) cited the importance of this location within the caravan route, emphasizing its great commercial activity from the sixteenth century, already, onwards, making it an obligatory stopping point for many travelers while being subject of claims from nomadic tribes, accustomed to looting.

Lastly, the writer Roger Mimó named the existence of this *ksar* in his work *Fortalezas de Barro* (Mimó 1996), even though he could not enter inside, since at that time it was still closed.

3 TERRITORIAL ESTABLISHMENT

Nowadays this *ksar* is situated on a sort of barren island within a vast oasis, bordering on the west with arable land although it originally lies outside the fertile region. Even though it is only just 250 m. away from the Ziz's riverbed, it is favoured by the existence of an extensive network of canals that run through the valley. One of them, in fact, runs parallel to the outer face of the west side of the *ksar*, a few meters from it.

The terrain does not aid its defense, as it lies on flat ground, at around the same level as the oasis, although it facilitates some visual control of the territory. It borders to the north with the Taurirt *ksar* and *Zaouia* Moulay Abdellah; to the south with the Titaf *ksar* and to the west with the Beni M'hali *ksar*, providing plenty of land for the population's food supplies.

4 URBAN MORPHOLOGY

4.1 *External shape*

Morphologically, the Sidi Bou Abdellah *ksar* follows a defensive structure model, recognizable in regular patterns and identifiable through simple geometric shapes. The floor plan displays a square morphology, which significantly must have been planned in response to the social structure formed by the members of the Chorfa and Qqebala families. Its four corners, coinciding with the four cardinal points, are limited by high towers nearly doubling the height of the wall; they also stand between the large canvases flanking the entrance. The western side of the *ksar* has a more irregular delineation, adapting to the limits of the territory.

4.2 *Urban layout and functional organization*

Its regular and orderly internal organization, as well as it square external morphology, may initially remind us of the classical Roman city, despite having many differences.

Complying with Islamic law, the city is intenally organized from two clearly defined, and even separate, spaces: the public area where social and commercial relations are established and the private one, the families' housing which is only accessed by inhabitants of the *ksar*. This part is in turn organized into three levels according to social structure. Firstly, the private family housing is organized from lower to higher level, and as a cell it is repeated throughout the *ksar*. By the successive repetition of houses, streets emerge, which eventually come to house families from the same lineage, resulting in neighborhoods inhabited by one single tribe.

The square, located next to the entrance, acts as the urban organizational backbone. In it, the social activities of the *ksar's* inhabitants take place, which explains the existence of a great open space to host parties and social events, where we also find a large public *hammam* equipped with a complex system of heated water, whose main entrance is located on the opposite facade of the square. On the north wall enclosing the square we find a door that leads to the residential area, which is structured around a main road, running from the southeast to the northwest, and which in turn is the basis from which secondary streets originate. In the first section where the mosque is located, we can observe a series of adobe pillars that are successively repeated along this stretch, supporting what originally would have been a structure of porticoes, which cannot be viewed today due to its collapse, giving this space a special relevance while protecting the street from the weather outside. From here on the street moves, in parallel to the south, allowing more privacy for the rest.

Secondary streets emerge, perpendicular to the main street, from four axes, extending in both directions up to the delimiting wall. These streets host the accesses to the houses which seek anew to comply with the precepts of the Koran, relating to two fundamental aspects: privacy of the family environment and equality before God. The first constraint is met by placing the doors of homes not facing each other, moving the axes of these secondary streets at the intersection with the axis of the main street and by trying not to exceed any neighboring property in height. Regarding the second precept, facades present no holes or anything else that stands out on their exterior, apart from a simple ornament around the door. Finally, a striking feature within the Sidi Bou Abdellah *ksar* is the existence of a road round the entire route of the wall, on the inside, allowing access to the defensive towers which unlike in other *ksour*, remain separate from private homes, indicating the possible existence of a body of guards in charge of surveillance and defense.

The width of streets is highly variable from one area to another, with widths of 4.2 m in the barns and *almácera* areas or 3 m in the residential area. In general, all of them are covered up to the first floor, surpassing this level when this space is added for houses that have gained this surface. They only remain open, with no cover, at street crosses, thus allowing light to enter and aiding ventilation. This system of covered streets facilitates life within the *ksar* as it isolates the interior from the harsh weather outside.

The area of political and administrative activity of the *ksar* lies to the east of the square. Here the urban structure changes, varying its outline and even its dimensional features. Next to the wall we find a large building, occupying a great surface, which stands out for its three floors, one more than the other homes in the *ksar*, and for the gaps capped by pointed arches that appear on the top floor, on the corresponding façade outside the *ksar*.

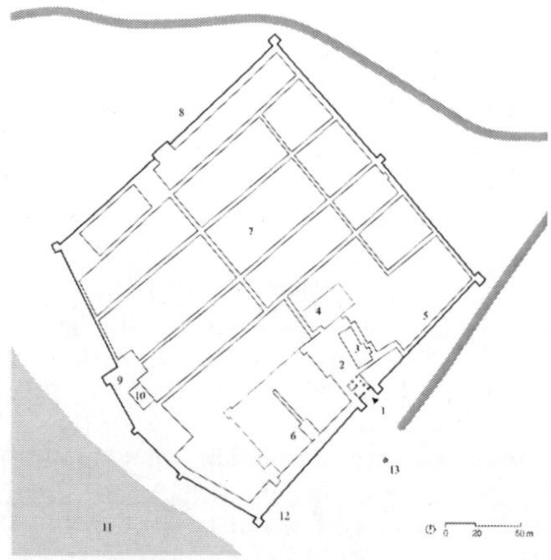

Figure 2. *Ksar* Sidi Bou Abdellah: 1 Entrance, 2 Square, 3 *Hamman*, 4 Mosque, 5 Political and administrative area, 6 *Almácera*, 7 Residential areas, 8 Parapet walk, 9 Cemetery, 10 *Marabout*, 11 Cultivable area, 12 Walls and Towers, 13 Well. (T. Gil Piqueras & P. Rodríguez-Navarro).

In the western part of the large square we encounter the area where the stables, barns and a large *almácera* are situated.

5 *KSAR* FACILITIES

Throughout our various expeditions we were able to inspect, measure, sketch, inventory and evaluate virtually all the existing urban elements within the *ksar*, leaving some to record due to the state of ruin in which they were in.

Afterwards we proceeded to classify them into four groups or units of study according to their functional characteristics, obtaining the following result:

1. Defensive units, which include the walled enclosure that protects the *ksar*, its watchtowers and the entrance door.
2. Residential units, formed by the study of the houses found.
3. Religious units, encompassing all the elements that are directly linked to religion, like the mosque, the cemetery and the *marabout*'s mausoleum.
4. Communitarian units. In this section we included all the elements related to everyday life in the *ksar*, like the square, the *hammam*, the orchard, stables, collective barns and the *almácera*.

5.1 *Defensive units*

5.1.1 *Walls and towers*

The *ksar* is equipped with its own defensive structure to protect its population from possible external attacks.

Thus the first noticeable defensive element is its great wall, which also acts as a boundary between the rural and urban spaces. It is displayed regularly, surrounding the entire urban layout and generating a more or less geometrically square structure. Like the rest of the *ksar* walls, the earthen wall was built on stone foundations in cases where the ground was not sufficiently stable. The total height of the wall is around 7 m, having a thickness, at ground level, of between 0.90 m and 1.50 m.

Slender towers fill the wall's corners, also repeating at half-length and flanking the *ksar*'s entrance. They have a height of 12 m, variable ground floor dimensions with the corner towers being the widest. In the last third of its parament, built of adobe, there are typically simplified crosses with irregular arms, attributed to the Berber culture. Both the towers and the walls are topped by triangle, shaped battlements.

5.1.2 *Entrance*

Entrances are elements that deserve special analysis because sometimes they manage to fulfill three different roles, such as contributing to the *ksar*'s defense system, facilitating social relationships by acting as a meeting point for its inhabitants and serving as a meeting place for the *Jmaâ*, a body that manages the *ksar*'s economic and political life.

The entrance to the Sidi Bou Abdellah *ksar* is located on the southeastern wall, overlooking non-arable land inside the *ksar*. The grandeur of its access is marked by two slender towers that guard the entrance door, as well as an *alfiz* topped by a small rectangular frame molded with tiling, followed by a set of niches closed at the top by lowered arches. Once we gain access we notice that the entrance shows a direct outline forming an enclosed space, delimited by four earthen walls, three hollow passages with their corresponding doors, all of them with a width of 2.60 m: the access door and two more on the way in, one towards the square and the other one towards the *almácera*, barns and stables.

The interior space, with a more or less square layout, is divided into three porticoes. The space generated on the right side is at a higher level than the rest. On the left, on the opposite side to the entrance door, there is a staircase leading up to the top floor.

5.2 *Residential units*

5.2.1 *Housing*

Inside the *ksar*, houses operate as independent cells, linked to a family unit, repeating a more or less common pattern over successive streets.

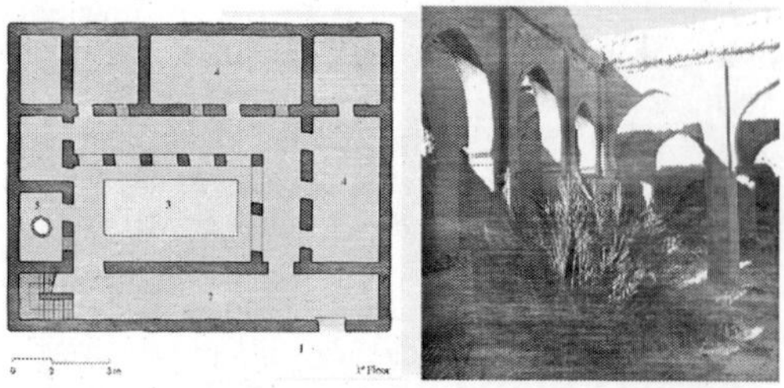

Figure 3. House, *ksar* Sidi Bou Abdellah: 1. Entrance, 2. Hall, 3. Courtyard and pool, 4. Rooms, 5. Well. (T. Gil Piqueras & P. Rodríguez-Navarro).

After studying and analyzing the housing we can determine that they are composed of a ground floor and an upper floor, generally presenting a regular ground floor of approximately 100 m^2, which is very striking compared to the data obtained after visiting other *ksour* in the area and which have smaller homes. Another element that stands out from the rest of *ksour* is the entrance door, which unlike the others has a thick decoration delimiting the entrance gap, making it easily recognizeable from within the, smooth and high walled, long and narrow streets.

Their structure is also different. Perpendicular to the main facade we find two earthen walls which work as partition walls, generating three interior spaces. In the center there is a large courtyard that includes a pool, as nucleus around which the rooms are organized and which are accessed by a series of gaps topped by pointed horseshoe-shaped arches, which in turn have bas reliefs topped by poly lobed pointed arches. This floor although resembling the layout of first period Roman houses (Laoust 1932), differs in that there are no windows towards the outside, not even on the first floor, following the classic pattern of Muslim housing oriented inwards. Thus the separation between the private área, for family use and the public one defined by the street is respected, as observed in the Koran, verses 4 and 5 of the *Sura* XLIX.

In the entrance hall, located before the square, there is a staircase that allows access to the upper floor and which leads to new rooms. Currently this floor is completely destroyed, making it impossible to recognize its spaces.

5.3 *Religious units*

5.3.1 *Mosque*

The mosque follows the classic structure found in other Mdagra *ksour*. It comprises a set of rooms among which we find an entrance hall located next to the main access, which in turn is linked to the prayer hall and the minor and major ablution halls.

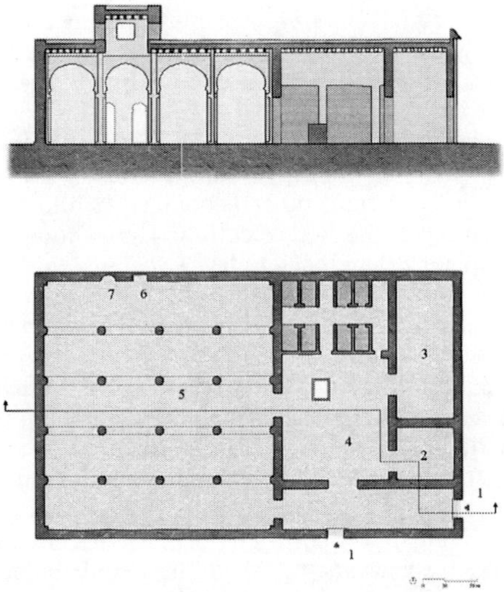

Figure 4. Mosque. *Ksar* Sidi Bou Abdellah: 1 Entrance, 2 Boiler, 3 Warehouse, 4 Ablutions room, 5 Prayer room, 6 *Mimbar*, 7 *Mihrad*. (T. Gil Piqueras & P. Rodríguez-Navarro).

Figure 5. Cemetery and marabout. (P. Rodríguez-Navarro, 2014).

There we find a well, a cauldron, a warehouse for firewood and a staircase from which the minaret could be accessed.

The prayer hall is composed of five naves separated by four porticoes formed by horseshoe arches. The pillars that support them display an octagonal shape and were built in adobe. On the *qibla* wall, to the east, we find the *mihrab*, formed by a large semicircular niche and the rectangular, shaped *mimbar*. Surrounding the *mihrab* we encounter refined decoration consisting of a series of bas-reliefs which by way of fascias, pilasters and niches, repeat geometric motifs, filling all the visible space.

Its location, unlike what usually happens in the Mdagra region, is found in the access area to the residential premises and has an area of 445 m^2. Today it is completely ruined.

5.3.2 *Cemetery*

The cemetery is located opposite the main entrance of the *ksar*. Overall, burials are carried out in the gound, with no headstones or any elements showing the tombs's situation beyond some simple stone slabs arranged vertically, limiting their length.

5.4 *Communitarian units*

5.4.1 *Square*

It is a space for public use, large and open at the top, which is located inside the *ksar* next to its entrance. From the square the main streets originate. Here is where the *ksar*'s community activities take place, so it is surrounded by all those elements that full a social role, like the *hammam* or the *fonduc*, although we do not know it there ever was one here. It sometimes also serves as a small souk. Architecturally, there are no relevant elements.

5.4.2 *Souk*

Due to the great caravans of merchants and the movement of goods that occurred across this region, among which there was also gold and slaves, the souk stood outside the *ksar*, thus being of an itinerant kind. This *ksar* was renowned before the sixteenth century for being, with *Ksar* Es Souk, one of the most important commercial centers in the region (Jacques-Meuniè 1982).

5.4.3 *Stable, poultry yard and collective barns*

Although it is quite common for animals to live with families occupying the ground floor of their houses, this is not the case here. The importance of this *ksar* means there is a clear separation between the rural and residential areas, possibly due to the existence of a marked social hierarchy in which some were responsible for land cultivation and animal care while others with higher social status were engaged in the *ksar*'s administrative and defensive management. The area designated for this usage is especially relevant, involving great agricultural and farming activity.

5.4.4 *Almácera*

In the Mdagra oasis there is a large of amount olive trees, so it is very common to find *almáceras* inside the *ksour*. In this case the *almácera* is located in the southwestern part of the *ksar* and its entrance is situated on the wall which runs parallel to the main wall. Through the wide door we enter a corridor that divides the space into three areas: on the left there is a large space where olives were collected upon arrival; to the right we find the processing area, where there is a large wheel and a weir. At the

Figure 6. *Almácera* (P. Rodríguez-Navarro, 2013).

back we see the main building with its numerous rooms and halls.

The characteristic that most distinguishes its location is the presence of a circular base finished in stone or masonry, in whose center there is a vertical log docking a horizontal one, which in turn passes through a cylindrical grindstone, a similar system to the Roman grinder.

6 CONSTRUCTION SYSTEM

The construction system is typical of the cities of southern Morocco. Walls are made of earth, resting on a base of stones embedded into a trench in the ground, which in this case, as in very few cases, surpasses ground level.

The floors are mainly composed of palm wood beams, cane gridding, an insulating layer made of palm leaves and a layer of compacted soil.

Vertical communication between floors is attained through earthen staircases built over a wooden frame, with adobe steps topped by a wooden nosing.

As for the decoration, although the *ksour* in this region do not particularly stand out for it unlike other valleys, like the Draâ valley, it is true that this *ksar* excels in comparison to others in the oasis, for its rich ornamentation and the variety of motifs among which we find simple geometric shapes, crosses, shamrocks, Fatima hands, moldings and archways.

7 CONCLUSIONS

Through the *ksar*, value is given to the concept of city as a sedentary setting for a population which, until its creation was nomadic, used to living in a rural environment. The earthen city, planned and projected, seeks to cover the needs of a Berber population, facilitating sociability among its members while protecting them from external dangers, which is justified in the case of the Bou Sidi Abdellah *ksar*.

This *ksar* known since the sixteenth century, presents a clearly organized structure, which distinguishes it from other Islamic cities and makes it typical of the Berber culture. Even though it is completely abandoned today, we can say that given its large size, rich ornamentation, singular configuration of homes, urban morphology and great amount of public facilities, this *ksar* was one of the most important ones within the caravan route linking Fez to Sijilmassa, noticeably standing out from the other *ksour* in the Mdagra oasis.

Overall, the *ksar's* urban planning illustrates an urban model adapted to the cultural needs of its inhabitants, as well as the climatic requirements of the environment, which has contributed to the development of an urban landscape that has today become a distinguishing feature of Moroccan culture.

NOTES

This research is included within the framework of the doctoral thesis title "Arquitectura de tierra en el Alto Atlas. Del oasis de Mdagra al valle del Outat. Marruecos" by Teresa Gil Piqueras and directed by professors Dr. Pablo Rodríguez-Navarro and Dr. Francisco Juan Vidal.

REFERENCES

Couvreur, G. 1968. La vie pastorale dans le Haute Atlas Central. *Revue de géographie du Maroc*, nº11. p. 14. en Peyron, M. 1984. Contribution à l' historie du Haut-Atlas Oriental: les Ayt Yafelman. *Reveu de l' Occident musulman et de la Méditerranée*. nº 38. p. 119.

Foucauld, Charles de. 1888. *Reconnaissance au Maroc 1883–84*, Challamel. Paris. p.352.

Jacques-Meuniè, D. 1982. *Le Maroc saharien des origines au XVIe siècle*. Librairie Klincksieck, Paris. p. 128.

Laoust, E. 1932. L'habitation chez les tranhumants du Maroc central. La Maison. *Hespéris Tamuda.* Tomo XIV. fasc. II. Larose. Paris. p. 138, p. 181.

Laoust, E. 1934. L'habitation chez les tranhumants du Maroc central. L'igerm. *Hespéris Tamuda.* Tomo XVII. Larose. Paris. pp. 154–155 y p. 159–161.

Mimó, R. 1996. *Fortalezas de barro en el sur de Marruecos*. Compañía Literaria. Madrid. pp. 127–130.

Peyron, M. 1984. Contribution à l' historie du Haut-Atlas Oriental: les Ayt Yafelman. *Reveu de l' Occident musulman et de la Méditerranée*. nº 38. p. 119.

Terrase, H. 1938. *Kasbas berbères de l'Atlas et des oasis*. Editions des Horizons de France. Paris. p. 160.

Earthen Architecture: Past, Present and Future – Mileto, Vegas, García Soriano & Cristini (Eds)
 ISBN 978-1-138-02711-4

Puzzolanic materials for restoration of earthen architecture

L.F. Guerrero & F.J. Soria
U.A.M.-Xochimilco, México D.F., México

R.S. Roux
U.A.T., Tampico, México

ABSTRACT: Lime has been used in earthen construction to make protective surfaces or to generate chemical stabilization. But these processes are slow which complicates its application in restoration. This is why people often chose to use cement as a plaster or stabilizer but, when time passes it causes unexpected reactions due to its high stiffness, impermeability and soluble salts developing. For this reason in the UAM-Xochimilco the use of puzzolanic substances in combination with lime it is testing in order to reduce the time for soil stabilization as well as improving its resistance, avoiding cement's collateral problems. This paper details the results of different comparative experiments focused on measuring the structural strength and moisture resistance of soil stabilized with lime, zeolites, volcanic ashes and powdered brick, in order to be used in earthen heritage architecture restoration and in contemporary building.

1 LIME AS PROTECTION OF EARTHEN BUILDINGS

Since ancient times most of Mexico's architecture was made with raw earth as the main building material. Combining rammed earth, cob, wattle and daub and adobe, pyramids, temples, palaces, fortresses, barns and most of the dwellings of the pre-Columbian civilizations were constructed.

When the quality of the soil was appropriate and the weather conditions suitable, this earthen architecture received maintenance through the simple use of soil in combination with natural substances that assured its superficial protection.

However, in regions with extreme climates or high rain downfall, earthen structures had to be protected through the use of stronger materials like stone and lime. This is the case, for example, of the great pyramids of Teotihuacan, Tula or Cholula; although their external appearance is that of stone monuments, this material only corresponds to the outer layer of the structure, having its nucleus made of compacted soil and adobe. The role of lime in pre-Columbian and Colonial architecture in most part of the Mexican territory was essential due to its use as binder of masonry as well as a protective substance for different built surfaces. These applications have allowed complex earth building systems to endure until our times, maintaining many of its physical, chemical and aesthetic qualities.

Despite that most research regarding the use of lime in architecture has focused mainly on stuccoes, mural paintings and polychrome surfaces that were part of temples and palaces, it is quite clear that whitewashed surfaces and mortars were part of practically all walls, roofs and floors built for ritual, productive, civil and residential functions on Mexican territory even until mid-20th century.

However, a less studied application concerning the use of lime in earthen architecture was as a stabilizing substance. Since ancient times, small proportions of lime were used in combination with earth in order to increase its mechanical strength as well as its resistance towards weather elements such as rain, wind or snow. The use of lime has been documented, for example, in walls and mud coatings from the archaeological zone of Paquimé (Fig. 1), as well as in most of the foundations of Tenochtitlan, in the center of Mexico City.

Figure 1. The use of lime has been recorded in the archaeological zone Paquimé, Chihuahua, Mexico (Photo L. Guerrero).

But this field is just beginning to be analyzed as a result of the scarce attention that has been given to the use of earth and lime as a building resource, even though still in the fifties and sixties of the 20th century their application in the construction realm was quite frequent. Unfortunately these materials have been gradually displaced by cement and other industrialized components. Two generations have been enough for the wisdom of thousands of years to become obsolete and the intangible heritage associated to the knowledge of their use and preparation procedures, to be forgotten.

Fortunately in recent years, as a result of research in the field of sustainable architecture, the use of lime is increasingly acquiring new dimensions, due to its evident economic, ecological and sanitary qualities.

This development is also linked to the broad level of dissemination that earthen building in general has gradually achieved, progress of which lime has been an inseparable complement. For centuries the effectiveness of calcium hydroxide as stabilizer, which limits the effects of water absorption, expansion and retraction of clays, among other attributes, has been constantly put to test.

This paper aims to explain several of the processes that have been followed in recent years at UAM-Xochimilco, focused toward the use of lime as stabilizer for existing soils in the surroundings of México City. This procedure seeks to determine not only the required amounts of stabilizing material, but also to explore the development of building components that can be used for the repair of missing parts in damaged heritage structures, as well as to build complete new elements such as walls.

Our more recent experimentation has evolved towards the implementation of a building procedure of poured earth, which uses soil, improved with lime, puzzolanic materials and a limited water dosage. The mixture is then emptied into a formwork or cast in several layers, applying a similar technique to that used in the construction of rammed earth walls.

Among the main findings that can be pointed out are de facts that this method has obtained earth components 40% more resistant to simple compression, it has also reduced capillary absorption of moisture and, above all, has an important time reduction of the building procedures due to the possibility of fast formwork retrieving.

2 SOIL STABILIZATION

Soil stabilization is known as the process that enables the control of dimensional changes that clays suffer when they come into contact with water. Different types of organic and inorganic materials, have been used throughout time to carry out these stabilizing procedures.

Although there are ancient documents that address the use of lime, such as the Prolegomena of Ibn Khaldun, written approximately in the 14th century (Monjo 1998: 33), it is evident that its application is much more distant in time and had spread all over the world in those cultures linked with the production and use of lime as a building material.

Lime has been used as a consolidating additive for earthen floors, foundations, water tanks, adobe bricks, rammed earth walls and more recently in Compressed Earth Blocks (CEB) (Oliveira 2005).

Since the Second World War, stabilized soils with lime have been widely used in construction, mainly in sub-bases of roads, railroad tracks and dams. From then, mainly in the field of road engineering, important research and practices of this technique were developed, thanks to which many standards and specifications have been tested for over more than forty years in countries such as the United States, France, Germany and Britain.

Even though there is historical data concerning this subject and the existence of scientific proof of lime's qualities as a soil stabilizer, associated with their physical behavior, there are still many facts to be discovered, mainly around the chemical phenomena that takes place internally in these mixtures.

It is well known that, depending on the dimension and laminar structure of mineral that compose the soil, the calcium cations get inserted between the clay micelles increasing their stability.

Some studies have shown that in these mixtures formations of neo-silicates and hydrated calcium aluminates are present, which improve the properties of soil, as well as the contribution of calcium hydroxide's carbonation (Oliveira et al. 1990).

However, the substitution of lime for Portland cement in the building industry, took place throughout the 20th century, and the latter was also incorporated into the field of soil stabilization.

Although "...it was in 1917 that H.E. Brooke-Bradley, (...) first constructed stabilized soil paving on Salisbury Plain" as the first "Scientific soil stabilization", the extensive application of cement as a consolidating additive developed from the 1960's onward (Cement and Concrete Association, 1953:2).

The preparation of CEB in those days was influenced by the use of cement opened as a component, which allowed raw earth very high mechanical strength and remarkable waterproofness.

However, as it is now recognized, cement production has a strong environmental impact due to the high temperatures needed for its calcination, which raises fuel consumption and therefore of greenhouse gas emissions. For these reasons, for almost twenty years the search for new alternatives in the concrete building field has aimed to decrease the quantities of cement needed and thus reduce the ecological footprint of such application. There have been many experiments that have proven that the inclusion of silico-aluminates, naturally present in volcanic ash, or ashes of vegetable origin, fly ash as well as some minerals such as zeolite, enhance the qualities of cement, allowing to reduce its percentage within the mixtures.

With the former background, the Autonomous Metropolitan University, in collaboration with the Autonomous University of Tamaulipas-Tampico, during almost five years has developed a series of investigations to recover the use of lime as a soil stabilizer (Guerrero, Roux & Soria 2011).

This research has had partial results, which were applied in the construction of a multipurpose space on a property belonging to the UAM-Xochimilco in the Southern part of Mexico City.

Experimental poured earth walls were made with soil stabilized by adding 5% of lime and 10% of cement. These percentages were determined by tests conducted in the laboratory of materials of the UAM-Xochimilco. Also the compressive strength was measured, reaching values of 29.35 kg/cm^2 after 7 days and 45.85 kg/cm^2 after 14 days (Soria, Guerrero & Roux 2013) (Fig. 2).

Nevertheless, for the purposes of our research, aimed towards the restoration of historic buildings, the use of cement was entirely dismissed, not only due to the above mentioned environmental impact issues, but also because of the evident series of pathologies that have been observed in heritage structures in which this building material has been incorporated (Guerrero 2008: 24).

Figure 2. Experimental poured earth walls built near the UAM-Xochimilco, Mexico (Photo L. Guerrero).

As it is well known both, sulfates present in cement as well as those who result of infiltrations, have the serious drawback that they, eventually in time, become expansive salts when in a humid state they interact with the tri-calcium aluminate phase existent in this material, which gradually destroy it.

With the former ideas in mind, we chose to recover information concerning the use of lime as a stabilizing substance, since it is a material that has ecological advantages over cement.

However, there are also references to the limitations of stabilizing with lime in the case of sandy soils for which the use of cement (Hoffmann, Negrini & Falleiros 2011) is usually recommended.

Furthermore, the soil stabilization with lime can take a long time, which complicates its use in restoration activities for heritage buildings that need to fix or support structural components in the shortest time possible, condition that has also boosted and recommended the use of cement due to its quicker reaction speed (Fernández 1982).

So then, with the idea of overcoming these limitations that soil stabilization with lime has, a series of experiments were proposed which allowed modifying the natural conditions of this substance, through puzzolanic processes.

3 PUZZOLANIC REACTIONS

The term "puzzolana" was originally used to name the sand of volcanic origin that exists around the Italian town of Pozzuoli, which has an important content of alumina and amorphous silica of high specific surface area (Sepulcre 2005), which in touch with calcium hydroxide in wet conditions, generates compounds that acquire a remarkable hardness at normal temperatures and even under water.

It is widely accepted that the knowledge of the reactions that these products generate was the basis for much of the advanced constructive technology that the Romans developed and that in the 18th century was rediscovered and optimized for the invention of Portland cement.

Physical-Chemical studies of these processes have enabled to show that there are many other substances—which are called generically puzzolanic materials—which have similar properties as those of the mentioned material from Puzzuoli.

These compounds may be of mineral origin as in the case of pyroclastic processes altered by weathering, or also be formed from organic residues as those present in sediments of ancient undersea deposits and even from anthropogenic activity, derived from the combustion of various industrial products such as iron slag, brick powder, fly ash, sugar cane or rice husk ash.

The concept of "puzzolanic activity" or "puzzolanicity" is defined as the property that amorphous or weakly crystallized silica and alumina compounds have to dissolve in a highly alkaline environment such as the one created by a solution of calcium hydroxide which, from an irreversible chemical reaction, generate di-calcium and tri-calcium aluminosilicates similar to those obtained in the setting of Portland cement (Quintana 2005:71). This reaction, involves the dissolution of glass or amorphous silica, which reacts with the hydrated calcium to form calcium silicate hydrate gel. The alumina will also be dissolved by the high pH conditions and react to form phases of hydrated calcium aluminates and hydrated calcium silico-aluminates normally crystalline, which contribute to the process of cementing and the increase in the strength of the whole.

The degree of solubility will depend on the specific surface area, which is the main factor within the different processes produced by diverse puzolanic materials. Eventually, however, most of the silica is dissolved and this leads to increase resistance in the long term (Sepulcre 2005:95).

4 METHODOLOGY AND RESULTS

Based on these premises, it was decided to carry on a series of studies in the Laboratory of Materials of the UAM-Xochimilco, aimed to compare the behavior of soil stabilization with lime in combination with three different puzzolanic materials available in the Valley of Mexico: zeolites, powdered brick and dust of an igneous rock known in the region as *tezontle*.

A series of samples and prototypes were developed aimed to extend the applicability of poured earth technology.

The main idea is to recover the wisdom of traditional building culture such as cob, adobe and rammed earth, but enhancing their properties through stabilization with aerial lime combined with puzzolanic materials.

The overall objective is to consider poured earth walls, not as a monolithic structure, but more as masonry of big blocks, that may develop flexible responses to earthquakes.

Also, small and lightweight molds or formwork can be used, which makes movements easier for non-specialized workers, as well as self-builders.

The key in order to obtain such goals depends on the possibility of having poured earth elements that can be unmolded in short periods of time. The building components that have been tested are made by filling the molds with successive layers of earth, similar to those used in rammed earth, but mixed with 18–25% of water.

For all testing the use of only one type of soil was determined, material that is marketed locally and that has relatively standardized characteristics. This soil is known as *tepetate* and is the product of the excavation, the mountains of volcanic origin that surround the Valley of Mexico.

Soil mechanics laboratory data showed that 100% of the coarse fraction of this soil passed through the No.4 mesh and 31% of its fine particles passed through sieve No. 200.

Subsequently the Atterberg Limits of the fine particles were determined, obtaining a 42.1 Liquid Limit and a Plastic Limit of 26.3; resulting in a Plasticity Index of 15.8 which led to the conclusion that this soil corresponds to the "SC" group or "Clayed sand" (USCS) (Juárez 2010).

The kind of lime used is aerial, of *Calidra* brand, with 82% of $Ca(OH)_2$ available, which was hydrated during three months.

Dosages for all specimens were performed in volume while maintaining 500 ml of soil, 50 ml of puzzolanic materials (sifted through mesh #100), 50 ml of lime (paste) and 200 ml of water.

These volumes allowed making sets of test cubes of 5 cm per side, which served for the completion of trials to simple compressive strength as well as moisture absorption tests.

The results obtained in tests of simple compressive strength (norm NMX-C-036-ONNCCE-2004) were very encouraging.

The average strength of cubic specimens of non-stabilized soil was 14.4 kg/cm^2, those stabilized only with lime 15.1 kg/cm^2, the ones with lime and zeolite 18.58 kg/cm^2, the ones with lime and *tezontle* 21.2 kg/cm^2 and finally those of lime and powdered bricks 21.7 kg/cm^2.

As far as capillary absorption is concerned, tests were made using the Mexican norm NMX-C-037-ONNCCE-2005.

The experiments revealed that after 10 minutes the cubic specimens of non-stabilized soil absorbed 8.45% of water, those stabilized only with lime 8.8%, the ones with lime and powdered bricks 8.75%, those with lime and zeolite 9.2%, and finally those of lime and *tezontle* 10.86%.

In total water immersion testing, soil without stabilizing specimens resisted 17 minutes before disintegrating (Fig. 3, right), while those containing volcanic rock and powdered brick remain intact under water after more than one month since initiating the test. (Fig. 3, left)

Research continues in order to make comparisons, now using lime in powder form, of similar proportions to the previous specimens, as well as a series of tests with increasing amounts of cement to count with reference data.

The findings that are presented in this text are very promising, because building elements can

Figure 3. Total water immersion testing (Photo L. Guerrero).

Figure 4. Restoration of an adobe monument using poured earth with lime and powdered brick, Colima (Photo L. Guerrero).

be ready in 40 minutes, with a better resistance to moisture and a 40% increase in their compression resistance when compared to non-stabilized earth.

5 CONCLUSIONS

The implementation of a poured earth technique using soil stabilized with lime and puzzolanic materials, may have important contributions in the restoration of earthen heritage buildings since the applications that are intended can follow two routes.

On the one hand it is expected to be able to use these components as a means to repair historical earth structures that have lost their unity (Fig. 4) On the other it is considered that earth, in plastic state, can be very useful for the intervention of walls or structural components made with poured mud, technique that aims to reduce the time to unmold the built elements.

One of the problems that frequently arise in the restoration of earthen heritage buildings, have to do with the undermining of the lower part of the walls either of adobe, rammed earth or wattle and daub. Mixtures in plastic state of soil stabilized with lime and puzzolanic materials may be poured in layers within the structural components with the support of an external cast or mold that serves as containment for the material.

Given the speed that is reached in this process, in less than one hour it is possible to remove molds and continue the restoration works. This same process can be used to replace missing elements derived from runoff of rainwater in walls, as well as for the filling of cracks.

A similar benefit is obtained in the case of contemporary structures designed to perform from anew (Fig. 5) Cast earth walls with lime and puzzolanic materials rather than cement can be unmolded quickly and, although it will probably not reach the compressive resistance of cement-soil, the ecological benefits of its application in structures subjected to limited efforts, can be an advantage.

Among the most notable findings from studies that are underway, is the fact that the best results of the puzzolanic materials used for stabilization with lime were obtained with powdered brick. The addition of this material increased the resistance of the soil by more than 40%.

Also, soil's vulnerability towards water was greatly decreased, without losing its porosity and permeability to water vapor, factors which have so much impact in their hydrothermal response and the adherence of other materials or plasters.

Figure 5. Poured earth students' workshop that took place in February in a public area near Xochimilco (Photo L. Guerrero).

From an economic point of view it is possible to consider that in Mexico lime has almost 50% less cost than that of the cement and powdered brick is a waste material. Because of this, it is possible to put in reach of communities with limited resources, a building material of easy and economic transformation, which can also be maintained and serviced by its own users.

It is essential to disseminate knowledge related to the use of lime and earth onto society, because these are not notions belonging to specialists in design, construction or restoration, moreover they are an integral part of the traditional and popular wisdom that should recover their place.

REFERENCES

Cement and Concrete Association. 1953. Soil-cement stabilization. *Concrete Quarterly* 20: 2–9, London.

De Oliveira, M.; Santiago, C. y D'Affonseca, S. 1990. The study of accelerated carbonation of lime-stabilized soils. In *International Conference on the Conservation on Earthen Architecture*: 166–170. Los Angeles: The Getty Conservation Institute-US/ICOMOS.

De Oliveira, M. 2005. O solo-cal: uma visão histórica e documental. In *Memorias del IV SIACOT-IIIATP*: 106–110. Monsaraz: Escola Superior Gallaecia.

Fernández, C. 1982. *Mejoramiento y estabilización de suelos*. México D.F.: LIMUSA.

Guerrero, L. 2007a. Tierra y cal. In *Anuario de Estudios de Arquitectura 2007*, UAM-Azcapotzalco, México D.F.

Guerrero, L. 2007b. Arquitectura en tierra. Hacia la recuperación de una cultura constructiva. *Apuntes* 20 (2): 182–201.

Guerrero, L. 2008. La cal y el patrimonio edificado. *La Gaceta del Instituto del Patrimonio Cultural* 12:19–29.

Guerrero, L.; Roux R. & Soria, F.J. 2011. Ventajas constructivas del uso de tierra comprimida y estabilizada con cal en México. *Palapa* Vol. V-1 (10): 45–57.

Hoffmann, M.; Negrini, F. & Falleiros, A. 2011. Tapia. In Neves, C. & Borges, O. (eds.) *Técnicas de construcción con tierra*: 46–61. Bauru, São Paulo: UNESP-Red Iberoamericana PROTERRA.

Juárez, E. & Rico A. 2010. *Mecánica de suelos* I. México D.F.: LIMUSA.

Monjo, J. 1998. La evolución histórica de la arquitectura de tierra en España. In *Arquitectura de tierra. Encuentros Internacionales*: 31–44. Madrid: Ministerio de Fomento.

Quintana, E. *Relación entre las propriedades geotécnicas y los componentes puzolánica de los sedimentos pampeanos*. PhD Thesis. Córdoba, Argentina. Universidad Nacional de Córdoba.

Sepulcre, A. 2005. Análisis comparativo de determinados aspectos sobre la hidraulicidad en los morteros de cal. In *Tratamientos y metodologías de conservación de pinturas murales*: 71–121. Palencia: Fundación Santa María la Real.

Soria, F.; Guerrero, L. & Roux, R. 2013. Alternative earth building techniques and their experimental applications. In Theimer C. & Wilson, Q. (eds.), *Proceedings-EARTH USA 2013*: 406–410. Santa Fe: Adobe in action.

Earthen Architecture: Past, Present and Future – Mileto, Vegas, García Soriano & Cristini (Eds)
© 2015 Taylor & Francis Group, London, ISBN 978-1-138-02711-4

The presaharian settlements of the Ziban: The persistence of Roman elements in native Islamic dwelling

S. Haoui
Institute of Urbanism and Architecture, University Saad Dahlab, Blida1, Algeria

ABSTRACT: Our study focuses on the problem of whether to consider the relationship between the ancient history of the Ziban region and Medieval Islamic architecture as, on the one hand, a mere re-use of existing building remains or, on the other hand, a manifestation of a blending of aspects of Islamic culture with a pre-existing Roman architectural order. In this article, we put forward our investigations concerning the characteristics of native dwellings in the presaharian settlements of the Ziban region—a region bordering the limes producing areas of Numidia corresponding to the agricultural territories of the Roman empire—and their relationship with the ancient historical past of the area. We bring to the forefront both elements of construction and typology which reveal the presence of an extinct building tradition and its influence on relative domestic architecture which has been often neglected in all other studies in favor of major architectural artefacts.

1 THE ZIBAN: DIVERSIFIED RESOURCES

1.1 *The territory: Situation and natural resources*

The Ziban (plural of Zab) is a sub-presaharian region situated at the north-eastern limit of the Algerian Sahara; an enormous territory separated from the north (tell) by the Atlas Saharan mountain range, 300 km south of the Mediterranean coast. It holds an important hydrographic position capturing the waters at the foot of mounts Zab and Aurès. Numerous rivers and springs, shooting out from the mountains and descending the valleys, run through the gently sloping lands of the Zab giving rise to oasis and the establishment of human settlements with their own dwellings and cultivated area. Because of their perennial nature, these hydrographic resources have guaranteed the permanency of occupation and urbanization in the region as well as its diachronic transformations (Trousset 1986) (Fig. 1).

Figure 1. Localization of the Ziban region in north-eastern limit of Algerian Sahara.

1.2 *The landscape: Organization and density of dwelling in the oases*

Throughout the two parts of the Ziban region, the oases settlements, with varying degrees of density, follow a north-east to south-west alignment. This alignment confirms the need of the settlements to follow the waterways which spilled into the Chott of this depression (Chott Melghir). The river Djeddi is one of the longest rivers in Algeria and flows 500 km in east-west direction amassing the waters from several rivers coming from the north. On reaching the Zab, its left bank is dotted with green zones of the oasis settlements of western Zab.

1.3 *The history: Ancient and Medieval archaeological richness*

If we take into consideration both the ancient history and the different archaeological studies of the region, we notice an alignment between the organization of the settlements and the borders of the ancient limes orchards of southern Numidia, Roman province in North Africa. Traces of this structure can be seen in different historical archaeological maps (Gsell 1911, Baradez 1949, Salama 1951).

One of the most important defensive sites of the region is the military camp at Gemellae, situated in the south of western Zab on the right bank of the river Djeddi. Built during the reign of the emperor Hadrian, the camp is surrounded by the remains of municipal city of Gemellae which several structures have been discovered and excavated (Fig. 2) (Baradez 1948). To the south, the camp is reinforced by a ditch (Euzennat 1990). This double and filter border (Trousset 2004) protected the oases by a system of forts and fortalices. From the outset, ancient civil and hydraulic structures were put into place to control the southern borders of these occupied territories and ensure agricultural production (Baradez 1948, Trousset 1986). The ancient settlements we see today in these countries are, perhaps, the consolidation and persistence, across the territory, of the antique borders of the cities, which are engendered by an "agricultural Romanization" (Leveau 1990).

Figure 2. Camp of Gemellae, view of the principia (Haoui S.).

In the Medieval Age, after the advent of Islam in the 8th century (Meynier 2007), the oasis of the Ziban region became a thorough fare through which the oriental caravans passed on their way to Agades in pursuit of their trans-Saharan commercial activities. The oases were also places of transit for pilgrims travelling from Laghouat to Gabès, passing by village of Sidi Okba (Trousset 1982), where numerous shrines, tombs, chapels and mosques show an incontestable architectural richness. From an urban point of view, Medieval historical literature relates the dense and fortified character of the Ziban cities, the names of sovereign towns as well as the richness and variety of the agricultural products (El Bekri 1859, Ibn Khaldoun 1856).

2 AREA OF STUDY: SELECTION GRID

Our research is concerned with both urban and architectural characteristics of historical towns which attest the continuity with the ancient past of southern Numidia. Our aim is to determine Roman influence on traditional domestic dwellings on their assimilation with late Roman foundations. To do this, several criteria were followed:

The proximity of the dwellings to the Roman camp: the remains of the military camp of Gemellae have been excavated by Baradez in 1947 and 1950. Our area of studies concerns the historical settlements situated on the left of bank of the river Djeddi and in a zone no more than 10 km as the crow flies from the camp site and its protected area.

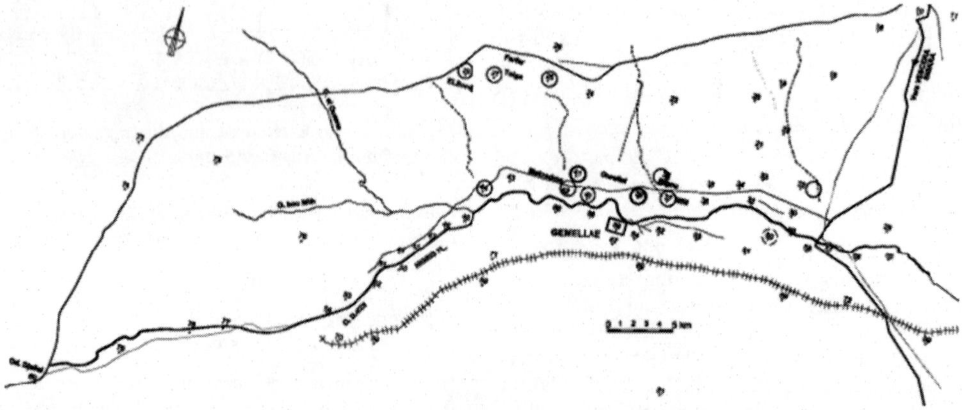

Figure 3. Location of settlements on the archaeological map (River *Djeddi* settlements—35 archaeological sites numbers—66 Roman camp of Gemellae) (Haoui S. from AAA and topographical map of Biskra).

Archaeological stratification: Medieval historical sources attest to the existence of settlements on these ancient sites during the Middle Ages. Gsell affirms that Arabic constructions probably rest on old foundations. Page 48 of the Archaeological Atlas of Algeria presents numbered archaeological sites of the western Zab. We can deduce from it the sites shown as dwellings. The density and concentration of these archaeological points are another indication of this stratification. Their verification with a modern typological map led us to identify zones of concentration (Fig. 3).

Building criteria: cartographical work means direct observation of the buildings within the chosen area. In the absence of deep excavation, however, which could have given us evidence of a vertical stratification and the presence of Roman foundations, we were forced to examine those dwellings where the ruins present the integrated structure of old building materials visible at the surface; for example, walls with bases made of cut stones or the use of engraved stonework. Nevertheless, our observations are based on clear and substantial evidence: the consistency and frequency of ancient elements in the construction of these buildings, especially the height and the continual use of stones in wall faces, The nature of the re-used items (cut stones, columns, capitals…), the uniformity and regularity of aspect as well as the quality of the masonry. These parameters of a co-existence between ancient and local traditional building materials are for us an indication of a possible topological stratification and motived us in our choice of this observation work.

3 TYPOLOGICAL STUDY OF TWO SAMPLE HOUSES

In our zone of study, the landscape is marked by the forms of a dense group of dwellings, deserted today and in a state of decay. They correspond to old urban centers.

3.1 *House 1: Lineal organization with portico*

It is situated in old village, 4.5 km North West of the military camp and 1.6 km from the river Djeddi. The Archaeological Atlas of Algeria indicated the existence there of an ancient fortress with Arabic houses built on Roman foundations. Few buildings remain standing but traces of carved stonework are numerous. We present here one of the rare houses we have managed to evaluate (Fig. 4).

3.2 *House 2: House with storey (Fig 5)*

This house is situated in old deserted village. It was abandoned in the autumn of 1969 when floods

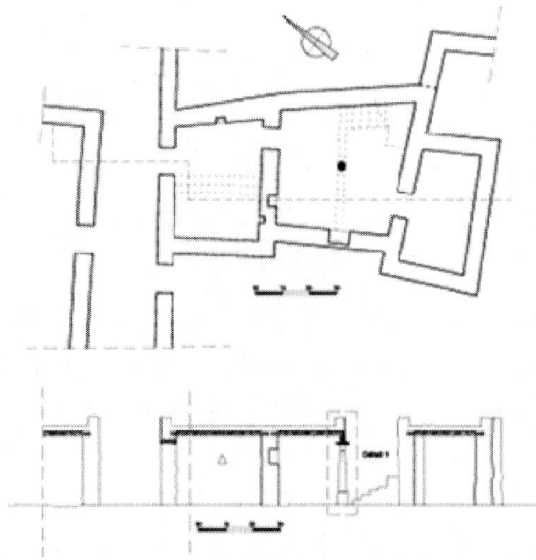

Figure 4. House 1, ground plan and section (survey and drawing Haoui S.).

afflicted the region and a new village was built. The settlement, situated 2.2 km North East of the military camp, Gemellae, is the nearest village to the antique municipality. Fortified in appearance, the village is accessible by portals which open onto a palm grove. The later has a Roman inscription in stonework and is supported by a brick archway which has completely collapsed today (Fig. 6).

The ground floor of our sample house is divided into two parts (one for guests, one for the family) each with its own staircase leading to the upper floor. Its uniqueness lies in the organization of the rooms around a spacious central area where the roof is supported by 6 palm trunks resting on stone disks. An opening in the roof covering allows light and air to penetrate the interior. All the earth brick walls of the house rest on bases made from huge stones and palm trunks. The outside walls are of 6.50 m height.

4 TECHNIQUE AND ORIGINAL FORMS OF CONSTRUCTION

The principal building materials in the dwellings under examination are cut stones from the Roman period, rough stones, baked earth bricks and palm trunks. We believe that most of the re-used stones and carved masonry comes from the ruins of Gemellae town. Stone is used principally in the foundations and base of the construction. Earth bricks are reserved for the body of walls. Timber from palm trees is present in both vertical

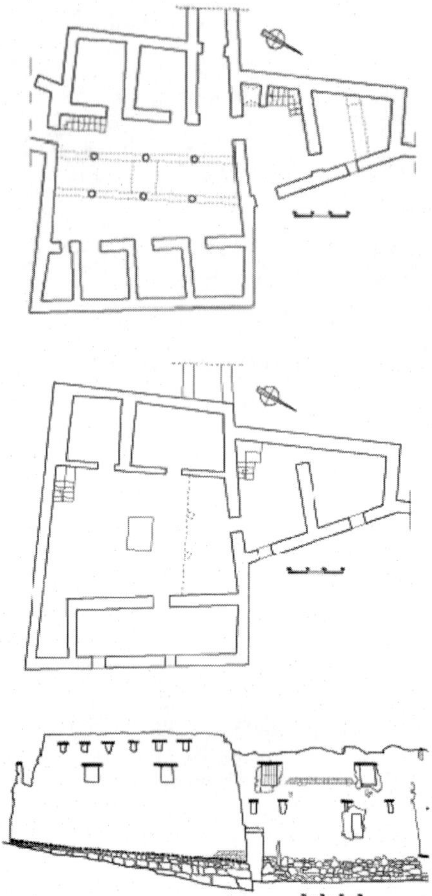

Figure 5. House 2, ground floor, floor and front elevation with the trunks wooden of date palm (survey and drawing Haoui S.).

Figure 6. Roman inscription, ancient entrance of settlement. (Haoui S.).

Figure 7. View of the front (house 2): The wall with blocks of roman stone and trunks of wooden palm date (Haoui S.).

and horizontal structures readily available in the area; the trunk was split into two or four parts and used in the framework of ceilings as rafters or supporting beams. Whole trunks were used as pillars to support the main rafters of the ceiling. In the various construction methods available using cut stones, earth bricks and palm wood, two details were brought to our attention:

4.1 *Hybrid wall*

A layer of palm trunks, split into two or four (according to the diameter of the trunk) was placed between the stone base of the wall and the upper layer of earth bricks, held in place by soil cement. This method is unique. Indeed, we know of trunks being used in walls to strengthen an opening or to prevent eventual cracking but in our case, the layer of wood is 80–100 cm from the ground. The trunks, whose lengths are equal to the thickness of the walls (60–70 cm) are disposed perpendicularly to the walls and carefully joined together by soil cement. The result is a hybrid structure in a perfect state of cohesion (Fig. 7).

4.2 *Antique columns in the settlement*

In certain cases, especially in the oldest buildings showing an advanced state of degradation, the trunks of antique columns were placed one on top of each other to act as secondary pillars in an extension. In the case of house 1, one such pillar supports the roof of the arcade. It should be noted that the join between the two pillar trunks was filled with plaster cement and wood shaving, all naturel fibers that stick better to the superimposed parts.

5 CONCLUSION

In this article, we have put forward the result of an investigation which is in itself just beginning. Our modest aim is to increase knowledge about traditional presaharian dwellings built in Roman-African rural areas within the provincial Roman boundary.

From an urban point of view, we notice the size of the establishments and the fortified nature of the ancient agglomerations. From an architectural point of view, we notice the recurrence of arcades (porticos) on only one side of the building leading to a modest courtyard. The door which opens onto the courtyard is offset from the outside door, something we usually attribute to Islamic buildings where really it belongs to a Mediterranean tradition. It can, in fact, be seen as the prolongation of Afro-Roman organization.

With reference to construction, the examples represented here attest to the frequent use of palm trunks. The importance of this natural material has been often neglected. Several studies have shown the use of different types of wood in building structures. In the traditional towns of the Maghreb, tree trunks are integrated into earth walls and provide a well-established antiseismic solution. In a territory closer to our zone of study, in Aurès, multi-layers of cedar logs are used in walls made totally of rough stones to form a horizontal system of support. However the method of construction that we have already described doesn't belong to any of these cited examples. Nevertheless, several hypotheses can be drawn. It could be an inferior horizontal system of support like the one seen in Aurès, but in this case, there is only one layer which would seem to refute this idea. Our second hypothesis takes into consideration that the layer of palm wood is between two different materials: stone and earthen bricks. This seems to suggest that the layer of wood acts as a kind of damp course to protect the earth walls from rising damp.

There are four reasons why this conclusion has been drawn. The first concerns the vulnerability of the earth to water which needs no further explication. The second is locality of this technique in areas where both ancient and modern histories have proved to be prone to flooding. The third concerns the horizontal position of the bark of trunks palm tree. It is rigid and almost impermeable which makes it a suitable barrier in the prevention of rising damp (Fig. 8).

The fourth is related to the physical characteristics of palm wood. It is porous and therefore, acts as sponge, absorbing water and preventing it reaching the baked earth bricks. The few studies made on this type of wood show that it is, indeed, hydrophilic to a high degree (Mekhermech 2012). To confirm the importance of this structure, we noticed that the walls with the timber were in a better state of conservation, even when partially decayed, than those without (Fig. 9).

We think then, that this last hypothesis is the most probable one. Palm wood could be therefore, an effective solution to reduce the vulnerability of architectural earth structures to rising water. To this hypothesis, further archaeometric research must be added to complete our doctorate study.

I can hardly close this subject without asking myself the question about the origins of this technique. Keeping to the debate about Roman continuity in traditional architecture, we know that wall bases made of blocks or rough stones and built of with earth bricks existed in the Roman era in various cultural provinces, especially Africa. As to the incorporation of palm wood as an intermediary band between earth bricks and stones, at this stage in our research, we possess no information.

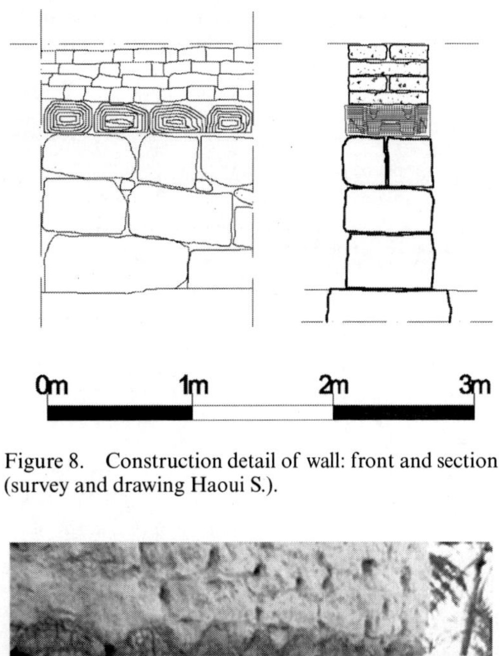

Figure 8. Construction detail of wall: front and section (survey and drawing Haoui S.).

Figure 9. View of an adobe wall in good condition on the degraded base (Haoui S.).

Could it be a building tradition of purely local origin in the period of transition between late Roman occupation and Medieval Islamic settlement or a lost African-Roman technique? This is, for us, another hypothesis which cannot be ignored since date palms existed well and truly throughout Gemellae and the Ziban region at the time of Roman era (Baradez 1949). This aspect of the persistence of elements from the Roman era in the Medieval Islamic architecture of southern Numidia and its influence in building techniques is a general question of study and touches on the role and influence of late antiquity in Islamic domestic dwellings as revealed in the works of P. Leveau (Leveau 1990 b).

The only certainty is that the disappearing skills and rural building techniques seen in these deserted settlements are the specific architectural expression of the unique nature of this presaharian landscape which combined both multicultural influences. The general objective of our research is to discover, within the relic of these vernacular dwellings, the reminders of an architecture incorporating old and ingenious skills.

REFERENCES

Baradez, L. 1948. Gemellae, camp d'Hadrien et ville des confins sahariens, in *Comptes rendus des séances de l'Académie des Inscriptions et Belles-Lettres,* 92e année, N.3: 390–395.

Baradez, L. 1949. *Fossatum Africae. Recherches Aé-riennes sur l'organisation des confins sahariens à l'époque romaine*. Paris: Éd. Arts et Métiers Graphiques.

Côte, M. 2005. L'urbanisation aujourd'hui au Bas-Sahara, in La ville et le désert, le *Bas Sahara algérien:* 13–26, Paris: Karthala.

El Bekri, A.O. 1859. *Description de l'Afrique septentrionale.* Paris: Ed. et trad. par William Mac Guckin de Slane.

Euzennat, M. 1990. La frontière romaine d'Afrique. In *Comptes rendus des séances de l'Académie des Inscriptions et Belles-Lettres*, 2: 565–580.

Gsell, S. 1911. *Atlas Archéologique de l'Algérie.* F°48. Al ger et Paris.

Ibn Khaldoun, 1856. *Histoire des Berbères et des dy- nasties musulmanes de l'Afrique Septentrionale*, trad. William Mac Guckin de Slane, vol. III. Alger.

Leveau, P. 1990a. L'organisation de l'espace agricole en Afrique à l'époque romaine, in *L'Afrique dans l'Occident romain (Ier siècle av. J.-C.—IVe siècle ap. J.-C.)*: 129–141. École Française de Rome.

Leveau, P. 1990b. Aspects de l'espace domestique dans l'habitat urbain antique des pays devenus musulmans autour de la Méditerranée, in *l'habitat tradi tionnel dans les pays musulmans autour de la médi terranée.* Volume 2: 333–366. Institut de recherches et d'études sur le monde arabe et musulman, IFAO, Collection: Patrimoine architectural.

Mekhermeche A. 2012. *Contribution à l'étude des pro priétés mécaniques et thermiques des briques en terre, en vue de leur utilisation dans la restauration des Ksours sahariennes.* Université d'Ouargla.

Meynier, G. 2007. *L'Algérie des origines, de la préhis toire à l'avènement de l'islam*, berzakh. Alger.

Trousset, P. 1982. Le franchissement des chotts du Sud tunisien dans l'Antiquité. In *Antiquités africaines*, n.18: 45–59.

Trousset, P. 1986. De la montagne au désert. Limes et maîtrise de l'eau, *Revue de l'Occident musulman et de la Méditerranée*, N°41–42: 90–115.

Trousset, P. 2004. Pénétration romaine et organisation de la zone frontière dans le prédésert tunisien. In *Africa Romana*, XV: 59–88. Rome.

Earthen Architecture: Past, Present and Future – Mileto, Vegas, García Soriano & Cristini (Eds)
© 2015 Taylor & Francis Group, London, ISBN 978-1-138-02711-4

Earthen architecture's on seismic areas: Costa Rica's case

I. Hernández Salazar
Chalíuk: Arquitectura y Restauración del Patrimonio, Heredia, Costa Rica

ABSTRACT: In Central America, Costa Rica is located in the zone of convergence of three tectonic plates: Cocos, Caribe and Nazca. This condition causes hundreds of earthquakes each year, some with higher and important magnitudes. As a result of these events, the legislation has hardened to the detriment of the traditional buildings of adobe and wattle and daub further aggravating their deterioration and contributing to its loss. This article is focuses on earthen architecture and its restoration in seismic areas, making aware of problems that include its maintenance under these conditions and showing some techniques for their conservation.

1 EARTHEN ARCHITECTURE: STATE ON SEISMIC AREAS

Earthen architecture has been present in the constructions of different civilizations and cultures for thousands of years. There are various designing and building techniques in which mud is used, from a resistive element as in a wall, to filling material or even finishing material as in wattle-and-daub houses. Just as with all existing construction systems, mud construction has great advantages as well as limitations, so its improvement and maintenance are a challenge to those professionals skilled in this type of construction system. This construction technique is usually associated with low cost buildings that are found in remote regions. These buildings are also perceived as marginal rather than high quality ones; however, mud constructions are part of the traditional landscape of many countries around the world.

In earthquake prone regions, mud seismic behavior has affected its credibility and popular acceptance due to ignorance of its capabilities and structural failure. As a result, mud construction is prohibited and its use is limited to repairs in existing buildings. Nowadays, some specialists, in many regions prone to earthquakes, have investigated about ways to prevent mud construction legacy. This has been done by setting suitable parameters for affected buildings, and providing economic, sustainable, and modern construction techniques.

Costa Rica is one of the most seismic countries in the world because it is located in the so-called "Pacific Ring of Fire"; consequently, Costa Rica, as well as many others countries, has limited mud constructions, and it is only allowed for building restorations. This Central America country has a long tradition in mud construction, and it has developed adobe and wattle-and-daub buildings that make up most of the architectural heritage of the country. However, this architectural heritage is threatened due to weather conditions, maintenance, and earthquakes.

In any case, traditional techniques have been replaced by modern materials and technologically improved, carrying implications for interventions in architectural heritage. In this situation is difficult to make the right choice of materials, techniques and criteria to be considered in a restoration. In addition, in recent years, some of the measures that have been taken in the field of conservation or intervention have been based on the use of "recipe book" of treatments to solve some pathologies.

There are different causes and agents of alteration in a building of adobe or wattle-and-daub. In this regard Costa Rica is not the best example of conservation due to some reasons like inappropriate technical skills, lack of a monumental architecture or skilled labor. On the other hand the earthquakes incidence and lack of interest of people for the heritage have, contributed to replace the historical mud constructions, with new cities, giving way to shopping malls, homes of reinforced concrete and even highways.

While the majority of declared heritage buildings are owned by descendants of the original owners, some have undergone severe changes to conform modern standards. Besides, the devaluation of the heritage has stimulated the conception that these buildings have undergone their useful life and they need to be replaced. Together with these intentions, several years ago it was prohibited to build with adobe and wattle and daub. This restriction stimulates the lack of maintenance of the buildings and emphasizes that these structures are not authorized.

Costa Rica makes severe interventions in heritage buildings for the safety of their inhabitants. While repairs or adjustments should be developed taking into account the characteristics of the original materials, there are new implementations with reinforced concrete beams, Portland cement mortars and sewn with steel rods (Fig. 1).

These elements contribute to the deterioration of buildings due to the materials and mechanicals incompatibilities which change completely the image and the character of mud constructions. The use of these materials has come from the hand of the erroneous recommendations of new builders or even associated to the rigidity of structures of concrete implemented in buildings as diverse as mud structures (Fig. 2).

Often, combinations of adobe and wattle and daub with rigid materials throughout the American continent have been developed since the 1940, mainly, reinforcing the use of concrete and brick for structural interventions or architectural modifications as a result of the damage that earthquakes left in the buildings. However, subsequent earthquakes have shown that the interaction between these materials produces new fissures in the area of contact and especially the width walls undue accelerates the collapse of these structures.

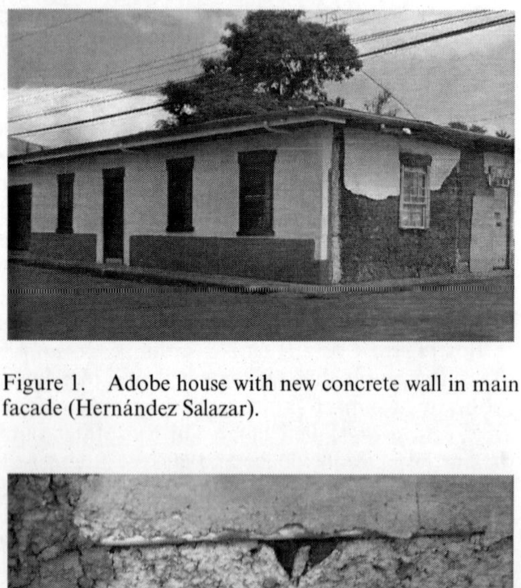

Figure 1. Adobe house with new concrete wall in main facade (Hernández Salazar).

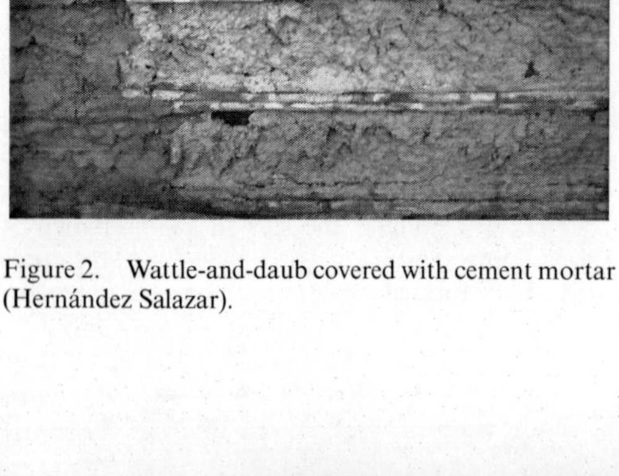

Figure 2. Wattle-and-daub covered with cement mortar (Hernández Salazar).

The great vulnerability of mud constructions to earthquakes is mainly based on its low resistance to traction, the poor quality of the material used for the walls or mortar, errors in the technique implementation, the lack of locking between the parts and the absence of tie-down items that fail to comply the structural unit of the work.

It is important to mention that in addition to these features, the excessive weight of the tiled roof generates an overload on the walls. This load is transmitted to the walls incorrectly, due to two main reasons: first, the absence of distribution beams generates point loads on the walls and second, by misconfiguring structure of composite roof by pieces of wood with little structural capability and insufficient dimensions.

2 ADOBE AND WATTLE-AND-DAUB

Adobe and wattle-and-daub are the most common mud constructions in the Central America area, while there are, around the world, more than one dozen types of earthen constructions. Both systems had a wide diffusion in the isthmus and particularly in Costa Rica, being responsible for the creation of its urban and rural image they allowed residential, institutional, religious and communal constructions.

The adobe began to be used after the arrival of the Spanish. However it had its apogee during the 18th and 19th centuries. The process to perform the adobes was laborious and was required enough time to carry out each one of the stages until the implementation of the material which consisted in a arduous and detailed process. The adobes were previously wet in a mixture of mud and water—that made the functions of mortar—and next they were placed interlaced following various types of rigging. Normally, the adobe wall didn't show huge openings, on the contrary, the walls were large earthen blocks with plastering that finally were terminated with a whitewash. On the other hand, when the opening of some vain was necessary, either for a window or a door, the walls were reinforced with wooden beams in their superior border. In different places of Costa Rica, much of the constructions were constituted by a framework and a cover of organic fibers and leaves. This technique was used along the entire American continent giving rise to the wattle and daub. Constructively, it's possible to identify a number of differences with other works using earth as raw material. These divergences are, basically, in the use of some materials of reinforcement in the wall, their structure system and layout or the behavior of the walls. Technically, the wattle and daub is a structure of wood, cane or bamboo, with a filling of ground mixed with organic fibers. In general, it consists of a primary structure made with wood above stones base. This structure is complemented by the installation of cane framework that is stuffed with a mixture of clay and dried organic fibers. That system has a higher resistance to seismic movements, at the same time allowing the construction of the buildings by stages and facilitates the growth of structures. Thanks to their constructive flexibility, it's possible to make partitions after the construction of the walls and the elimination of existing sections if necessary.

The wattle and daub technique is classified generally as a traditional building system to be used by people without any kind of specialization in constructive terms; however it constitute, a sustainable system with widespread presence along the Costa Rican territory. This kind of construction allowed the evolution of new architectural styles in the country and facilitated their adaption to the needs of people at the time.

3 EARTHQUAKES AND EARTHEN ARCHITECTURE: EFFECTS

Mud constructions have a very low resistance to traction; consequently the seismic forces produce fissures and major cracks that subsequently lead to ruin. Although in some cases, the buildings have withstood the onslaught of movements during a certain period of time, it may be sufficient a single telluric event to make them collapse. This is one of the reasons why this type of architecture presents further deterioration compared to other buildings of the same period, built with different materials.

At the time that an earth building is shaken by an earthquake, it is subjected to horizontal forces product of the vibrations, hauling flexion stress on the walls. Curvatures, deformations, collapse, breaking, rockfalls and landslides are some of the problems these solicitations produce (Fig. 3).

On the other hand, the vertical forces that originate from an earthquake also must be taken into consideration, since they affect directly and negatively, all massive or heavy building elements or those that are attached to the principal structure and have their only support in this (structures in projection, balconies). Also, vertical forces cause oscillations, deformations, landslides or collapses of another element like arches, columns, roofs structures and ladders nearby to walls, among others.

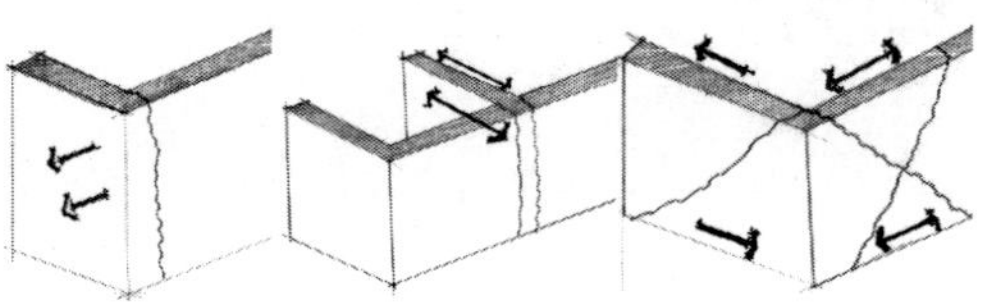

Figure 3. Principal fissures and cracks produced by horizontal forces on buildings (Hernández Salazar).

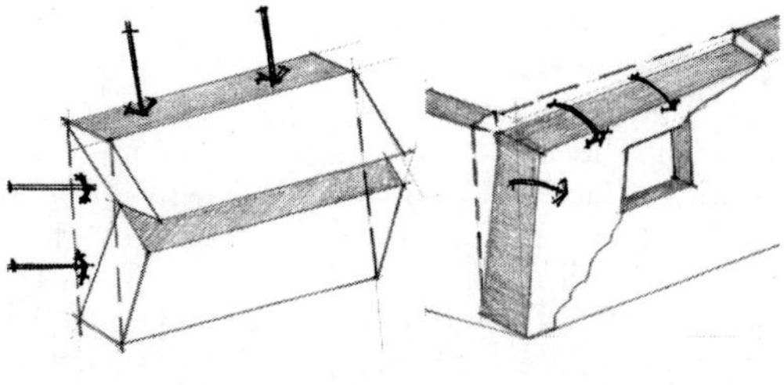

Figure 4. Cracks and deformations produced by vertical and torsion forces on buildings (Hernández Salazar).

Finally, there are the forces of torsion. These are originated from the combination of the horizontal movements of the terrain and vertical forces. Their consequences may be very important especially in constructions with very irregular distributions (Fig. 4).

Another important aspect is that according to form in plant, regularity in the distribution of the structure and height. The buildings may be affected differently depending on the materials used. These provisions in conjunction with the use of the land make even more vulnerable this type of constructions. In the case of buildings with walls of various thicknesses, they will behave in different ways, facilitate the collapse of structure. Long walls without intermediate supports or in their weakest points such as corners also undermine their resistance against earthquakes, so it can be vulnerable to fall in an imminent ruin.

Another aspect that affects a proper seismic behavior in mud constructions are the high walls and with little thickness, since they're often weaker and therefore these have a lower resistance to perpendicular forces. Simple measures can prevent the construction of buildings with irregular shapes, roof structures without adequate support or with excessive weight, buildings on inappropriate areas, the creation of large openings in the walls or poorly executed, can collaborate with a best performance of structures on future telluric events.

4 SEISMIC REINFORCEMENT TECHNIQUES

There are new procedures that are being implemented in several countries, mainly South Americans, that are intended to, minimize the consequences that have earthquakes on the patrimonial structures built with earth. They have been developed several alternatives, ranging from the installation of cast beams, vertical stiffeners inside the factories of adobe, placement of pieces of wood to reinforce corners and encounters of the walls, realization of abutments to minimize the collapse of walls, among other techniques, which could be reduce the damages in mud structures according to the newest researches.

However, the works haven't stopped with the implementation of the above-mentioned reinforcements; on the other hand, efforts have continued, to implement innovative alternatives, such as the placement of nets of synthetic fibers bordering the walls or placing pieces of wood for to try a monolithic behavior. This behavior is the most appropriate for that kind of constructions.

The following are the solutions most recently used in Costa Rica with the purpose of facilitating the seismic adaptation of land constructions. This study is an indicative of solutions that exist today for approach this problem.

One of the initial measurements to achieve an improved behavior in earth constructions is the placement of horizontal elements cast at the top of the walls. The installation of these components, tries to achieve a suitable anchorage with the structure of the roof, a uniform distribution of load on the walls and subsequently, your optimal transmission to the masonry. They're one of the most important elements in the seismic adaptation of earthen buildings, because the most of the failures in these structures occurs when the walls are cracked vertically and overloaded. Moreover, the placement of elements—usually wooden—pieces at the end of the walls, gives strength and stiffness. At the same time that facilitates the continuity of the level, avoiding disintegration of segments out and preventing its detachment (Fig. 5).

An important consideration with respect to the clamping elements is material that you are going to choose, it must to be compatible with the masonry, either physical or structurally. The most common material for this use is the wood because is able for traction forces absorbed although also the steel is used.

On the other hand, the connections between wall and roof with the beams should be carried out by means of elements of bracing and bindings. These pieces tend to be steel pins or wooden stakes. Also the use of vertical rods embedded in the walls connected with the principal beam is another solution that ensures the link between different elements.

While the wood is the material most used for beams, the stainless metal or the metal with non-corrosive treatment, they're another alternative that can be put into practice for these uses.

The connections between the walls that intersect can possess sufficient strength to withstand moderates earthquakes, the same as the original mud construction walls. Is important improve connections and for to achieve this objective, there are different techniques like fill the cracks, the insertion of rods or the installation of wooden parts, these measures allow the structure keep (Fig. 6).

There are some solutions to improve encounters between the walls, applied mainly to the top of them. These supports are intended to prevent that at the time of a seismic movement, they separate the masonries because they move according to the direction of the forces. Frequently, in historical buildings of adobe, wood elements braces the corners, in order to avoid the overturning of the walls are installed.

Other solutions that have been used to solve that problem of weaknesses in earthen structures, is to reinforce these points not only at the top, if not, creating intermediate sites and install other elements. This alternative is commonly used when the collapse that has occurred throughout all corner, making intervention necessary in the entire section.

The use of buttresses and pilasters is presented as a solution to recover the structural stability of the buildings, increasing their resistance to horizontal movements or torque. The buttresses act as supports to prevent the turning of the wall outward, also contribute to improve the integration of the walls that converge in the corners. In this sense,

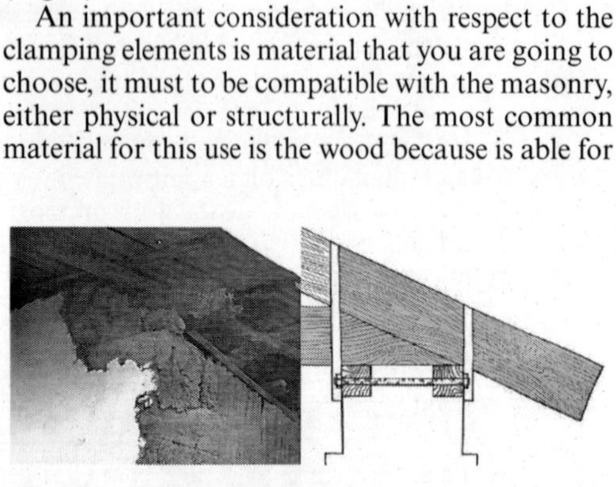

Figure 5. Adobe building. Placement of wooden beam at the top (Hernández Salazar).

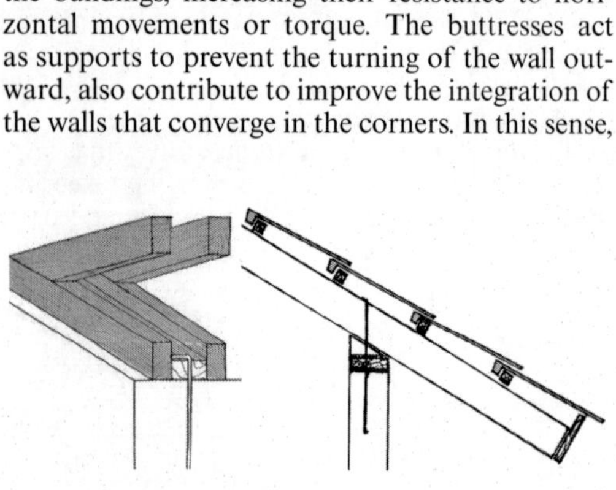

Figure 6. Different connections between the walls and roof's structure (Hernández Salazar).

it's common for buildings, arise buttresses in areas such as the vertices or intermediate areas in the walls perpendicular to then integrate them into the structure of the wall. It's a structural element used in adobe buildings to stabilize walls of low-rise or great slenderness (Fig. 7).

These construction elements must be compatible with the building material and it must be consider, the space available to build them.

There are different systems for to repair the cracks in walls. For earthen structures there are various kinds of techniques to improve these cracks; for example, injecting or sewing them. In the first case, it should attack first the origin of the injury. If there is a crack caused by foundation movements, for example, first must to eliminate the displacement and subsequently repair the cracking. Where cracks are generated by beams breaks, in the same way, it's necessary to eliminate the problem and finally fill the fissure.

It is important take into consideration that the cracks in the walls reflect movements of the building and in others occasions, can hide major pathologies.

Another restoration technique is make sewn on the walls, procedure performed when building movements will be occur once repaired the fissures. It tends to be one of the most commonly used options in the case of constructions subjected to constant seismic events, since it allows oscillation and keeps tied the walls. For the implementation of this technique is recommended use the same materials of the original construction to ensure the continuity of structural walls, although sometimes the sew is made with elements of wood like a keys and even with rods of fiberglass or steel with anticorrosive treatment. Where you opt for this option it should be considered that may appear little injuries in an area close to the repaired. This is because many times is necessary sewn a larger area of the own lesion for example, by to introduce long keys or insert surrounding new adobes to improve the rigging on the wall.

Moreover, has been implemented the system of rehabilitation with elements of wood bordering the walls. This system consists in the installation of wooden pieces in horizontal and vertical direction, in order to increase the resistance of the walls against horizontal movements, at the same time contributing to increase the stability and the joint work of the structure. The tables are placed on both sides of the walls and they're fixing by galvanized steel bolts or threaded steel rods. Also, wooden parts are interconnected by rods or thru bolts that go through the walls. It's recommended that the wood pieces are less than the separation obtained by dividing the height of the wall between 15 cm, and it's necessary to provided that the tables do not have less than 0.20 m thick.

This system is intended to assimilate the structural behavior of adobe's buildings, with the constructions of wattle and daub, trying to make the walls behave as a unit, resisting the perpendicular forces and movements of torque. It has been developed by the Ministry of Culture in Costa Rica, who has implemented it in some buildings throughout the territory. The results have been satisfactory, since you can get a monolithic system. However, it's important to consider that incorporate pieces of wood inside the walls, in the long time could present pathologies related to moisture (Fig. 8).

Notwithstanding the number of techniques and solutions that currently exist, in addition with use of materials and modern methods applied to restoration processes, would generate some problems, mainly, by the incompatibility between the constructive systems and the architectural heritage.

The kind of resistance, the thermal behavior, chemical or mechanical and the physical characteristics are only some of the variables involved in the degradation process of historical structures when they're immersed in this type of modern treatments.

Although there have been important advances in the field of restoration and successful solutions, from the point of view of seismic behavior applied in earthen constructions, there is still popular beliefs and even professional, believing that in the use of

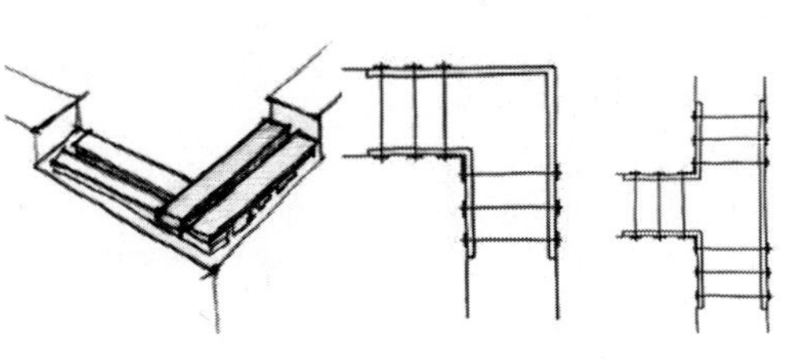

Figure 7. Solutions to improve encounters between the walls (Hernández Salazar).

Figure 8. Adobe building with wooden elements bordering the walls (Hernández Salazar).

contemporary materials found the strength, stability and survival over time, reason why, still continue applying contemporaries materials in heritage conservation projects. Responses to the pathologies that have this type of constructions, often respond to criteria customized for each approach in particular, generating solutions to continue implementing in buildings with similar characteristics. This type of behavior, has contributed to the slow diffusion of good solutions in constructive terms, working harmoniously with the mud buildings.

The efforts should be continued to focus on the multiple possibilities of the earth as a building material and especially the historical value of the works produced with it. The identity of the inhabitants of a country is also in its architecture, so they must have special interest in the rescue of it, doing it in the most appropriate way and under the utilization of the original materials.

Eradicate the implementation of solutions that alter the physiognomy of the buildings, the awareness of the importance of the historical value and to believe that there are more than a simple aesthetic character; are the new challenges for professionals in the field of heritage restoration.

5 CONCLUSIONS

The consequences of the seismic events on the earthen buildings over the years, suggest that their lack of stabilization and fragile behavior have contributed to multiple and repeated losses of this heritage around the world. Also, has come to identify the land as an unstable and unsuitable material to build up in areas with high seismic levels and even its use has been banned in many countries, under the criteria of human security. However the fragility of the land, the progressive loss of construction techniques, lack of maintenance of the built historic buildings with it, the improper techniques using contemporary materials, all these aspects, among others, contribute to that at the time of a quake, the mud buildings are even more inconsistent in this conception and therefore more prone to crashes and collapses.

The challenge of seismically adapt this buildings is a task essential and convenient, since it facilitates the preservation of historic resources of a country, at the same time that strengthening the security of its users. The loss of traditional adobe and wattle and daub constructions should not become a normal consequence after a tectonic event, nor should the implementation of seismic adequacy intrusive or inefficient be the trend.

Also is important to prevent the conception of mud like a dangerous and unsafe material continue consolidating in the collective memory of the citizens of a country, where most of its traditional buildings that have given identity, were drafted with this raw.

The current generation and the future have the responsibility to conserve and safeguard the structures that still survive, through the use of appropriate supportive measures and little invasive, so are preserved the roots of vernacular architecture built with earth.

It's possible to achieve a balance which guarantees the authenticity of a historical construction and the stability against seismic forces. However, responds to any construction in the most appropriate manner to such efforts, it should not rely of only the application of one of the proposals that have been considered formerly at structural level, on the contrary, optimize the earthen constructions behavior should be the ideal and the consequence of a series of joint actions to achieve results after a long process of research.

Only exploring new initiatives for the conservation of the architectural heritage built with earth and making it the priority from the research point of view, they may continue finding new design clues that establish the guidelines to follow and implement for future interventions.

NOTE

This article is a part of a personal research project about *"Restauración de la arquitectura de tierra en Costa Rica. Criterios y técnicas de intervención en zonas sísmicas"* development by Chalíuk: Arquitectura y Restauración del Patrimonio.

REFERENCES

Blondet, M. et al. 2013. *La Intervención del Patrimonio Edificado en Tierra en Áreas Sísmicas y las Cartas de Conservación.* Lisboa: Universidad de Coimbra y Porto.

Blondet, M. et al. 2003. *Construcciones de Adobe Resistentes a los Terremotos: Tutor.* California: Earthquake Engineering Research Institute.

Carranzas Aedo, W. & Rivero Olmos, A. 2002. *Bahareque: Guía de construcción parasísmica.* Villefontaine Cedex: Ediciones CRATerre.

Corporación de Desarrollo Tecnológico. 2012. *Evaluación de Daños y Soluciones Para Construcciones en Tierra Cruda. Manual de Terreno.* Santiago de Chile: Ediciones Raizfutura.

Getty Conservation Institute. 2006. *Proceedings of the Getty Seismic Adobe Project 2006 Colloquium.* Los Ángeles: Getty Center.

Ginell, W. et al. 2006. *Guías de Planeamiento e Ingeniería Para la Estabilización Sismorresistente de Estructuras Históricas de Adobe.* Los Ángeles: Getty Conservation Institute.

Webster, F. 1999. *Seismic design of adobe structures in Costa Rica.* San José: Earth building.

Earthen Architecture: Past, Present and Future – Mileto, Vegas, García Soriano & Cristini (Eds)
 ISBN 978-1-138-02711-4

Rammed earth consolidation techniques: Restoration of Talamanca del Jarama city wall

E. Herrero García
Heritage Specialist Architect, Madrid, Spain

ABSTRACT: Preserving the remains of rammed earth exposed to weathering is still an outstanding problem in the field of historical conservation in Spain. Current efforts are mainly focused on the recovery of former volumes by reconstruction (using more or less compatible materials) rather than on conserving the material. The choice is often between pulling structures down or simply letting them die. But sometimes, those little ruins are the actual remains of a whole age. This is the case of the Christian earth wall of Talamanca del Jarama, now reduced to a stretch of 15 meters. Very valuable information about materials and construction techniques used in the central area of the Castilian Plateau in the 12th and 13th centuries can be still inferred from these remains. It is therefore important to preserve them. Studies on rammed earth stabilization techniques and decisions regarding restoration are presented in this paper.

1 TALAMANCA DEL JARAMA CITY WALL

1.1 *Historical introduction*

Several archaeological excavations have revealed that the Talamanca area has been inhabited since pre-Roman times. However, historical data on population are scarce and isolated until the ninth century—the time when Talamanca began to gain importance following the increase of military pressure of the northern Christian kingdoms. Its geostrategic situation in the valley of the Jarama River made this place the first defensive bastion on the road connecting the Guadarrama Mountains with Toledo.

This made Talamanca del Jarama suffer continual strife, during which their own walls suffered damage from these invasions. There are records of destruction in the years 861, 878, 939, 1047, 1050 and 1062.

In 1086 Talamanca and its subsidiary towns and villages passed definitely into the hands of the Christians. After the Battle of Navas de Tolosa in 1212, the final defeat of Islamic forces occurred. Talamanca would not suffer more attacks from this point on, and a period free of Muslim incursions, protected by the Archbishop of Toledo, began. During this period many important works were conducted on the wall under the orders of Archbishop Jiménez de Rada.

In 1574, Talamanca ceased to belong to the archbishopric of Toledo and passed to the Crown.

1.2 *The walled enclosure*

The fortified area of Talamanca is located on the east bank of the Jarama River, along the Valdejudíos Creek. It stands on a flat plane of low height, with the exception of its southeast flank, closest to the stream, where the terrain is steep. The plant enclosure forms an irregular pentagon of about 310 m from north to south, and 430 m from east to west, and with a perimeter of around 1200 m.

Today the wall only retains 30% of its layout. Only some loose sections are preserved, many of which are covered by existing buildings or overgrown with vegetation.

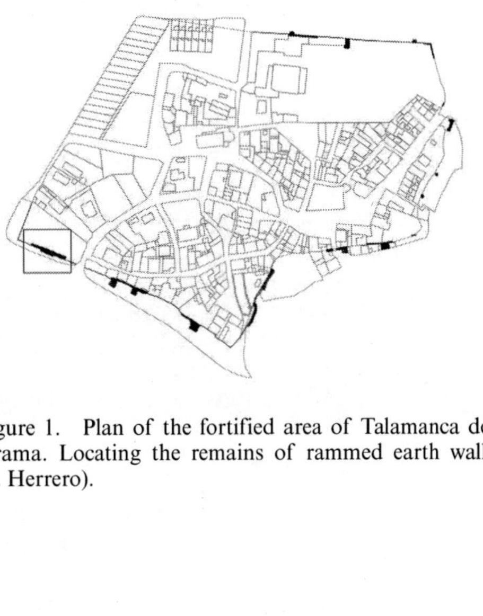

Figure 1. Plan of the fortified area of Talamanca del Jarama. Locating the remains of rammed earth walls (E. Herrero).

Figure 2. Extrados of the preserved 12th–13th century wall. This section is located in the southwest corner of the enclosure (E. Herrero).

As far as we know, the origin of the wall of Talamanca del Jarama is Arabic, and investigation has shown that it was formed by a rammed earth wall over a masonry base. However, this first Islamic wall underwent numerous destructions and repairs which can still be read on some of the stone foundations that remain. For this reason, it is very difficult to easily date the remains of the wall.

Few traces of earth wall are preserved, but among them, one (which during the current restoration works has been dated by thermoluminescence from the 12th to 13th century) stands out. It would correspond with the rebuilding of the walls after passing into Christian hands and despite its small size, the construction system that used to build it up could still be read in the different wall elements. This is why its conservation was so important.

2 RESEARCH ON INTERVENTION TECHNIQUES ON RAMMED EARTH

In 2010 the General Direction for Historic Heritage of the Autonomous Region of Madrid commissioned the restoration project of a stretch of the wall. This stretch included the section made of rammed earth mentioned above. Due to the difficulty of consolidating earth materials, it was decided to investigate the different stabilization techniques for earth walls and their possible applications to this specific case.

2.1 *Techniques for structural stabilization*

2.1.1 *Repairs with one side formwork using traditional rammed earth technique*

This system consists of recovering the lost wall volumetry by using a one face formwork and filling it with a mixture similar to the existing one, by rammed layers. The last two layers, which cannot be compacted, are filled up with a more plastic mixture. The earth mixture should not contain organic matter and its clay content should not be too high, otherwise retraction cracks may appear. If the earth from the fallen wall is still retained on the ground, it is the most suitable to be used, because the proportion of clay is already adequate.

Figure 3. Reconstruction with new wall sections after cleaning up and regulating the original wall. Bounou. Morocco (E. Herrero).

The biggest problem with this technique is linking together the old and the new material. It is essential to brush and prepare the old surface, removing loose material. It is also necessary to introduce linking elements. These elements should be non-metallic, such as pieces of brick or reeds. When filling with new material, the surface must be moist so that water ensures the connection between the clays contained in both materials (existing and replacement ones).

It is not suitable to wedge and repair minor faults (such as it is in the case study of this wall). Recovering important volumes is a good solution because for minor repairs it is difficult to have access in order to compact the new layers of material.

This type of repair is one of the most commonly used today, as it involves the reuse of the traditional techniques that built the original item. However, it is not actually a traditional solution: no examples of this type of repair being used have been found. Repairs of old items with new rammed earth used to be done, but completely eliminating unstable areas of the wall (regularizing their support and unions) and building them up from scratch. It does not seem to have been used to cover small gaps of a material, which was more commonly done using stone masonry, brick or adobe.

2.1.2 *Repairs using concrete-encasement*

This technique consists of recovering the volume and stability of the earth work by introducing a new element made of a different formwork material such as concrete, with very different characteristics of strength and permeability.

Cement concrete-encasement was heavily used in the twentieth century for repairing rammed earth walls since it assured stability while keeping the new intervention clearly differentiated. These

Figures 4 & 5. Emergency intervention conducted by the General Direction of Cultural Heritage under the supervision of architect D. José Ballesta Juste in 1987, consisting of a concrete-encasement. Talamanca del Jarama (E. Herrero).

Figures 6 & 7. Left: Masonry repairs in the base. Talamanca del Jarama. Right: Specific repairs with adobe. M'Hamid. Morocco (E. Herrero).

actions were necessary in many cases to prevent some monuments from collapsing, but today, when knowledge about the use of cement on historic buildings is much greater, it is not an advisable option. Today cement concretes have been replaced by natural hydraulic lime concretes, which are much more flexible and breathable.

The main characteristic of this method is that it is not reversible and it changes the original load system. The new and the former structures do not work together: the concrete-encasement becomes the actual structure. A large supporting area underneath this new structure is needed as it has to be self-stable.

2.1.3 *Repairs using similar natural materials*

This system consists of repairing the damaged areas of the wall with adobe elements (or CEB), which have a similar physical composition. This technique has been traditionally used to repair earth walls.

Its advantage over the repair with the technique of one side formwork is that, as a compatible system, it is more flexible and locks well with the original earth work, allowing it to cover the lack of material quickly and easily (and less expensively). As with the formwork, you need to thoroughly wet the original factory at the time of filling to favor the connection between the clays contained in both materials.

It is recommended to use a mud mortar for bonding adobe bricks. The mortar has to be less resistant than the adobe itself in order to avoid faster weathering in the adobe bricks than in the bonding.

2.1.4 *Repairs using different natural materials*

This repair consists of wedging or filling the lack of material with a material other than earth, such as stone, brick or concrete.

Its biggest problem is that these materials have different strength and permeability than earth, and they usually suffer less damage, remaining unchanged while the earth deteriorates. For this reason, this solution can be hazardous if used to fill the surface at upper levels, as you may end up causing the entire block of surrounding rammed earth to fall. It works best for fillings in depth or for repairs at the base of the wall, where it can even work as a protective base against rainwater.

These repairs should not be confused with the brick or stone belted building systems, because these work in a different way. These systems, for example, work well for controlling the expansions of earth walls and to prevent cracking.

2.1.5 *Projected earth*

This process consists of projecting earth at high speed onto a surface, which promotes adhesion. The projection of material as a system has its origin in the projection of concrete, but recent studies have shown good performance when using earth (Fuentes García 2010).

Although it is still under investigation and it has not been tested enough, in the future it may be a system to consider, as long as a material compatible with the original wall is used. Cement compositions should be avoided.

2.2 *Techniques of surface consolidation*

2.2.1 *Ethyl silicate surface consolidation*

Ethyl silicate is a silicate-based compound which uses highly volatile ethyl alcohol as a solvent. Ethyl alcohol makes the pores remain open allowing evaporation of the alcohol, improving breathability. The product creates strong chemical bonds between the clay particles, preventing separation of the clay layers in the presence of water, forming thereby a protective and impermeable film.

It is necessary to apply it in dry conditions, and on several layers to ensure good penetration. The ethyl silicate has a strong penetrating and consolidation performance (which makes it irreversible), and has therefore been widely used for the consolidation and waterproofing of rammed earth monuments.

However it is important to use a very diluted product, as it can create rigid crusts several centimeters thick, eventually provoking the collapse of the whole block of earth.

2.2.2 *Consolidation by lime washing*
Lime water is a natural consolidant composed of calcium hydroxide diluted in H_2O, i.e. the water found naturally on slaking rafts. This water contains consolidating properties, as it forms crystals that harden when in contact with ambient CO_2.

It is used as follows: for cleaning up and preparing the wall surface, it is saturated with water to provide a deeper, more intense entrance of lime water. Then lime water is applied nebulized, stopping whenever the surface shows a faint glow. Application must be made by steaming to allow greater penetration (no brushing or jet). At least four applications are needed with a gap between them of about 8 days (it depends greatly on the environmental conditions). Each application may proceed while the surface absorbs. Excess unabsorbed water should be removed immediately after each application, to avoid carbonation on the surface and the consequent reduction of penetration in subsequent applications.

This technique is rarely used, mainly due to low penetration rate even if many layers are applied.

2.2.3 *Oil addition*
The addition of some oil (mainly olive or linseed oil) on the wall to make it waterproof manages to reduce the absorption of some water but not hydric erosion, therefore it does not greatly improve its behavior (Madonado Ramos 1997).

2.2.4 *Surface coating*
Earth is a material with low surface resistance to impact and weathering, so it is important that the walls formed with this material have good surface protection against wind, rain, vegetation or insect attack. It is therefore advisable to use coatings.

To coat appropriately and avoid detachment, there should not be an excessive difference of resistance between the new material and the structural base of the wall. Therefore, traditionally, the coatings consisted of applying a thin layer of mud mortar with a clay soil (to improve waterproofing), which was mixed with straw (to improve cohesion and avoid cracking). The best way to ensure a smooth operation is to preferably apply it in several layers (at least two) with a thickness not exceeding 20 mm. The base layer should contain more clay and coarse aggregate than the last one. The ideal quantity of straw is difficult to determine because traditionally it was estimated roughly and depends greatly on the composition of the earth.

Lime based coatings are also common, as their mechanical behavior is very harmonious with that of earth. Lime mortars used are normally 1 part of lime and sand 3 or 4 (Castilla Pascual 2004).

Cement mortar is not to be used, mainly because of its excessive resistance and its waterproof properties which favor the condensation of water within the material (the wall needs to breath and be permeable to water vapor migration), thus generating loss of cohesion of the material.

Many times for strengthening and speeding the setting of coatings, bastard mortar (a mixture of lime and cement to form an artificial hydraulic lime) is used. However it would be more appropriate to use a natural hydraulic lime which is more compatible with the earth and has better breathability and cohesion.

3 INTERVENTION MADE ON THE RAMMED EARTH WALL IN TALAMANCA DEL JARAMA

At the time of the restoration works, the rammed-earth wall had lost much of its protection systems: its base in the inner face was underground, it had no coronation and the coating was almost completely lost (only one small piece is preserved in the external side). The wall was therefore fully exposed to the direct action of rain and wind, which were generating significant loss of volume.

As the intrados base of the mud wall was buried and in direct contact with the ground, the masonry basement could not properly fulfill its function, and moisture from the soil had eroded the earth of the base to the point that its stability was compromised. Therefore, as a first step, it was necessary to lower the ground in the inner side until the stone plinth could be detected and regain stability, repairing the eroded area.

Recovering its volume was rejected mainly due to two reasons:

1. The wall was completely out of context, not having a clear link with the rest of the walled enclosure. Centuries of neglect of the Talamanca city walls led them to lose almost all of their original layout, so that retrieving the overview was conceptually and economically infeasible.
2. There was insufficient documental and stratigraphic information, to allow a clear volumetric delimitation without giving rise to a false history. How should length and height be rebuilt? How should the coronation be finished?

The interest of preserving this small stretch was mostly historical and documentary. Therefore, minimal intervention was desired, mainly based on restoration of stability (using adobe brick), and regularization of the surface to reduce water ingress (putting a mud and lime coating).

3.1 *Stabilization using adobe brick*

Once the stabilization system for the wall had been studied it was decided that the best solution for this case was to repair using adobe bricks. The bricks were produced on site. As it is a traditional technique that is no longer used, the job was supervised by two specialists in adobe construction. Mr. Jesús Ortega and Mr. Pablo Demetrio explained *in situ* the method they used to employ back in the 40s and 50s in the area of Guadalajara, when adobe was commonly used to build houses and fences.

3.1.1 *The composition of adobe is*

- Clayey earth from the former wall. When selecting the soil the first layer of the ground, which is high in organic matter, was removed as the earth for the production of adobe cannot have any. After gathering the earth suitable for the production of bricks, we proceeded to sieving it to remove coarse content and obtain homogeneous granulometry.
- Threshed rye straw. The straw fibres should not exceed 5–6 cm.
- Water.

3.1.2 *Preparation of dough and moulding the bricks*

After sieving the earth, straw was gradually added and mixed. Once earth and straw were well blended, water was added, and mingling was continued with shovels and rakes until the correct texture of the dough was obtained. The amount of straw that was added was approximately 15–20% of the volume of mud used in each kneading.

To mould the bricks bottomless wooden moulds, called *adoberas,* were used. They were filled with a mixture of mud and straw. Afterwards they were pulled out directly from the ground for drying. To cut the excess of material and to match the surface, a wooden ruler was used. The removal from the casts was performed by soft vertical shake. The mould should be kept moist to facilitate removal of mixture so it was immersed in water after each use.

Adobe was left to dry for a first period of at least 24 hours. Once proven dry, bricks were placed on edge to facilitate drying on all sides. Bricks were left to dry for a minimum of 7 days until used.

Figures 8 & 9. Photographs of the underpinning work with adobe in the soffit of the wall. The bricks were laid on headers and a mixture of mud and straw was used as a binder mortar. Brick dimensions are 29 × 13 × 6.5 cm (E. Herrero).

3.1.3 *Recovery of putlog holes*

The conservation of this stretch of wall had a primarily didactic purpose, therefore, the location and layout of the wall putlog holes, which were completely covered by landslides, was essential. These holes were reconstructed where underpinning adobe was made and simply consolidated in all the other areas. The recovery of the putlog holes was important, because these elements help us to study the dimensions of the original sections of the rammed earth wall and the methodology used in its construction.

3.2 *Coating of mud and lime*

Once the composition and condition of mud was studied, it was decided that it was necessary to coat the wall to protect it from rain and wind, and regularize the surface in order to avoid water accumulation.

The earthy texture of the coating was desired in order to change the image of the monument as little as possible, which is why mud mortar was used. But as the durability of a mud coating is very short, lime was added to improve it.

After making various samples of mortar, the best result obtained was the following composition: 1 Kg of natural hydraulic lime mortar, 600 grams of selected earth (from the original mud wall on the surrounding ground), 600 grams of gravel and 150 grams of straw. The ideal composition of these mortars is difficult to obtain and only time will tell if this was the right choice. However, it is important

Figures 10 & 11. Detail of a putlog hole before and after reconstruction (E. Herrero).

Figure 12. Detail underpinning adobe base (E. Herrero).

Figure 13. State of the earth rammed due to the action of rain and wind, which were generating significant erosion on it (E. Herrero).

Figures 14 & 15. Left: After the consolidation works on the wall Right: The mud coating enriched with lime that was placed (E. Herrero).

Figures 16 & 17. Intrados side. Comparative pictures before and after restoration (E. Herrero).

Figures 18 & 19. Extrados side. Comparative pictures before and after restoration (E. Herrero).

to note that any such coating would need a minimum of maintenance, something that is currently forgotten after conservation works.

The mortar obtained was applied as traditional lime mortar, taking special care in wetting the entire area to ensure a good bond between rammed earth wall and the mortar.

4 CONCLUSIONS

The difficulty of maintaining a wall of earth that has lost its protection systems means that the actions carried out in most fortified enclosures of rammed earth consist mainly of the full recovery of the volumes or in the absence of intervention. But in cases such as Talamanca, where reconstruction is not conceptually or financially feasible, but it is necessary to preserve the remains because of the information they can still provide, it is important to consider a third option based on structurally and superficially consolidating the rammed earth, with minimal intervention.

Regarding structural consolidation, the most suitable solution with the criteria of minimum intervention, reversibility and compatibility of materials, is repair using adobe bricks. This technique is well suited to restore walls of earth because it clearly marks the new intervention while fitting perfectly, as the base material is the same.

According to tests carried out in Talamanca, using silicates or lime washing to consolidate the surface is not very effective. This is mostly because they do not prevent the accumulation of water in the cavities and irregularities of the surface and it ends up entering inside. The best way to avoid this water ingress when the wall does not have coronation, is the use of small pieces of adobe to fill important voids and mud mortar to regularize and coat this wall. In this way, the water drains quickly, preventing it from penetrating inside.

So although at first it may seem that the use of a coating may lead to a reduction of information, ultimately it is the best way to preserve it. This coating will extend the life of the earth wall and may be reversible in the future. Furthermore, once the first rains fall and the coating starts to get dirty, the earth mortar quickly loses the appearance of being new and integrates well with the surroundings.

REFERENCES

Castilla Pascual, F.J. 2004. *Estabilización de morteros de barro para la protección de muros de tierra*. Madrid: Tesis doctoral. Universidad Politécnica de Madrid.

Fuentes García, R.M. 2010. *Construcciones de tierra. El tapial. Nuevo sistema de construcción y restauración mediante la técnica de Tierra Proyectada*. Granada: Tesis doctoral. Universidad Granada.

Maldonado Ramos, L.; Castilla Pascual, F.J. & Vela Cossío, F. 1997. La técnica de tapial en la Comunidad autónoma de Madrid. Aplicación de nuevos materiales para la consolidación de muros de tapia. *Informes de la construcción* N° 452. Madrid: CSIC.

Earthen Architecture: Past, Present and Future – Mileto, Vegas, García Soriano & Cristini (Eds)
© 2015 Taylor & Francis Group, London, ISBN 978-1-138-02711-4

From stone to rammed earth: The open plan architecture in Sinyent Farm (Valencia, Spain)

F. Iborra Bernad
Departamento de Composición Arquitectónica, Universidad Politécnica de Valencia, Valencia, Spain

I. Matoses Ortells
Consellería de Educación Cultura y Deporte, Generalitat Valenciana, Valencia, Spain

ABSTRACT: The rammed earth construction is often associated with heavy walls and small openings, at least in the lower part of the building. However, the assimilation of a typology thought to be executed in stone could lead to surprising solutions. This is the case of the medieval farm in the site of Sinyent (Valencia) built probably around 1330 A.D. as head of a holding controlled from the monastery of Santa María de la Valldigna. The building, now abandoned, is a unique example of synthesis among the solution of a medieval farm and a manor house of Andalusian inspiration. The first refers to the model observed in Sangüesa by Vicente Lampérez, with ground floor opened through arches and a closed upper floor for the dwelling. The second reference follows the model of the early Christian palaces of Valencia and probably reflects an iconic image, perhaps that of the house of the priory of the monastic order in the city of Valencia.

1 OLD "ALQUERÍA" OF SINYENT

The Muslim word *alquería* nowadays refers to a farmhouse, but it is also used for small rural towns, often located in the vicinity of a defensive construction, and not very far from a point of water supply. The entire territory was controlled by a *hisn* (a fortress) in this case the castle of Corbera, which offered shelter and defense to the population of nearby farms and towns in times of insecurity.

In the case of Sinyent, we found a place situated very close to the Júcar River in a strategic position, probably the confluence of the roman Via Augusta and Via Sucro-Saetabis. In addition, from immemorial times, this area was one of the main crossing over the Júcar River. When crossing on foot or horseback was not possible, it was done by boat. It is known that in 1263 one of the main benefits of Corbera dominion was precisely the *barcha de Corbera* (the boat of Corbera) and, when the counts of Carlet acquired Albalat (a town next to Sinyent) in 1494, the sale document included the privilege of boat crossing (Matoses 2012).

About the settlement of Sinyent, we know that since 1328 it belonged to the Cistercian monastery of Santa María de la Valldigna, founded by James II in 1298. The former owner had been Pedro Ferrer de Campredó, who acquired the lands to various individuals. We know that properties sold by his executors were the farmhouse of Sinyent with adjoining orchard and a dovecote, three oil mills with his workshop and a cauldron for the oil, and four houses next to the main building. These latter were situated in territories of Corbera dominion, bordered with two public paths (still existing) and lands of the same testator. A vineyard and an oven to bake bread located on the settlement of Sinyent, among other properties, completed the property acquired (Gascon 1996: 135–136).

2 THE MEDIEVAL FARMHOUSE PROPERTY OF THE MONASTERY OF VALLDIGNA

The word *Granja* (Farmhouse) normally refers to a set of buildings for agricultural uses, including housing for the owner, the managers and the settlers, stables, barns, etc. Buildings are generally grouped around a courtyard, sometimes surrounded by a fortified wall. The meaning of *alquería* and *granja* is virtually the same, but we prefer to keep the two terms in order to differentiate Muslim and Christian phase. The site of Sinyent, called "La Granja", remained in the possession of the monastery of Valldigna until the confiscation of 1837, when it was sold by the State to an individual.

2.1 *Main façade*

"La Granja" appears to the visitor as an imposing mass of rammed earth and brick walls, whose antiquity is immediately betrayed by the presence of pointed arches in doors and windows. The ground floor consists of a portico with arches, from which only the central, built with stone and slightly larger in size than the rest, has remained open. There also appear arches in the side walls and the rear façade, opened over a large courtyard where other newer buildings stand around. Upstairs there are three medieval windows, but only the central remains intact (AA.VV. 1996: 623, Soler 2002: 206). The arched shape of these windows serves us to establish a probable chronology in the first half of the fourteenth century, before the trefoil arches were widespread. The presence of a small Gerona stone column with capital of palms is also significant. The manufacture and marketing of standardized shafts and elements of Gerona stone began to acquire importance in the first half of the fourteenth century (Español 1999).

The main façade of "La Granja", with an open ground floor and few windows at the top seems similar to the Italian municipal palaces of the thirteenth century, like the famous Broletto of Como (1215). The function of this open space, within a rural setting, could be holding agricultural markets. In this sense, we find a similar arched façade, now blocked off, in the "Alquería del Moro" complex, in Benicalap (Valencia), but it only has the arches and not an upper floor. Similarly the "Molino de los Frailes" (Friars Mill) in Campanar (Valencia) presents arches in a part of its façade, but neither

Figure 1. Main façade (real state) and hypothetical restitution (Iborra).

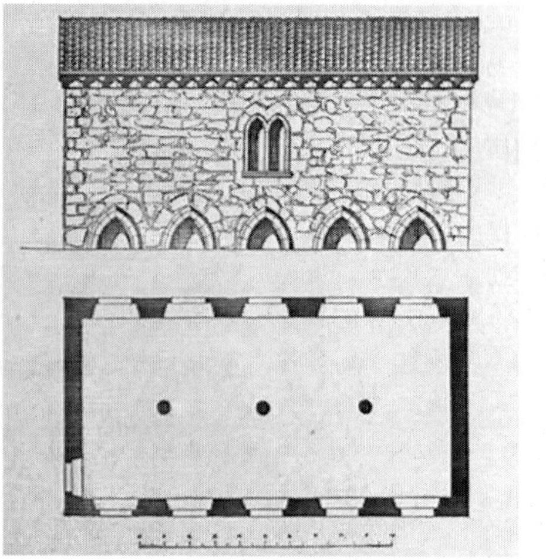

Figure 2. "Granja de Sangüesa". Drawing by V. Lampérez.

of these examples has the rotundity of the house in Sinyent.

Greater similarity presents another medieval building that existed until the early twentieth century in the vicinity of Sangüesa (Navarra). It was also a farmhouse dependent on a monastery and was documented by Vicente Lampérez before it was destroyed to get the stone in a railway station. It was included in his book *Arquitectura Civil Española* (1922) as the only example of this type.

As described by the author: "It was a rectangular building, 16 × 8 meters, of Gothic style, built with masonry and a gable roof. On the façades it had several pointed arch windows, some of them twins, corbels cornice, and at the bottom, very underground, several Gothic arches like gates or a porch. Inside, already dismantled, were the remains of a room with eight vaults over central pillars (of the type of the Poblet cellar) and, if I remember correctly, another wooden floor. At the level of this the voids of fireplaces and chimneys were seen in the walls. It belonged to a *Claveria* or farm dependent of the Royal House of Roncesvalles, and it was already mentioned in documents of the twelfth century, and described in the sixteenth. The building belonged to the thirteenth or fourteenth centuries. A great pity was the destruction of this interesting building, executed to get the stone in a railway station" (Lampérez 1922 I: 64–65).

2.2 *Rear façade*

If the main front is interesting, much more unique is the rear façade. It has a tripartite composition, with lateral bodies that seems protrude slightly, as it were old towers. These "towers" have all four fronts perforated on the ground floor and have large arches opening onto the courtyard at the first floor. Between them stands a portico of three unequal arches on the ground floor and an upper gallery with five openings, including a narrower pass that could serve as a door linked to an external staircase, now missing.

The pointed arches in this upper gallery are reminiscent of those in the interior of the "Alqueria Fonda" in Campanar (Valencia), dated in the fourteenth century. They also bear a certain resemblance to the arches of the ground floor in the Admiral Palace of Valencia, a building also constructed in the early fourteenth century, but very intervened in later periods.

The tripartite plan and its inhomogeneous double bay (the back bay is slightly smaller) looks remarkably like many examples of Andalusian architecture. Some unique details very significant, as the unequal width of the towers, which was present in the Real Palace (Valencia), Galiana Palace (Toledo) or Alja-fería Palace (Zaragoza). Another particularity is the hall with a great arch and two smaller gates flanking it in the intermediate wall, present in the three examples cited (Iborra 2012 y 2013).

These particularities give us a first impression that the building might be erected on the southern hall of a Muslim palace or recreational estate. The particular metric of walls and rooms refers to Valentian palms of about 22.65 cm, but the total measures fits better to the ma'muní cubit. This is very evident in the rooms flanking the internal gate. Specifically Eastern measured 3.77 meters (8 cubits) and Western 3.51 meters (7.5 cubits). The total length of the gallery, including these rooms, would be 19.68 meters (42 cubits). The outer theoretical total size would be 45 cubits, but this measure and that of the central room are distorted because of the thickness of the walls, which have been materialized in Valencian palms, losing about 5–6 centimetres over the ideal total of each wall.

Figure 3. Rear façade (real state) and hypothetical restitution (Iborra).

Regarding the rear bay, where the position of the arches does not respond to any Islamic prototype, it is fully dimensioned in the Christian metric. Staking appears to have started on the west side, with a dimension of 6.86 meters (30 palms). The central section measured 4.55 meters (20 palms) and the eastern absorbed four inches more of the overall length, conditioned by the Andalusian metric. Perhaps most representative are the totals of 20.98 × 8.67 meters, equivalent to 92.5 × 38.25 palms, which are very close to 45 × 18.5 ma'muní cubits. All this suggests that the stake has been made in Valencian palms, but reproducing a prototype executed in Islamic cubits, executed exactly in the same scale.

Reuse of an earlier Muslim building is impossible with a fully perforated ground floor. Neither the use of earlier foundations has any sense, because the piers in the portals would require a larger area. In addition, the building to which its plan refers seems an urban palace of certain sumptuousness and not a mere farmhouse houses formed by free peasants.

Another possible explanation is the copy of a particular building prototype, which could be a transformed Andalusian palace. But in this case it would have been more logical to choose the scheme of a rural building. The last alternative is the intentional copy of a representative image of a building common to other properties of the monastery, but modified by introducing the necessary arcs to perform the rural mercantile function. Since the monastery of Valldigna was a new religious foundation, it has no sense the survival of a Muslim palatine scheme, but it could be logical thinking in another building, like the priory house of the order in the city of Valencia.

This latter building was located in the current Landerer street, historically called Valldigna street. Orellana situated it opposite a square that exists on it, opened in the second half of the eighteenth century after the demolition of several houses (Orellana 1924: II, 673–674). The site of the building is nowadays a part of the plot of the Patronato de la Juventud Obrera, where it is currently established the Sala Escalante theatre. The large plot encompasses se-veral older buildings. The main one was the palace of the Barons of Cortes and Ruaya, in which the institution moved in 1895, although later other properties were acquired. One of these would be the old priory house Valldigna or rather, the housing block built in the nineteenth century.

What most interests us about this is that the façade of this building has a length of about 20 meters and the eighteenth century urban plan of T.V. Tosca (especially in the engraved revision made by Fortea in 1738) represents it as a double bay building with a large backyard. The first priory house was donated by King James II by privilege of July 5, 1319. The Cistercians of Valldigna also founded there the Priory of St. Bernard "de las Casas de Valencia" (of the houses in Valencia), which allowed them to establish a representation with the rank of embassy. Here was also the residence of the abbot, where economic, cultural, artistic and literary activities were centralized (Gascón 1996).

The hypothesis that the plan of Sinyent reproduces the primitive priory house in Valencia enters, therefore, as far as plausible. This strange building has a number of elements that really are not working well in a rural environment, but would make sense in an urban context.

2.3 *Internal layout*

A good proof of the symbolic and representative character of the building is that the internal layout does not respond to the apparent external schema. Alleged rear façade towers were never really implemented.

The space above the brick gallery is filled with a medieval rammed earth wall, as in some way occurs in the courtyard of the Royal Palace in Perpignan. Moreover, arrow slits turning towards the gallery and a free space corresponding to the passage from the primitive stairs are preserved in the top of the central wall. This disappeared staircase leading up to the terrace would be attached to the east wall, where two corbels are preserved at an intermediate level of the wall. A small window, opened to hoist a banner, is also at this same level. Moreover, the absence of stone corbels for the roof of the alleged gallery confirms that it was uncovered.

The south bay was originally covered with a flat roof situated over the level of the current cornice, as deduced from the presence of corbels, but the area of the gallery seems to have been uncovered in origin, as said above. The stone corbels and distribution beams over them are preserved, as well the marks of the primitive pavement with its slope and remains of the four original gargoyles on the main façade, consisting of tiles. Above this level we found the remains of a low parapet with square arrow slits approximately 45 × 45 cm located about 20 cm over the floor. The reduced height of the parapet and the presence of the arrow slits suggest that it would be a crenellated parapet, although no remains of crenellations had been yet identified.

Inside, the medieval floors have been replaced with wooden beams and brick vaults, probably contemporary with the implementation of the new gabled roof in the late eighteenth or early nineteenth centuries. These floors are in a poor condition due to continuous filtrations of water, with partial collapses, large deformations and even decomposition of gypsum in many of the vaults.

The ground floor is much transformed, being used as a residence until the mid-twentieth century. The humidity has affected both walls and partitions, so their condition is deplorable and is not worth restoring them only by ethnological values, given the importance of the medieval phase of the building. Taking into account that it originally was a open plan space and that the ground floor arches are well preserved, releasing of the original space seems the most suitable solution.

On the main floor, at least two very interesting Gothic brick doors are preserved. In one of this the fine plaster border can be perfectly seen, contrasting to the rest of the wall with a much rougher

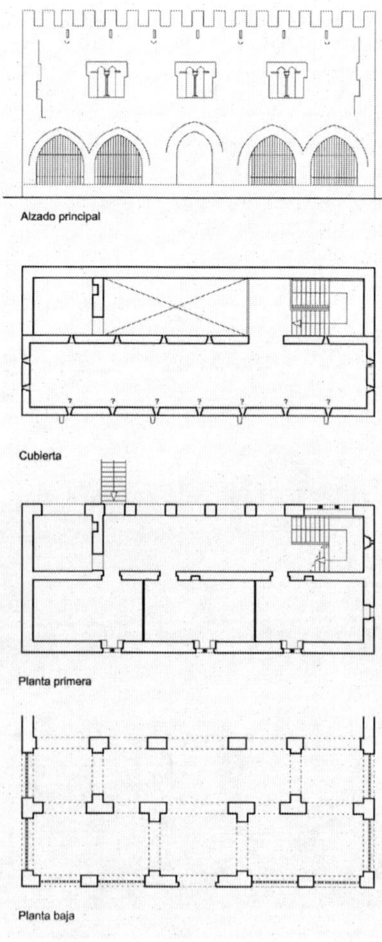

Figure 4. Hypothetical restitution of the main façade and plans of roof, first floor and ground floor (Iborra).

plaster. Perhaps they are among the best preserved examples of this era.

There is a third old door walled up and a square opening of more modern appearance, linked to the plateau of the stairs in the gallery, and two small credenzas or niches in the wall. The original layout seems to have had three bedrooms, with equidistant partitions now missing, that had nothing to do with the position of the arches on the ground floor. The latter were part of the inherited model and make sense only in the type of manor house in which the plan could be inspired. Supporting a wall on one of these arcs it would be easy getting the prototype layout of a medieval palace with a *sala* or main room and a *cambra de paraments* or secondary room, used as a representative bedroom.

The rear bay should have been, at least in the ideal model, a large hall with a gallery opened to the courtyard and two small alcoves in the ends, rising a little more in the form of towers. However, this was only in theory, as explained above. It would be originally a false gallery, which could be reached by an external staircase that would provide access to this main floor, having another staircase to reach the roof.

3 CONSIDERATIONS FOR FUTURE INTERVENTIONS

All that has come down to us is genuine, except the floors and partitions. The medieval building is perfectly readable and even rebuildable. But a huge risk of losing the authenticity of this image runs in a intervention seeking to return to its original state rebuilding what is now lost, even if the new addition is made with traditional forms and materials. Equally dangerous is an act of excessive zeal to preserve only the original parts of the building with the addition of contemporary elements, which too often ends up being an operation of personal glory of the architect. What we think accurate is acting in a conservative way, although there are some elements that are unrecoverable and must be replaced.

Despite the initial temptation to anyone who begins to understand the building, it seems unwise attempting to recover the original volume. Aside from altering a consolidated external image, the execution of a new parapet wall (still using traditional techniques) inevitably would present a different aspect, much newer. Even patina processes with earth tints are no longer than a superficial dyeing that never reaches the rough texture of aged material. Equally problematic would be rebuilding the crenellations. Making new crenellations without any example preserved could be considered a false historicism, but not recovering it would also give a false impression of the primitive building.

Moreover, any attempt to restore the original upper part with the same technique of rammed earth will have the appearance of a "hat" on the historic building, given the difficulty of aging this material properly. A practical alternative would be that of making a brick parapet covered with rough plaster to imitate the texture of the wall. However, and despite the care by matching tones, difference between new and old finishes is always appreciable. The alternative would be renewing the entire surface of the main façade, operation which could be perhaps debatable on stone walls but is not recommended at all in a rammed earth wall covering with plaster, as shows the experience in Betera Castle (Valencia).

For these reasons and given the easy readability of the original elements, we consider wise keeping the current roof. It is the most sensible option considering the materiality of the building itself. Otherwise, it connects within the modern criteria of restoration understood as conservation and valuation of the entire history of the building. A model or a drawing in an explanatory panel inside the complex could help to understand the original configuration.

The rear façade is more problematic, due to the current state of advanced degradation. But a brick wall is more grateful than a rammed earth wall in the case of partial reconstructions, as it is apparent in the tower of the Casa de la Serena in Alfara del Patriarca (Valencia) or Forna Castle (Alicante). A type of brick similar to the original would be used to maintain a harmonious image of the complex and may, in any case, be subtly marked with a profile of the area rebuilt with a recessed level, or some element identifiable only from a close proximity, as a small brass band. The date of the reconstruction recorded in some of the new bricks or additions would document it perfectly. The arches of the windows in the main façade also might be completed within the very principles of the scientific or philological restoration, widely used by archaeologists.

More questionable could be the reposition of the columns in the windows. As far as we are concerned, a clearly identifiable new element would break unnecessarily the harmony of the façade, and a volumetric solution could be too rough on a south-facing location, where noon light marks the shadows of his decoration. Otherwise, the Gerona stone columns, with its bases and capitals, were produced serially, and there is one of the original pieces preserved in the building. It would therefore be reasonable making a reproduction of it, identifying new parts with a date to be easily identifiable in future.

Regarding the arches walled up in the ground floor and rear galleries, the opening of them would be well supported by various international agreements on restoration. But indiscriminate opening in the exterior walls in a building located on the outskirts of the town, will raise the need for expensive safety glass or metal fences. They allow an external vision of the ground floor, but also alter significantly the established image of the building. In addition, the suppressed walls are part of the history of the building and even could have been executed in late medieval times.

Moreover, opening the arches would force to work again the brick pillars, made with a very poor mortar and damaged by moisture. These pillars should be reinforced with a hidden structure or replacing its deteriorated bricks.

Finally, the main problem is the stability of the building and its walls of rammed earth, which could be quite degraded by rain infiltration. The main façade maintains the brick cornice, despite having lost a part of the roof which protected the building. The same occurs in the intermediate wall. Injections of lime in rubblework may be more useful, but here we consider preferable pouring water slurries of lime from the top, which is in any case the most affected and discouraged area. This solution, used during the Middle Ages to consolidate vaults, would allow calcification and hardening of the top of the wall, which is the area that will receive the new roof. In any case, it must be avoid at all costs the replacement of the core of the rammed earth wall by modern brick wall, a practice that has unfortunately been common until recently.

It is also worrying the inclination of the eastern end wall. The possibilities of stitching the corner, or a foundation underpinning should be studied in detail. It could be a level tied in correspondence with the intermediate floor, controlling much visual impact of external perforations. In this case, the bars would be anchored on a rammed earth wall, material which presents adherence problems and the danger of sliding, as was noted during the restoration of the Castle of Alaquás (Valencia). In that building, it was decided the execution of a mortise and a ring-beam of concrete. However, the experiences of recent earthquakes in Italy have shown the disastrous results of interrupting the continuity of a wall introducing elements of greater rigidity. Therefore, we believe sufficient and recommended a tied from inside with crossing bars of fiberglass or carbon fiber, connecting with the wall at the roof and floor levels.

Figure 5. Arched window in the first floor (Iborra).

Regarding the horizontal structures, its state is virtually unrecoverable. Since they re to be replaced, it makes no sense keeping the system of beams and vaults, built after the rest of the house. We could chose between returning to the original design of floors or the introduction of a modern and contemporary solution. Paradoxically, most modern criteria seems to be reproducing the original traditional solution, which could be identified observing the footprints preserved, although it would be necessary to adapt it to current overloads and fire regulations, which could force to increase its section. In this way have been restored the Casa del Mestre Penya in Sagunto (Valencia), reproducing the original floor of which there were some physical remains, but increasing the section of beams and raising the level of the pavement about 10 or 15 cm.

REFERENCES

AA.VV. 1996. *Castillos, torres y fortalezas de la Comunidad Valenciana*. Valencia: Levante.

Español, F. 1999. Los materiales prefabricados gerundenses de aplicación arquitectónica, (siglos XIII–XV), In *L'artista-artesà medieval en la Corona d'Aragó*: 77–119. Lérida: Universidad de Lleida.

Gascón, V. 1996. *Florones del Señorío Valldignense*, Tabernes de Valldigna.

Iborra, F. 2012. La imagen y la arquitectura del palacio del monarca en los reinos de Valencia y Mallorca durante el siglo XIII, In *Las Artes y la Arquitectura del Poder. Actas del XIX Congreso del CEHA*. Castellón de la Plana: Universidad Jaime I.

Iborra, F. 2013. Corte y cortesanos: evolución tipológica residencial y ecos del palacio del monarca en el Reino de Valencia entre los siglos XIII y XV. *Anales de Historia del Arte* 23(II): 469–485.

Lampérez, V. 1922. *Arquitectura civil española de los siglos I al XVIII*. Madrid: Editorial Saturnino Calleja.

Matoses, I. 2012. Desvelando el arcano de la ribera del Júcar. La casa de Sinyent, In *Actas del IX Congreso Internacional de Arquitectura en Tierra. Tradición e innovación*. Valladolid: Universidad de Valladolid.

Orellana, M.A. 1924. *Valencia Antigua y moderna*. Valencia: Acción Bibliográfica Valenciana.

Soler, P. (dir.) 2002. *Guía de arquitectura de la provincia de Valencia*. Valencia: Colegio Oficial de Arquitectos.

Zaragozá, A. & Iborra, F. 2008. Una aproximación a arquitecturas desaparecidas: el palacio episcopal, el palacio de En Bou y la capilla del Real Viejo de Valencia. In Zaragozá, A. (ed.), *Jaime I (1208–2008) Arquitectura año cero*: 134–156. Valencia: Generalitat Valenciana.

Earthen Architecture: Past, Present and Future – Mileto, Vegas, García Soriano & Cristini (Eds)
© 2015 Taylor & Francis Group, London, ISBN 978-1-138-02711-4

Contemporary rammed earth construction. Alexandre Bastos—creativity and maturity

F. Jorge
ARGUMENTUM Edições, Lisbon, Portugal

ABSTRACT: Rammed earth has been a very popular traditional building technique, without change since early times. Low material costs and raw technique pleases the housing needs of poor populations without resources for complex technologies. In the past, rammed earth was not a choice but the outcome of the economic and cultural context of countryside. Today, rammed earth is no longer a technology of the majority but an option of a few selected elite.

> *"The times when earth was the basic building material for the poor are now over. Today it is used to build houses for the rich. But unfortunately, the State is still unable to come up with the earth to build social housing."*
>
> —Elie Mouyal, Architect, Marrakech

Rammed earth construction illustrates the uses of a material that is in abundant supply in the territory. It is a traditional building technique that has been repeated without change since early times.

It is perhaps one of the oldest and most widespread building techniques in the world and is still used today in the same way it was in the past.

The material earth is readily available on site. The method has been used to develop a self-building method by the users of the built structure themselves under the guidance of a master. Low material costs and rudimentary techniques were the two-fold response to the housing needs of poor population groups lacking the resources for complex technologies. In the past, rammed earth was not a choice but simply a consequence of the economic and cultural reality in the rural world. Today, rammed earth is no longer a technology of the "masses" but the choice of minority groups.

In addition to its advantages in political, economic, social, technical and ecological terms, there is also a cultural and architectural interest in 'earth' as a material. The diversity of earthen architectures and the building methods make possible a kind of commitment to counter cultural imperialism and against a return to the uniform and character less norms of the 'International Style', which has been difficult to escape from for a number of years now.

Thus, the use of earthen architecture could facilitate the vital reintegration of architecture into diverse cultural and popular traditions that are inherent to the communities and reconcile it with the purposes and uses of local engineering, recreating at the same time a dynamic cohesion and a lasting link between the history, the present and the future."

This quotation from Jean Dethier, taken from his introduction to the catalogue for the exhibition *Des Architectures de Terre ou l'Avenir d'une Tradition Millénaire* [Earthen Architecture or the Future of a Thousand-Year-Old Tradition], illustrates how the return to the tradition was a way of giving a response to the situation as it was experienced in the latter half of the 20th century, providing a great impulse for a generation that sought solutions to modern-day problems to look to the past.

In the early nineties—curiously enough, the decade in which the 7th International Earth Conference was held in Portugal—a new generation of architects became inspired by a growing awareness of vernacular architecture and traditional building techniques.

At that time the first examples of single-family homes (for the urban middle classes) built entirely using the rammed earth technique and designed by a generation that had consolidated knowledge of the traditional building methods could already found.

Towards the end of the 20th century it was the new generation of architects, engineers and builders who inspired, and influenced each other, towards the re-learning and reutilisation of the traditional rammed earth technique—both in new works and, in particular, in the renovation and rebuilding of ruins or abandoned houses that had been built with rammed earth decades earlier.

The merging of the knowledge of the traditional techniques and the development of building practices thanks to new technological advances opened up a path of optimism and hope for a professional practice that was closer to the people and the master builders and, thus, a more mature practice that absorbed and protected the local culture and sought to give it a lasting future.

This new generation of architects, influenced by traditional techniques, turned to rammed earth as a way of continuing traditions and safeguarding a more utopian and more sustainable future.

The architect Alexandre Bastos stands out of this generation. His professional work has become a manifesto for, and defence of, the use of rammed earth as a valid building technique based on informed and balanced choice. This choice was based on a vast body of international experience reflecting ecological, creative and conceptual options.

Rammed earth was chosen as a traditional building method, a method that was also seen as offering technological rigour and sustainable versatility and one that could be used with conceptual maturity, based on a pluralistic approach to artistic freedom of expression.

Alexandre Bastos, who settled in the Odemira area in 1984, could have no idea that when he began work on designing and building his own design studio in 1993 he was setting out on a professional trajectory that was to include dozens of rammed earth buildings. He had just completed his first five years of working in the profession in the company of some good architects of the day. His background also included attendance of a Fine Arts/Painting course before taking up architecture.

To his academic training Bastos added a particular sensitivity to vernacular architecture and good knowledge of the territory.

In Odemira his knowledge of the local culture was consolidated through his friendship and association with masters of the rammed earth technique, who were the last remaining carriers of the expertise on this method that had been handed down through generations.

Driven by a desire to work on his own terms, he designed and built his own studio working with these masters in rammed earth, with whose help he developed his own technical knowledge.

"In contrast to the past, using earth as a building material today is an option and not a necessity. This difference highlights the utopian aspect, all the more so because rammed earth walls (which are normally just the exterior walls) make up 10% to 14% of a whole work, and as the costs are less than half those for a conventional building project (there are no pillars, foundation footings or beams or normal lintels; there are no double brick walls with insulation or gutters for moisture; nor does plaster have to be applied to one side of the wall), it is clear that the conventional method takes a lot more time and involves much greater costs (...)."

(...) "Never before had architects who work with earth thought so much about construction and its destruction. Tradition had it that houses were born, lived and then died. This utopia of the permanent construction, the choice between being ephemeral and being eternal, alters the concept of construction and, above all, that of architecture. It brings with it others utopias: the full freedom and great flexibility that rammed earth, which can be modified for the coming generations, offers—making the ephemeral with an eternal material and r-using the same material at zero cost, giving it new life."

Alexandre Bastos followed his path, conducting ongoing investigation into the technologies that the masters revealed to him whilst building his works and also drawing on the constant disquiet of his professional practice, with his questioning of the way in which he conceived his designs and the coherence of the expression theoretical reflections inherent to the act of designing.

His studio building was and is a paradigmatic work not only for his own career, but also in terms of the contemporary development of the rammed earth method in Portugal.

Figure 1. Atelier building—exterior (F. Jorge).

Figure 2. Atelier building—interior (F. Jorge).

First and foremost, for his conscious choice of the technique (for which it was difficult to get permits at the time), but also for the objectivity of the building's programme, a painting, photography and architecture studio, a space with a double ceiling height and a mezzanine, that included a laboratory and sleeping quarters for occasional use.

The building, wholly built in exposed rammed earth, establishes an extraordinary relationship with the exterior through its arched door span, which bathes the area reserved for painting in natural light, and through the window, which has been designed to frame the pond on the exterior, bringing the magic of a reflecting pond into the building's interior.

Likewise, in the house in Carvalhal/Troviscais, a relationship with the exterior is established in a daring and imaginative way, emphasising the fluidity of indoor and outdoor spaces and enhancing each space, recreating their scales.

Wherever possible, Alexandre Bastos incorporates figurative suggestions and sculptural interventions in the architectural elements of his buildings, be it through the creation of pictorial silhouettes in the whitewashed plaster on the rammed earth walls, or the incorporation of frescoes painted on the plaster, or the combination of material textures, as is the case in the house in Zambujeira Velha de Santa Maria, or bas-reliefs on the walls, as is the case in the São Luís Market, also in the vicinity of Odemira.

In other cases, the conceptual originality reveals itself in the detachment of the structural elements from the architectural components, as in the house in Caldeirinha/Troviscais, where a powerful brick arch is freed from the walls and ceilings, reflecting a deconstructivist poeticness that marks the image and organisation of the interior space into living and kitchen areas—which are merged but separate at the same time.

As is the case in the Wally Makepeace house in Troviscais, sometimes it is the typology of the commission that allows for the conception of volumetrically independent, but interlinked spaces which thus take on an unmistakeable identity.

Based on the five works presented, which were chosen randomly form a total of several dozen rammed earth dwellings that Alexandre Bastos designed and built over a period of 25 years of intensive and dedicated work, it becomes very clear that the choice of this building technique is a determined, informed and well thought out one that contributes to the continuity of a building tradition and architectural originality and is inherent to someone with such a unique and coherent professional trajectory.

Alexandre Bastos has chosen a path that fuses two professions, those of the architect and the painter, enriching his works with his technical and artistic contributions and enhancing the quality of the architectural object with a creativity and maturity derived from a pluralistic approach to artistic expression.

Figure 3. Residence in Carvalhal/Troviscais (F. Jorge).

Figure 4. Residence in Carvalhal/Troviscais (F. Jorge).

Figure 5. Residence in Zambujeira Velha de Santa Maria (F. Jorge).

Figure 6. Residence in Zambujeira Velha de Santa Maria (F. Jorge).

Figure 7. Residence in Troviscais (F. Jorge).

Figure 8. Residence in Troviscais (F. Jorge).

He is a prime example of how an architect with a background in the Fine Arts can bring added value to the architectural creative process.

Furthermore, his admiration of the vernacular in architecture, and his understanding of its models and the construction systems, have allowed him

Figure 9. Residence of Wally Makepeace in Troviscais (F. Jorge).

Figure 10. Residence of Wally Makepeace in Troviscais (F. Jorge).

to develop a deep knowledge of the practice, which is so ably demonstrated in his solutions that combine the best of tradition and modernity.

His works are superb examples of the typological diversity and aesthetic value that is possible in the building process when working with a material that offers outstanding opportunities in terms of colour, texture and form. As Alexandre himself puts it:

"Rammed earth offers all the creative possibilities of painting…"

REFERENCES

AAVV. 2005. *Arquitectura de terra em Portugal—Earth architecture in Portugal*, Lisbon: Argumentum.

BASTOS, A. 2013. Liberdade—terra/taipa in *International seminar on earth architecture*. Universidade Do Porto.

Des architectures de terre; ou l'avenir d'une tradition millénaire Catalogue to an International Travelling Exhibition.

Dethier, J. 1986. *Histoire, actualité et future*. Paris: ed. Centre Pompidou.

Mileto, C., Vegas, F. & Cristini, V. (ed). 2012. *Rammed Earth Conservation.* Crc Press/Balkema.

Earthen Architecture: Past, Present and Future – Mileto, Vegas, García Soriano & Cristini (Eds)
© 2015 Taylor & Francis Group, London, ISBN 978-1-138-02711-4

The fitness of earthen architecture

C. Kraus
University of Kansas, Lawrence, KS, USA

ABSTRACT: Fitness in phenotypic evolution is a concept that describes the relative ability of certain traits to contribute to the successful reproduction of a thing, e.g. how well a thing is adapted or *fit* to its circumstances. Typically associated with biological evolution, this concept can equally be applied to the evolution of culture—e.g. the concept of the meme, as a cultural equivalent to the biological gene. In this paper, earthen architecture is viewed through the dual evolutionary lenses of gradual speciation and punctuated equilibrium. Its fitness—or it cultural currency—is explored through a brief survey of ideological constructs that have succeeded in sustaining this time-tested way of building since nearly the beginning of human history to our current age of environmental consciousness.

1 FITNESS FOR SURVIVAL

Sustainability, in its purest form, is the fitness to endure. Throughout nature one observes the instinctive function to endure, for at least as long as is required to successfully pass on one's genetic or memetic imprint. Contrary to contemporary rhetoric, sustainability of the built environment is less about the planet than it is about people. The planet, in whatever altered state, is likely to outlive humanity regardless of how badly we behave. Yet, what is good for the planet is usually good for people...in the long run. Proponents and opponents of sustainability alike want the same thing, to provide their progeny the opportunity for healthy, happy lives. Proponents of sustainability, however, possess the gift of foresight.

1.1 *Vernacular architecture as gradual speciation*

The evolution of vernacular architecture, relative to non-vernacular traditions, can be understood analogously to *gradual speciation*, where fitness incrementally changes over long periods of time. Vernacular architecture encompasses the empirically designed built environment, typically by owners, whole communities, or craftspeople practicing an inherited craft rather than professional architects or master builders explicitly applying theoretical and philosophical constructs.

The inherent function to endure is the hallmark of vernacular architecture. In much the same way as living organisms slowly evolve over time in response to a unique set of conditions, vernacular architecture slowly and incrementally adapts to climate, socio-cultural conditions, economic forces, and security needs. Motives of a particular society are drawn from the intimacy of personal experiences and cultural ideologies.

A society's traditions and customs contain the particular memes responsible for past preservation. Conservative impulses privilege this fact, while progressive impulses privilege the equally, and increasingly, necessary rapid adaptation that allows a society to continue to respond to its mutable environment.

1.2 *Non-vernacular architecture as punctuated equilibrium*

One of the most celebrated, criticized, and misunderstood tenets of Modern architecture was the valorization of function. Modernists tended to cast off vestiges of tradition that no longer served their original purposes. This was believed to be necessary in response to societies that had developed an exaggerated historical awareness, as evidenced in the heritage of nineteenth century architecture. Modernists were compelled to purge these 'junk' memes, illustrated by frequent appeals to architectural essences.

In vernacular societies, no such intentional cultural reformations were needed because, like its biological corollary, these societies typically had slowly, almost imperceptibly, adapted in lockstep to equally slow changing external and internal forces. However, in non-vernacular architecture, relative to vernacular traditions, the rates of change and the intentionality behind these changes more closely aligns with the biological concept of *punctuated equilibrium*. Non-vernacular architecture tends to change rapidly yet sporadically, accompanied by explicit, self-conscious reflection, in the best of cases. Motives are drawn from current symbolic, societal, and technological forces as well as idiosyncratic theoretical constructs. These rapid changes, however, are as likely to result in regression as progression.

Recognizing the inherent sustainability of vernacular architecture, it is not surprising to witness in our *age of environmental consciousness*, increased scrutiny of this time-tested way of building. Paradoxically, the very awareness of vernacular architecture forever bars its authentic practice.

A rapidly changing, global and interdependent contemporary society vastly complicates a vernacular approach to sustainability. No longer can we be solely concerned with striking a concinnity with local contexts. Contemporary society—no longer plural as it once was—is now obligated to consider the complex whole. Yet, now more then ever, the essential lessons of vernacular architecture hold much promise for a more sustainable society.

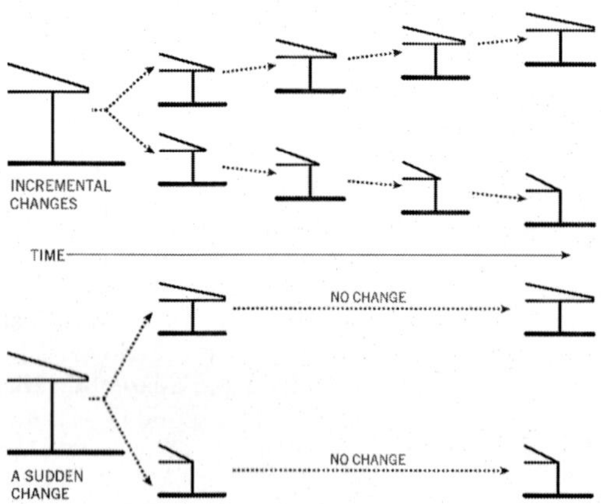

Figure 1. Vernacular architecture as 'gradual speciation' and non-vernacular architecture as 'punctuated equilibrium' (Diagram by author).

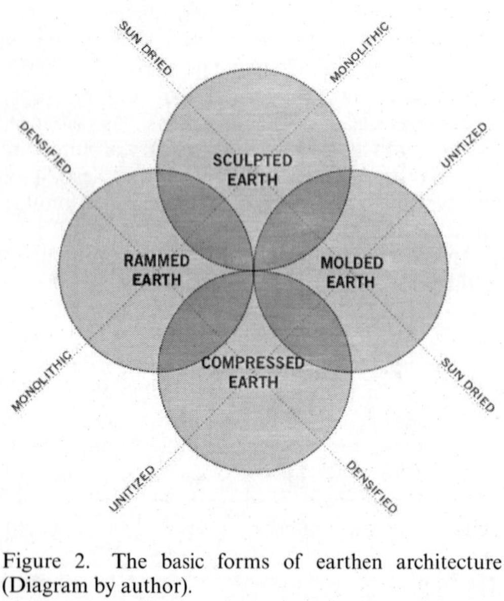

Figure 2. The basic forms of earthen architecture (Diagram by author).

To possess fitness in architecture is to achieve a sufficiently high combination of symbolic, social, and technological acceptance. When cultural ideologies shift slowly, a gradual speciation in architecture will tend to maintain fitness. However, when cultural ideologies experience massive change and/or significant heterogeneity, a punctuated equilibrium model of architecture emerges, historically with an equal tendency to flourish or fail, e.g. big consequences.

Lessons of a vernacular architecture may serve as steady guideposts even during periods of punctuated equilibrium. One material, earth, closely associated with vernacular architecture, appears to possess a simultaneous contextual specificity as well as near-universal applicability. Its variability and adaptability has historically given it the capacity to remain fit in nearly every region and in many cultures. Today, these qualities are seen as a liability in societies that value predictability and standardization.

1.3 *The techniques of earthen architecture*

Earthen architecture has arisen independently and in various forms across space and time, further enriched through complex patterns of cultural transference, so that today a multitude of techniques exist. The majority of these earth-building techniques can be grouped into four basic classifications, here referred to as *Sculpted Earth*, *Molded Earth*, *Rammed Earth*, and *Compressed Earth*. Each of these classifications concentrates on the manner of forming and strengthening the earthen material.

The incredible variety of earthen building techniques, arising from the diversity of cultures and the heterogeneity of available soils, has enabled earthen architecture to endure for nearly the whole of human history–until the industrialization of societies. Industrialized materials have now largely replaced earthen ones, at first in industrialized nations and later exported to quasi-industrialized nations. Yet, throughout the last few centuries, periodic and fragile resurgences in earthen architecture have emerged in response to political, humanitarian, and ecological concerns.

2 SYMBOLIC IDEOLOGIES IN EARTH

Earthen architecture arose independently in many of the world's civilizations, particularly in areas with a lack of good alternative building materials such as stone or wood. Beyond necessity, however, earthen material has frequently held powerful, symbolic connotations. Early works of earthen architecture often transcended utilitarian function to embody spiritual, symbolic, and emotional functions, e.g. burial mounds and temples. These symbolic connotations were conferred on the earthen material itself.

The mud brick, a seemingly unremarkable artifact, was believed by ancient Egyptians to be capable of possessing magical properties. Recent archeological evidence has discovered, among valuable artifacts found in royal tombs, collections of four unbaked mud bricks painted with iconography bound to the four forms of Meskhenet, the goddess of birth (Wegner 2010). These *birth bricks* have long been a source of mystery and wonder. Evidence suggests they were used during childbirth for their utility as well as their spiritual significance. The expectant mother would stand upon the double-stacked bricks and crouch over an open pit, positioning the midwife to receive the child. These bricks evoked the birth goddess and were "intimately related to the creation of the world" (Roth & Roehrig 2002, p. 131).

The ancient pyramids of Egypt were commonly built of stone. There were, however, according to the ancient Greek historian Herodotus, at least four Egyptian pyramids built of unbaked mud brick, such as the one attributed to the Egyptian king Asychis (Goodhugh 1845). Herodotus claims that upon his uncharacteristic pyramid, Asychis inscribed the following passage, "Compare me not with the stone pyramids, for I am as superior to them as Jove is to the other gods. Thus was I made. Men probing with poles the bottom of a lake, drew forth the mud which adhered to them, and formed it into bricks" (Goodhugh, p. 1117). From this inscription, it is clear Asychis more highly esteemed earthen material over stone (Urquhart 1850). Although it is not fully understood why Asychis felt this way, it does appear to reinforce the reverence bestowed on the material.

Figure 3. An ancient Egyptian birth brick. Photograph courtesy of Josef Wegner. Used with permission.

Earth, in these instances, appears to possess a pre-reflective human significance, with associations of fertility, origin, and creation. In the *Book of Genesis*, humankind was formed from soil. "And the lord God formed man of the dust of the ground, and breathed into his nostrils the breath of life" (Holy Bible, Genesis 2:7 1769). According to the *Huainanzi*, the goddess Nüwa created humankind from yellow clay and mud (Luo 2004). According to the *Qur'an*, Allah shaped the first human from clay and breathed a soul into it (Qur'an, Ṣād 38:71–72).

In nearly every major religion, these types of origin stories are recounted. In the twentieth century, a scientific basis arose for the origins of life originating in an earthy soup, with theories based on, or similar to, the work of Alexander Oparin's *The Origin of Life* (1953). In the absence of oxygen, life spontaneously metabolizes from a 'primeval soup'. In the 1950's, researchers conducted experiments on inorganic, earthy, compounds sealed in glass tubes and subjected to sparks simulating lightning strikes to spontaneously create amino acids, the basic building blocks of life (Miller & Urey 1959). Further research has shown montmorillonite clay can be used as a base material to simulate primitive cell division (Hanczyc et al. 2003). In each case, an earthy substance holds the potential for life.

Eleven years after the fall of the Berlin Wall, located along the former border between West and East Germany, and on the site of a destroyed church, the Chapel of Reconciliation arose as a rammed earth oval enveloped by a timber screen (Minke 2006). The massive walls, built from earthen material mixed with rubble from the remains of the church, symbolize the rebirth of a nation. However, today earthen architecture is more likely to possess symbolic significance in relation to its connotations with the planet earth and its implicit associations with a growing environmental consciousness. Whether from birth to life, from life to death, or from death to resurrection, earthen material has served as a manifestation of deeply held symbolic beliefs.

3 SOCIAL IDEOLOGIES IN EARTH

Coinciding with the political and social upheavals ushering in the Modern mentality in Europe, French architect François Cointereaux championed a resurgence of earthen architecture. Its simultaneous rise with the issue of human rights in France appears to be no mere coincidence. Historian Paula Young Lee recognizes "the politically-charged" nature of rammed earth in 18th century France (2008, p. 58).

Cointereaux's political leanings were in line with Physiocracy, an emerging philosophy emphasizing a *government of nature*, where values of laissez-faire, self-sufficiency, and a dedication to the agrarian roots of society were held in high regard. Pisé de terre became the ideal political symbol as well as an effective utilitarian material. The abundant and inexpensive earth was capable of providing the individual with the means to ensure his or her own freedom, liberty, and prosperity, with the added benefit that *dirt* was seen as the antithesis of everything the revolution despised.

Cointereaux believed pisé could improve rural living conditions by providing rural dwellers with high quality, inexpensive, incombustible, and readily abundant building materials. As Cellauro and Richaud observe, "For small and large landholders alike, who had survived recessions and droughts, there were considerable advantages in constructing buildings that were cheap, durable, and which could be erected with one's own labour from a material available on one's own property (2006, p. 141).

Despite failing to establish pisé as a lasting and broadly accepted form of building, Cointereaux made considerable impacts during the 1790's throughout Germany, Denmark, Finland, and Russia, as well as in the United States and Australia (Ibid). It is little surprise then, that individuals such as United States Supreme Court Justice Bushrod Washington, General John Hartwell Cocke, and Virginia Supreme Court of Appeals Justice St. George Tucker, would all be so deeply compelled by translations of his works. These Virginian gentlemen were prominent members of the American Colonization Society and each were social activists advocating for slave's rights (Hallock 2004). Justice Washington, for instance, enthusiastically embraced rammed earth, building at least seven earthen structures—residential and agricultural—with a primary aim to reform slave housing practices (Ibid). Beyond cost savings, Washington appreciated the increased indoor comfort resulting from rammed earth's thermal mass.

Social values would similarly inspire singer-songwriter Woody Guthrie to write the novel 'House of Earth'– an ode to earthen architecture—during the Dust Bowl that devastated the American Midwest (Guthrie 2013). Guthrie saw first-hand how poorly timber frame homes, relative to earthen homes, withstood the unforgiving dust storms that terrorized rural communities in Kansas, Oklahoma, and Texas. Convinced that earthen building would be the salvation of these communities, Guthrie became an ardent promoter. Here too, social ideologies underpin his advocacy. "In Guthrie's fierce proletarian worldview, the rural poor are thereby shafted by the iron boot heel of capitalist greed merchants, and he finger-points everyone from bankers to lumbermen to termite-like real estate brokers, enemies of the little guy" (Brinkley & Depp 2012).

Guthrie wasn't the only one eyeing potential salvation in earthen architecture. Thomas Hibben, an architect-engineer with the United States Resettlement Agency, was given the task of designing 75 homes in Gardendale, Alabama. His enthusiasm for rammed earth was enough to convince his bosses to allow him to build seven 'experimental' rammed earth homes in a decidedly Modern style as part of the Gardendale Homestead project (Carpenter 2010). Hibben could be described as an individual with socially minded ideologies. "I believe architecture to be an expression in three-dimensions of the social, economic, intellectual, and spiritual state of civilization" he once stated (Carpenter, p. 45). Hibben's experimental work in rammed earth continues to make a potent argument for the feasibility of rammed earth as a contemporary building technology. More than 75 years later, of the original rammed earth homes at Gardendale, all seven remain. Hibben, himself, noted that, several years after their construction, "of all of the building types erected there [Gardendale], those

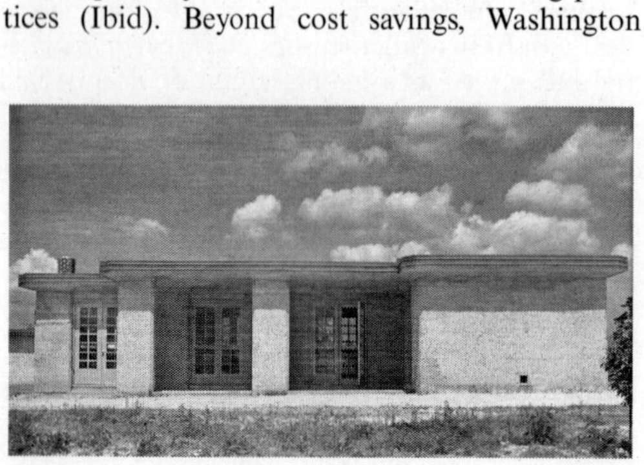

Figure 4. A rammed earth home at the Gardendale Homestead, 1937. Photograph courtesy of the Library of Congress.

Figure 5. Tsunami reconstruction project, Kirinda, Sri Lanka, Shigeru Ban Architects, 2007. Photograph courtesy of Shigeru Ban Architects. Used with permission.

of Rammed Earth construction have cost least for maintenance" (Carpenter, p. 56).

Since the invention of the CINVA Ram in 1952, compressed earth blocks have been used extensively to house some of the world's poorest communities. This relatively new earthen building technology was "anchored in an initial concern to provide a new, economically and socially relevant response to housing production for the very poor, and has continued to focus on this as its area of application has developed" (Guillaud et al. 1985, p. 8).

The recurring theme of community empowerment through do-it-yourself mud brick construction is central to the mission of the Adobe for Women project, which seeks to empower a community largely composed of women and children, whose husbands and fathers have migrated to other areas in hopes of making a better life for their families (Minner 2011).

After a tsunami devastated the fishing village of Kirinda, Sri Lanka in 2004, Japanese architect Shigeru Ban turned to the earth to help the local community rebuild. Ban chose compressed earth brick as the primary structure for 100 new prototype homes in an effort to stimulate the local economy as well as provide high quality, affordable housing (Ban 2014).

To this day issues of economic equality, freedom, liberty, and social justice often inspire the use of earthen architecture.

4 ENVIRONMENTAL IDEOLOGIES IN EARTH

Earthen architecture, particularly in the United States, has been primarily associated with social groups considered marginal or alternative, such as the poor—and their advocates. In the 1970's, earthen material experienced yet another fragile resurgence, this time due to it environmentally sustainable attributes, such as its low embodied energy, high recyclability, potentially long lifespan, and low toxicity while having little impact on biodiversity and virtually no depletion of biological nutrients. However, a sustained environmental consciousness would not be established beyond marginal groups until several decades later.

Instead, these building technologies have become synonymous, in Western societies, with obsolesce or regression. Defying this scarlet letter, late-twentieth century advocates of rammed earth experimented with industrial stabilizers such as Portland cement, mechanical mixing and compaction, and engineered hydrophobic admixtures.

Additional advances in this long-dormant technology have concentrated on several of its most notable weaknesses, namely poor thermal resistance and lack of standardization. While many of these modern improvements, imported from concrete construction, resulted in increased structural and moisture performance, they have also succeeded in compromising some of the earth's most admired qualities—e.g. its naturalness, affordability, and sustainability.

As the allure of rammed earth grows, the impact of labor costs are of prime concern. Due to the physical compaction of the soil, rammed earth is inherently labor intensive relative to concrete or light wood frame construction. Recently, David Easton has pioneered a new method of earth building cleverly termed pneumatically impacted stabilized earth (PISE). Once again borrowing from the concrete industry, PISE uses a rotary gun and hose to deliver a mix of dry soil, cement, and water into a one-sided form (Easton 2012). The impact force from the hose and the industrial binder combine to form strong durable walls at a fraction of the labor required in other earthen building techniques. The open side of the wall is later shaved or troweled to achieve a pleasing surface not entirely unlike a stucco finish, with the ability to significantly vary the surface texture.

In 1992, Meror Krayenhoff, developed an insulated version of stabilized rammed earth, building upon the work of his mentor David Easton. Unlike Easton, Krayenhoff was building with rammed earth in cold climates. While earthen building techniques possess high thermal mass, effectively equalizing diurnal temperature swings, they possesses low resistance to heat transfer. To alleviate this shortcoming, rigid insulation is inserted into the interior of the form, around which the earthen material is rammed. The typical stabilized insulated rammed earth (SIRE) wall sandwiches a 4" thick insulating panel between an 8" thick exterior wythe of rammed earth and a 12" interior one (Krayenhoff 2012). While this assembly is able to achieve a high thermal resistance factor akin to insulated concrete, when dynamic considerations are factored in such as solar insolation and the effects of thermal mass on internal heat and humidity sources, the dynamic R-value of the coupled thermal mass and thermal resistance is even greater. In fact, research has shown that a composite envelope, such as that pioneered by Krayenhoff, can be greater than the sum of its parts in terms of thermal resistance, humidity regulation, and air tightness, resulting in high performing, comfortable buildings in even extremely cold climates (Fix 2009, Hall 2012).

Despite advances in strength, durability, thermal resistance, and speed, the in situ nature of earthen building limits its application in instances where the permissible building season is short or in cases where site limitations or construction schedules prevent its adoption. The advent of compressed earth block had provided earth builders with a new unitized system of construction at a small scale, however, not until recently has the notion of pre-fabrication been applied to rammed earth wall paneling systems. Martin Rauch has teamed up with architects Jacques Herzog and Pierre de Meuron for the New Ricola Herb Center using massive pre-rammed panels. In addition to enabling the project to be erected quickly and beyond the limitations of weather, the pre-fabrication process begins to cast rammed earth as a standardized building technology. According to Rauch, "Many people think that nowadays loam is only used in developing countries to build the most basic huts. However, the Ricola Herb Center makes a strong statement about the future perspectives and the positive qualities which loam construction can have in Europe as well" (Ricola 2013).

The next chapter in earthen architecture may shift from the massive to the microscopic, from borrowed concrete technology to technology native to the soil. In recent years, researchers have drawn from timely advances in geomicrobiology and civil engineering to explore microbially induced calcite precipitation as a means to biostabilize earthen material (DeJong et al. 2006). In 2009, Magnus Larsson proposed his wild idea to stop the desertification of the Sahara by using a biomineralization process to lock sand dunes in place, forming architectural spaces along this new edge condition (2009). Around the same time, Ginger Krieg Dosier has been developing a more practical, scaled down application, the bioMason sand brick using soil bacteria as a cementing agent (Matus 2013). Collaborators Daniel Hirmas, Jennifer Roberts, and I have explored microbial biomineralization to enhance the material properties of rammed earth using a process we call *microbially indurated rammed earth*—MIRE (Kraus et al. 2013). In each case, advances in technology have been employed to rescue an environmentally sustainable yet fading craft. Despite challenges on the basis of performance, labor costs, lack of standardization, and general misconceptions, a heightened ecological sensitivity has ensured its continued fitness.

5 FITNESS OF EARTHEN ARCHITECTURE

Once upon a time, the fitness of earthen architecture was rooted in its symbolic power. With the rise of Positivism and the Industrial Revolutions, its fitness began to ebb. Yet the survival of earth building depends on its continued significance. In a post-vernacular world, earthen architecture, if it is to experience a lasting resurgence, must be capable of evolving rapidly yet authentically to remain *fit*, and will be increasingly rooted in political, social, humanitarian, technological, and ecological concerns.

Already, intensified interest is causing this timeless way of building to experience a paradigm shift. Compressed earth block and rammed earth, in particular, have begun to slip beyond the confines of the margins on the strength of our awakened environmental consciousness. Beyond passive vernacular qualities, the future of earthen architecture will likely be inextricably intertwined with advances in nanotechnology, digital fabrication, and other contemporary technologies. Yet, it is equally important to recognize the role that a phenomenological appeal to our senses plays in the continued fitness of earthen architecture.

REFERENCES

Ban, S. 2009. From conversations with the author.

Brinkley, D. & Depp, J. 2012. This land was his land. *The New York Times Sunday Book Review* (July 9).

Carpenter, J.L. 2010. Dirt cheap: the Gardendale Experiment and rammed earth home construction in the United States. *PhD dissertation, University of Maryland.* College Park: Digital Repository at the University of Maryland.

Cellauro, L. & Richaud, G. 2006. François Cointeraux's École d'Architecture Rurale (1790–91) and its influence in Europe and the colonies. *Architectural History* 49 pp. 129–148.

DeJong, J.T., Fritzges, M.B. and Nüsslein, K. 2006. Microbially Induced Cementation to Control Sand Response to Undrained Shear. *Journal of Geotechnical and Geoenvironmental Engineering* 132.11 pp. 1381–92.

Easton, D. 2012. "Pneumatically impacted stabilized earth (PISE) construction techniques." *Modern earth buildings*. By M. Hall, R. Lindsay, and M. Krayenhoff. Oxford: Woodhead Publishing pp. 385–400.

Fix, S. & Richman, R. 2009. Viability of rammed earth building construction in cold climates. Toronto: Ryerson Univ.

Goodhugh, W. 1845. *The pictorial dictionary of the Holy Bible.* London: H.G. Bohn.

Guillaud, H, Joffroy, T. & Odul, P. 1985. *Compressed earth blocks: manual of design and construction, volume II.* Deutsches Zentrum für Entwicklungs technologien—GATE.

Guthrie, W. 2013. *House of earth.* New York: Infinitum Nihil/Harper.

Hall, M.R. 2012. "Hygrothermal behavior and occupant comfort in modern earth buildings." *Modern earth buildings*. By M. Hall, R. Lindsay, and M. Krayenhoff. Oxford: Woodhead Publishing pp. 17–40.

Hallock, G. 2004. Pisé construction in early nineteenth-century Virginia. *Perspectives in Vernacular Architecture* 11 pp. 40–53.

Hanczyc, M., Fujikawa, S., & Szostak, J. 2003. Experimental models of primitive cellular compartments: encapsulation, growth, and division. *Science* 302 (no. 5645) pp. 618–622.

King James Version. 1769. *The Holy Bible.* New York: Oxford Edition.

Kraus, C., Hirmas, D., & Roberts, J. 2013. Microbially indurated rammed earth: a long awaited next phase of earthen architecture. *ARCC: The Visibility of Research* pp. 58–65.

Krayenhoff, M. 2012. "North American modern earth construction." *Modern earth buildings*. By M. Hall, R. Lindsay, and M. Krayenhoff. Oxford: Woodhead Publishing pp. 561–608.

Larsson, M. 2009. Turning Dunes into Architecture. *TED: Ideas worth Spreading* (December 2).

Lee, P.Y. 2008. Pisé and the peasantry. *Journal of the Society of Architectural Historians* 67 (no. 1) pp. 58–77.

Luo, J. 2004. *Over a cup of tea: an introduction to Chinese life and culture.* Dallas: University of America Press.

Matus, M. 2013. Award-winning biomason grows bricks from sand and bacteria to reduce CO_2 emissions. *Inhabitat* (September 16).

Miller, S. & Urey, H. 1959. Organic compound synthesis on the primitive earth. *Science* 130 (no. 3370) pp. 245–51.

Minke, G. 2006. *Building with earth: design and technology of a sustainable architecture.* Basel: Birkhauser.

Minner, K. 2011. Adobe for women/blaanc borderless architecture and Caeiro Capurso. *ArchDaily.* (July 21).

Oparin, A. 1953. *The origin of life.* New York: Dover Publications.

Ricola AG. 2013. The new Ricola Herb Center – sustainability and leading architecture. *Press Release* (January 23).

Roth, A.M. & Roehrig, C.H. 2002. Magical bricks and the bricks of birth. *The Journal of Egyptian Archaeology* 88 pp. 121–139.

Saheeh International. 2011. *The Qur'an: english meanings and notes.* Riyadh: Al-Muntada Al-Islami Trust.

Urquhart, D. 1850. *The pillars of Hercules.* London: R. Bentley.

Wegner, J. 2010. "A decorated birth-brick from South Abydos." *Archaism and innovation* by D. Silverman, W. Simpson, & J. Wegner. New Haven: Department of Near Eastern Languages and Civilizations, Yale University.

Earthen Architecture: Past, Present and Future – Mileto, Vegas, García Soriano & Cristini (Eds)
© 2015 Taylor & Francis Group, London, ISBN 978-1-138-02711-4

Compressive strength of blood stabilized earthen architecture

C. Kraus, D. Hirmas & J. Roberts
University of Kansas, Lawrence, KS, USA

ABSTRACT: Throughout the literature on earthen architecture, references to the historical use of animal byproducts such as blood, manure, and urine to increase compressive strength and moisture resistance are fairly common. However, little empirical research has been done to substantiate these practices. In this paper, we present the results of an experiment testing the compressive strength of blood stabilized rammed earth. Triplicate unstabilized and blood stabilized rammed earth cylinders were tested for compressive strength at 7, 14, and 28 days. Results show the compressive strength of blood-stabilized cylinders to exceed the control cylinders by approximately 36% at 28 days. Biostabilization of rammed earth with blood shows promise for a more sustainable built environment.

1 BLOOD STABILIZATION AS A VERNACULAR PRACTICE

The ancient use of blood as a stabilizer for earthen architecture most likely arose due to the spiritual or symbolic significance placed on these materials, with earth acting as a *prima materia* and the blood infusing the earth with life. "In the belief that Mithras' slaying of the bull brought life to the earth, bull's blood became a symbol of the sap of life with a power to protect houses (of clay) from the attack of Ahriman, the evil power" (Winkler 1956 p.157).

Over time, through the gradual memetic evolution of vernacular architecture—whether these symbolic meanings persisted or not—blood-stabilized earthen material did appear to possess increased strength, moisture resistance, and durability (Eires et al. 2013, Beas 1991). Its perceived benefits where obvious enough for it to have arisen independently in many vernacular societies.

Today it is common for researchers interested in earthen building to acknowledge the role that blood stabilization has had in historical and, in a few cases, contemporary architecture. "Animal products like blood, urine, manure, casein and animal glue have been used through the centuries to stabilize loam" (Minke 2009 p.41). "Cactus mucilage, ox blood, paper, corn husks, and manure are examples of binding agents that increase the durability of the material (Rael 2009 p.114). While bio-stabilizing practices such as animal's blood are well known, the "precise effect has not been examined scientifically and is unknown" (McHenry 1984 p.76).

2 SPECULATIONS ON THE STRENGTHENING MECHANISM OF BLOOD STABILIZATION

Adhesives derived from animal blood have been used in applications such as plywood and have been shown to achieve a dry bonding strength equivalent to their conventional, environmentally damaging, phenol formaldehyde-based counterparts (Lin and Gunasekaran 2010). The bonding strength of blood-based adhesives is a result of the serum albumin protein found in fresh animal blood as well as dried blood powder. Albumin has also been shown to significantly increase the moisture resistance of earthen materials due to its dual hydrophilic/hydrophobic properties (El Azreq et al. 2013).

3 A METHOD OF BLOOD STABILIZATION

3.1 *Pre-compaction soil analysis*

A reasonably well-graded, unmodified course sandy limestone-based soil was selected as the base for the experiment and used for both the control (unstabilized rammed earth) and the treatment samples (blood stabilized rammed earth). This soil was chosen due to its suitability for unstabilized rammed earth as well as it being readily available. The particle-size distribution was determined according to ASTM (2009) standard D6913–04 (Fig. 1).

Following particle-size analysis of the soil, a modified Proctor compaction test was performed according to ASTM (2012) standard D1557–12 to determine the optimum moisture content—6.71%—to achieve the maximum dry density (Fig. 2).

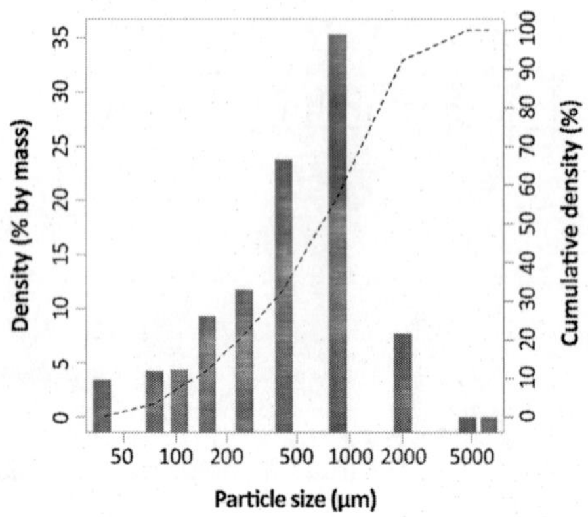

Figure 1. Soil particle-size distribution showing the percent by mass of each size fraction (bars; left axis) as well as the cumulative density function (dashed line; right axis) (Graph by authors).

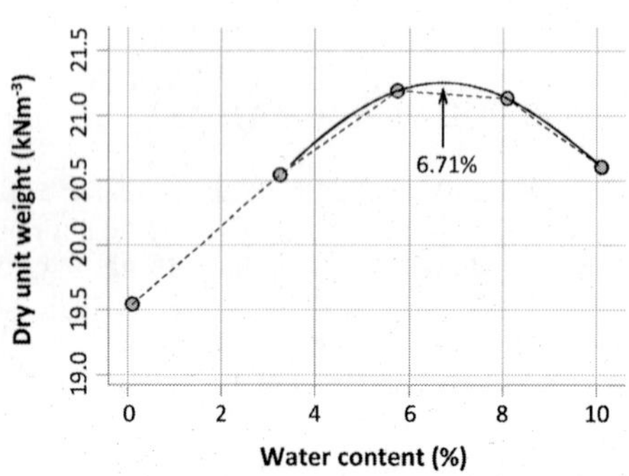

Figure 2. Modified Proctor compaction curve used to determine an optimum gravimetric moisture content of 6.71% (Graph by authors).

Based on increased compressive strength results in previous experiments, the optimum moisture content was increased by 10%, resulting in total moisture content of 7.38%. Since the soil used for this experiment contained 1.30% residual water content by mass, an additional 6.08% artificial ground water was added to bring the sample moisture content up to 7.38%. Thus, for the control samples, artificial ground water was added such that the final mixture contained 73.8 mL per 1 kg of soil. Similarly, for the treatment samples, blood was added in the same volume-to-mass ratio as the control.

3.2 *Preparation of test cylinders*

Following pre-compaction analysis, the base soil was sterilized using an autoclave and dried to a low moisture content (see above) prior to cylinder preparation. In addition, all equipment and surfaces were sterilized to prevent contamination. For both the control and treatment specimen, triplicate 4 × 8 inch (10.16 × 20.32 cm) cylinders where prepared for 7-day, 14-day, and 28-day compression tests using standard steel molds and a pneumatic rammer with a 3 inch (7.62 cm) steel butt attached to an air compressor delivering 11.3 CFM (19.2 $m^3 s^{-1}$) at 90 psi (621 kPa). An additional cylinder was prepared for each specimen type for thin section analysis. Each cylinder was rammed with four lifts of equal mass, scarified between lifts to increase lift-to-lift bonding. The bottom and top surfaces of the cylinders were made as uniform and level as practical to achieve accurate compression testing results.

For the treatment cylinders, freshly acquired bovine blood was thoroughly and mechanically agitated to achieve a homogenous consistency before being mixed into the soil in a manner consistent with the artificial ground water that had been added to the control specimen (Fig. 3). The test cylinders were allowed to dry slowly in a controlled environment prior to compression testing and thin section analysis (Fig 4).

3.3 *Compressive strength testing*

Compressive strength tests were conducted on the rammed earth cylinders according to ASTM (2007) standard D1633–00, modified using 4 × 8 inch

Figure 3. Bovine blood was used in place of an artificial ground water in the treatment specimen (Photograph by authors).

Figure 4. Non-stabilized and blood-stabilized rammed earth cylinders prepared and ready for compression testing (Photograph by authors).

Figure 5. Typical compression testing of blood-stabilized rammed earth cylinders (Photograph by authors).

(10.16 × 20.32 cm) steel molds and neoprene pad caps at 7, 14, and 28 days after initial cylinder preparation. Peak force was recorded in pounds and compressive strength was calculated based on the cross sectional area of the cylinders (Fig 5).

3.4 *Thin section analysis*

A pilot experiment was conducted prior to the current study where several blood stabilized rammed earth cylinders were prepared to determine best practices for stabilizing soil with blood. One of these cylinders was selected for thin section analysis to provide an opportunity to observe the material fabric after the addition of blood and to infer a mechanism for increases in compressive strength. Samples were impregnated with a clear epoxy and a mounted thin section was prepared on a 1 × 2 inch (2.54 × 5.08 cm) slide and stained for calcite with alizarin red. Thin sections were prepared by Texas Petrographic Services, Inc. (Houston, TX).

4 THE RESULTS OF BLOOD STABILIZATION

4.1 *Compressive strength*

After seven days from initial cylinder preparation, the control cylinders achieved an average compressive strength of 257 psi (1.77 MPa); exceeding the 187 psi (1.29 MPa) compressive strength of the blood stabilized cylinders (Fig. 6). After fourteen days, the blood-stabilized cylinders had closed the gap, achieving an average compressive strength of 518 psi (3.57 MPa), compared to the control's

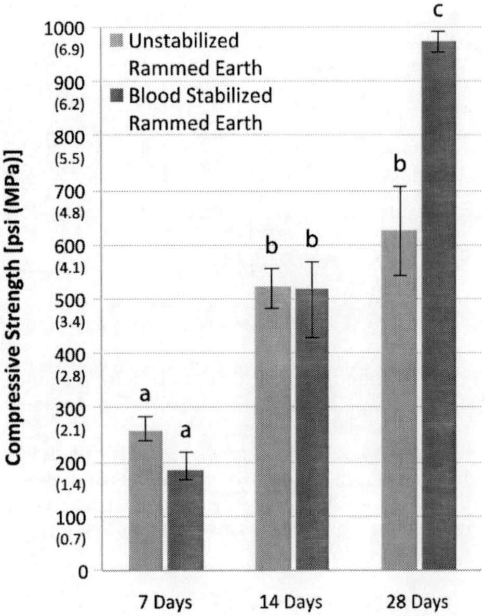

Figure 6. Comparison of compressive strengths for unstabilized and blood stabilized rammed earth. Bars with the same letters correspond to mean compressive strengths that are not significantly different from each other at an α-level of 0.05. Error bars represent the range of the triplicate strength values (Graph by authors).

521 psi (3.59 MPa), which were not significantly different at an α-level of 0.05.

By twenty-eight days, the blood-stabilized rammed earth cylinders had exceeded the control cylinders by approximately 36%, achieving 973 psi (6.71 MPa) compared to 625 psi (4.31 MPa). The increased compressive strength of the blood stabilized rammed earth over the control reinforced the earlier pilot experiment using a lower performing soil in which the blood stabilized samples exceeded the control samples by more than 200%.

4.2 *Blood-stabilized soil matrix*

An impregnated thin section from the pilot experiment is shown in Figure 7. The blood appeared to be staining the inorganic micromass with a reddish brown-black color indicating that the blood is likely aggregating some of the clay and silt fraction of the sample and the larger framework grains that are in contact with the micromass. This is likely the mechanism adding strength to the material. As the blood dries it coagulates; this coagulation appears to be hardening the surrounding matrix and increasing the overall strength of the samples. It has been proposed that proteins and hemoglobin in the blood bond the soil matrix after air-drying (Winkler, 1956).

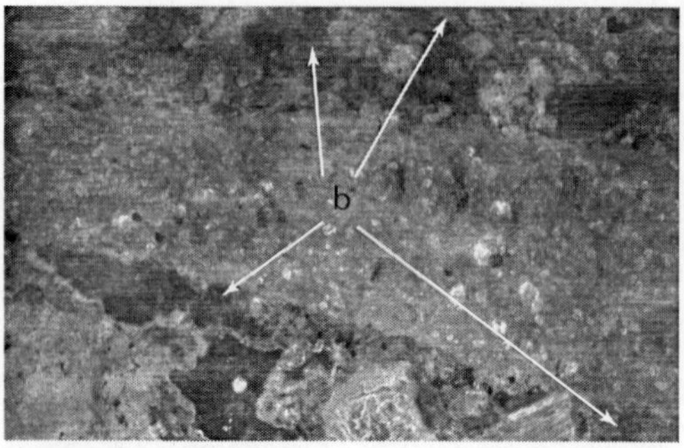

Figure 7. Blood-stabilized thin section from an unpublished pilot experiment. The scale bar in the bottom right corner of the photomicrograph has a length of 10 μm. The arrows point to blood residue that appears as a dark reddish brown-black stain of the inorganic micromass. This photomicrograph was taken under phase contrast (Photomicrograph by Texas Petrographic Services).

5 THE POTENTIAL OF BIOSTABILIZATION

Contemporary advances in rammed earth material have largely been borrowed from concrete construction—e.g. cement stabilization, hydrophobic admixtures, insulated rammed earth, and pneumatically impacted stabilized earth (Krayenhoff 2012; Easton 2012). While these advances are commendable, there appears to be tremendous potential in lesser processed, lower polluting technologies—such as biostabilization—that promise to maintain the low embodied energy of earthen materials while increasing its strength and moisture resistance to meet contemporary building standards. Additionally, biostabilizers may be able to be extracted from waste streams as industrial byproducts, further contributing to its lighter ecological footprint.

In 1978, in the context of an earlier era of heightened environmental consciousness, architect Buckminster Fuller asked his colleague Norman Foster an insightful question, 'How much does your building weigh?'(Foster 2001). Fuller understood that to lessen one's negative impact on the world, efficient use of material and resources was critical. Ultimately, as Foster aptly points out, today we are able to measure the impact of our building in more sophisticated ways, relying less on pounds of steel and more on pounds of CO_2 emissions. The weight of a building is not always a good predictor of ecological impact, as rammed earth can attest to. In our current era of environmental consciousness, Fuller's question ought to be reframed; *how deep is your ecological footprint?* In this respect, bio-stabilized earthen architecture has a promising future.

ACKNOWLEDGEMENTS

We would like to acknowledge our students who graciously contributed to the research: Alyson Bents, Timothy Bents, Joshua Boling, Zachary Dawson, Brittany Hodges, Anna Johnson, McKenzie Liebl, Chloe Lockman, Jared Pechauer, Ben Peek, and David Versteeg.

REFERENCES

ASTM. 2007. *Standard Test Methods for Laboratory Compaction Characteristics of Soil Using Modified Effort* (56,000 ft-lbf/ft3 (2,700 kN-m/m3)). ASTM Standard D1557–12, West Conshohocken, PA: ASTM International.

ASTM. 2007. *Standard Test Methods for Compressive Strength of Molded Soil-Cement Cylinders*. ASTM Standard D1633–00, West Conshohocken, PA: ASTM International.

ASTM. 2009. Standard Test Methods for Particle-Size Distribution (Gradation) of Soils Using Sieve Analysis. ASTM Standard D6913–04, West Conshohocken, PA: ASTM International.

Beas, M.I.G. 1991. Traditional architectural renders on earthen surfaces. Master Thesis, Historic Preservation, University of Pennsylvania.

Easton, D. 2012. Pneumatically impacted stabilized earth (PISE) construction techniques. *Modern earth buildings*. By M. Hall, R. Lindsay, and M. Krayenhoff. Oxford: Woodhead Publishing pp. 385–400.

Eires, R., Camões, A., and Jalali, S. 2013. Earth Architecture: Ancient and New Methods for Durability Improvement. *Structures and Architecture. Concepts, Applications and Challenges*. Ed. Paulo Cruz. Boca Raton, FL: CRC. 962–70.

El Azreq, M.A., El Hammoumi, A., Iben Brahim, A., Kerroum, M., Gueraoui, K., El Mouraouah, A., Lbadaoui, A., Kasmi, M., Birouk, A., and El Harrouni, K. 2013. SANSWALL: Sandblasted armed non-stabilized wall of rammed earth. *Advanced Studies in Theoretical physics* Vol. 7, no.1 pp. 25–56.

Foster, N. 2001. Richard Buckminster Fuller. *Buckminster Fuller: Anthology for the New Millennium*. Ed. Thomas T.K. Zung. New York: St. Martin's, pp.1–8.

Krayenhoff, M. 2012. "North American modern earth construction." *Modern earth buildings*. By M. Hall, R. Lindsay, and M. Krayenhoff. Oxford: Woodhead Publishing pp. 561–608.

Lin, H and Gunasekaran, S. 2010. Cow blood adhesive: Characterization of physicochemical and adhesion properties. *International Journal of Adhesion and Adhesives* 30 pp. 139–44.

McHenry, P. 1984. *Adobe and rammed earth buildings: design and construction*. New York: Wiley.

Minke, G. 2009. *Building with earth: design and technology of a sustainable architecture*. Basel: Birkhauser.

Rael, R. 2009. *Earth architecture*. New York: Princeton Architectural Press.

Winkler, E.M. 1956. The Effect Of Blood On Clays. *Soil Science* 82.2 pp. 157–72.

Earthen Architecture: Past, Present and Future – Mileto, Vegas, García Soriano & Cristini (Eds)
© 2015 Taylor & Francis Group, London, ISBN 978-1-138-02711-4

Location and conservation of rammed earth in Chelva (Spain)

M.Á. Lloría Cosín
Colegio Oficial de Arquitectos de la Comunidad Valenciana, Chelva, Valencia, Spain

J.A. Cantó Vañó
Colegio Oficial de Arquitectos de la Comunidad Valenciana, Bocairent, Valencia, Spain

S.D. Lozano Esteban
Colegio Oficial de Arquitectos de la Comunidad Valenciana, Castellón, Altura, Spain

ABSTRACT: The city of Chelva, located within the province of Valencia, is characterized by an architecturally emblematic historical city center where rammed-earth construction predominates. These structures are built with locally available soil, barely processed single rammed walls, which not only provide character to the buildings, but also reveal a vast variety of compressing and finishing techniques. This variety has resulted in the compilation of a catalog of construction solutions, based on the same principle, rammed earth, but with multiple and diverse applications and results. This building technique was initially used for imposing defensive constructions such as the city walls that could be the origin of the village, but it is especially present in modest domestic buildings due to its simplicity and affordability for those with limited economic means. The study of these past constructions will ensure the future survival of the rammed-earth walls which are still standing in the narrow streets of Chelva.

1 INTRODUCTION

Chelva is a small town in the interior of the province of Valencia. Currently on a progressive population and economic decline, it has been a place for transition and settlement for the various cultures that have come and populated the Iberian Peninsula. Nowadays, no remains of Iberian or Roman settlements can be traced in the village itself, but relevant remains of an Iberian settlement have been located on a small hill in the northern part of the town where the existence of the Roman aqueduct Peña Cortada, among other minor findings, attest to the settlement of these cultures in the immediate environment of the current location of Chelva. Chelva's most historic neighborhoods have been listed by the regional government, where importance is given to its urban traces and the materials used to construct buildings, mostly private homes (Decree 168/2012). For this reason, it is considered appropriate to review the state of conservation of the buildings and the possible actions to be performed on them. Historically the materials used for construction are those which are in the immediate environment and soil is one of the most abundant elements.

Many houses, today located in the narrow winding streets in the historical city center with a strong Islamic origin, were built with rammed earth. There are also monumental buildings erected with rammed earth, as the existing Castle-Palace in the city center and the defense structure called La Torrecilla, which are also the oldest buildings in town with the exception of the Roman aqueduct.

2 HISTORICAL EVOLUTION

Although the origin of the city is not confirmed, there are remains of an Iberian settlement, nearby the urban settlement, and the remains of the Roman aqueduct also reveal the passage of this civilization. The first town written records date back to the 11th century as endorsed by the existence of the Almohad walls (Benavent & Magro 1998), found in the Castle-Palace, at the current Plaza Mayor. This enclosure would have continuity with the current neighborhood of Benacacira, where narrow winding streets built during the Muslim occupation, would also be surrounded by ramparts to protect the enclave from its elevated position. Prior to the formation of this neighborhood, the first residents settled in a natural shelter from rough stone that exists in the present district of Petrosa, literally "stony thing" (Mares 1681), almost under the current Plaza Mayor, which opens to the south and is protected from the north winds by the form of the slope where Chelva sits.

During the 13th Century, with the Christian reconquest, different neighborhoods were built around the Islamic area of Benacacira, considered the oldest neighborhood, and located on the south side. This period also saw the creation of an area called the Jewry, which began to form the Moorish quarter outside of the walled Christian part and along the old access road to the villa. The old town hall is also from this era, although it was renovated later in the medieval period when the Renaissance façade was added (Torralba 2003). All these neighborhoods were adapted to the topography with some sloping streets that follow the contours and others, some with stairs, following the slope of the hillside. In 12th Century the villa evolved northward with the construction of the Archbishopric Church of "Nuestra Señora de los Angeles," the construction of the current town hall and the current setting of Plaza Mayor. Around these elements new streets began to develop in an orthogonal mode expansion in the 19th century in a flat area. In the 20th Century the town continued to grow northward, with new and wider streets, which do not exceed ten meters, and higher buildings were constructed along the current CV-35, main access road to the town. The gradual expansion of the town to the north has allowed the historical framework to remain intact until the present day.

Figure 1. Hypothesis urban evolution (MA Lloria).

3 CONSTRUCTION TECHNIQUES AND LOCAL MATERIALS

Constructions in Chelva used local resources: the stone is very porous and soft and its possible to build with it the lower floors and areas in contact with the ground; and in the upper floors, the rammed earth is a very present element throughout the town, combined with other materials. The floors were built of wooden logs and plaster or brick vaults, depending on the season. The floors of the rooms were of rammed earth itself at low levels, fine plaster floor in many floors and, over the years, clay tiles or more contemporary materials. The roofs, usually gabled, are covered with curved tiles, Arabic, on a bundle of clay to regularize, extending over a hurdle that relies on pine logs from the nearby forests, usually without timber, and with lights up to six meters. The woodworks did not normally exist on the top floor of the house or *cambra*, because they need to be ventilated. The façades are white washed in many cases in white and jambs with hollow blue. Façades also painted in blue or green, but are currently limited. In some cases, different layers of lime reflect these colors. Often the rammed earth comes into view, especially in the back of the houses that fall in orchards or private areas that can observe the essence of building this element, so characteristic of the village of Chelva.

At first glance, when wandering the streets of Chelva, different types of rammed earth walls can be observed. Looking for representative elements, there are walls of rammed earth, lime-crusted, gypsum-reinforced brencas, brick reinforced, etc, which belong to various domestic buildings. As per monumental buildings, defense constructions in this case, there are walls of the 11th and 12th centuries, with great certainty, as well as the Castle-Palace and Torrecilla, a watchtower surrounded by a wall located one kilometer north of the town.

4 WORKING METHODOLOGY

To characterize and locate the different types of rammed earth walls in Chelva, this study began with a process of data collection based on the observation of the walls and private homes, especially those which can be seen from the street. Sometimes private owners have allowed observation of interior courtyards or rear façades that were not visible from the street, but these were isolated cases. Photographs were taken of all those elements that at first glance seemed to fit the construction technique of the rammed earth wall. There were also many elements covered by mortar or plaster, which complicate the verification of their rammed earth origin. All data from the overall construction and building system have been entered into a database. In most cases walls were made with rammed earth starting on the first floor and it was not easy to measure its most characteristic features, such as height tables, drawer height, distance between needle and wall thickness. In later phases of the research study, detailed specific cases were analyzed to exemplify different types of existing rammed earth walls in town. For this purpose, photogrammetry techniques were used in addition to direct measurement of the components. This whole process was implemented to examine domestic buildings, abundant and varied within the historic areas. There were also two immediate defensive enclaves, Castle-Palace and Torrecilla, which would require detailed studies to characterize all existing rammed earth walls in each of them and thus corroborate whether or not these were built with the same technique. These two elements were examined using the same methodology for data collection than the rest, but the wide surface of walls of rammed earth and the complexity of these constructions provided a considerable breadth of information that allowed the description of the various types of rammed earth structures observed in the rest of the town. All data were grouped, labeled and classified according to the different constructions in town, which in turn were grouped into various historical neighborhoods considered and defined administratively. A code was also added to each building with a rammed earth wall element and to each façades built with this technique.

5 EXISTING TYPES OF RAMMED EARTH

The process to classify each of the rammed earth wall found in the town consisted in describing each of the types and how they were recognized by observing each element (Font & Hidalgo 2009). There are many elements that were covered, usually with lime, which complicated the identification of the material, and were classified as indeterminate, and there were many walls which could be rammed earth wall but could not be identified as such, and which were recorded to be analyzed in subsequent study phases. The types of rammed earth wall found in Chelva and pertinent descriptions follow.

- Earthen type. Walls that do not have additives in their mass, and which show brown colors, similar to the color of the land soil from the surrounding gardens. In addition to the color, this type is also characterized by the very texture of the earth walls, with imprints of the tables, needles, etc.
- Lime-crusted type. Walls with mortar crust made of earth mixed with lime or limestone, which shuttering boards about before tamping mass. This outer coating produces whitish spots features within the mass that can be seen when the earth wall is cut. These elements have a lighter brown color than the walls of soil and are more uniform than before.

It must be said that many times in the same wall the crust on the outside of the wall facing and dirt inside combined, are supposed to economize by reducing the amount of lime to be used in construction. In this study the elements are described by their appearance alone counted data from both sides of the walls.

- Gypsum-crusted type. Walls considered as a variant of the wall above, it is identified because the crust or wall coating presents plaster mortar, in its interior may be ground or stones, but there are not certain data of the execution. Such walls are best known for the color causing the outer face, white to gray, and the texture of the facing where the imprints of boards show a different texture than the ones shown on the lime-crusted wall.
- Gypsum-reinforced. Walls seen in surfaces made of a wide range of materials and in curved lines produced by the combination of layers of gypsum mortar and ground which blend together at the edges of the higher footings but not in the center, resulting in spots (Maldonado & Vela 2001). This type is sometimes combined with plaster buttresses—*rafas*—inserted at the ends of the walls. Such walls result in striking geometric forms which are often hidden under painted façades, but which have a color variation due to the different construction materials used when exposed.

Figure 2. Rammed earth wall (MA Lloria).

- Brick-reinforced type. Walls which usually combined solid brick or sometimes stone, with layers of rammed earth. This type of bearing walls has lime scale on its exterior side, which confers greater resistance and improves adhesion of the mass of the brick wall reinforcement. These walls have a whitish tint and are only found in a limited number structures in town.
- Stone and earthen type. This type of wall has only been seen in those elements that are impaired and have lost their crust or surface layer. It is possible that many of the walls are constructed with this kind of reinforcement or improvement, especially in the lower parts where the action of water is more aggressive but has not been shown on many occasions. On the other hand is a good sign not to see the inside of the walls of earth because it means that the existing elements are not badly damaged.

Figure 3. Gypsum-reinforced wall "brencas" and/or "rafas" type (MA Lloria).

Figure 4. Brick-reinforced wall type (MA Lloria).

6 LOCATION AND CONDITION

The distribution of rammed earth constructions in town is not uniform, given that these are located in different neighborhoods without an apparent order. Sometimes different and contiguous houses share the same wall but are part of different properties, as a result of the successive segregations that have been made over time. In other cases the technique of the wall covered cannot be appreciated, but the continuity of cornices and termination elements lead to the conclusion that the construction could be coeval, even the same between different properties. Overall the condition of the walls of rammed earth is acceptable; most of them are in homes and are repaired, usually with cement or plaster to cover the cracks or holes of the needles. Unfortunately there are many empty homes where the deterioration of the wall is more pronounced and sometimes close to ruin and there are specific cases of total ruin. In cases in which the rammed earth wall is ruined, it is possible to observe the composition of these walls and how the different techniques and materials for the various parts of the structure are combined. The façades on the exterior sides are often made of lime crust, while the inside of the wall is earthen. This effect is very likely due to cost savings: constructors did not use lime where the action of the water is less invasive, but did use lime in the most sensitive to deterioration outdoor areas. Another important group is the one composed by walls that have a coating or plaster finish around them, executed prior to tamping or pouring the material into the footings, joining the mass of the wall in the drying process of each of the phases. However, with the current data it has not been feasible to describe the composition of various types of existing walls in Chelva. Nevertheless, it has been observed that most of the needles used in domestic buildings are round. In many cases, it is not possible to exactly determine the shape of the needles due to the deterioration of the weep holes by the action of animals or by the water damage that has distorted these elements so characteristic of this construction system. This study is intended to obtain data and characterize the most abundant types of rammed earth walls in town to provide knowledge to act on them and preserve its essence.

For the specific cases of defensive buildings, and most monumental buildings in town, the condition and characteristics are variable within each of the enclosures. Sometimes there are walls in perfect condition and in other cases there has been a considerable loss of coating or finishing materials that partially allows the identification of the techniques and materials used in its construction inside the wall. These constructions combine the use of soil with stones and lime crust in occasions. Elements

Figure 5. Ruin rammed earth wall (MA Lloria).

Figure 6. Location of rammed earth walls (MA Lloria).

Figure 7. Monumental rammed earth wall (MA Lloria).

in contact with the ground are complemented with stones and lime-crusted wall, to give it greater strength where it is most needed. In the case of the tower of the Castle-Palace, the rammed earth wall seems only made of soil at the coronation, but the loss of volume does not support this statement categorically. The main feature of these constructions, differing the domestic, is the form of needles. In these monumental buildings, the needles holes that have survived until today are rectangular in shape, approximately 7 × 2 cm, and ascribe its construction to the period of the Muslim occupation, specifically to the Almohad period.

7 CONCLUSIONS

Although the study is not yet complete, it is confirmed that there are many rammed earth buildings in Chelva, particularly private homes. This construction system uses local resources in a sustainable manner, and its preservation is fundamental. This knowledge of how buildings were built in the past is essential to conduct the necessary repairs in the present in order to preserve them for the future. Valuing traditional construction techniques, knowing them and respecting them allow us to act in the most appropriate way to keep its characteristic building technique linked to the place where it has been made. This study provides a basis to future proposals aiming at preserving rammed earth wall constructions as representative of Chelva's historical, architectonic, and aesthetic values.

REFERENCES

Benavent Ávila, Fernando; Magro y Moro, Julián V. 1998. Construir sobre lo construido: El caso de la posada de Chelva. En Actas del Segundo Congreso Nacional de Historia de la Construcción, A Coruña, 22–24 octubre 1998, eds. F.Bores, J. Fernández, S. Huerta, E. Rabasa, Madrid: I. Juan de Herrera, SEdHC, U. Coruña, CEHOPU, 1998.

DECREE 168/2012, de 2 de noviembre, del Consell, por el que se declara Bien de Interés Cultural, con la categoría de Conjunto Histórico, la Villa de Chelva y sus huertas, en el municipio de Chelva. [2012/10176]. Conselleria de Turismo, Cultura y Deporte de la Generalitat Valenciana.

Fermín Font y Pere Hidalgo. 2009. Arquitectures de Tàpia. Castellón: Colegio Oficial de Aparejadores y Arquitectos Técnicos de Castellón.

Maldonado Ramos, Luis; Vela Cossío, Fernando. 2001 Curso de construcción con tierra I, II y III. Madrid: Instituto Juan de Herrera, 2001.

Mares, Vicente. 1681. La Fénix Troyana. Teruel: Edita Teruel Imprenta "La Federació" 1931. 2ª edición.

Torralba Rull, Jerónimo. 2003. Las casas consistoriales de Chelva. Chelva (Valencia): Ed. La Fénix Troyana. Asociación Cultural.

Earthen Architecture: Past, Present and Future – Mileto, Vegas, García Soriano & Cristini (Eds)
 ISBN 978-1-138-02711-4

The walls construction system in Muza Tower (Benifaio, Spain)

V. López-Mateu & S. Tormo-Esteve
Departamento de Construcciones Arquitectónicas, Universitat Politècnica de Valencia, Valencia, Spain

T.M. Pellicer
Departamento de Ingeniería de la Construcción y de Proyectos de Ingeniería Civil, Universitat Politècnica de Valencia, Valencia, Spain

ABSTRACT: The study of the Islamic defensive architecture is very important due to their great extension in the Iberian Peninsula. This issue presents a direct connection with the construction system commonly used in their architecture at that time: the rammed earth. The studies carried out so far show common features and some peculiarities that could relate the technique from the geographical situation and the available resources, to different periods of construction or building technology evolution. Because of that, it is necessary to analyze in depth this construction system in each case study. It is the key matter to recognize their characteristics in order to layout and to apply the adequate conservation and restoration works that preserve the heritage significance. This paper presents the results of technical analysis related to the Muza Tower, focused on its walls building technique. The methodology used and the results obtained reveal some characteristics about this system that can be used as a reference to similar constructions. They can also be relevant for future research works and to increase the knowledge about the structural historical technique.

1 CASE STUDY DESCRIPTION

1.1 *Main data and location*

The Muza Tower is one of the Islamic towers that were built in the Iberian Peninsula, in the geographic area known as Al-Andalus, between the 11th and the 13th Century. The accurate chronology referred to its construction and its uses are still uncertain.

This tower is placed in Benifaió, a town located just about 20 km far from Valencia. Despite its proximity to the capital, it is sited in an agricultural area, close to an industrial park and it remains isolated from the main civic center. This situation is mainly due to the 19th Century layout of the railway, which runs close to the tower and cut the pedestrian path with the urban area. The previous studies carried out so far have considered that this defensive structure correspond to one of the earlier Islamic settlements in the area; and it is also considered as part of the defensive belt in the southern part of the city of Valencia (Bazzana & Guichard 1978). Nevertheless, the specific site related to the territory was unknown when the research started.

1.2 *Building condition: Previous approach*

Nowadays, the tower shows specific problems, especially in its top level, where the walls present severe erosion and the merlons have disappeared in some cases.

That is very usual in these defensive constructions, but it still remains some of them due to the strength of its masonry work.

Nevertheless, the rest of the walls are in a pretty good condition. But this is except for the northern corner that needs urgent and complex repair due to a crack that could produce the break of the wall in that corner.

By a naked eye it is also possible to appreciate different repairs, especially at the base of the tower. And looking in more detail, they appear around the front door, which is also quite common in this type of construction.

Inside of the tower, as in other similar buildings, there are different levels of floors. These levels would correspond with the access gate and several openings that it is possible to recognize. Which is not very regular as well as very impressive of this tower is the existence of a diaphragmatic wall. In the inner part of the tower walls there are some holes, cavities and horizontal tears. These horizontal tears appear in some positions on the outside of the walls.

Few more data were known about the tower; although there were many assumptions or explanations of all these tracks on the inner tower walls, as possible ladders, or other building elements.

1.3 *Methodology followed*

Due to the reasons exposed, before to undertake the restoration process, it was needed to elaborate complete preliminary studies of the tower. The preliminary studies are also required by the local Heritage regulations, in this case the *Ley de Patrimonio Cultural Valenciano* (art. 35.4).

However, in Spain there are only few general regulations for the contents of these preliminary studies, including the particular study of rammed earth walls. The standard UNE 41805:2009, indicates generically how to face up to the preliminary studies in a Heritage building.

This standard establishes the study of the constructive aspects to analyze the different types of materials and their compatibility, including the in situ masonry or rammed earth walls. It recommends considering the different external agents (biological, physical, chemical, environmental, etc.) as main responsible for the deterioration process.

Nevertheless, it does not establish how to carry out specific analysis of the rammed earth walls. Therefore, the general methodology was adapted to this specific case using the following steps:

1. To review the scientific literature about similar study cases.
2. To directly compare with other close or nearby towers (Espioca, Aledua, Bofilla, etc.).
3. To accomplish the visual inspection of the tower and the direct measurement tasks.
4. To undertake a complete photographic report and advanced graphical surveys.
5. To use the non-destructive testing techniques (NDT), such as infrared thermography.
6. To take small walls samplings to analyze their main characteristics (physical, chemical, etc.).

Figure 1. Muza Tower. View from the eastern corner, showing the SE and NE sides and the main entrance of the tower (López-Mateu & Pellicer 2013).

Nowadays, all these tasks are considered essential in a rigorous combined multi-disciplinary analysis, which can provide accurate data and results.

2 PRELIMINARY STUDIES

2.1 *Background*

According to the previous experience in Sot de Chera's Tower restoration (López-Mateu et al. 2012) a multidisciplinary group of UPV researches was formed in September 2012. The *Diputación de Valencia* put them in charge of the preliminary studies task to determine the tower condition and its possible restoration.

The studies carried out have been an effort of previous research in several fields, from the historical and documentary to botanical, technical drawings and material analysis as well as the construction system. They have provided the major data to determine technical characteristics and special features of the tower walls. From these special features we can underline the type and dimensions of the rammed earth walls, as well as the formwork components.

In these walls, it should be noted the particularity of the dimensions of the rammed earth formwork and the shape and the arrangement of the elements. These peculiarities have been studied and known through the tracks obtained from the walls after the first technical drawings of the exterior and interior walls.

However, in this case as in other defensive constructions, one can perceive that they are not only earth walls (build only with earth), but also stone walls molded with a mass containing stones of various features, joined by lime mortar, with an accurate grain-size distribution, similar to a concrete wall, but using lime instead of cement.

2.2 *Historical studies*

The historical sources indicate that are next settlements usually linked to a spring of fountain (as in the area called *Font d'Almaguer)* during the Iberian period, as well as placed in locations close to the tower from the Roman times. There have been archaeological excavations in different campaigns between the years 1922 and 2002 with quite remarkable finds (Bonet Rosado 2004).

The origin of the Islamic site is unknown, and there is a controversy because of the existence of a very near tower, at the center of the urban area next to the city hall. This tower is known as *Torre de la Plaza*, and it was restored in 1994 (Beltran i López 2006).

The first references to the present name of the town are from the medieval period. These quotations appear in the *Llibre del Repartiment* (Ferrando i Francés 1978); although there have not found precise allusions to the Muza Tower. Therefore, there are still unknowns about the exact date of the construction and its possible transformations or the subsequent historical changes it has undergo.

Nevertheless, some of these uncertainties have been explained, such as the reason for its peculiar location, in a farming area with different crops, apparently unrelated to the surrounding territory.

The location of some towers were linked to a strategic position as a road, an elevation in the field (a hill or a mountain) or the existence of a river or permanent water stream. Otherwise, in the Muza Tower case, the presence or the location of the tower is explained by the existence of a number of sources or springs in the area. They were originated in the underground streams between a mountain system and a lake or lagoon nearby.

This fact has confirmed the historical data of the location of fountains and springs in the area since ancient times. The existence of multiple fountains is also mentioned in several articles, although the particular situation of them is not known at this moment. One of these springs was located close to this tower. That is maybe one of the main reasons of its location in this specific area. Consequently, it has been crucial the investigation of the historical mapping of the area. This study has revealed the particular shape of the lot around the tower, which appears to have a ring around it. The trace of this shape has remained over time. Due to this research, it has been possible to discover the existence of a spring in the northeastern part of the Tower. That appears in the 1879 parcel map of the area. It is also peculiar in this tower the use of the building as a dovecote during some years. The numerous holes or caves in the inner walls illustrate this usage. These holes were the nest of the doves, as it is possible to distinguish in other dovecote buildings.

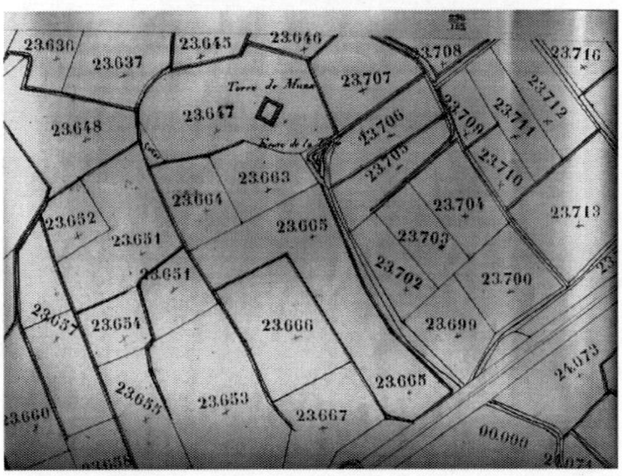

Figure 2. Detail of the parcel map from *Acequia Real del Xúquer*, 1879 (*Acequia Real del Xúquer* Archive).

3 CONSTRUTIVE RESEARCH

3.1 *Dimensional analysis*

First of all, it was carried out the study of the constructive technique of the rammed earth walls. This investigation include two aspects: the dimensional and the formwork arrangement where the materials were spur inside.

It was possible to carry out the general and particular measurements due to the relative good condition of the tower.

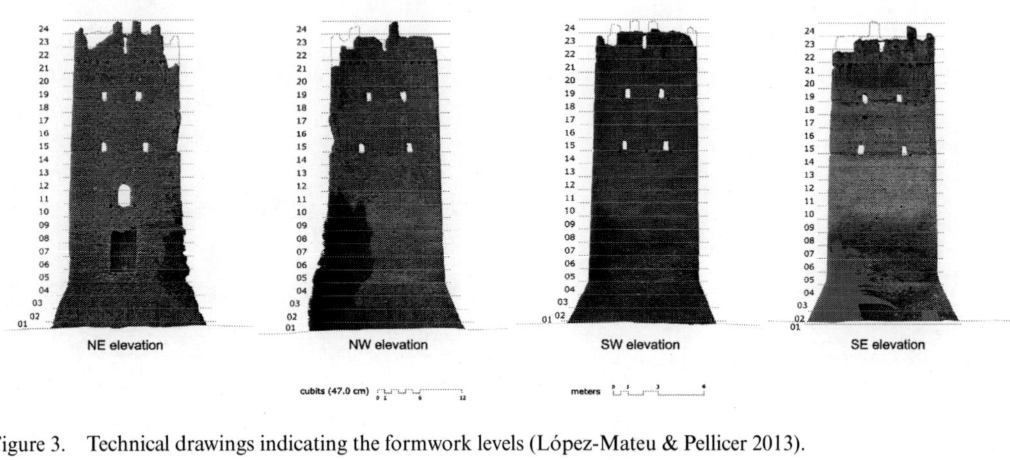

Figure 3. Technical drawings indicating the formwork levels (López-Mateu & Pellicer 2013).

Otherwise, in other towers the walls are more damaged and it is only possible to find a few remnants of the battlements or even the walls. In these cases, it is not possible to find the formwork traces.

The technique of laser scanner was used for the technical drawings as well as the photogrammetric image rectification, combined with measurement and in situ data and topographic survey. From this technical drawing were obtained the accurate (general and partial) dimensions of the wall. Without these techniques, it would be very difficult to carry out this work. That is because the walls have a lot of irregularities, some of them due to its own construction and some others by its own deterioration.

The work allowed obtaining a point cloud, both from the outside of the tower and from the interior. It permitted to perform the necessary cut of the 3D-model by points or areas of study.

Using this photogrammetric technique, it was obtained a few uprisings and very detailed photogrammetric rectifications, on which has been possible to "read" the document of the building.

The most relevant aspect respect to the dimensions of the tower has been the investigation of the sections identifying the module used to build the formwork. In this case, it is the Arabic cubit called black cubit (ma'muni), of approximately 47 cm.

In this tower, the whole construction is adjusted according to this module or pattern. It is possible to verify not only in elevation (as so far has been studied and considered), but also on the ground level. It also can be measured in the cross section of the wall, following the different floors.

According to the present research, the module for the formwork is two Arabic cubits (94 cm). It should be borne in mind that to check this measure in the traces of the wall, the distances would be less than that, between 90 and 92 cm in average.

This difference is due to the construction process: it is not possible to fill and level completely the mold. The top layer is necessarily compacted against the walls of the formwork, decreasing the height a few centimeters.

Moreover, the top layer should support and hold the sticks of the next level. These sticks would create incisions on the wall, when the mass is still fresh. This will allow embedding and fitting of the next mold. Following this module, it

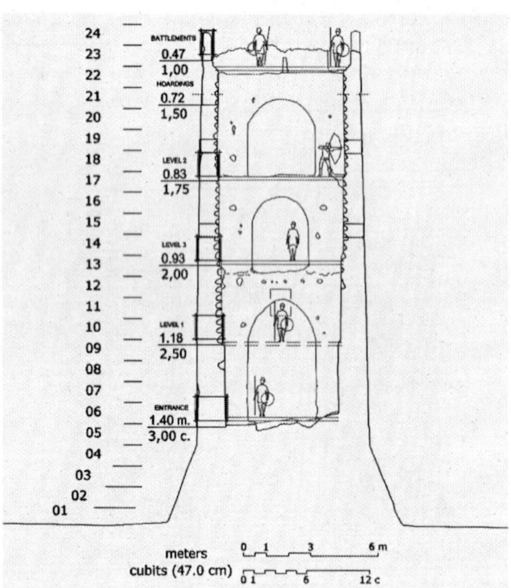

Figure 4. Muza Tower cross section showing the hypothetical floor levels (López-Mateu & Pellicer 2013).

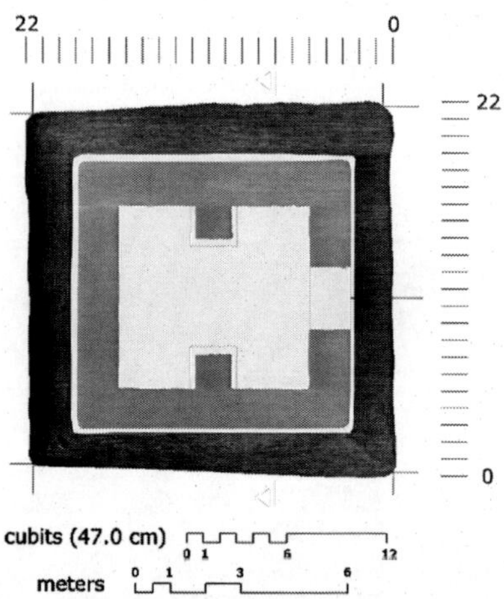

Figure 5. Muza Tower floor plan showing the cubit units (López-Mateu & Pellicer 2013).

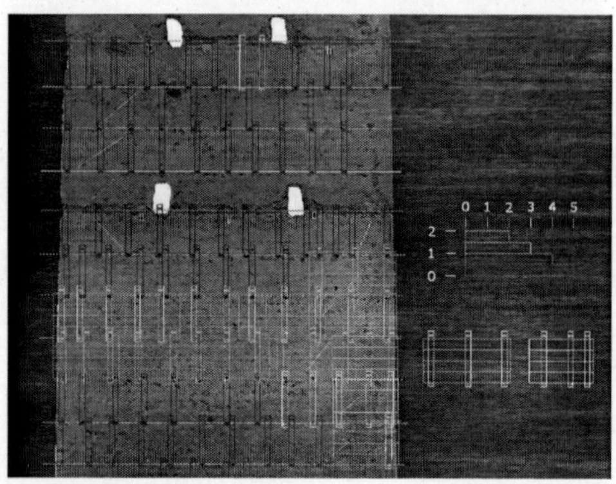

Figure 6. Partial view of the SE facade, with the formwork models superposed over the photogrammetric elevation (López-Mateu & Pellicer 2013).

has been established that between different levels (or changes in interior section) there is a constant height of 3.60 m.

That is the height of the base level as well as the first three floors, while the height changes to 4.50 m in the last level. That means the floors are located at four and five modules of rammed earth respectively.

The location and size of the loopholes used by ancient archers are directly related to the floor levels obtained. The last floor was the roof that would complete an upper level formed by a parapet surrounding the terrace.

This breastwork disappears in the corner to build the wooden hoard. On the top of the parapet there were the battlements, some of them still remain in the tower. This height corresponds with two modules that would complete the complete height of 24 modules (21.60 m). All these measures are linked to the construction system technique and the building layout.

3.2 *Construction technique study*

According to several papers and treatises consulted (López Elum 2002) the materials used for the construction of these walls were those relating to the environment in which they were built. This is a logical hypothesis due to the economy of means in that period. Nevertheless, in this case it should be noted that the materials used were sought on more remote sites due to the precise selection of them.

The area is flat, with few landforms and the natural waterways are not far away. The nearest ravine is the *Barranc de Tramusser,* located 1 km far north of the tower. Most of the network of the irrigation canals is subsequent to the realization of the *Acequia Real del Xúquer* (13th Century dated).

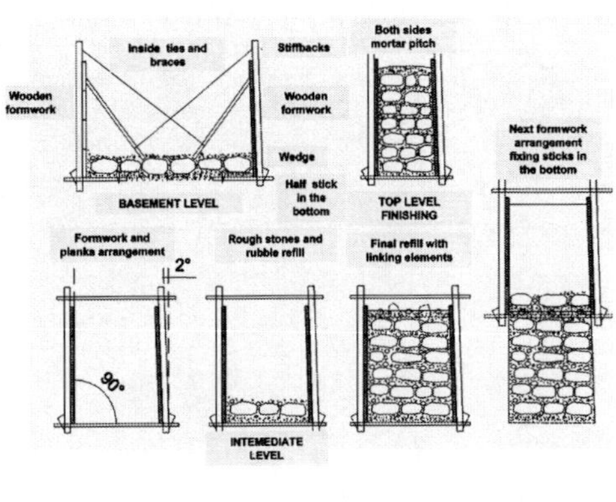

Figure 7. Cross sections showing the construction process with the formwork and the wall components (López-Mateu & Pellicer 2013).

The nearest mountains are the Llombai anticline and the Alginet Mountains. They are situated 5 km west of the tower. The lands that surround the tower are mainly agricultural lots. The appearance of these fields is clay soil. Nevertheless, the tower is slightly sunken, thus the land of these crops may come from fillings. For this reason, the first works was to investigate the type of dominant soil in the area through geological and geomorphologic studies. For this work it was consulted some manuals as the one published by *Instituto Valenciano de la Edificación* (IVE 2006), as well as the Geological map of Spain (IGME 1980).

These sources indicate that the field where the tower is located is an area of soft clay. The lithology of the area indicates that it is an area of colluvium. This means that the ground is formed by a type of barrage (detritus material transported and deposited in transient or permanent way by a current of water), consisting of very fine grains of silt and sand.

These formations have their origins in streams with small slope. In this case they are next to a marsh or lagoon areas and they are reinforced by the action of human beings due to the cultivation of the fields, which is predominant in the area.

Despite the foregoing, in the case of the Muza Tower it is found a masonry consisting mainly of stonework more than rammed earth work. The stones used are of limestone type according to the analyses carried out. They are of two types: angular and rounded, which also indicates its different origin.

At the bottom of the tower, there is an area where the more superficial part of the wall has been damaged due to the action of the Ivy. In this place, one can appreciate the courses of stones, especially those of medium size, around 15 × 20 cm. They are forming something like a masonry wall rather than a wall made of earth layers.

Nevertheless, the placement of the stones is regular, forming a kind of courses which would allow an easy placement and compaction into the mold. Thus, according to several authors (López Martínez 1999, Martín Civantos & Martín Garcia 2012, Mileto et al. 2012) it could be called masonry wall (*tapial de calicantos*), cofret masonry with lime mortar or cyclopean concrete wall. Consequently, it is possible to use some of the features of these materials to evaluate the behavior of the wall.

According to the granulometric analysis, compared with Fuller Grading Curves, it can be seen that the distribution of the different fractions of aggregates makes the mass very compact. This shows that the selection of materials and its gradation was specifically studied by the builders of the tower.

The selection of these materials is also particularly suitable for the defensive function of the tower by its strength, although the traces of having suffered attacks are not appreciated. It can be also considered that it is related to the durability and stability of the building in an area with high humidity and torrential rainfalls in the autumn and the spring seasons. This type of material and its realization with a very compact mixture, as well as its richness in lime mortar has allowed their preservation until our time despite being unused.

All these design features mean a great knowledge and experience in the use of the most suitable materials for the realization of this type of walls and a precise selection of its components.

4 CONCLUSIONS

It is necessary to recognize the historical background, the location and the environmental conditions to understand the building as an integral system as well as the origin of rammed earth construction system.

This technique is commonly used in ancient buildings and the materials used are related to the particular location. However, in some cases the ancient builders had chosen very specific components for the walls among their possibilities. This aspect indicates that they were already concern about the durability and withstand of their work.

The use of the same constructive system with a module as a pattern and a selection of materials guarantee the success of this purposes.

The results obtained in this tower show that it is possible to recognize that the module (in this case, the black Arabic cubit of 47 cm) affects not only to the height of the formwork, but also to the entire building (floor plan, cross sections, and the establishment of different levels).

The advanced technical drawings give a great advantage to deal with an accurate analysis of these dimensions and the construction process or sequence. These data are considered very relevant, since they could open the way for the verification of the general and specific dimensions of other towers in relation to its construction technique.

Nevertheless, there is still missing the connection between the dimensions of the formwork of the earth walls and buildings and the exact date of their construction.

REFERENCES

Bazzana, A. & Guichard, P. 1978. Les tours de defense de la Huerta de Valence au XIII^{e}S. *Melanges de la Casa de Velázquez*, tomo XIV: 73–106. París.

Beltran & López, F. (2nd ed.) 2006. *La Torre de Benifaió*. Valencia: Ajuntament de Benifaió.

Bonet Rosado, H. 2004. El yacimiento romano de Font de Mussa de Benifaió y el servicio de investigación prehistórica de Valencia. *El mosaico romano de los orígenes de Roma*: 29–45. Barcelona: Ediciones Gas Natural SDG, SA.

Ferrando & Francés, A. 1978. *Llibre del Repartiment*. Valencia: Vicent García Editors.

Font, F. & Hidalgo, P. 2009. *Arquitecturas de tapia*. Valencia: Puchades, S.L.

Font, F. & Hidalgo, P. 2011. La tapia en España. Técnicas actuales y ejemplos. *Informes de la Construcción* 63 (523): 21–34.

Instituto Geológico y Minero de España (IGME). 1980. *Mapa geológico de España: Sueca*. Madrid: Servicio de Publicaciones Ministerio de Industria y Energía.

Instituto Valenciano de la Edificación (IVE). 2006. *Guía de estudios geotécnicos para Cimentación de Edificios y Urbanización*. Valencia: Instituto Valenciano de la Edificación.

López Elum, P. 2002. *Los castillos valencianos en la Edad Media: materiales y técnicas constructivas*. Valencia: Biblioteca Valencia, D.L.

López Martínez, F.J. 1999. Tapias y tapiales. Loggia *Arquitectura y Retauración*. 8: 74–89.

López-Mateu, V. et al. 2012. Restoration of the Castle of Sot de Chera's Tower (Valencia, Spain). In Mileto, Vegas & Cristini (eds), *Rammed Earth Conservation*: 375–380. *Proc. 1st Int. Conf. Valencia, 21–23 June 2012*. London: Taylor & Francis Group.

Lopez-Mateu, V. & Pellicer, T.M. 2013. Rehabilitación de la Torre Muza de Benifaió: Estudios Previos y Primera Fase del Proyecto de Intervención urgente de consolidación.

Martín Civantos, J.M. & Martín García, M. 2012. Analysis parameters for systematization of rammed earth walls in Granada and Almeria (Andalucía, Spain). In Mileto, Vegas & Cristini (eds), *Rammed Earth Conservation*: 175–180. *Proc. 1st Int. Conf. Valencia, 21–23 June 2012*. London: Taylor & Francis Group.

Mileto, C. et al. 2012. Restoration of rammed earth architecture in the Iberian Peninsula: Ongoing research. In Mileto, Vegas & Cristini (eds), *Rammed Earth Conservation*: 369–374. *Proc. 1st Int. Conf. Valencia, 21–23 June 2012*. London: Taylor & Francis Group.

Earthen Architecture: Past, Present and Future – Mileto, Vegas, García Soriano & Cristini (Eds)
 ISBN 978-1-138-02711-4

Design in Jericho: Contemporary earthen architecture and local cultural landscape

F.M. Lorusso & S. Mecca
DIDA Department of Architecture, INN-LINKS Research Unit, University of Florence, Florence, Italy

ABSTRACT: The paper will analyse the design of a Cultural Centre in Jericho, Palestine. The project develops an integrated centre for the production, representation and cultural exchange together with hostel in collaboration with the Palestinian administration. Based on a partnership between Italian and Palestinian professors and students, the project and design goal is to build a sustainable Mediterranean architecture as evolution of a specific cultural landscape induced by interaction with contemporary architectural language and environmental needs. The final project was prepared after a Palestinian and Italian students workshop as requested by the Ministry of Public Works & Housing and the local authorities. The local culture of earthen architecture in Tel el Sultan, Ottoman adobe and timber frame buildings, together with the splendour of the mosaic floors of the remains of the Hisham Palace, reverberate immaterial and material echoes, evoking substances, forms and signs.

1 INTRODUCTION

The School of Architecture of Birzeit University (Palestine) and of University of Florence organized on May 2009 in Jericho a workshop, promoted by the Ministry of Public Works & Housing, the Jericho Governorship and the Jericho Municipal Government. The main objective of the workshop was an experimental project based on the study and valorisation of traditional earthen architecture (adobe), local materials and local building cultures that would satisfy the principles of eco-sustainability and bio-compatibility.

Figure 1. Adobe wall of ancient city of Jericho at Tell el Sultan (S. Mecca).

2 LOCAL TRADITIONAL ARCHITECTURE

The proposed design solution is based on a contemporary reinterpretation of the fundamental elements of the local architecture, with particular reference to the culture of mud brick buildings present in the Jericho Valley (Fig. 1). The local traditional architecture was mainly earthen architecture; there are many examples of adobe brick constructions in Jericho: from the walls and dwellings of ancient city of Jericho at Tell el Sultan, some of which date back to the 8th millennium BC, to the traditional dome house units (Fig. 2) or the elegant dwellings and accommodation facilities built during the British Mandate in the first part of the 20th century (Fig. 3). At the end of the workshop the students and lecturers of the two universities drew up a preliminary project for the construction of an innovative architectural complex in the Jericho municipality, granted by the Palestinian National Authority and strategically positioned vis-à-vis the town centre. The project was officially presented to the local authorities involved.

3 SOCIO-CULTURAL SUSTAINABILITY

The fieldwork was based on the interpretation of the traditional and local buildings and settlements, which were considered as appropriate to meet functional needs, environmental and cultural historically have consolidated.

The analysis of the architectural heritage has identified as characterizing the high density and

Figure 2. Traditional earthen dome building (S. Mecca).

Figure 3. A house built during the British Mandate, beginning of 20th century (S. Mecca).

geometric irregularity of the settlements, in which irregular public spaces combine the residences of one or two floors built around an enclosed patio full of plants. It identified as characterizing the mitigation systems of sun and heat through wise management of the shadow and the interception of the fresh winds from North-West, as well the role of trees and vegetables, of porches, of the courtyards and gardens, of the external walls without openings over the entrance door, of the screens of *mashrabiya* and of the flat roof with high parapets.

The project defines a new Cultural Centre composed of two functionally distinct parts that are, however, closely integrated: a multi-purpose building for the production and representation of cultural events (theatre, dance, music, conventions) and a hostel consisting of 42 double rooms and 8 small apartments for four people, each with their own patio and garden (Fig. 6).

An open-air theatre is hollowed into the ground; it is preceded by a square that acts as a physical hinge between the two parts and is the barycentre for social exchanges for town residents and visitors alike (Fig. 4).

A small square with a pomegranate tree, "tree of life" (represented symbolically in the Qasr Mafjar diwan, one of the most important artistic historical buildings of Jericho) acts as the city gate of entry from the road and the car parks, for both the accommodation area and the Arts Factory gallery-loggia. There are also two parking areas: for visitors and for the employees of the two buildings.

Figure 4. Aerial view of open-air theatre and Arts Factory (F. M. Lorusso).

Figure 5. View of internal plaza between hostel units (F. M. Lorusso).

Figure 6. Aerial view of Jericho Project (F.M. Lorusso).

The plan is organised around the central "urban" node of the theatre-square, and is deliberately laid out transversally across the site in an east-west direction so as to enjoy the view of the sunrise and the sunset and to exalt the visual relation with the two most important landscape and panorama references: the Jericho Valley and the Jordan River with the desert and the mountains in the background and the table lands of the Mount of Temptations.

The hostel on the opposite side is characterised by a complex morphology: four separate parallel residential blocks are placed in a NW-SE direction and between them are public spaces and paths that converge on the theatre-square and channel the cool northern winds. The scheme creates connective spaces that provide transversal walkways and small shaded squares with seating or fountains (Fig. 5). The monolithic volume of the Arts Factory is on the south side with reserved service access and parking.

4 THE ARCHITECTURAL CHARACTERS

The project is based on the valorisation and strengthening at the same time the traditional architectural culture of heavy adobe masonry, the local economy based on local resources and traditional building practice and of traditional and sustainable architectural language. In particular the theme of shade, especially of shaded spaces that act as intermediaries, as a visual and luminous filter between the outside and the inside, synthesized in the extraordinary idea-shape of the fretwork of the traditional *mashrabiya*, has given shape to the strong and dominant architectural theme of the large fretwork walls placed as screens in front of the facades of the accommodation facility and Arts Factory.

The distribution of the wide openings is reinterpreted: their size varies; their distribution is not uniform; some openings are unencumbered, others screened by brick grills; some are plastered others embellished with ceramic or mosaic borders; this reinterpretation up-dates the analogical reference at the conceptual and formal level.

The buildings are plastered with a traditional earth and straw mortar; the base perimeter resting on the ground consists of a course of blocks of pale stone that link up with the external paving. This paving is also in stone, with different shaped flagstones separated by strips of stone of varying widths.

This paving is also used in the theatre square and for the tiers of seating.

There are two different residential typologies:

- Small two-roomed apartments for two/four people with a little kitchen and bathroom, on the ground floor, 60 m^2, each with a small, paved patio with an underground water tank and a small private garden with a palm tree and/or a fruit tree (orange, lemon, pomegranate, banana).
- Double rooms with private bathroom, 30 m^2, on the ground floor of the two central blocks and on the first floor of all four blocks. These rooms have a private balcony, protected by the perforated screen wall on the elevation. The rooms on the first floor are reached by an external stairway.

The hall of the hostel is sited at the northern end of the central block: it includes the reception area, sitting-room, breakfast room and bar on the ground floor and a meeting room and offices

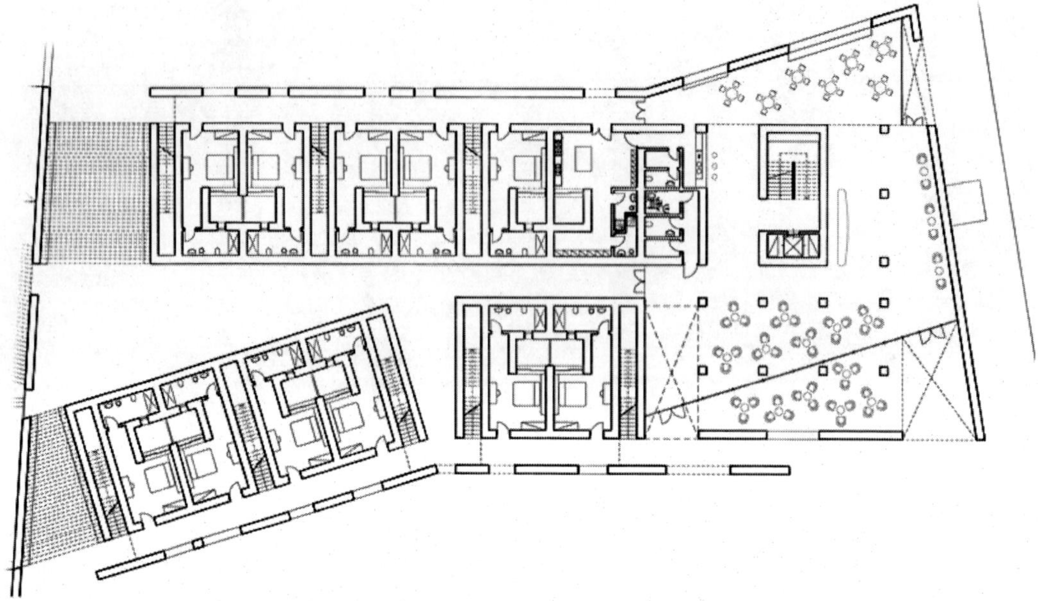

Figure 7. Plan of part of the hostel apartments (F.M. Lorusso).

on the first floor. A bar and a shop are planned for the ends of the two lateral blocks where they will be in direct contact with the two squares (Fig. 7).

The *Performing Arts Factory* building has been planned as a multipurpose space: it consists of two halls of different sizes separated by a three storey internal tower that houses the lavatories, storage areas, a goods lift, the stairs and the lifts, administrative offices and the dressing rooms for the open-air theatre (one floor is below ground level). At the west end the dressing rooms and lavatories are in the basement, the lavatories for the public visiting the open-air theatre-square are on the ground floor and a smaller multi-purpose hall is on the first floor.

The larger hall on the ground floor has space for about 400 seats; these can be removed so the space can be adapted to different uses. The eastern hall can be used for accessing the other hall or it can provide extra space for rehearsals and set designs. The eastern front facing the city and the northern front facing the theatre-square are both characterised by a large fret-worked wall that screens them and forms a gallery-loggia that welcomes and distributes visitors and spectators and, like a balcony or theatre box, looks out onto the open-air theatre below.

Most of the basement area is given over to storage and a stagecraft laboratory and it connects directly with the stage of the open-air theatre.

5 THE STRUCTURAL SYSTEM

The structural system of the buildings is based on the use of load-bearing walls made of raw earth elements assembled using classic adobe technology. The standard dimensions of these elements are $10 \times 20 \times 40$ cm, and they are laid in such a way as to ensure a resistant thickness of 60 cm. Structural continuity at each floor level is guaranteed by the presence of continuous girders in laminated wood, appropriately positioned within the thickness of the walls and connected to each other in the corners and other points of intersection, with double or single sections. Whether the sections are double or single depends on whether the particular wall is internal and supports two floors with beams running in the same direction, or whether it is a perimeter wall with beams of only one floor resting on it. The foundations consist of a continuous row of squared stone elements that widen out so as to increase the area that bears the vertical load transmitted and any eccentricity it might have (Fig. 8).

The load-bearing floor structures are created by aligning a series of wooden joists in parallel; the joists are about 40 cm apart. Other transversal connections placed in staggered rows protect the sections (30×20 cm) from twisting. The resistance of the walls to seismic activity is ensured by the structural continuity of the building, the bonding between floors and perimeter walls and the general uniformity in plan and elevation of the distribu-

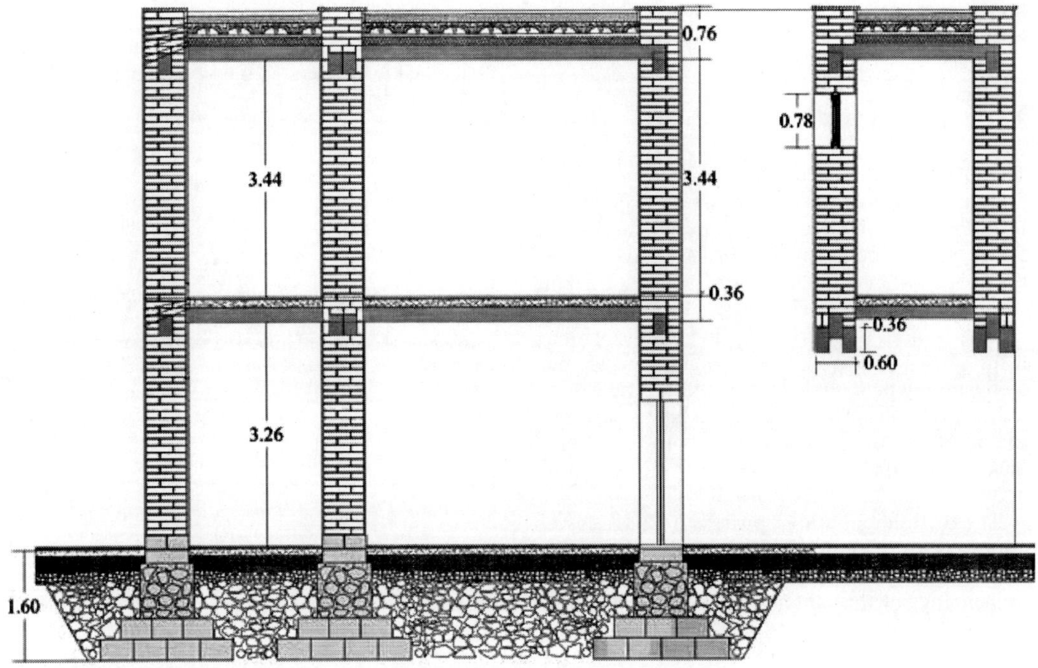

Figure 8. Building detail of masonry and timber structure. (F.M. Lorusso).

tion. Where there are higher walls, the over-all rigidity is compensated by inserting a laminated wood framework into the wall thickness.

6 MECHANICAL SYSTEMS AND ENVIRONMENTAL SUSTAINABILITY

The overall concept of the mechanical systems is based on the following fundamental goals:

- The lowest consumption of external material resources, as cement, steel, chemical material, etc.
- Hygrothermal comfort within the buildings independently of the external climatic conditions, the number of occupants and their activities.
- A lowest energy consumption target for the complex: the energy required for air-conditioning, hot water production, mechanical ventilation and artificial lighting.
- The complex should produce enough energy to be self-sufficient and to have a low environmental impact thanks to the use of the primary resource in the area: the sun.
- The different systems should all be integrated, flexible, easy to access and easy to maintain.
- Their optimal management should be ensured by adopting control systems for managing and regulating, that make it possible to respond to climatic changes and the variability of energy requirements in order to avoid a costly waste of energy.

Air-conditioning and hot water systems are diversified systems in order to meet the requirements of the different activities that will take place in the building complex.

The hostel will have an air-conditioning system that uses an air/water heat pump and radiant floor panels integrated with dehumidifiers. The combined system made by massive walls, shading walls and small courtyards will assure the basic hygrothermal comfort for users.

To ensure a minimum water supply and its sustainable use there will be a reserve water tank with a 10.000 litre capacity. Rainwater will be recuperated and wastewater will also be used for the lavatory flushes or for irrigating the green areas; the entire water plant will be integrated with systems for saving water (timed taps, double flush lavatories etc.) in all the bathrooms. To ensure the building complex generates enough energy to be self-sufficient 2 photovoltaic generators with a combined power capacity of approx. 130 kW will be installed: one on the roof of the theatre and the other on the roof of the hall.

7 CONCLUSION

The local culture of earthen architecture in Tel el Sultan, of Ottoman adobe and timber frame buildings, together with the splendour of the mosaic floors of the remains of the Hisham Palace, reverberate immaterial and material echoes, evoking substances, forms and signs.

This project is the fruit of a multidisciplinary team: the principles of sustainability and bio-compatibility have guided this project so new material and technologies have been used only if necessary to improve the quality and comfort of the complex, making it compatible with modern living standards, and to facilitate maintenance and management.

These differentiating process expresses an approach to "responsive understanding", as natural and indispensable necessity of update the phenomenology of the contemporary world, whatever it is, according an ethical and aesthetic innovative re-generation of the historical memory and its constructive knowledge, with no picturesque and local vernacularism.

REFERENCES

Breschi A, Roggio, Lorusso F.M. & Vallerini L. 2006. *Parco delle Cave. Seminario di progettazione di un ambito periurbano nella città di Cagliari.* Milano: Electa.

Fathy, H. 1986. *Natural energy and vernacular architecture: Principles and examples, with reference to hot arid climates.* Chicago: University of Chicago Press.

Fathy, H. 2008. *Costruire con la gente. Storia di un villaggio d'Egitto: Gourna.* Milan: Editoriale Jaca Book SpA.

Lorusso F.M. 2005. *Mediterraneo, paradigmi di architettura nuova.* Convegno internazionale sull'architettura mediterranea, Ordine degli Architetti di Nuoro.

Lorusso F.M, 2005. *"Progettare i luoghi" in Architettura 'naturale'. Origine e diffusione dell'architettura vegetale e in terra cruda.* Cabras (OR).

Lorusso, F.M. & Mecca, 2010. *"...un'immagine fugace con tutte le sue ombre" L'architettura mediterranea di Álvaro Siza.* Hiart, n°4.

Earthen Architecture: Past, Present and Future – Mileto, Vegas, García Soriano & Cristini (Eds)
© 2015 Taylor & Francis Group, London, ISBN 978-1-138-02711-4

The *karsheef*: Problems of cultural and environmental conservation

G. Martines
Polytechnic of Bari, Italy

ABSTRACT: The paper will report the results of a research project conducted by the Polytechnic of Bari about the earthen construction in clay and salt in Siwa. This construction technique is known since the 14th Century thanks to the journey report by Abu Abd Allah Muhammad Ibn Battuta. The research group applied to the study of the technique of *karsheef* (mud and salt mixed) in the aim to propose some intervention methods to prevent its decay. The three main themes in this case were: first, to understand the variation of the earthen techniques at the time when a different culture was aggregated to the original Siwa inhabitants; second: to analyze the variation of the climatic environment in the oasis in the last few decades and how it may have affected the earthen construction; and last, to study the behaviour of the population in the use of the traditional material. It is very important to say that in a close environmental system like an oasis, a traumatic change of the climatic aspects, joined to the change of the construction techniques and of the material, and eventually of the use of the ground, in only a few years, could make enormous damage to the balance of the oasis system. For the restoration of the *karsheef* technique, it is proposed to use the traditional methods, in the line of continuous maintenance, which is the only way to preserve these buildings: renovating the original material with intervention of partial rebuild. But the new settled areas are in a different position from the historical site: they are placed in the valley instead of on the top of the hill, with a very important problem of foundation, and the traditional buildings are no more used, and moreover their roofs are ripped to reuse the wood in new construction.

1 THE EARTHEN CONSTRUCTION IN SIWA

The Polytechnic of Bari, started in 2009 a line of research about the conservation of the earthen construction, in particular the study found an application in the Regional Cooperation Support Program financed by Italian State and Regions, project DIARCHEO—line 2.4—dialogue and culture of Mediterranean APQ—sub Project Siwa.

Siwa oasis is located in the Egyptian desert, in a depression under the sea level, next to the Lybic boundary, at a crossroad of important caravan routes.

The earthen construction techniques are diffuse in different manner throughout the whole world.

Siwa oasis, and in particular Shali (the hill to which belongs the first settlement) presents a very particular building and constructive material. Here there are no materials suitable for building, with the exception of a small quantity of a very resistant red sedimentary stone. The main problem, however, is not the bad quality and the scarcity of stone, but the absolute lack in the area of materials suitable to make the traditional binding mortar. Probably these are the factors that led to the use of a building technique with *karsheef* (*Hirir* in the dialect of Siwa).

The *karsheef* technique, however, differs from the earthen techniques traditionally systemized:

- Dug of materials
- Cut blocks
- Blocks compressed
- Extruded blocks
- Adobe
- Unformed blocks (compressed)
- Straw clay
- Daubed earth on frame
- Poured clay (compressed)
- And their various composition.

In some cases, in the Rammed earth technique, or in the Poured Clay, it is possible to find some inert materials like stones to give more strength to the lower part of the wall.

In Shali we found a very interesting form of earthen construction that mixes salt blocks, coming out from the deposits left by the drying of the salt lake that generates the oasis, and the salt mud created by the process of sweetening of land for the cultivation of palm trees.

Even the palm trees wood is used for the construction (for roof, ceilings, even doors) for the lack of other suitable materials. In short, the traditional building technique is achieved through the

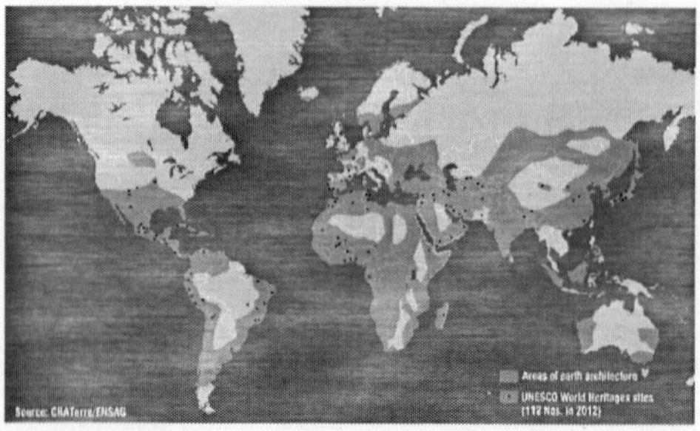

Figure 1. Distribution of earthen sites in the world; (CRATerre-EAG—UNESCO - "Earthen Architecture—Constructive Cultures and Sustainable Development").

Figure 2. The realization of a wall in *karsheef* (Martines).

use of materials produced in the 'closed system' of the oasis.

The wall construction technique, today still in use, implied the execution of masonries made of salt blocks arranged in a not different way from an *opus incertum*: the blocks are located without cutting or regularizations, as arriving from the piking-up sites, with the only attention of arranging frequent bondstones.

The salt blocks are joined with a salty mud: *tilakt* (*Tiin* in the dialect of Siwa); this mud is made by the clay token from the farming sites and is desalted in freshwater pools.

2 THE KARSHEEF TECHNIQUE

This construction technique is described in the report of the "journey" of Abu Abd Allah Muhammad Ibn Battuta, a rich Berber that in the 14th century has his *siyaha* (as he call that—that would mean to flow like water), the *hajr*: the religious pilgrimage through the Egypt desert. In his report he says: "[…] we arrived at Taghaza, a town without resources who has a curious feature: houses and mosque are made of blocks of rock salt and have the roof covered in camel leather There are no trees, just sand and a salt mine where they are found by digging huge slabs—so much so that a camel does not bear more than two at a time—arranged one above the other as if they had been cut and put on purpose underground." in fact the crusts of salt from the dried seabed of the lake, which break spontaneously as a result of crystallization, appears similar to the specially made plates.

The walls realized with salt blocks are very small: 20 to 60 cm even for building of three or four levels. This was possible thanks to the very strong conjunction between salt blocks and salty mortar.

In fact, when the wall was composed, the salty blocks surfaces, connected by the humid mud, left some salt starting a slow process of binding that in few years granted a monolithic behaviour of the masonries. In the composition of the wall were left some spaces that, in the connection process, could give place to the development of salt crystallization that binds the materials. Therefore was generated, around salt blocks, a boundary of ligature in which the new generated crystals binds chemically and mechanically the mortars to the blocks. An incredible process of indifference between "inert" and binders makes these structures behaviour like monoliths, but in each downpour (or in each masonries intervention with the flooding) the joints between materials are re-made.

In the ancient part of the city, on the urban wall, which is exposed to wind and rainfall, it is possible to observe a particular condition for the creation of crystals. By analysis (conducted by Fratini F., Rovero L., Tonietti U.), the *Tiin* have proved to be composed of sodium chloride, magnesium, gypsum and other minerals.

Figure 3. The cropping of *karsheef* slabs from the dried seabed of the salty lake of the oasis (Martines).

Figure 4. Remains of *karsheef* after the liquefaction of the mortar due to the loss of shells on the head of a wall (Martines).

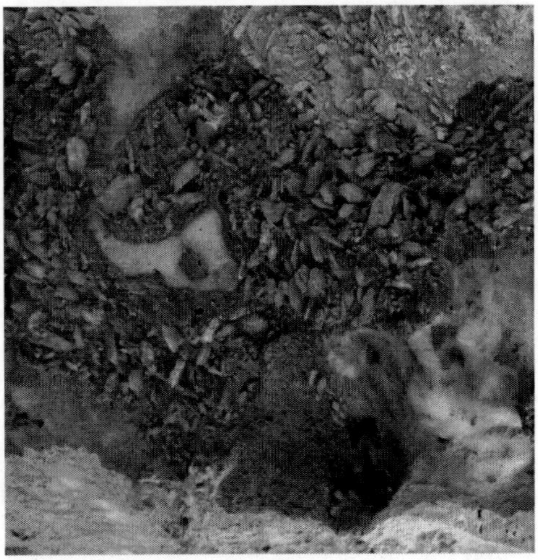

Figure 5. Effect of the 'ligature crown' between salt blocks and clay mortar (Martines).

Tiin mud is composed by 40–60% of Clay mineral, 10–15% by Quartz, 20–30% by Gypsum, till 20–40% Halite. *Karsheef* is composed by 40–86% of Halite with few traces of other elements.

3 THE TECHNIQUES THROUGH THE TIME

In the Ottoman period this settlement was enlarged, as another one was created on the overlooking hill of Adrar; it was originally created by the Bedouin

Figure 6. Crystals taken from the north side of the Shali urban wall near the door of the North Mosque (Martines).

Berbers that settled in the oasis, creating a dual urban structure founded on two overlooking hills, each of them with independent paths and defensive systems.

There are strong typological and technological differences between the two settlements. Only since the 19th century, the urban fabric of the oasis was enlarged downhill, linking again the two poles and expanding in the underlying tableland.

On the Hill of Adrar is possible to find different construction typologies and a different use of the same materials, linked to the use in the southern Sahara region: in this area is possible to find some stone used in place of the salt blocks. It prevents the 'ligature crown' process and in fact today we can see isolated elements instead of the common monolithic behaviour of masonries. Another 'strange' element we could find in Adrar is the use of the palm tree: the Siwa palms are very stubby and have a very low density wood, easily attacked by pests. In fact Siwan people use to left them tied to stones on the bottom of the lake for at least one year to sterilize them. After that they employ the palm tree divided in half, to use as beams or decks. In Adrar is possible to find same palm tree cut into wedges (about 60°) as it is used in the south of the Sahara, where the palm trees are much taller and have a much denser wood.

In the last century the technique of the *karsheef* is not lost, even if it is now accompanied by the use of *tobe* (gypsum blocks) and reinforced concrete, that is used as well as the *karsheef* in new buildings.

Figure 7. Detail of siwan ceiling realized using palm tree beams (Martines).

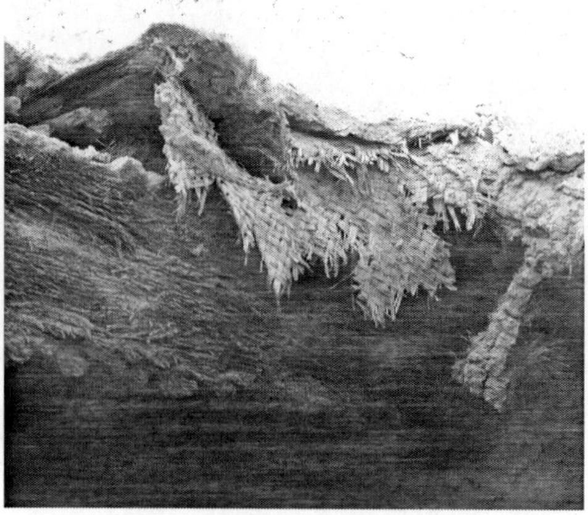

Figure 8. Mats made with woven palm leaves on the ceiling of a collapsing building in Shali (Martines).

Today, the use of the traditional techniques registered few differences from the past. Today *karsheef* is taken from other sites for the exhaustion of the original 'cave', and are often re-used the materials from collapsed buildings. The new collected salty blocks are not so pure and contain a great part of clay. It makes more simple the process of joint between salt and mud, but reduce the compressive strength of the overall wall, so are now used ticker walls and their resistance to rainfalls is minor.

For this reason, as well as for the climatic change of the last half century and for the extension of the city in the valley down the hills, today are used several elements to protect the masonries from the water: the most used traditional method is to apply mats, made with woven palm leaves, on the upper side of walls, as well as to cover the ceilings and to lie beneath the foundations.

4 THE CONSERVATION MATTER

Preventing the decay of earth walls, regardless the type and material, represents a serious technical and methodological problem.

First of all, it is not possible to define a "normal state" or an "original state" of the form of a building, because the particular type of material imposes a continuous work of repair, renovation and adaptation of the structure. So it is not possible to have a "point zero" to aim to in the restoration. On the other hand, the missing material from the walls usually is the main theme of the restoration, and the typology of missing is to determine the type of intervention. But in *karsheef*, as in the most of earthen techniques, the mass *is* the structure, so, for the conservation, we *need* to reintegrate the original material, against the most well-established rules on restoration in the European culture.

Again in contrast to our established rules, we should expect a very poor durability of the intervention, according to the type of material. The alternative would be to change the material or forcing it with some chemical alteration that would distort the concept of the intervention.

Therefore we can find a solution only if we are allowed not to observe the strict conservation of the form, but the conservation of the history of the constructive techniques that, in the time, have spontaneously adapted to the change of the needs.

The philosophy adopted by the Bari Polytechnic for the mosque restoration provides for the respect of every addition or change the monument has undergone in time, in order to preserve the use and building typology of this geographical area; it sees in the process use-maintenance-adjustment of the structure, an untiring continuity of the tradition of the religious use and of the building techniques. A right respect of the building techniques has been noticed even in the overlapping of maintenance and adjustment-change works that the structure underwent and our intervention is intended to confirm it.

Even in the restoration techniques the local tradition proposes optimal resolution for most of the problems, like underground humidity, cantonal fractures and bindings.

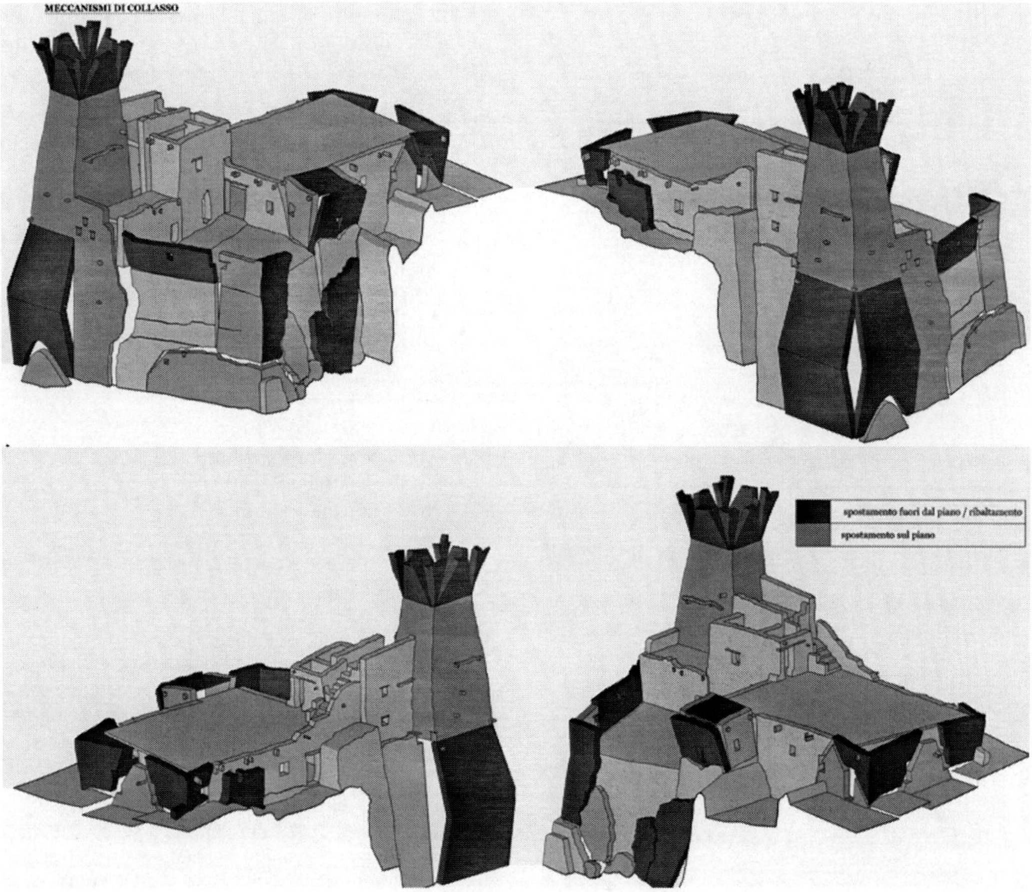

Figure 9. Schematic view of the static and seismic problems of the Shali North Mosque (Martines).

It has already been specified the entry in the use of mats of woven palm leaves for the purpose of dehumidification that have begun to be present in the buildings from the last century, used under the wall to separate the *karsheef* from the ground upcoming humidity.

5 CONCLUSIONS

Our Program was expected to provide a sort of guidelines for the conservation of traditional structures. In particular, we conducted a study on the Shali North Mosque to identify the conservation problems and their causes, and to verify if the traditional techniques could give well-grounded solutions to the matter.

Following the intentions to respect the 'closed system' of the oasis, we analysed and collected a detailed list of traditional solutions, which seemed able to correctly response to the conservation problems, and we suggested in which case and how to use them.

We only introduced a new factor, which is an extensive use of wooden binders and palm mats, already presents in some areas of the oasis, but unfortunately abandoned in new buildings, in favour of materials from the global market, not so suitable for such a precarious balance.

NOTES

The paper comes from the work of the Support Program for the Regional Cooperation—Framework Programme Agreements: Countries of the Mediterranean south shore—Line of Intervention 2.4 "Dialogue and Culture" Integrated Project Diarcheo | Sub-Project Siwa.

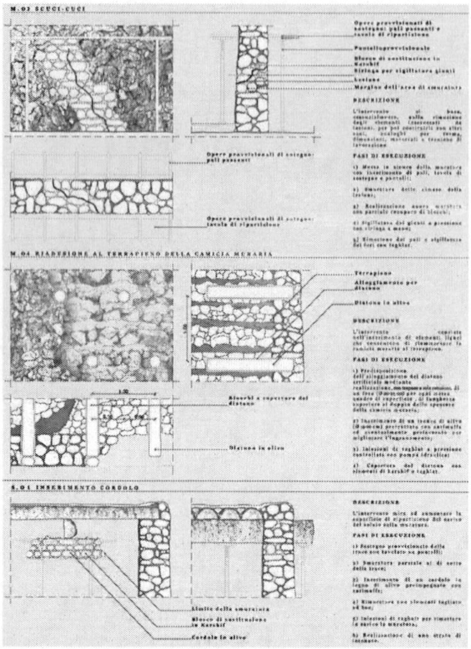

Figure 10. Some schedule of the typed intervention developed for the case of the Mosque (Martines).

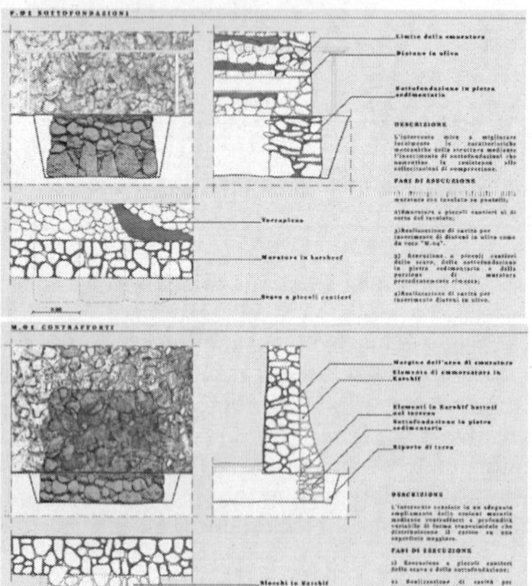

Figure 11. Some schedule of the typed intervention developed for the Recovery plan of Shali (Martines).

The photographs in the present relation were taken by the author. The drawings were elaborated by the work-team of the Polytechnic of Bari.

- Research team:
  - Prof. Arch. Attilio Petruccioli (coordination)
  - Arch. Calogero Montalbano
  - Arch. Giacomo Martines
- Tutors:
  - Arch. Paolo Perfido
  - Arch. Valentina Castagnolo
  - Ph.D. Arch. Domenico Catania
  - Arch. Claudio Rubini
  - Arch. Adriano Spada
- Work-team:
  - Arch. Sabina Calvano
  - Arch. Piergiorgio Loconte
  - Arch. Stefania Pizzolante
  - Arch. Pasquale Ricupero
  - Arch. Antonio Vitulli

REFERENCES

Fratini, F., Rovero, L. & Tonietti U. 2009. The Salt Architecture in Siwa Oasis—Egypt (XII–XX Centuries). In *Proceedings of the 4th International Congress on "Science and Technology for the safeguard of Cultural Heritage in the Mediterranean Basin*, Caire.

Houben, H. & Guilland H. 1996. Earthen architecture. In *The Courier n°159 Dossier, Investing in People Country*.

Mileto, C., Vegas, F. & López Osorio J.M. 2011. Criteria and intervention techniques in rammed earth structures. The restoration of Bofilla Tower at Bétera (Valencia). In *Informes de la Construcción* 523. Madrid: CSIC.

Petruccioli, A., Martines, G., Montalbano, C., Bertagnin, M. & De Antoni, D. 2012. The Siwa Oasis: The art of building and sustainable development. In *Le Patrimoine, moteur de développement—Heritage, a driver of development—Actes du symposium de la XVIIème Assemblee Generale de l'ICOMOS—proceedings of the 17th ICOMOS General Assembly.* 1099–1110. Paris: UNESCO—ICOMOS.

Terra, 2008. *10th International Conference on the study and conservation of earthen architectural heritage.* ICOMOS—ISCEAH—International Scientific Committee on Earthen Architectural Heritage.

Earthen Architecture: Past, Present and Future – Mileto, Vegas, García Soriano & Cristini (Eds)
 ISBN 978-1-138-02711-4

Characterization of external renderings of rammed earth constructions in the Algarve

L. Mateus & J. de Brito
Instituto Superior Técnico, University of Lisbon, Portugal

M.R. Veiga
National Laboratory of Civil Engineering, Lisbon, Portugal

ABSTRACT: The Algarve region in southern Portugal has one of the most significant rammed earth construction legacies in the country. This construction method is characterized by its high susceptibility to water damage and external renderings are essential to the longevity of these constructions. The search for mechanical compatibility between external renderings and rammed earth is also important to enhance the longevity of these constructions. Some of their commonest anomalies occur due to the interaction between the rammed earth substrate and the rendering materials, because of their different characteristics and performance. Both hydric behaviour and mechanical strength of the rammed earth renderings were determined through laboratory tests. The mechanical characterization of the rammed earth walls was also carried out. The characteristics of these two materials and their behaviour as an assembly were analysed and discussed.

1 INTRODUCTION

This paper is based on a survey and sample collection of exterior lime renderings from traditional buildings in western Algarve and of their rammed earth backgrounds.

The use of protective renderings is common in Portugal, unlike in other regions of the world. These constructions have external renderings made of coats of mortar with lime, clay and sand, made on site (Margalha et al. 2011).

At this preliminary stage a set of samples was assembled for their physical, mechanical and chemical characterization in laboratory.

Within a Ph.D. study at Instituto Superior Técnico, characterization tests were performed by the first author at the National Laboratory of Civil Engineering (LNEC), allowing the establishment of analysis criteria aiming at the understanding of the degradation phenomena of these renderings.

2 OBJECTIVES AND SAMPLES

Renderings are the first physical barrier inhibiting the direct contact of water with rammed earth substrates, preventing leaching of the fine aggregates and loss of load bearing cross-section.

The main degradation phenomena are related to the poor resistance of the earth substrate to horizontal actions and the high susceptibility of earth to water damage (Magalhães & Veiga 2009). The main objective of the characterization process of the renderings is to understand the natural degradation phenomena and how they are associated with the background characterization (Mateus et al. 2005). As a result of this analysis, a search was undertook for clues on the adequacy of applying certain rendering compositions, both in rehabilitation scenarios and in new construction.

The collection of samples for the research, aiming at physically and mechanically characterizing rammed earth construction, focused on the following aspects: 1. type of substrate: rammed earth; 2. type of coating: exterior; 3. geographical area: western Algarve; 4. type of construction: current dwelling 1 to 2 storeys above ground; 5. construction period: from 1850 to 1950.

Fifteen buildings (Fig. 1) were selected to test their renderings. Rendering samples from all the fifteen buildings were submitted to water absorption and drying tests; seven of the buildings were tested regarding compression strength of external renderings and rammed earth substrates. These constructions exhibited degradation of the external renderings, especially lamellar detachment, loss

Figure 1. One of the constructions analyzed in western Algarve (L. Mateus).

of adherence and cracking related to movements of the earth substrate.

3 CHARACTERIZATION METHODOLOGY

The methodology used to characterize the renderings comprised the following major steps:

1. Field survey: Buildings selection and systematic collection of renderings and earth backgrounds samples;
2. Laboratorial stage (performed at LNEC—National Laboratory of Civil Engineering, Buildings Department): Execution of tests for physical and mechanical characterization of the collected samples: capillarity (water absorption) and subsequent drying; compressive strength (this test was performed both on renderings and earth substrates).

Since samples were irregular in shape and disintegrate easily, standard test methods could not be applied to determine water absorption and compression strength, which demand well defined shapes and some cohesion. Therefore, tests developed at LNEC for ancient renderings within other research projects (Valek & Veiga 2005; Veiga et al. 2004) were used.

4 TESTS CAMPAIGN

4.1 *Capillary absorption and drying—renderings*

The first stage effectively established response patterns to water contact, by simulating a natural scenario of rain exposition. It defined the level and rate of water absorption of applied renderings and it correlated them with the drying rate. This analysis is especially relevant given the high susceptibility to water of rammed earth substrates. It also evaluated the expected water retention period within the rendering layer after rain water absorption and in what way it can affect the integrity of the earth substrate. The technique used, described and validated in previous work (Veiga et al. 2004), is visualized in Figures 2 and 3.

The 45 samples - 3 for each of 15 buildings selected—were weighted at the absorption stage and after drying, under similar temperature and humidity conditions. The mass of the specimens was measured at given time intervals to determine the mass increase resulting from absorbed water. A similar process occurred at the drying stage.

This test aimed at establishing direct relationships between the strength capacity of existing coatings in the buildings selected and strength capacity of the rammed earth substrate (Veiga et al. 2004).

4.2 *Compression strength—renderings and rammed earth*

If the mechanical incompatibility between the rendering and the earth substrate is confirmed, it represents a contribution to the degradation phenomena of external renderings. The compressive strength is one of the relevant characteristics to be

Figure 2. Measurement of specimens mass during the water absorption process—renderings (L. Mateus).

Figure 3. Drying process of the specimens—renderings (L. Mateus).

Figure 4. Samples for compressive strength tests—renderings and earth substrate (L. Mateus).

evaluated within that scope as an overall indication of the mechanical characteristics.

So, the study was focused on determining the compressive strength both of coatings and rammed earth. For each of the seven buildings, 3 rendering samples and 5 substrate samples were tested (Fig. 4). Since the samples are irregular in shape and disintegrate easily, the standard test methods could not be used. Instead, methods developed at LNEC within earlier research projects on old renderings were used.

5 RESULTS AND DISCUSSION

One of the main aspects concerning renderings degradation processes is their susceptibility to water. The render function is to delay the water absorption and promote quick drying. It should have a low capillary coefficient combined with high drying rate.

The results obtained are translated into capillary absorption and drying diagrams (Fig. 5). The contact capillary coefficient mean values are presented in Table 1.

The analysis of the water absorption and drying test results leads to the following conclusions:

1. The water absorption process is quick. The highest absorption rate occurs in the initial water absorption period (first 5 minutes), implying a significant capillarity coefficient of the renderings analyzed;
2. In most cases, almost the whole absorption of the coating is reached after a short period (under 25 minutes); the remaining water absorption process occurs at a very low rate;
3. Only two buildings have slower absorption processes—Ferragudo (Fig. 5) and Sargaçal;
4. The drying process is slower, drying periods are significantly larger than absorption periods and rates are smoother over time;

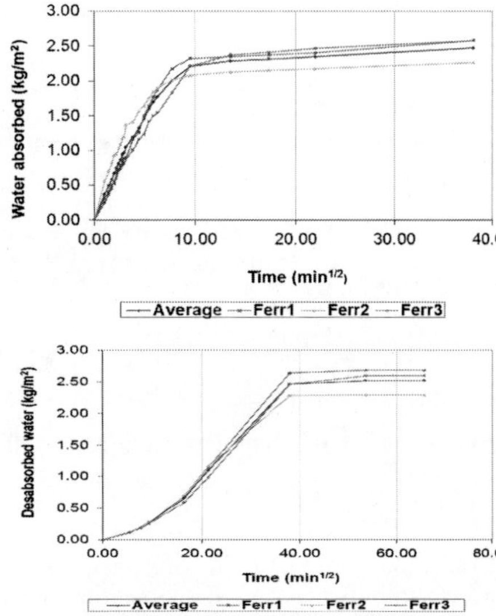

Figure 5. Water absorption (up) and drying (down) diagrams—renderings (L. Mateus).

Table 1. Contact capillary coefficients—renderings.

Case study	Coeffcient of capilarity by contact (kg/m^2 · min1/2) – Average			Overall absorption (kg/m^2)
	Cc 5	Cc (90–10)	Cc24	
Ferragudo	0.33	0.18	0.07	2.47
Vale da Pinta	0.84	0.02	0.07	2.67
Monte Vidal	0.67	0.15	0.09	3.24
Barragem	0.77	0.09	0.10	3.87
Monte Bravura	0.60	0.05	0.07	2.78
Sargaçal	0.10	0.20	0.12	4.70
Monte Ruivo	0.46	0.06	0.05	1.95
Vale Figueira	0.72	0.04	0.07	2.50
Aljezur A	0.82	0.20	0.12	4.37
Bensafrim	0.43	0.26	0.09	3.38
Benagil	0.58	0.00	0.04	1.58
Silves	0.48	0.19	0.08	2.90
Praia Benagil	0.68	0.05	0.07	2.80
Aljezur B	1.46	0.01	0.10	3.88
Aljezur C	0.77	0.06	0.08	2.97

5. Therefore, there is a very significant difference between the period during which saturation of the coating occurs throughout its thickness and that corresponding to total drying of the

previously saturated coating; this implies, during the saturation/drying cycle, a water retention period in the interface between the coating and the rammed earth substrate;

6. The high capillarity of the renderings promotes the contact of infiltrated water with the rammed earth substrates and their progressive degradation by leaching of the fine aggregates, which progressively weakens the adherence of the renderings to their substrates.

The results of compressive tests are summarized in Figure 6 and Table 2. They led to the following conclusions:

1. In most buildings, there is a significant difference of the average compressive strength of renderings and substrates. In some cases, the rendering's strength is 2 to 3.5 times higher than the substrate's strength, showing the mechanical incompatibility between these materials. Exceptions are Monte Vidal and Sargaçal buildings, which have similar rendering and substrate compressive strength values (slightly higher in renderings);
2. Some external coatings show high compression strength (around 3.1 MPa) compared with current air lime mortars (Veiga et al. 2004), resulting from the carbonation of the lime in time, that slowly increases the strength of applied lime mortars; this generates excessive stresses in the interface with the rammed earth substrate, as a result of different deformation capacities (Binda et al. 2000), promoting loss of adherence between these materials and frequent spalling in this type of building;
3. There are significant differences between the compressive strength of the various renderings, ranging from 0.7 MPa (Monte Vidal) to 3.1 MPa (Ferragudo); by comparison, rammed earth's compressive strength is more homogeneous, with values ranging from 0.6 (Monte Bravura) to 1.4 MPa (Ferragudo).

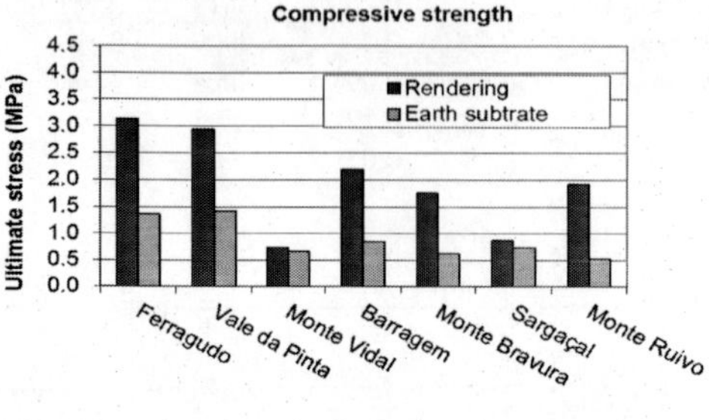

Figure 6. Compressive strength of renderings *versus* earth substrates, for each construction tested (L. Mateus).

Table 2. Compressive strength of specimens (average and standard deviation) [MPa]—renderings and earth substrates.

Coating		Substrate	
Coating Ferragudo Rendering		Substrate Ferragudo Rammed earth	
R_m	3.1	R_m	1.4
Δ	0.89	Δ	0.08
Coating Vale da Pinta Rendering		Substrate Vale da Pinta Rammed earth	
R_m	2.9	R_m	1.4
Δ	0.03	Δ	0.06
Coating Monte Vidal Rendering		Substrate Monte Vidal Rammed earth	
R_m	0.7	R_m	0.7
Δ	0.01	Δ	0.02
Coating Barragem Rendering		Substrate Barragem Rammed earth	
R_m	2.2	R_m	0.9
Δ	0.83	Δ	0.01
Coating Monte Bravura Rendering		Substrate Monte Bravura Rammed earth	
R_m	1.8	R_m	0.6
Δ	0.06	Δ	0.02
Coating Sargaçal Rendering		Substrate Sargaçal Rammed earth	
R_m	0.9	R_m	0.7
Δ	0.07	Δ	0.01
Coating Monte Ruivo Rendering		Substrate Monte Ruivo Rammed earth	
R_m	1.9	R_m	0.5
Δ	0.28	Δ	0.02

6 CONCLUSIONS

The present rebirth of rammed earth construction has various reasons, linked to the acknowledgement of its advantages in terms of traditional construction identity to be maintained. However, it also reveals some drawbacks, namely concerning their low resistance to earthquakes and susceptibility to water action.

The group of buildings analysed is widespread in Western Algarve. This is associated to intended variations in the characterization of the substrate and coating materials, namely in terms of the grading curve of the earth mixes. Even though these variations are noticeable, tests performed in laboratory revealed the existence of response patterns common to all the buildings tested, either in terms of water presence (and subsequent drying) or materials compression strength. The following are underlined:

1. Water absorption is quick, revealing great susceptibility to water of the renderings analysed;
2. Drying is much slower, implying long water retention periods in the coating thickness, promoting its absorption by the background material (rammed earth) and leading to a progressive loss of compactness and cohesion of this material;
3. In most buildings, the compressive strength of the coatings is much higher than that of the substrate material, leading to stresses in the interface that promote precocious loss of adherence, cracking and spalling of the coatings.

ACKNOWLEDGEMENTS

The authors acknowledge the support of LNEC—National Laboratory of Civil Engineering, especially the team from the Renderings and Insulations Group of the Buildings Department, for their help and guidance in the execution of some of the lab tests. The authors also acknowledge the support of ICIST (Instituto de Engenharia de Estruturas, Território e Construção), IST (Instituto Superior Técnico).

REFERENCES

Binda, L. et al. 2000. Investigation procedures for the diagnosis of historic masonries. *Construction and Building Materials* 14.

Magalhães A. & Veiga M. R. 2009. Physical and mechanical characterization of ancient mortars. Application to the evaluation of the state of conservation. *Materiales de Construcción* 59 (295).

Margalha G. et al. 2011. Traditional methods of mortar preparation: the hot lime mix method. *Cement and Concrete Composites* 33 (8).

Mateus L. et al. 2005. Methodology of laboratory and in situ characterization of traditional renderings in rammed earth constructions (in Portuguese). *IV SIACOT—Iberian-American Seminar of Earth Construction, Monsaraz.*

Valek, J. & Veiga, M. R. 2005. Characterization of mechanical properties of historic mortars. *STREMAH 2005—Ninth International Conference on Structural Studies, Repairs and Maintenance of Heritage Architecture, 22 / 24 June 2005, Malta.*

Veiga, M. R. et al. 2004. Capillarity tests on Historic mortar samples extracted from site. Methodology and compared results. *13th International Masonry Conference, Amsterdam.*

Earthen Architecture: Past, Present and Future – Mileto, Vegas, García Soriano & Cristini (Eds)
© 2015 Taylor & Francis Group, London, ISBN 978-1-138-02711-4

Masked defensive architecture. El portal de Massalavés

I. Matoses Ortells
Architect and Quantity Surveyor, Valencia, Spain

ABSTRACT: Sometimes defensive architecture, over time, was adapted so the restorations would comply with any new functional requirements. Such interventions could modify the existing typology, producing superimposed typologies that obviated the nature of the monuments. This happened to the doorway of Massalavés, a defensive construction that was refurbished integrating its typological characteristics to those of a traditional home, masking the characteristic elements of its nature. The building was used during the 20th century as the Town Hall, which was left in a state of neglect after being left vacant. A recent intervention, aimed at repairing the building in order to give it a new use, has favoured the study of the original construction techniques employed to build it, the masonry, the wall made of limestone and masonry mortar in a wood formwork, and the rammed earth, during the process discovering concealed elements of high heritage value. The article will try to undertake a historical and constructive approach of the building by explaining the interventions that were employed, to foster their knowledge and dissemination.

1 HISTORICAL APPROACH

The Christian conquest of Valencian lands in the first half of 13th century was a relatively short process, due to, among other reasons, the scarce resistance of the Islamic forces during most of the sieges and also to the strategy followed by James I, who chose agreements over invasion. (Gómez-Bayarri 2003, Vol. III, 175).

The integration of the Arab population in the configuration of the Kingdom of Valencia is considered one of the main reasons for the success of the Reconquest, at the same time favouring the survival of the Muslim workforce, until their definitive expulsion, in the first decade of the 17th century.

However, after the crusade there were constant uprisings, especially in the region of the Júcar river, the Islamic defensive limit in the conquest of the northern area of the kingdom and the main focus of riots and insecurities in the years following the land distribution to the Christians.

In times of Islamic rule, the hamlet of Massalavés, Maçal-hauézin or inn of the Hauézin (Llorens 1976,55), had a population fundamentally dedicated to agriculture, manned by a defensive construction, from which there are still remains.

After its recovery by the Christians it was given to the Milá's lineage by royal privilege, thus the barony of Massalavés, a Christian manor of mainly Arab population, was created. However, it didn't have a stable governance, being passed onto several feudal lords over the subsequent three centuries, until the news of the claim of the Milá in 1493 (Llorens 1976, 64–78).

Such social and governance instabilities, together with fortuitous natural disasters, such as the regular overflows of the Júcar river, justified the construction of the new Massalavés's fortifications, using the remains of the Arab tower, as well as the various interventions of repair and refurbishment which the building went through during the first times of its existence.

Fortifications, being constructions of singular characteristics with regards to their implementation and design comparing to the rest of architectonical types, were uncomfortable and difficult to maintain when there was no longer a sense for its usage. They would be abandoned, becoming quarries that hastened their deterioration, only taking advantage of the materials they offered. In other cases, they were reused, given their convenient location, as base for other buildings, including becoming new fortifications.

There are a huge number of cases of fortified elements that have been transformed into buildings for other uses, either defensive towers into homes, castles restored and reformed into shrines, watchtowers into belfries (Fig. 1), walls used as dividing walls, etc. The alterations could be of such a magnitude that they sometimes hid any elements that were characteristic of the fortifications, making them unrecognizable, hidden under the architectonic circumstances of their new use.

The castle of Massalavés also went through several construction stages throughout its history.

Figure 1. Jérica, a Roman tower transformed into a Mudejar bell tower. Photo F. Sánchez.

Figure 2. The *Casa de la Vila*. Photograph from the book: Villa y Baronía de Masalavés. Unknown author.

Its permanence now hides under an unrecognizable appearance of a traditional house, masking the signs of its defensive past (Fig. 2).

As we previously mentioned, it was originally a tower of a Muslim hamlet, adapting its structures to build fortified spaces after the Reconquest, the doorway forming part of it. Later it was used as the palace of the Barony of the Milá's lineage, which achieved social relevance thanks to its contributions to the crown and to links, by marriage, to the Holy See (Llorens 1976, 86).

After acquiring the title of Villa by royal privilege in the 16th century and after the municipal charter in the 17th century, there are several references to the use of the castle as a place for meetings of the Vila's General Council (Madoz 1848, Vol XI, 278; Llorens 1976, 133), the prelude to its subsequent refurbishment into Town Hall, in 1889, when the Barony of Massalavés died out with the abolition of feudal lords in Spain in 1838.

2 DESCRIPTION OF THE BUILDING

The state of the building at the moment of the intervention already foretold of the existence of the original building. With the loss of the coating on the first level on the front facade, a wall of limestone ashlars became visible, with a slightly pointed arched access built from ashlar, that was blocked for the insertion of a trabeated hole. These elements were strange for a building whose parameters were those of a traditional house between dividing walls: a house of two levels, with a gabled roof, a façade of short length perforated with gaps of symmetrical composition and built using simple techniques and cheap materials (Fig. 3).

Inside, on the lower level, there is a pointed barrelled vault, outside of the building's symmetry axis and resting on a wall made of rammed earth, reinforced by a dividing wall of modern construction. On the back facade there is another ashlar pointed arch, bigger in size than the one described on the main facade and also partially blocked with reused ashlars, allowing for a trabeated gap that connects to an exterior staircase, of modern manufacture, that connects with the upper level rooms. These rooms on the upper level, all had independent access from the staircase, and their roofs were covered in wood, wattle and plaster. The structure, resolved with wooden trusses resting on the walls, had partially collapsed due to a lack of conservation works.

The archaeological intervention answered many of the questions raised during the initial visits. On the inferior level, the original paving and the

Figure 3. Elevation of the Casa de la Vila before the beginning of the works. Photograph I. Matoses.

Figure 4. Inside after the removal of the coating of the wall, loophole sectioned by window. Photograph I. Matoses.

auxiliary access elements, such as the portcullis, were discovered. But more interesting was the finding of a small space, between the wall of rammed earth and the dividing wall. It had a staircase providing access to the superior level from a semicircular arched door located inner elevation, blocked with reused masonry.

Together with another door from which the footings of the jambs are known, it very probably delimited the room that Pascual Madoz mentions that was utilised as a dungeon during the times in which the building was used as the headquarters of the General Council of the Vila (Madoz 1848, Vol XI, 278).

Unfortunately, an unwise decision was made that justified the search for lower layers, which led to the removal of the medieval staircase, which had survived until now.

On the upper level the archaeological intervention discovered, against an initial hypothesis, that the medieval building reached said level, compromising a wall made of limestone mortar and masonry with the rammed earth technique. Loopholes, that had been blocked, appeared in its walls, two of them having been sectioned during the opening of the modern windows (Fig. 4).

The door of the balcony and the upper level access door, both located in the symmetry axes of the inner and outer facades, were built over pre-existing holes resolved using semi-circular arches of brick rowlocks.

Beneath the remains of the uncharacteristic roof, on the faces of the walls the putlog holes for forgings that had disappeared could be seen and the footings of a timbrel vault, probably a roof, more typical of these type of constructions.

The construction technique most used by the Arab workforce in the Júcar river area (Hidalgo and Matoses 2012) were walls made using limestone and masonry mortar in a wood formwork. In this case, its use would correspond to the reconstruction of a previous space, or to the conclusion of an unfinished one, with less financial resources.

With regards to the layout of the modules, the wall is made of limestone and masonry walls of approximately 2.80 m in length, equivalent to six cubits *ma'muní* and one metre in height, 2 cubits *ma'muní*, joined to the lower and upper rowlocks but not to the limestone walls of the perpendicular walls, in whose joints the traces of the top end of the timber formwork, called *barzón,* alternatively appear. The distance between the putlogs of the formwork is approximately one metre and it is square in shape, 0.10 m each side, with a wall thickness of around 0.80 m. This data shows a somehow rudimentary work, with a formwork made of less elaborated pieces (the wooden laths used just needed small cuts) compared to those used in the same area in times of Islamic rule.

Besides great differences in the measurements of the elements of the wall, an over-simple displacement of the formwork at the wall joining points occurs, all of these signs of the technique's empovering.

3 ANALYSIS OF THE INTERVENTION

After being left vacant following the building of the new Massalavés Town Hall in 1975, the undertaking of a rehabilitation of the building as a history museum was recently decided. The project lacked previous studies, a fundamental tool in these kinds of interventions for the analysis of the building from formal, historical, constructive, archaeological perspectives, etc., so there was not enough of a knowledge base to establish the appropriate intervention criteria for the analysed data (González 1999, 54–57). The administrative situation wasn't favourable either, as usually happens with most historical heritage restorations, the funding was obtained through public administration grants, with very tight deadlines and resources and very inconvenient requirements. The initial plans were modified, beginning with the objective of the intervention, finally finding that, due to the findings at the archaeological stage, it was more appropriate for the building to become a museum of the site.

The objective of the intervention in a building of historical heritage greatly conditions the architectonical project and the effect of the works on it. That is why the decision about its final intended use it is of huge importance, evaluating the compatibility of said use with the proper conservation of the heritage.

So interventions open to unspecified use must be avoided; interventions where there is no evaluation of the minimum impact required for the implementation of the desired activity or for the convenience of its co-existence with the building.

In our case, the objective triggered the creation of a museographical project, elaborated by the architect Javier Hidalgo, that would explain the phases that the building had gone through and the value of the material remnants of each of them, had they survived or not, to provide visitors with a better understanding. The intervention no longer intended the conditioning of a building containing a museum, but the valuing of the building as a museum.

One of the decisions that further conditioned the intervention was the reconstruction of the roof, which was in a state of total disrepair, due to the abandonment that the building had suffered.

It had to be rebuilt and it was deemed that the construction of a flat roof, typical of the structure's defensive element (Fig. 5), was more appropriate for the building, rather than a sloping roof. If the sloping roof had been kept, reflections about its replacement or not could have been opened, as it had value as an element of traditional architecture and for being part of the history of the building, but that was not the case.

For its materialisation the walls had to be raised to level their height, and this was achieved with a wall made of limestone mortar and masonry that was recovered from the excavations (Fig. 6). To do so samples of the mix were produced, allowing checks of texture and tones. The battlements that were probably part of the doorway cresting were not built, as there was no hard evidence about their

Figure 5. Hypothetical reconstruction of the doorway. Image from the museographical project by J. Hidalgo.

Figure 6. Image of the pouring of limestone mortar on the wall built in the doorway. Photograph I. Matoses.

Figure 7. Image of the building after the intervention, with the recovery of the flat roof. Photograph I. Matoses.

existence nor any information on their appearance, so that the interpretation of their inclusion could be confusing.

The museum intended to highlight its defensive stage, more attractive for the visitors, but without neglecting the period during which it was the Town Hall (Fig. 7). In this sense, the windows of the main façade were blocked, to recover the loopholes that had been sectioned when they were opened, but elements that witnessed their prior use were kept, such as the forged elements of the balcony, the oculus with the date when the refurbished building became the Town Hall, the embedded safe on the first level, etc.

Even then, the lack of specialization of the contractor company and of the technicians and directors responsible for the intervention, contributed to the loss of remnants of undeniable value, such as the footing of the timbrel vault on the upper level or the medieval staircase previously mentioned.

Other aspects of the intervention that were important for its proper understanding include the opening of the back arch, whose blockage prevents the vision of the access to the space, fundamental for the understanding of the doorway (Fig 8). It couldn't be opened, as the adjacent space belonged to a different owner, leaving its acquisition and opening for the future.

Figure 8. Image of the inside of the building, with the arch blocked. Photograph I. Matoses.

4 CONCLUSIONS

This article, a synthesis of a deep study of the building, intends to disseminate the experience of the Portal de Massalavés restoration, a building of a defensive character concealed behind the typical image of a traditional home. Besides the historical approach and the formal and constructive descriptions of the building, a critical analysis of the intervention is made, with the objective that the information can work as reference point for similar future interventions.

REFERENCES

Bazzana, A. 1983. Les structures: fortification et habitat, Castrum 1. *Habitats fortifiés et organisation de l'espace en Méditerranée médiévale, Travaux de la Maison de l'Orient, Maison de l'Orient et de la Méditerranée*, Jean Pouilloux. Lyon.

Cuchí i Burgos, A. 1996. La técnica tradicional del tapial, *Actas del primer Congreso Nacional de Historia de la Construcción*, Madrid, 19–21 septiembre de 1996, pp. 159–165, Instituto Juan de Herrera, CEHOPU. Madrid.

Escalona Molina, M., 2009. *Estadal, una aproximación al universo de la Mensura*, Servicio de publicaciones de la Junta de Andalucía.

Font Mezquita, F. & Hidalgo Chulio, P., 2009. *Arquitecturas de Tapia*, Colegio oficial de Aparejadores y Arquitectos Técnicos de Castellón. Castellón.

Gómez-Bayarri, J.V., 2003. *La Valencia Medieval, en la colección Historia general del Reino de Valencia, vol. 3*, Real Academia de Cultura Valenciana. Valencia.

González Moreno-Navarro, A.1999. *La restauración objetiva. Método SCCM de restauración monumental*, Diputación de Barcelona. Barcelona.

Hidalgo Mora, J. y Matoses Ortells, I., *Rammed earth architecture in the middle course of the river Júcar*, en actas de Restapia 2012, International Conference on rammed earth conservation, Taylor & Francis. Valencia.

Lloréns y Raga, P.L., 1976. *Villa y Baronía de Masalavés*, Quiles. Valencia.

López Elum, P., 2002. *Los castillos valencianos en la Edad Media. (Materiales y técnicas constructivas)*, Biblioteca Valenciana. Valencia.

Madoz, P., 1848. *Diccionario Geográfico-Estadístico-Histórico de España y sus posesiones de ultramar Vol XI*. Establecimiento tipográfico de P. Madoz. Madrid.

Matoses Ortells, I., 2013. *Desvelando el arcano de la ribera del Júcar, la casa de Sinyent*, en construcción con tierra, pasado presente y futuro, actas del IX Congreso de arquitectura de tierra de Valladolid, UVA. Valladolid.

Sarthou Carreres, C. 1992. *Castillos de España*, Espasa-Calpe. Madrid.

Soler Molina, A., 2009. *Massalavés: geografía, historia, patrimonio*, Ayuntamiento de Massalavés. Valencia.

Earthen Architecture: Past, Present and Future – Mileto, Vegas, García Soriano & Cristini (Eds)
 ISBN 978-1-138-02711-4

Sequential organization for self-built earthen construction site

S. Mecca & M. Achilli
DIDA Department of Architecture, INN-LINKS Research Unit, University of Florence, Florence, Italy

ABSTRACT: A self-built process needs, as all building processes, a specific and appropriate organization for assuring performances and reliability. Conditions in which constructors operate on site closely interact with the performance of the construction process and particularly with the risk of non-conformance of a building. Based on case studies conducted on sequential organizational structures of the building process, a self-building construction plan is proposed as application of organizational rules as at suitable planning approach for self building earthen construction sites. The co-farming project in Cenaia in the province of Pisa was demanded by a local group of citizens named Ecocity in order to reorganize the urban structure and give a new identity to the settlement. The goal of the Ecocity group is to realize an experimental cohousing and cofarming unit according a self-building process.

1 FOREWORD

Earthen architecture, which finds its main value as expression of cultural intangible heritage, at the same time needs to assure technical and economic performances and reliability: advanced organizational models are more needed for self-building processes applied to traditional vernacular building techniques: not only communities must be assured about quality of building, but also self-building site have to efficient in using resources, mostly human resources.

The reliability of building processes is related not only to the level of performance required but also to their organizational structure. Conditions of uncertainty in which building operators work on site closely interact with the performance of the construction process, and particularly increase the risk of non-conformance of a building to the performances required, and therefore reduce the efficiency of processes.

A technical and organizational risk analysis of construction processes can lead construction management to design organizational structures capable of reacting to environmental conditions in which "on site" production operations develop, and also able to increase the reliability of the building process. Working on organizational interfaces can lead not only to changes in detail design, but also in performances required to each operator, in order to obtain higher efficiency levels.

Based on case studies conducted on sequential organizational structures of the building process, a set of organizational rules can be applied to self-building for implementing experimental "lean construction" processes. Organizational rules based on risk organizational factors analysis can lead to the designing of a "lean" organizational structure of construction processes which can better fit the needs of self-building operators.

2 THE SEQUENTIAL MODEL

The sequential model is a partial application of concurrent engineering and partnering concepts and practices (Austin 1994, Hall 1991). The model is called *sequential* because is characterised by a careful analysis of the organizational and technical interfaces and the interdependence links that derive from them.

The elements that characterize a sequential organisation are:

- Risk management of interfaces between building process operators;
- Flow control of the process on the interfaces between the sequences;
- Formal phase of training and preparation of the site activities, with time and adequate resources.

The sequence becomes the reference for identifying the overall specifics of the final and intermediate performances.The most important part of sequential organizational work consists in:

- Analyzing the activities and the interfaces.
- Identifying the appropriate intermediate specifications and related responsibilities for each operator.

- Redistributing and aggregating tasks and related responsibilities for each operator to form integrated teams of operators.

A sequence is a set of building operations fully carried out in a continuous way by only one team (or firm) and without other operators working at the same time in the same place. The rules to create a sequential organisation can be as follows:

- Each building phase or sequence should construct a coherent and autonomous subsystem of the construction process.
- Each sequence should allow the team of operators in charge the chance to work fully and continuously, without any time-wasting.
- In the interfaces between the different sequences there should be space for the control of sequence completion, which represents a tool of conflict-solving between operators, in conformity with the ISO 9000 norms.

The sequential organisation of a self-building construction site, given the same materials and products, leads to an increase in the reliability of the construction process due to, for instance, the use of a more efficient conflict solving model and a heightened motivation of the workers.

The organizational structure consists of two levels:

- The whole process organisation based on a limited number of sequences and on interface milestones where the technical and process advancement can be checked.
- The organisation of each sequence based on the internal detailed scheduling of the construction process.

The rhythms of the sequences have to be harmonised and verified in the identified place units, to avoid operational overlapping and the physical over-staffing of operators, causes of safety risks and risk of poor quality.

3 PARTICIPATING IN PLANNING

Action aimed at reducing and eliminating the critical points allows a progressive definition of a basic organisation structure and an operative programme.

The splitting into sequences (Fig. 1) facilitates a progressive definition and checking at each building phase of the project through sketches, scheme design, production drawings, specifications, traditional and visual scheduling of work. (Fig. 2).

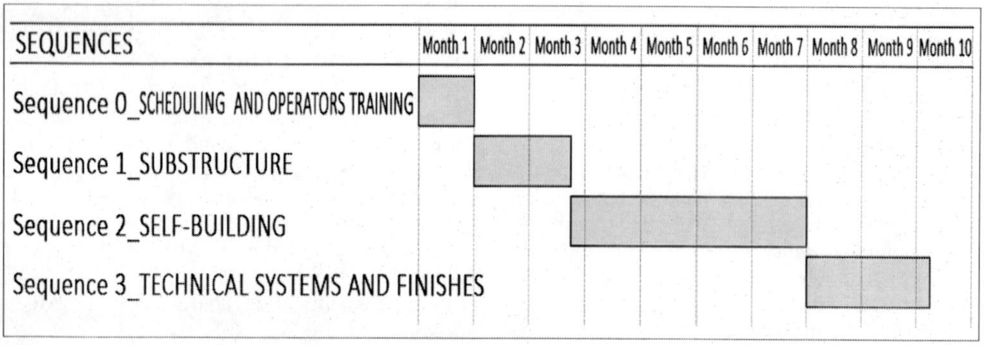

Figure 1. General scheduling of sequential construction site (M. Achilli).

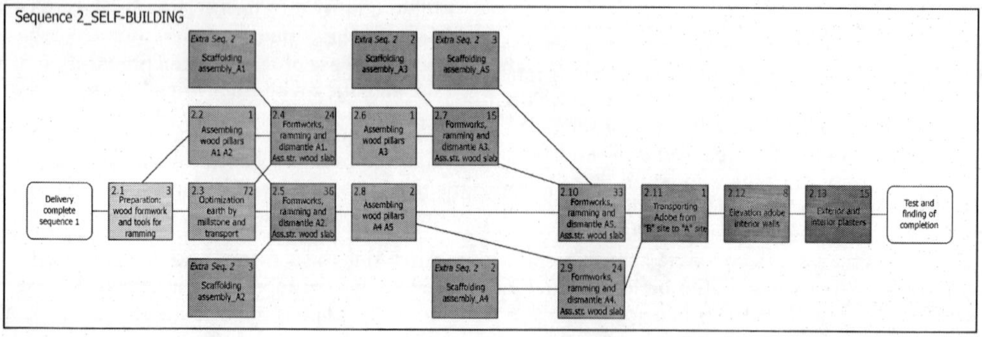

Figure 2. CPM scheduling of sequence 2 (M. Achilli).

The participation of the operators responsible for the sequences in the drafting of the plan and schedule, as far as the parts they are competent for and the technical and organizational interfaces are concerned, can assure feasibility of the general program and of the sequence programmes, responsible autonomy, participation, training and motivation necessary for the assumption of joint responsibility in achieving the final result.

This sequential procedure requires operators be involved as early as possible, since the scheduling of the organizational structure already begins with the first sketches and the global project.

The traditional plan and mostly the visual plan becomes the main tool for communication between construction operators, the basis for detection of weak points and for determining task transfer or regrouping.

This phase indicates how the sequence 0 of building site preparation is aimed at:

- Careful examination of the project, programme of sequences, technical and organizational interfaces;
- Contextual verification of its workability and its possible modification to adapt it to the technical specifications of the operators' competence;
- Shared organization of scheduling with attention given to the harmonisation of production rhythms;

Formal acceptance of the building programme.

- Training of operators on the base of the working programme.

4 OPERATORS TRAINING

The preparation period of the building site is the ideal period for the training of operators who will have to carry out new tasks for traditional trades; training can continue during work on the site, but is more effective if begun in "zero" sequence. Several specific surveys have revealed the interest shown by building operators for processes that enrich and widen their competence and improve working conditions. On sequential building sites there has been an evolution of skills in this direction, whether or not the operators had specific training for the process.

During the development of the building site the control of completion can sustain a parallel quality and productivity control and the exploitation of the operators professional competence.

5 SELF-BUILDING SITE ORGANISATION

The co-farming project in Cenaia near Pisa (Italy) was demanded by a local association of citizens named Ecocity who aimed to develop a project to be included in the existing village of Cenaia financed by private funds and capable to reorganize the urban structure and give a new identity to the settlement.

The goal of the Ecocity association was to realize an experimental cohousing and cofarming village by creating a group of residential units with the spaces for productive, cultural and touristic facilities related with lands for farming activities and using local materials.

The Cenaia-Ecocity project choosed as a mode of realization of the residences a self-building process, as an ethic and economic choice.

The self-building process not only can reduce the cost of building, but is adding value to a earthen architecture, is coherently a "short production chain", and is creating the competence for a self-maintenance of buildings by the community.

The planning of the work according to an organizational and advanced method is necessary for self-building processes which need integrate specialized and non-professional workers, especially if when new (traditional) construction techniques as ramed earth are applied.

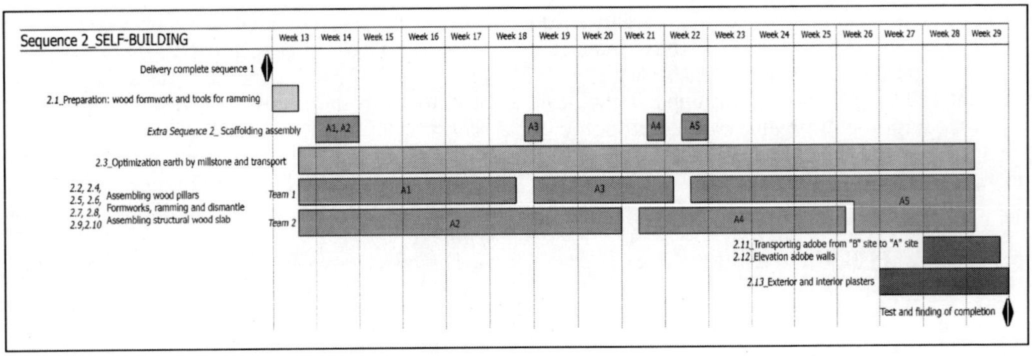

Figure 3. Bar chart scheduling of sequence 2 (M. Achilli).

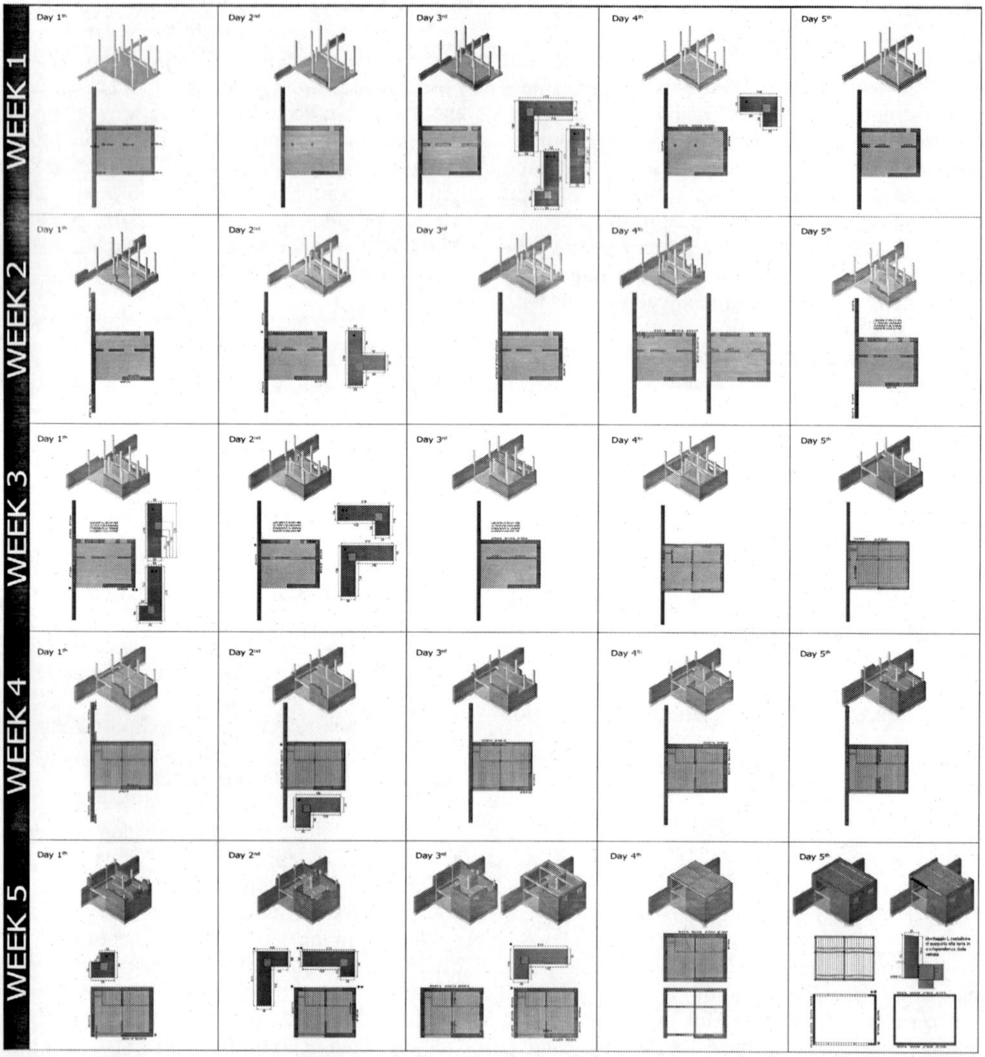

Figure 4. Visual scheduling of self-building activities of sequence 2 (M. Achilli).

The experimental sequential organization allowed a shared schedule of works, materials, time, cost and quality control and to achieve significant and tangible results through a higher level of participation and organizational autonomy of all operators and a maximum reduction in environmental uncertainty.

6 THE SELF-BUILDING PROCESS SCHEDULING

The building process is a chain of actions and relationships that lead from the project to an accomplished work through its management. The building process is divided into stages, operators or actors, and a set of procedures that must be taken to ensure that the final work is done in a manner responsive to the needs identified in the application, respecting the predictions of the timing and costs of execution.

The approach obliges building operators to think and work in an integrated manner to achieve common goals.

The efficiency of the construction process in self is determined by the correct use and exchange of information, the adoption of appropriate and organizational models that enable the coordination and the integration of the subsystems, the establishment of stable relationships between

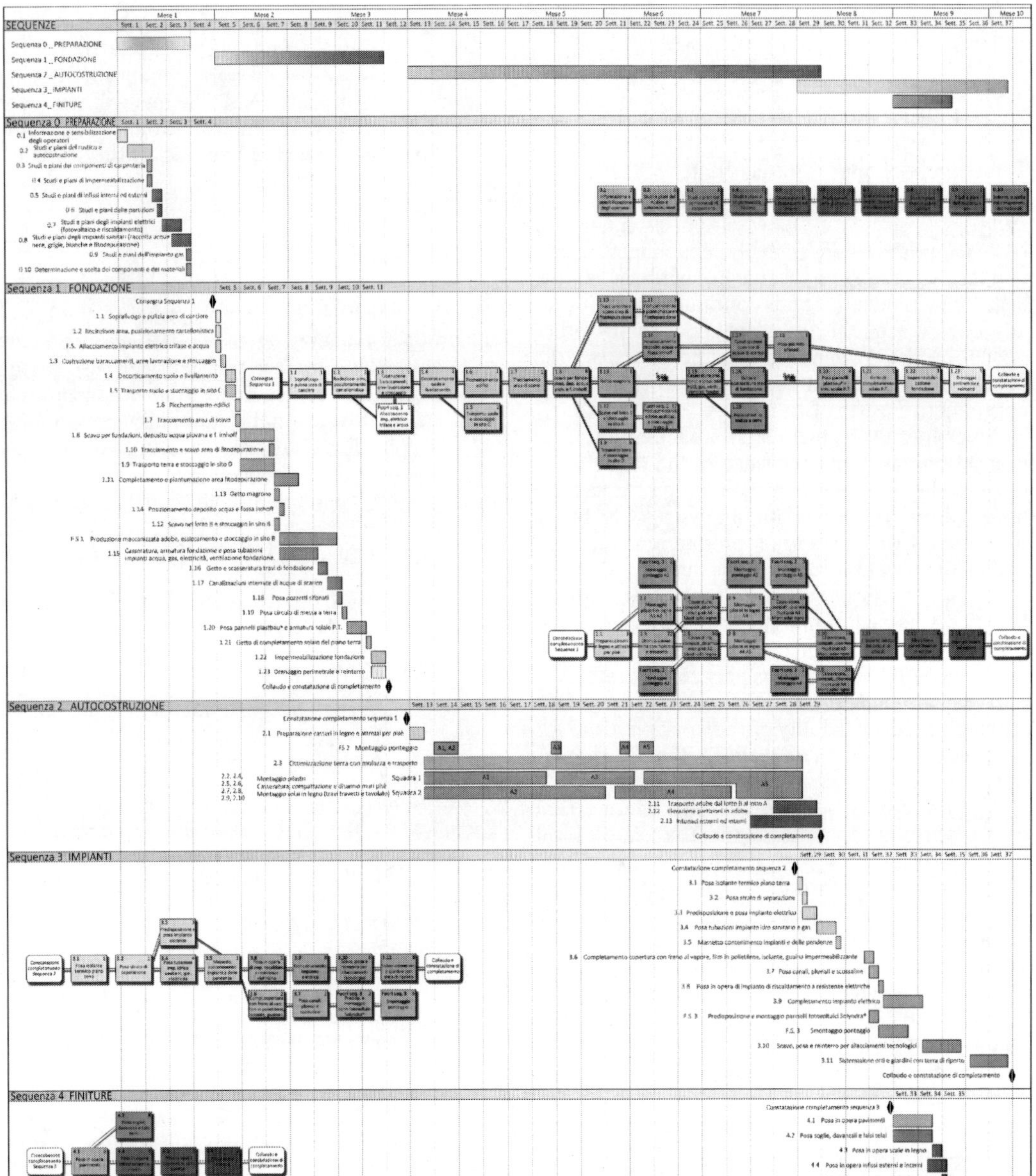

Figure 5. CPM and bar chart scheduling of site activity (M. Achilli).

actors in the process building. The self-building requires to predict and consider the dynamics and the involvement of the various stakeholders: the main enemies are improvisation and carelessness.

Compared with the traditional building process, where each operator has professional competence and structured experience, the sequential approach for the Ecocity-Cenaia is able to manage non—professional operators and their need for coordination.

A self-building site means to actively participate and share the process of housing production: the work is not done by professionals, but by people with different backgrounds who provide their professional work already in other fields of employment and that they build their home on their "free time", providing their manual skills.

7 THE SELF-BUILDING SEQUENCE

The sequential approach has been applied to the central masonry phase, the sequence 2 (Fig. 1, 2 and 3): the sequence 2 is a self-consistent phase, both temporally and spatially, enabling the self-building operators in charge to perform all the tasks independently and continuously, without any kind of mutual interdependence with other operators; the sequence allows the operators the possibility of full and continuous work, without unproductive waste of time.

According the sequential approach an integration action has been preformed in the sequence 0, through the movement and grouping of operative tasks, through a systematic analysis of architectural design documents, of production drawings and of site organisation and scheduling.

The interfaces between the sequences 1 and 3 have been defined as milestone where to check the completion of the sequences, the quality performances and solve the conflicts between operators.

A visual scheduling (Fig. 4) has been produced for checking the reliability of the work plan and for training the self-builders team by the 3D simulation of building process (Figs. 2, 3).

It is the central sequence of site, charaterized by the self-building process and the use of local resources as the earth excavated for the foundation. The 6 house units to built require approximately 650 cubic meters of earth. The excavation of the foundation produces 950 cubic meters of soil. The excavated earth then has to be compacted reducing about 30% the volume, and therefore: $650 \times 1.3 = 845$ mc < 950 mc. The soil of the site is sufficient in quantity to build. A sequential design approach of a site is suitable when we are dealing with unskilled workers who provide their free time to this common goal. Whereas Ecocity is for 26 families, it was determined that all persons should contribute manpower right from the start with at least 1 component that ensures 4 consecutive hours for each daily work.

The daily goal is to produce 12 cubic meters of rammed earth walls, and it is possible avoiding the confusion caused by inexperience and organizing 6 work teams of 2 persons; 3 teams will work for each unit, then two houses will be built at the same time, and will succeed in working as the sequence shown in the graph (Fig. 3).

8 CONCLUSIONS

The experimental planning proposed to the Ecocity group for managing the construction process of a co-housing project can confirm, also at the planning phase, that a sequential organizational approach can be a suitable planning approach for self-build earthen construction sites, as the Cenaia experiment can show. The sequential organization of a building site enhance the participation and sharing of building activities program, integrating planning, training and scheduling and information sharing.

REFERENCES

AA. VV. 2008. *Terra Incognita. European earthen architecture: discovering, preserving.* Lisboa/Bruxelles: Argumentum & Culture Lab.

Bertoni, M. Cantini, A. 2008. *Autocostruzione associate ed assistita in Italia.* Roma: Editrice Dedalo.

Forlani, M.C. 2001. *Costruzione e uso della terra.* Roma: Maggioli.

Foti, M. 1991. *Progettare per l'autocostruzione.* Torino: Ed. Clut.

Mecca, S. 1999. *As sequences flow: proposal of Organizational rules for 'lean construction' management, in: Proceedings of IGLC-7, Seventh Conference of the International Group for Lean Construction.* University of California, Berkeley, California, USA.

Mecca, S. 2002. *Comprendere il cantiere: verso nuovi paradigmi per l'organizzazione del cantiere edile.* Pisa: Edizioni ETS.

Mecca, S. & Natticchia, B. 1995. *Costruire per sequenze: Efficienza e affidabilità nel cantiere edile.* Firenze: Alinea.

Scudo, G. & Narici, B. & Talamo, C. & Tubi, N. 2001. *Costruire con la terra: Tecniche costruttive, campi di utilizzo e prestazioni.* Napoli: Sistemi Editoriali.

Earthen Architecture: Past, Present and Future – Mileto, Vegas, García Soriano & Cristini (Eds)
© 2015 Taylor & Francis Group, London, ISBN 978-1-138-02711-4

Restoration of rammed-earth monumental buildings 1980–2011, the Valencian Community (Spain)

M. Mestre Sabater, L. García Soriano & M. Diodato
Instituto de Restauración del Patrimonio, Universitat Politècnica de València, Valencia, Spain

V. La Spina
Departamento de Arquitectura y Tecnología de la Edificación, Universidad Politécnica de Cartagena, Spain

ABSTRACT: The main objective of this study is to carry out an analysis of the interventions made through the last thirty years on rammed earth buildings in Valencian Community and sponsored by the Generalitat Valenciana (Valencian regional government). The research aims to analyze the techniques and materials used in the interventions on rammed earth and how they were performed underlining how the experience and knowledge of a primitive construction technique, as rammed earth, is or is not incorporated in new practices, new materials and even new technologies. Similarities and differences between various restorations are described showing also the pathologies that originated after the restoration works. Finally, practices that are adequate and respectful of historic architecture are described.

1 FOREWORD

1.1 *Aims of the study*

The fundamental aim of the study is to carry out a detailed analysis of the construction techniques and materials used in interventions on rammed-earth monumental buildings that the Generalitat Valenciana (Valencian regional government) financed from 1980 to 2011. From the information found in the Historical Archive of the Valencian Community it is also possible to deduce the conservation criteria applied in each case by the author of the restoration project.

Considered that almost thirty years have passed since some of the interventions included in the study were concluded it was possible to evaluate the final result and to see how they have endured the passing of time.

Therefore each intervention was analyzed from a documental point of view and, secondly, the current condition of the buildings and the presence of pathologies which may have developed since the interventions were identified.

The research collected information and experiences from the interventions in monumental rammed earth architecture in order to generate knowledge based on past experiences that can be available and easily shared. As a consequence it is possible to generate a series of guidelines and criteria that will provide valuable knowledge to the experts who need to intervene in monumental architecture within the geographic framework of the Valencian Community.

The final objective of this research is to determine which types of intervention are more suitable than others, which show more respect to the original building, and which resist more to the passing of time.

1.2 *Methodology of the study*

The method used in the research has several clear steps that involved the files found in the Historical Archive of the Valencian Community which allowed to perform a systematic analysis of the intervention projects carried out over the decades, and therefore to draw some global conclusions.

The methodology used consists of four fundamental phases:

- preliminary analysis
- typological search
- recording data
- detailed analysis and drawing conclusions

1.2.1 *Preliminary analysis*

The first step was to complete an historical study of each building regarding the different restoration works that were carried out. This analytical phase is performed by using direct and indirect sources: published material, inquires made to the agencies involved in the restorations, project documentation, etc.

1.2.2 *Typological search*

The second stage entailed an exhaustive search in the Historical Archive of the Valencian Community to identify the different construction phases and the various interventions that sometimes overlapped on each of the selected buildings.

To proceed to this search, the buildings, military, civic and religious, were chosen knowing, from the previous step, that they were restored between 1980 and 2011.

1.2.3 *Recording data*

For each of the analyzed interventions a record sheet was produced with all the relevant data regarding the intervention techniques, criteria and the current state of the building.

The current state of preservation was discovered with thorough technical survey of the buildings to register the conditions in which can be found today: deterioration, evolution of the material, use, integration with the surroundings, etc.

1.2.4 *Detailed analysis and drawing conclusions*

The most important work is the final synthesis of each case studies and the subsequent comparison between them. Moreover, it was fundamental to understand and analyze the original construction technique and its diverse variants to fully understand the interventions done.

The following chapters contain a systematic analysis of the mentioned interventions: firstly, in each case the original construction technique, its variant, the material used for the proposed works and criteria are examined; secondly, the focus passes on the actual restoration and current condition of the building.

2 THE RESTORATION OF RAMMED EARTH ARCHITECTURE IN THE PROJECTS FUNDED BY GENERALITAT VALENCIANA

2.1 *Research in the Historical Archive of the Valencian Community*

The Law of the Statute of Autonomy of the Valencian Community (May 1982) grants exclusive responsibility to the Generalitat Valenciana for matters concerning regional archives which are not owned by the state. So in 1983, the responsibilities for matters of Culture, which previously was held by the Council of Culture, Education and Science, were transferred to the Directorate-General of Culture.

The Historical Archive of the Valencian Community was founded according to the article 88 of the Law of December 2009, (this law changed the old denomination of "Central Archive of the Valencian Community") and is part of the Valencian Council of Culture, Education and Science, whose mission is to collect, preserve and make available the documentation of the Generalitat Valenciana.

2.2 *Selection of the study cases*

Due to the management and organization changes currently being implemented in the internal system of the Historical Archive of the Valencian Community, the phase of searching and compiling information encountered some difficulties. First of all, it was not possible to obtain the list of "grants and investment in cultural infrastructure and rehabilitation of artistic and architectural heritage", which meant that it was not possible to obtain the total number of records of restoration and rehabilitation works done on monumental buildings between 1980 and 2011.

In order to obtain this information and to locate the case studies, multiple sources were explored. Firstly a list of monumental architecture built with rammed earth in the Valencian Community was drawn up. A bibliographic consultation allowed identifying which of these buildings had undergone interventions during the last few decades. The second step was to visit the historical archives and request the lists of registers belonging to each of the selected buildings divided in the categories of "town" or "name of building", and to consult those which were included in the initial selection.

In this way, a definitive list was obtained and was created a personal catalogue with the records of "Monumental rammed earth architecture in the Valencian Community that have undergone interventions funded by the Generalitat Valencia in the last 30 years".

It was not possible to consult all the records requested in the Historic Archive of the Valencian Community because some of them were located in the Territorial Service, others were lent out during the period of the research, and others, the most recent registers, had not yet been transferred to the archive. However, it was possible to access the documentation of 72% of the located interventions in monumental rammed earth architecture in the Valencian Community.

A total of 436 records belonging to 72 buildings were consulted. 40 cases are in the province of Valencia, 22 cases in the province of Alicante, and only 10 cases in the province of Castellón (Fig. 1, 2).

However the total number of analyzed interventions is greater than the number of the buildings because some of them could have been the object of more than one intervention, as the restoration works could be delimited to a specific part of the building. On the other hand, not every restoration work concerns an intervention on rammed earth

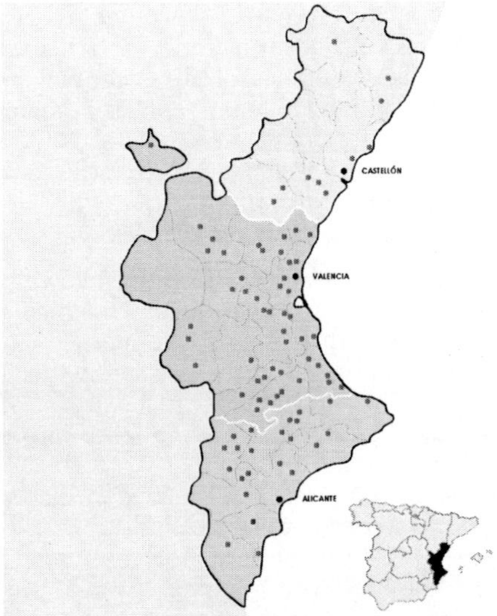

Figure 1. Map of the Valencian Community and the towns with restored rammed earth architecture (Authors).

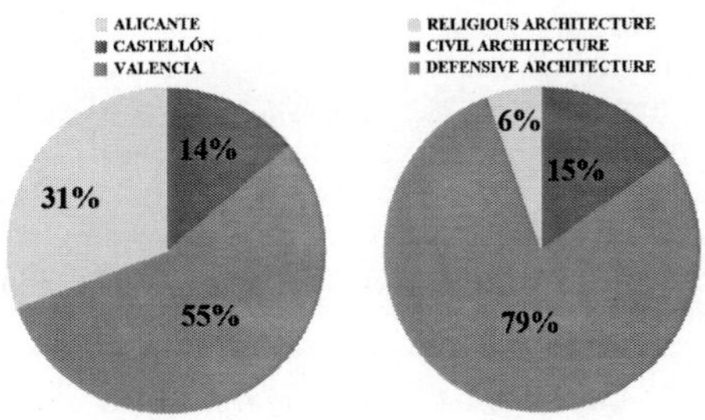

Figure 2. Diagrams of the distribution of catalogued buildings. Provinces and architectural typologies (Mestre Sabater).

walls, given that some interventions involved different sorts of works such as adapting a building to its environment, installing lighting or paving. These interventions were not included in the study.

3 ANALYSIS OF THE INTERVENTIONS

3.1 *Restored rammed-earth buildings*

3.1.1 *Typological analysis of the case studies*

After compiling the data, the typological distribution of the case studies was deduced (Fig. 2). The restoration works were done on a various range of rammed earth materials located in castles, city walls and towers (57 buildings). They were also found, in much lower proportions, in religious and civic buildings (4 and 11 buildings respectively). The majority of rammed earth constructions studied belong to defensive buildings of Islamic origin which are spread over the territory of what used to be Sharq Al-Andalus. Most of these rammed earth fortifications were built by the Almohads, and were subsequently used by Christians, albeit sporadically rather than continuously.

The defensive architecture is placed on the edges of towns (the castles in Gandia, Corbera, Cullera, and the towers in Benifallim or Moixent) as well as integrated into the urban fabric (the towers of Alfarp castle, Albal, and Torrent). Among civic buildings, there are examples of manor houses (the palaces at Aielo de Malferit, Llutxent or Betxí) while religious architecture includes chapels and monastery (La Cartuja de la Vall de Christ or the monastery at Simat de la Valldigna).

3.1.2 *Analysis of the original variants of rammed earth used in the buildings*

Regarding the original rammed earth construction technique, a fundamental piece of data to better understand the reasons that could have led to a particular restoration techniques or materials was obtained.

The passing of time caused transformations and alterations to the original structure of many of the buildings, making it a complicated task to identify the variant of rammed earth in an already restored wall even looking closely to it or consulting the descriptions provided by the architects of the intervention.

The designation of the traditional rammed earth technique found in the archive documentation was usually simply "rammed earth wall" without going into deeper descriptions. In contrast, in some cases the definitions found in the reports attached to the restoration projects do not exactly match with the real construction technique and are lacking of scientific consistency. In general, even if the reports were written by professionals, there is confusion over the terminology of rammed earth techniques especially in the old texts. For example, formed masonry with lime mortar or cyclopean concrete (rammed earth formed by a mortar consisting of lime, sand and gravel of different sizes, including large stones) which has lost its crust could have been confused with ordinary masonry. Given this situation, several research groups have been working during the years to sort out these concepts and develop a lexicon of the diverse variants of rammed earth. (AA. VV. 2001b, www.restapia.es).

It is essential to learn the exact construction variant used in the building when proposing an intervention, and therefore it is very important that this variant is correctly defined. In addition, there are many examples where more than one construction technique is used, and various rammed earth materials are combined. For example, in the Corbera castle, lime concrete, formed masonry with lime mortar and rammed earth reinforced with bricks on the outer surface were found. The Gestalgar castle displays canvases of rammed earth reinforced with layers of lime mortar, rammed earth with stones and lime mortar infill and rammed earth mixed with earth and stone.

In most cases, the original construction technique of the buildings, based on both written and photographic evidence, corresponds to different variants of rammed earth improved with lime: rammed earth reinforced with layers of lime mortar and rammed earth with masonry and lime; while less frequently richer variants, such as lime concrete and Valencian rammed earth, were found.

3.2 *Analysis of construction techniques and intervention criteria*

The exact details of the intervention techniques as the procedures and materials used for each intervention were detailed in the breakdown of the budget for each restoration work found in the archive records. Faced with the varied nature of the projects different solutions were proposed based on the particular needs of each building. However, there are many aspects which do coincide in the majority of the interventions, in terms of criteria, techniques and materials used.

Intervention criteria are of great importance for any project involving the restoration of architecture, but in most of the analyzed documentation, they are not mentioned explicitly and in those projects where the criteria are described, they are not always followed. Therefore, in many interventions, the priority was to find solutions to the existing problems and pathologies rather than making sure that the works were governed by clear and specific intervention criteria.

In order to carry out the analysis of the construction techniques and the intervention criteria, the detailed area of the wall where each intervention was done were considered. The projects are differently structured according to the different parts of the building: the volume of the walls (making a distinction between the crowning of the walls and the volume) or their surface. This allowed separating the interventions criteria in relation to the different parts of the architecture (walls, crowning, and surface).

3.2.1 *Crowning: Criteria and techniques*

In 70% of the analyzed cases, the top part of towers or castles was totally or partially reconstructed, and the reasons for this reconstruction are the same for both partial and total substitution. The purpose was to clarify the architectural shape and simplify the identification of the monument which is considered an aesthetic and typological singularity. A second purpose is to cover and protect the material from water and atmospheric agents, which is a functional and structural reason.

In some cases, reconstructions took place with a lack of sufficient documented information that could clearly define the original profile of the building.

The intervention technique applied in volumetric reconstruction, in most cases, is actually the original rammed earth technique which means a reconstruction using a formwork similar to the original.

Examples of new rammed earth are:

- Sot de Chera tower, in which the following type of rammed earth is defined: "rammed earth reinforced with layers of lime mortar, on both sides, composed of natural earth, with a high clay and lime mortar content at a dosage of 1:2, with the paste poured and flattened into layers not higher than 10–15 cm"
- Palace of Alaquàs: "lime rammed earth, cement, sand, gravel and bricks"
- Tower in Albal: "bastard mortar rammed earth and masonry with a surface covered with an opaque mortar finish". This intervention left no visible traces of the construction technique that was used.

There are also cases where totally different construction techniques were used, but where it was attempted to use similar materials, such as the Cofrentes castle (Fig. 3c), where solid bricks and cement mortar were used. In others cases the technique, materials and coloring are totally different from the original ones, such as the crowning of the Beselga castle with lightweight reinforced concrete with expanded clay.

3.2.2 *Materials: Criteria and techniques*

In 65% of the cases, the criteria of the works was "total recovery of the original volume"; in 25% the criteria was "completion or consolidation of the original volume", and in 10%, no work was done to the rammed earth material (Palau de l'Ardiaca and Major Tower at Banyeres).

In addition, the main objectives for a "total recovery of the original volume" are the same as for the crowning and can be summarized as follow:

- The recovery of the profiles of the walls, in order to clarify the architectural shape and the identification of the monument (an aesthetic and typological reason).
- The protection of the material and core of rammed earth from atmospheric agents and degradation (a functional and structural reason).

When the intervention was the "consolidation of the original volume", this would be applied in order to guarantee the stability of the material and to protect and reinforce the most eroded sections of the walls, without intervening more than what was strictly necessary.

To achieve the "total recovery of the original volume" of the material, the most commonly used

Figure 3. a. Alcalá de Xivert Castle; b. Castle in Castellón; c. Cofrentes Castle; d. Simat de la Valldigna Monastery (Mestre Sabater).

technique was again rammed earth. This technique was applied on existing walls by installing a formwork on one side and filling the space between the old wall and the formwork with earth (Beneixama tower at or Castell Vell in Castellón) (Fig. 3b).

It is usually possible to see traces of the rammed earth technique, although there are exceptions where the material has been completely covered with a layer of mortar.

However, in other cases, to achieve the "total recovery of the original volume" different techniques and materials were used, for example "a mixed mortar of cement, lime and sand with a dosage of 1:1:6" at the Elda castle, or "a new wall made from solid brick and cement mortar with a rendering of lime mortar with a dosage of 1:4" at the Cofrentes castle.

Lastly, the techniques used for the "consolidation of the original volume" would have involved reintegrating the original holes in the material, or putlog holes, "with a mixed mortar of cement, lime and sand, with a dosage of 1:1:6, and with the addition of natural colorants" (Beselga castle and Sot de Chera tower).

3.2.3 *Surface: Criteria and techniques*

The analysis of the documentation permitted to define three principal criteria of surface intervention: "total replacement of the surface", "partial replacement of the surface" and "preservation and protection of the surface".

The most commonly used was "partial replacement". Its main objective was an aesthetic one as to restore the décor and uniform image of the monument. It was also done for functional and structural reasons, in order to protect the old surface and, at the same time, fill of eroded rammed-earth wall.

"Preservation and protection" was used with the purpose of maintain the original rammed earth material, which imply the acceptance of certain levels of deterioration, in order to obtain the only possible outcome: to preserve the material authenticity of the monument which means not only to preserve the general appearance of the construction but also to preserve the marks done on the surface of the monument by the passing of time.

There is no clear justification for the "total replacement of the surface", as it would totally hide the old matter, which goes against the idea of preserving the material authenticity and doesn't allow the building to be considered as an authentic historical document. There are cases where the intervention was rather unfortunate, resulting in the original material being covered with a layer of white cement mortar (Castell de Guadalest, Albal or Almussafes towers).

In most cases, the technique used for "partial replacement" was to apply a finish with a base of cement mixed with lime and river sand, with a dosage of 1:1:4 and with the addition of natural pigments. The only difference between these cases was how the finish was applied, which allowed the old surface to be distinguished to a greater or lesser degree. There are also cases where the finish was applied differently depending whether the surface was old or new.

The techniques used for "preservation of the surface" were mostly to clean and consolidate the surfaces by filling in any no-structural cracks with lime wash, sealing superficial cracks and fissures with lime mortar with a dosage of 1:2, applying a new coat of plaster, removing any loose element and smoothing over with colored lime mortar applied with a trowel, etc.

4 RESULTS OBTAINED AFTER THE INTERVENTIONS AND THE CURRENT CONDITION

By surveying and inspecting the sites, an attempt to identify and recognize the damages and pathologies appeared after the intervention was done. Generally speaking, by considering the mechanisms that most commonly produce problems in earth constructions is clear that the main weakness is the rapid deterioration due to the action of environmental agents (AA.VV. 2011). In the cases studies, pathologies associated fundamentally with extrinsic factors such as environmental, biological, man-made or mechanical agents were found. The most abundant damages in the inspected building consists of damp stains of a blackish color, efflorescence or whitish stains, crumbling and decomposition of the surfaces as well as cracks.

Due to the passing of time, around 85% of the analyzed constructions displayed alterations related with the presence of water, either from dampness caused by capillary rise or filtration, or from condensation and rain. Other common alterations were caused by climactic agents such as wind, rapid changes in temperature, and air and environmental pollution.

To a lesser extent, 20% of cases displayed structural deficiencies due to intrinsic factors: weakness associated with the components of the material, or defects in its implementation, such as lack of adhesion between the original earth and the rendering. These correspond principally to alterations caused by inadequate treatments or interventions, with the use of materials which are incompatible with the original ones and have created more problems than solutions. In 15% of cases, the damage was caused by human factors, such as vandalism. Regular maintenance, if planned would have contribute to the preservation of this pieces of heritage.

5 CONCLUSIONS

After studying the restoration works done on monumental rammed earth architecture financed by the Generalitat Valenciana, the results were not always satisfactory. In many cases, the correct techniques and compatible materials were not used, and the demanding criteria of the discipline were not met. The construction technique of rammed earth requires specific knowledge, both about its original construction and the intervention techniques, which is fundamental for its preservation.

Faced with the different preservation condition of these structures (Fig 3), a range of intervention options become available, both in terms of the criteria to apply and the techniques to use. Therefore it is essential to identify the material, constructive and aesthetic impact of the interventions, and their durability over time in order to choose the most compatible and suitable techniques to reconcile at once the needs to act on the progressive deterioration of the structures and the demands of the basic preservation criteria.

NOTE

This work is part of the research project "The restoration of rammed earth architecture in the Iberian Peninsula. Criteria, techniques, results and prospects", which was financed by the Ministry of Science and Innovation (ref: BIA 2010–18921). It is a project approved in 2011 which attempts to research the diverse criteria and intervention techniques carried out in rammed earth constructions over the last 30 years.

The study entitled 'Rammed Earth with Rammed Earth: Restoration in the Historic Archives of the Valencian Community from 1980 to 2011' consists of research and analysis of interventions of monumental rammed earth architecture in the Valencian Community from the 1980s up until 2001, which were financed by the Generalitat Valenciana.

REFERENCES

Arxiu Històric de la Comunitat Valenciana: 9/1958-4; 9/27687-4; A-007710; 9/27823; 9/27971-2; Rel. 19661; Rel. 6639; 9/25342-21; 9/27732-11; Rel. 18915; 9/24313; Rel 19615; 9/22516-8; 9/22449-8; 9/11379-2; Rel. 15516; Rel. 19886.

AA.VV. 2011a. Aproximación a la patología presentada en las construcciones de tierra. Algunas recomendaciones de intervención, en *Informes de la Construcción, Vol. 63*, 523, 97–106. Julio–septiembre 2011.

AA.VV. 2001b. *Terra Europae. Earthen architecture in the European Union*. Pisa: Edizioni ETS.

Carbonara, G. 1997. *Avvicinamento al restauro*, Liguori, Nápoles. K.

García Soriano, L., Mileto, C. & Vegas López-Manzanares, F. 2012. Restoration of monumental rammed earth buildings in Spain between 1980 and 2011 according to the Archives of the IPCE en C. Mileto, F. Vegas, V. Cristini (ed.), *Rammed Earth Conservation*, CRC-Balkema/Taylor & Francis Group, Londres, 339–344.

La Spina, V., Martella, F., Mileto, C. & Vegas López-Manzanares, F. 2012. Restoration of rammed earth architecture: the projects funded by the Diputación Provincial de Valencia (Spain) en C. Mileto, F. Vegas, V. Cristini (ed.), *Rammed Earth Conservation*, CRC-Balkema/Taylor & Francis Group, Londres. 369–374.

Sanna, S., Mileto, C., Gulli, R. & La Spina, V. 2012. La restauración de la tapia en la Comunidad Valenciana a través de las intervenciones financiadas por las diputaciones. Criterios, técnicas, resultados y perspectivas, *Arché*. Núms. 6 y 7, UPV. 439–448.

Earthen Architecture: Past, Present and Future – Mileto, Vegas, García Soriano & Cristini (Eds)
© 2015 Taylor & Francis Group, London, ISBN 978-1-138-02711-4

Conservation of earthen architecture in the United States of America: Some case studies

C. Mileto & F. Vegas
Universitat Politècnica de València, Valencia, Spain

ABSTRACT: The United States has a broad and varied built heritage using different architectural earthen techniques. Diverse conservation works have been performed on this monumental and domestic heritage, regarding both the criteria and techniques used. The text attempts to present a range of different criteria and intervention techniques by examining some case studies, representative because of their different construction dates, location, constructive technique and the type of intervention carried out. A compared analysis of these cases allows us to take a preliminary look at the conservation of earthen architecture in this country and understand it also as a useful experience in other geographic and cultural contexts.

1 INTRODUCTION

The US has an important architectonic and archaeological heritage of buildings made of earth both in large monuments and traditional architecture. The extension of the country and the differences in climate and territories have had a marked influence on the variety of techniques used. Furthermore, the heterogeneity of constructive cultures that exist in the country and the autochthonous cultural influences and those imported throughout history can also be seen in the amount of building variants. The study of these techniques and their cultural mixtures is of enormous interest not only for American researchers but for all researchers interested in earthen architecture, built earthen heritage and traditional architecture in general.

On the other hand, the criteria and techniques used in the conservation of this heritage are also interesting both in their own context and in comparison with the criteria and techniques used in other parts of the world. The necessary contextualisation of the interventions in the panorama of US conservation allows us to understand fully the criteria and objectives of the interventions and outline the historic and cultural values identified in the country's earth heritage. This second part of the research belongs within the research projects that the authors have been developing over the last few years. The study of these aspects in the context of the United States made it possible to broaden the cultural and technical perspective in view of the possible comparison with the many interventions carried out in the last thirty years, especially in the Iberian Peninsula.

2 METHODOLOGY USED IN THE RESEARCH

The research presented in this text is based on multifarious methodology that uses different registers for data collection: bibliography consultation, case studies, interviews and queries to experts in the subjects of the research, field visits, case analysis, debate of themes and results with experts in the field. The bibliographic research focused chiefly on two major points: the conservation of earthen architecture in the United States, the main subject of this text, and earthen architecture as such in the United States. Examining earthen architecture in the United States permitted us to understand fully the cases of conservation and restoration that we found, since both familiarity with the construction techniques of earthen architecture and the cultural and constructive context provide the grounds for its conservation. Thanks to the broad view in the bibliography, we chose the most interesting cases to make an in-depth study of earthen architecture in US territory and the most important and representative cases to analyse the criteria and techniques used in conserving these structures.

3 EARTHEN ARCHITECTURE IN THE UNITED STATES

The variety of techniques used for the construction of earthen architecture in the United States depends on many geographic, climatic, cultural and historic factors. In the first place, the huge size of the country and its geological and climatic diversity clearly influences both the use of earth as a construction material and the technique used for building. The distribution of the techniques in US territory has been dealt with in some publications (Noble 2000) that show us the scope of earth architecture in the country. Apart from the abundant use of this material from coast to coast and from north to south, it is important to point out the variety of techniques used according to the location: cob, manual adobe, moulded adobe, rammed earth, half-timber, wattle and daub, sod, mud plasters and mortars on stone masonry, etc. So we can find examples both in traditional and monumental architecture and archaeological sites all over the country.

4 CONSERVATION OF EARTHEN ARCHITECTURE IN THE UNITED STATES: CASE STUDIES

The cases studied to examine more profoundly the criteria and techniques used in the conservation of earthen architecture in the United States were selected, on the one hand, from the bibliography and, on the other, from the possibility of having direct access to the information about the interventions and the people involved in them. Some institutions like the Getty Conservation Institute and the Architectural Conservation Laboratory of the University of Pennsylvania deserve mention, as does the constant work of the National Park Service. It is also worth underscoring the activity carried out by some associations like Cornerstones Community Partnerships, consisting above all in cooperating and counselling local communities. On the other hand, there have been many collaborations between entities, as is the case of the GAIA Project, a collaboration between the University of Pennsylvania, CRATerre, from Grenoble, and ICCROM, from Rome, which has given rise to interesting research since the nineties about earth plasters and their conservation, and the cooperation between the University of Pennsylvania and the Getty Conservation Institute. The cases presented below do not assume to be a comprehensive list of the conservation of earthen architecture in the United States, but a useful sample thanks to the variety of techniques used in the initial construction, the geographic and climatic diversity and the variety of criteria and techniques employed in the interventions.

4.1 *Mesa Verde National Park, Colorado*

Mesa Verde is a huge archaeological complex with houses and cliff dwellings made of masonry with earth mortar. Most of these masonry constructions were plastered with mud, polychromed in some cases (Rohn 2006). For over ten years the Architectural Conservation Laboratory of the University of Pennsylvania, with the partial collaboration of the Getty Conservation Institute, has been carrying out research and conservation here, one of the most ambitious and important projects for the consolidation of earthen structures in the United States. The classification and conservation works on the exterior surfaces focused on different parts of the site: several façades (Area J & Area 26) of the Cliff Palace in the Speaker Chief Complex (Collum 2008), Kiva F of Long House (Ferron 2007), Kiva E of Long House (Reynolds 2007), Spruce Tree House (Adler 2009; Krieger 2011), Far View House (Byungwook 2009). At the present time the Architectural Conservation Laboratory of the University of Pennsylvania is working on the Fire House. In these places, the earth renderings were classified and a detailed study of the compatibility of the consolidation products to be used on the mud surfaces was carried out, yielding results about the variability of colour, penetration, cohesiveness and durability. An important part of the project was the use of gelatines (consolidants made of animal glue) to reinforce the earth renderings (Aloiz 2011). This type of product was used with the idea of conserving the renderings with the least possible impact and the greatest possible material compatibility or of seeking a counterbalance between the cost and the results obtained. The use of this sort of degradable consolidant also makes it possible to reach an interesting compromise between native culture and the demands of contemporary conservation. Native Americans, who repair their nearby dwellings on a yearly basis, feel that a neglected place like Mesa Verde should slowly go back to the earth it came from. On the other hand, the National Park Service, an agency in charge of the upkeep of monuments, is obviously in favour of maintaining the monument as it stands today. The use of animal consolidants permits it to be conserved in its current state, but allows its slow degradation over time and therefore its return to the earth it came from (Aloiz 2011).

4.2 *Three national monuments in the south-west: Chaco Canyon, Salinas Pueblo Missions and Bandelier*

These are three enormous archaeological sites built by Native Americans out of stone masonry with earth mortar and mud and straw rendering (Russo 2005), polychromed in the case of Bandelier, which have repeatedly been subjected to conservation works from the late 19th century until the present day (Crosby 1986; Matero 1998). In the last few years, on the basis of a series of studies carried out beforehand (Beas 1993; Matero & Bass 1995), an investigation and a chemical physical and mechanical analysis of the treatments applied since the end of the 19th century to consolidate the earth renderings present in these three places. The products used in this type of intervention throughout history in the three places were Portland cement, acrylic resins, polyvinyl acetates, etc. The effects of the products used were compared as regards variability of colour, penetration, cohesiveness, durability, steam transmission, freeze-thaw sensitivity, erosionability towards water, module breakage and mechanical resistance. The compared analysis led to interesting conclusions about the compatibility of the different materials used in the case of consolidation of the earth plasters and the most suitable formulation for the new mortars to be used to fill in gaps.

4.3 *Casa Grande Ruins National Monument in Arizona*

Casa Grande is one of the oldest archaeological sites in the United States with a spectacular building made of earth in the 14th century (Houk 1996).

The archaeological site has been protected by a preservation order since 1889 and subjected to continual conservation works (Matero et al., 2000). The study and intervention proposed by a multidisciplinary team from the Architectural Conservation Laboratory of the University of Pennsylvania in collaboration with the Getty Institute, since 1996, has focused on classifying the earthen building and establishing guidelines to protect and conserve it. In the first place, the constructive technique of the building itself was studied: for many years different authors have been debating about whether it was adobe (Shapiro 1999), rammed earth (Mindeleff 1898; Lumpkins 1972) or some other technique. Thanks to the observation of degradation phenomena, they have managed to determine that it is a cob structure (Matero 1999).

On the other hand, precise methodology was drawn up to study the degradation of the earthen construction by identifying the phenomena typical of this material, and a glossary of specific terms was established. Within these studies a detailed analysis was carried out on the earth that the building is made of, which brought to light a relationship between the characteristics of the surface and the nucleus of the earthen structure with regard to the degradation phenomena that have taken place and brought about some physicochemical transformations (Matero 1999).

This phenomenon consists in a migration of the natural lime present in the earth inside the wall towards the surface in such a way that it creates a very fine film of calcium carbonate that protects the earth in the wall. The identification of this degradation phenomenon, studied specifically in the case of Casa Grande, is particularly relevant because it could throw some light on other cases of earthen structures with similar characteristics.

4.4 *Fort Union in New Mexico and Fort Davis in Texas*

The case study and intervention on Fort Union (Garder 2005) and Fort Davis (Thybony 1990) enters the world of adobe architecture conservation, especially common in the American Southwest. There have been many adobe constructions or reconstructions in historic sites, but there are fewer cases of conservation of the original adobes and fewer still of mud plasters on adobe. The study of the degradation of the adobe carried out on Fort Davis in Texas (Burt, Wood & Warden 1998), and the conservation of the adobe of Fort Selden, where specific studies have been carried out on the degradation of adobe caused by rainwater (Romero 1990) in order to find a suitable mixture for producing adobes for the conservation of historic adobes (Selwitz 1995) is of great interest. Within the GAIA Project, during the period going from 1992 to 1997, the state of conservation of the renderings and finishes of adobe structures in Fort Union and Fort Davis was studied in order to draw up some guidelines for the conservation of forts that were later put into practice. In the first phase, the history of both forts was studied along with the history of the interventions carried out since they were declared national monuments: 1959 in the case of Fort Union and 1961 in that of Fort Davis, after which they became units of the National Park Service. This study yielded conclusions regarding the compatibility and durability of the treatment applied earlier to conserve the sites (Matero 1995). Once the studies had been completed, two representative areas of the two forts were chosen (Mechanics' Corral in Fort Union and Officers' Row in Fort Davis) to carry out a pilot project for the conservation of plasters on adobe surfaces. The pilot experiment consisted in documenting the coatings, emergency consolidation works, interventions consisting in putting back the rendering on the surface with hydraulic lime, cleaning with a soft brush, consolidating the surface of the plasters with diluted acrylic resins. The materials to be used were selected for their compatibility, stability and durability, after studies in laboratory and tests on site (Matero & Bass 1994).

4.5 *Adobe houses at Fingers Lake, New York*

This example was chosen as a case study because, unlike the previous ones, it represents traditional non-monumental architecture and is a paradigm of how buildings can be cared for and maintained in good condition by their owners and the local community without being taken under the aegis of the National Park Service. The literature about adobe houses in the north of the United States (Piper 1990; Piper 1999; Dassler 1990) is very scant and we would even go so far as to say practically non-existent; most of the literature about adobe in the States completely overlooks these interesting mid 19th century buildings (Noble 2000). However, people living in the area and especially the inhabitants of Geneva (NY) are familiar with these buildings and treasure them, and they have been catalogued, studied and protected thanks to the Geneva Historical Society. In Geneva itself a dozen adobe houses dating from the second quarter of the 18th century are in a perfect state of conservation. These buildings, rendered with mud and straw mortars, are so well conserved thanks to their owners' endeavours.

4.6 *High Hills of Santee, Stateburg, South Carolina*

In 1821 Dr William Anderson replaced the old timber wings of his 1760 mansion in Stateburg (Borough House) with new rammed earth wings and added another seven outbuildings made with the same technique on the grounds of his plantation (kitchen, laundry room, cabins for slaves, library, medical centre) (Golebiowski 2009). Influenced by him, between 1852 and 1852 the architect Edward C Jones built the nearby Church of the Holy Cross out of rammed earth. For almost two centuries, Anderson's buildings have required no restoration works whatsoever because they have constantly undergone maintenance: both on their bases, their roofs and their surfaces, the three most important points for the conservation of rammed earth (Golebiowski 2009). The same cannot be said of Church of the Holy Cross, although the damage suffered is mainly due to external agents and has nothing to do with the fact that it is made of rammed earth: in 1886 a serious earthquake affected the top of the bell tower; in 1902 a hurricane knocked down a tree that destroyed the north wall of the church, which was reconstructed out of reinforced concrete in 1950. In 1974 the top of the tower was rebuilt out of reinforced concrete, along with the south wall of the nave. At the same time, the church was plastered with cement mortar, which, in time, has come loose due to a lack of transpirability. These cement plasters may be partly responsible for the serious termite attack the church has suffered, which made it necessary to perform a great deal of work on the timber structures (Golebiowski 2009). In view of this data, it is interesting to observe that all the pathologies present in these buildings were caused by natural catastrophes or human-induced agents and that the rammed earth has resisted perfectly for almost two centuries a damp climate such as that of South Carolina.

4.7 *Restoration of adobe architecture in the US southwest*

Adobe architecture is quite commonly found in all the southwest of the United States (Lumpkins 1972; Smith 1982; Austin 1990; Garrison 1990). Apart from the abundance of vernacular adobe architecture that can be found in the towns in the area (Bunting 1964; Seth 1988; Spears 1986), there are also many monumental architectures built with this constructive technique and which have been restored over the years (Iowa 1985; Crosby 1986; Arnon 1990): San José de Gracia Mission Church in Las Trampas; San Antonio Chapel in Chacon; San Francisco de Assisi Mission Church in Ranchos de Taos; Santa Cruz Church in Arroyo Seco; Santa Teresa del Niño Jesús Church in La Puebla; San Miguel Mission Church in Santa Fe, etc. Among the most outstanding examples of adobe constructions are some of the missions built by the Spanish in the 17th and 18th centuries (Lee 1990; Hannaford 1990). San José Mission in Tumacacori (late 18th, early 19th century) is one of the most interesting cases, not only because it is built out of adobe but because of the restoration works practised on the church and the ruins of the mission. The mission was abandoned in 1848 and declared a national monument in 1908. The walls of the church, made of adobe, presented serious damage caused by damp, so after 1975studies and interventions directed by Antony Crosby were carried out (Crosby 1978). The study of the church's fabrics revealed that the problems caused by damp, probably in existence since the building was erected, had worsened considerably after 1920, when interventions using cement mortars began; these had not got any better in 1973, when whole sections of the walls were stabilised with cement (Crosby 1987). Finally, the intervention was minimal, just enough to ensure the future of the building: drainage and maintenance works on the original earthen plasters in the exterior (Crosby 1987). On the other hand, the case of the Franciscan Mission in Pecos (18th century) is important because of the presence of different types and sizes of adobe, which reveal the stratification of the different construction stages (White 1996), and because of implications of interventions to fill in the gaps in the adobe. These gaps filled in at different times with different types of adobes, mortars imitating the texture of adobe, flat plaster, etc., caused different impacts on the historic structures that give rise to an interesting reflection about criteria for the reintegration of adobe fabrics. The current intervention on the east walls of the church is being carried out by Cornerstones Community Partnerships, a cooperative that has been working since 1986 on the preservation and restoration of architecture in Southwest US, especially focusing on earthen architecture and its restoration in the region. This organisation has collaborated many times with local communities in the restoration of churches, parish churches and little communal buildings in many towns in the southeast. Most of these projects consisted in removing the cement rendering and other additions to historic adobe buildings because they were incompatible with the original adobe walls, and they reverted to traditional techniques. The organisation published a very interesting manual on this subject (Cornerstones Community, 2006). Among the cases of traditional architecture, Taos Pueblo, whose inhabitants have practised annual maintenance works on it since it was founded around the year 1000, is deemed to be of special interest (Iowa 1985). This exceptionally well-conserved traditional town was listed as a World Heritage Site by UNESCO in 1992. Unlike the case of Taos, there are numerous other towns built by Native Americans that have not received so much attention over the years. In many of these places, new materials (cement mortars, cement blocks, aluminium windows, etc.) have been used instead of the traditional materials, thus causing serious problems for their conservation. Even some interesting experiments, as is the case of the Sky City Cultural Center in Acoma, have failed because new materials such as asphalt-stabilised adobe, incompatible with traditional earth renderings, were used (McHenry 1990).

5 CONCLUSIONS

The cases discussed here have been chosen to illustrate a wide range of construction periods and different construction traditions: from Native Americans in Mesa Verde, Casa Grande and the other monuments in the Southwest to the houses built out of adobe in the English fashion in the north of New York State, without forgetting the group of dwellings built at the end of the 18th century in South Carolina or the architecture of the Spanish missions in New Mexico. On the other hand, each case has given rise to diverse reflections about conservation criteria and techniques. So, for example, the conservation project for the Mesa Verde complex provides the opportunity to address the criteria for intervention on architectonic heritage in general and the conservation of mud rendering in particular. Both the case of Mesa Verde and that of Chaco Canyon, Salina Pueblos Mission or Bandelier are an example of the conservation criteria and techniques of mud rendering on different surfaces (stone or adobe walls) and with different disintegration or adhesion issues. On the other hand, this study of earth plasters is complemented with the studies carried out in Fort Union and Fort Davis, where the plasters on the adobe walls are lime and gypsum stucco. A detailed scientific study of these plasters made it possible to identify suitable products for their conservation. These studies about the conservation of the plasters are completed with the studies performed (Burt, Wood & Warden 1998) on the conservation of the adobe structures themselves. The case study and conservation of the main building at Casa Grande gave rise to interesting reflections about the knowledge and inspection of the construction technique and the relationship between the materials used for construction, the constructive technique, degradation phenomena and physicochemical processes created by degradation. The studies carried out on this matter point to new possibilities for the interpretation of both the constructive techniques and the consequences of the degradation mechanisms. The buildings at the High Hills of Santee, Stateburg, South Carolina, have demonstrated the durability of rammed earth construction even in humid climates. The cases of adobe architecture in the Southwest and its conservation led to reflections about its degradation mechanisms (Tumacacori), the incompatibility with industrial materials and the difficulties arising from the intervention both at a technical level (Tumacacori) and as regards intervention criteria (Pecos). Not only monumental architecture has been addressed, but also traditional and residential architecture, both in the adobe dwellings in the north of New York State and in the towns of the Pueblo peoples in New Mexico. In both cases, the importance of the maintenance and continued use of these dwellings is evident (Geneva and Taos Pueblo) as the only way of guaranteeing their survival, although the dangers of interventions and manipulations often exist, without clear guidelines based mainly on appreciation and respect for these traditional architectures (New Mexico towns).

ACKNOWLEDGEMENTS

This research was possible thanks to the grant for Spanish residency professors and researchers abroad of the Ministry of Education, Culture & Sport (Salvador de Madariaga programme) awarded to Camilla Mileto (ref: PRX12/00225). On the other hand, the authors of the article would like to thank Frank Matero, Randall Mason, Aaron Wunsch, William Whitaker, Catherine Myers from the University of Pennsylvania; architects Richard Piper, Maribel Beas; Robin Jones and Jake Barrow (Cornerstones Community Partnership); Kent Diebolt (founder of Vertical Access) for their help, support and advice throughout this research.

REFERENCES

Adler, R.S. 2009. *Weathering The Storm: Diagnostic Monitoring For Preventive Conservation At Spruce Tree House, Mesa Verde National Park, Colorado*. University of Pennsylvania.

Aloiz, E.M. 2011. Investigations into a Historic Treatment: The Efficacy of Gelatin as an Adhesive for Earthen Finishes at Mesa Verde National Park. University of Pennsylvania.

Arnon, N. & Baca, S. 1990. Churches, symbols of community: the preservation of New Mexico's adobe churches. *6th International Conference on the Conservation of Earthen Architecture.* Los Angeles: The Getty Conservation Institute: 143–148.

Austin, G.S. 1990. Adobe and related building materials in New Mexico, U.S.A. *6th International Conference on the Conservation of Earthen Architecture.* Los Angeles: The Getty Conservation Institute: 417–423.

Beas Guerrero, M.I. 1993. Consolidation of traditional plasters: a laboratory research. *7ª Conferência Internacional sobre o Estudo e Conservação da Arquitectura de Terra.* Lisboa: Direcção Geral dos Edifícios e Monumentos Nacionais: 410–416.

Bunting, B. 1964. *Taos Adobes. Spanish Colonial and Territorial Architecture of the Taos Valley*. Albunquerque: University of New Mexico Press.

Burt, R.A., Woods, P.K. & Warden, R. 1998. Three-Dimensional Modeling of Erosion to Historic Adobe Wall Ruins at Fort Davis, Texas. *APT Bulletin, Vol. 29, No. 2*: 35–38.

Byungwook Lim, A. 2009. *Soft Capping of Archaeological Masonry Walls: Far View House, Mesa Verde National Park*. University of Pennsylvania, Philadelphia.

Collum, N.S. 2008. *A Comparative Study of Earthen Surface Finishes of the Eastern Façade of Open Area J and the Northern Façade of Open Area 26 at Cliff Palace, Mesa Verde National Park, Colorado.* University of Pennsylvania.

Cornerstones Communyty, 2006. *Adobe Conservation. A Preservation Handbook*. Santa Fe, Sunstone Press.

Crosby, A. 1978. A Preservation Monitoring System at Tumacacori National Monument. *Bulletin of the Association for Preservation Technology, Vol. 10, No. 2*, U.S. National Park Service Issue: 47–76.

Crosby, A. 1986. Preservation Maintenance in the Southwest. *Bulletin of the Association for Preservation Technology, Vol. 18, No. 3*: 48–55.

Crosby, A. 1987. Monitoring Moisture at Tumacacori. *APT Bulletin, Vol. 19, No. 4*: 32–44.

Dassler, L. 1990. *Nineteenth century New York state earthen homes. 6th International Conference on the Conservation of Earthen Architecture*. Los Angeles: The Getty Conservation Institute: 430–437.

Ferron, A 2007. *The Consolidation of Earthen Surface Finishes: A Study of Disaggregating Plasters at Mesa Verde National Park.* University of Pennsylvania, Philadelphia.

Gardner, M.L. 2005. *Fort Union National Monument.* Tucson: Western National Parks Association.

Garrison, J.W. 1990. *The evolution of adobe construction systems in the Southwest (U.S.A.) and related conservation issues. 6th International Conference on the Conservation of Earthen Architecture*. Los Angeles: The Getty Conservation Institute: 53–56.

Golebiowski, J. 2009. *Rammed Earth Architecture's Journey to the High Hills of The Santee and its Role as an Early Concrete.* Graduate Schools of Clemson University & The College of Charleston.

Hannaford, D. & Edwards, R. 1990. *Spanish colonial or adobe architecture of California, 1800–1850*. Stamford: Architectural Book Publishing Company.

Houk, R. 1996. *Casa Grande Ruins National Monument*. Tucson: Western National Parks Association.

Iowa, J. 1985. *Ageless Adobe. History and Preservation in Southwestern Architecture*. Santa Fe: Sunstone Press.

Krieger, R. 2011. *Architectural Finishes at Spruce Tree House, Mesa Verde National Park: Characterization and Interpretation of Room 115(2)*. Univ. of Pennsylvania, Philadelphia.

Lee, A.J. 1990. Spanish Missions. *APT Bulletin, Vol. 22, No. 3*: 42–54.

Lumpkins, W. 1972. *Adobe Past and Present*. Santa Fe: Museum of New Mexico,

Matero, F.G. 1995. A programme for the conservation of architectural plasters in earthen ruins in the American Southwest: Fort Union National Monument, New Mexico, U.S.A. *Conservation and management of archaeological sites.* London: James & James, 1: 5–24.

Matero, F.G. 1999. The conservation of plasters in earthen archeological sites. Cultural resource management: conserving earthen architecture. *Halifax: Cultural Resource Management, Volume 22, 6*: 59–62.

Matero, F.G. & Bass, A. 1994. Orphans of the storm: the preservation of architectural plasters in earthen ruins. Cultural resource management: conserving earthen architecture. *Halifax: Cultural Resource Management, Volume 17, 4*: 21–26.

Matero, F.G. & Bass, A. 1995. Design and evaluation of hydraulic lime grouts for the reattachment of lime plasters on earthen walls. *Conservation and management of archaeological sites.* London: James & James, 1, No. 2: 97–108.

Matero, F.G., Del Bono, E., Fong, K.L., Johansen & R., Barrow, J. 2000. Condition and treatment history as prologue to site conservation at Casa Grande Ruins National Monument. *Terra 2000: 8th international conference on the study and conservation of earthen architecture*. London: James & James: 52–64.

Mchenry, P.G. 1990. Rebuilding Ácoma Sky City. *APT Bulletin, Vol. 22, No. 3*: 55–64.

Mindeleff, C. 1898. Aboriginal Architecture in the United States. *Journal of the American Geographical Society of New York, Vol. 30, No. 5*: 414–427.

Noble, A.G. 2000. Patterns of Building Material as Revealed in the Farm Housing Survey of 1934. *Material Culture, Vol. 32, No. 1*: 57–77.

Piper, R. 1990. Earthen architecture of New York State: adobe construction in a non arid climate. *6th International Conference on the Conservation of Earthen Architecture*. Los Angeles: The Getty Conservation Institute: 117–121.

Piper, R. 1999. Earthen Architecture in the Northern United States European Traditions in Earthen Construction. Cultural resource management: conserving earthen architecture. *Halifax: Cultural Resource Management, Volume 22, 6*: 30–33.

Reynolds Hall, L. 2007. *Characterization, Analysis and Interpretation of the Surface Finishes of Kiva E, Long House, Mesa Verde National Park*. University of Pennsylvania.

Rohn A.H. & Ferguson W. 2006. *Puebloan Ruins of the Southwest*. Albuquerque: University of New Mexico Press.

Romero Taylor, M. 1990. The Fort Selden Adobe Test Wall Project. *APT Bulletin, Vol. 22, No. 3*: 35–41.

Russo Zinn, W.A. 2005. *Cement Modified Earthen Mortar—An Investigation Of Soil—Cement Performance Characteristics At Three Southwestern National Monuments.* (Master's Thesis). University of Pennsylvania, Philadelphia, PA.

Selwitz C. & Caperton T.J. 1995. Chemical Stabilization of Adobe in the Restoration of the Montaño Store. *APT Bulletin, Vol. 26, No. 2/3*: 37–41.

Seth, S. & Seth, L. 1988. *Adobe! Homes and interiors of Taos, Santa Fe and the Southwest*. Stamford: Architectural Book Publishing Company.

Shapiro, J.S. 1999. New Light on Old Adobe: A Space Syntax Analysis of the Casa Grande. *Kiva, Vol. 64, No. 4*: 419–446.

Smith, E.W. 1982. *Adobe bricks in New Mexico*. Socorro: New Mexico Bureau of Mines & Mineral Resources.

Spears, B. 1986. *American Adobes. Rural Houses of Northen New Mexico*. Albunquerque: University of New Mexico Press.

Thybony, S. 1990. *Fort Davis National Monument*. Tucson: Western National Parks Association.

White, C. 1996. Adobe Typology and Site Chronology: A Case Study from Pecos National Historical Park. *Kiva, Vol. 61, No. 4*: 347–363.

Earthen Architecture: Past, Present and Future – Mileto, Vegas, García Soriano & Cristini (Eds)
© 2015 Taylor & Francis Group, London, ISBN 978-1-138-02711-4

Innovation through a public construction using rammed earth in Alsace

L. Piccon
Nunc Architectes, Eckbolsheim, France

A. Mesbah
ENTPE-LGCB, Université de Lyon, Vaulx en Velin, France

ABSTRACT: Traditionally, earth has always been used for construction in Alsace, in the north-east of France, typically by slotting mud between timber framing. The project to build a gallo-roman archeology museum was an opportunity to innovate with rammed-earth techniques. This three-floor museum is a high-energy-performance rammed-earth building. The external skin is composed of an inside monolithic rammed-earth wall and an outside prefabricated rammed-earth masonry wall with cork insulation granules placed between the two. To the South, a Trombe wall covered with glass provides effective solar gains. To obtain administrative permission to erect this building, it has been necessary to develop an appropriate experimental approach. The text will detail the mechanical and thermal experimental studies conducted on rammed earth and the building strategies that have resulted from this research.

1 ORIGIN OF THE CONSTRUCTION CHOICES

Dehlingen's Archaeological Interpretation Center (AIC) is situated in the heart of the village. Surrounded with existing houses, it is therefore relatively protected from bad weather.

This project is the result of an architectural competition. It consists of a building divided into two parts facing each other on either side of a narrow strip of light.

The first part is the Koeppel House, dating from the seventeenth century and renovated in its original spirit. The new part has come to replace the former barn, which burnt down in 2001. (Fig. 1).

This extension was completed in rammed earth. The choice of this construction technique, in which earth is compacted in batches, is a reference to the stratigraphic layers of archaeological excavations that reveal the layers of history.

Moreover, Gallo-Roman buildings were constructions of great simplicity and constructive efficiency. Today, concerns about sustainable development and energy efficiency have architects strive for the same aims.

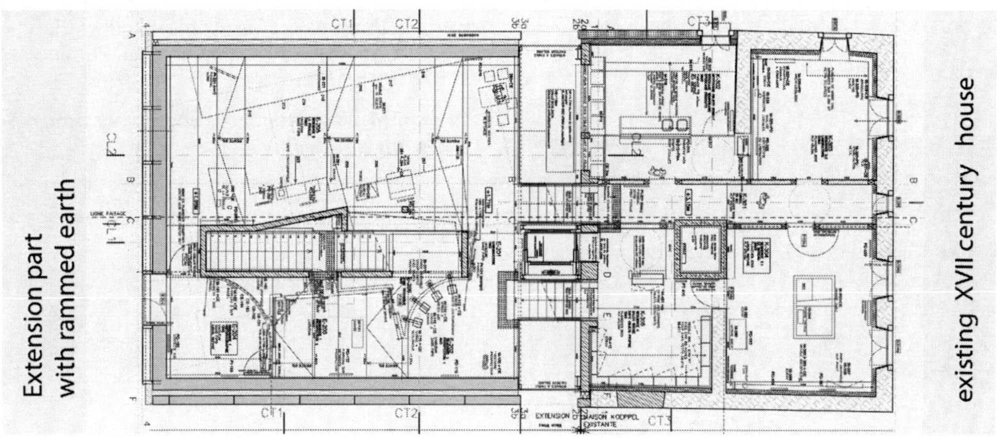

Figure 1. First level plan (Nunc architectes).

The new building was designed as a contemporary construction; however, it was decided from the start to resort to local materials and secular techniques close to the ones used in Gallo-Roman villas.

The aesthetic strength of this building comes from the raw expressiveness of the materials used. The structural material is also the finishing material. The contribution of this project to modernity lies in the current knowledge of materials and in the ability to combine them into high performance applications.

The energy concept of AIC is based both on the thermal inertia of the building and an efficient management of solar gains to reduce heating needs.

The thickness of the rammed-earth walls, the massiveness of the wooden floors, the quality of the roof insulation all contribute to the high thermal stability of the building.

A glazed double skin has been placed in front of the south earthen wall, leaving an air gap of 36 cm maximum.

The objective of the double skin is to preheat incoming air in the winter and mid-season (Fig.2).

In the summer, the opening of the double skin provides a significant depression, which ensures the circulation of air necessary for the extraction of natural indoor air, and prevents overheating.

The aim to make the most efficient building possible has led us to develop a wall composed of two earthen walls coupled with insulating granular cork. Beyond the energetic performance, the point of this triple wall is to separate the supporting function from the protective function in case of bad weather: the load-bearing wall located inside has good hygrometric regulation properties and is fully protected from humidity. The outer wall is aimed at being protective. At the same time, this solution allows to enjoy the aesthetic quality and expressive power of earth both on the inside and the outside of the building.

Figure 2. South façade (Nunc architects/L. Boegly).

2 TECHNICAL INNOVATION

The project is innovative from several points of view:

- a rammed earth wall is used as a load-bearing element in a three-floor museum (i.e. a public building) in a region where mud filling is traditionally used in timber-framed buildings.
- a complex "triple-skin" system including two distinct rammed-earth walls.
- the thermal performance of the complex.

Specific research was needed to validate these choices.

2.1 *Soil analysis*

The earth was selected from a quarry located less than 5 km from the site. It was extracted, screened and stored for several months on the extraction site. 8 tons of stored soil were collected to conduct analyses, tests, and to build a prototype to validate the quality of the earth, its physical and chemical characteristics, strengths and possible applications.

Atterberg limits:

The test was carried out on a 0/400 μm fraction of the material. The following results were obtained:

- Liquid limit: WL = 45.21%
- Plastic limit: WP = 36.94%
- Plasticity Index: PI = 8.27%

A Methylene Blue test was conducted on a sample of fine-grained soil. The value obtained was VB = 4.5. Hence, we concluded that the clay present in Dehlingen is very sensitive to water.

2.2 *Mechanical characteristics in compression and tensile strength of earth material*

A series of tests was conducted [Mesbah, A 2011] by the "Laboratoire de Génie Civil et Bâtiment (LGCB)" of the National School of Public Works (ENTPE), to determine:

- the influence of binders (lime and/or cement) on Dehlingen's soil

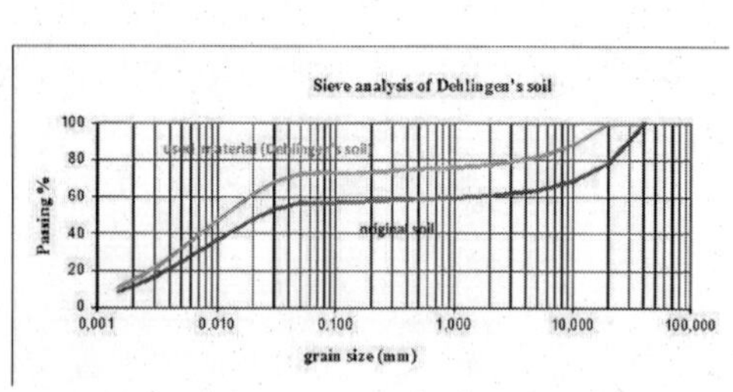

Figure 3. Particle size analysis (A. Mesbah, ENTPE).

– the most optimal composition
– the optimal parameters for our applications
– the implied curing conditions

More than 100 specimens were produced with raw material and then stabilized with cement, lime and a lime-cement mixture by static compaction approaching standard proctor density. For mechanical characteristics, a compressive tensile strength test (splitting test) was used. It was carried out on samples aged 30 and 60 days, to determine the most appropriate mixture.

We deliberately chose to produce our specimens at very low compaction force and to do our research in unfavorable conditions to determine the minimum performance that Dehlingen's soil could offer. We found that the performance of the local soil was good enough. If the material was to be used in good conditions, it would be even more efficient.

The cure of the specimens was performed in a chamber at 21°C and 65% humidity. At the end of the curing, all the specimens were partially wet (15% residual water content). Each series was tested immediately or after the erosion or capillary process.

During the manufacturing process of the structure, the soil's water content was approximately 15%. Then the water content decreased to 5 or 6% for a few months to finally stabilize at 1 or 2% after a few years. It is to be reminded that compression strength increases as earth cures.

The tests were carried out upon leaving the preservation chamber, i.e. when the specimens were still partially wet and their resistance therefore quite low. As a matter of fact, the mechanical tests underperform reality; the final structure actually proves to get stronger throughout its existence. Observing the influence of the binder, we noticed that the 6% treatment with cement got the best compressive strength in a saturated state. Therefore, it was decided to use this composition for the exterior wall exposed to rainy conditions. The treatment of the inner wall will be a mixture of lime 2% + cement 2% which is more ecological than cement 6%. The selected values are shown in Fig. 5.

Further trials performed by the manufacturing company at the beginning of the construction gave the same results. However, sand was added to avoid excessive cracking and limit shrinkage.

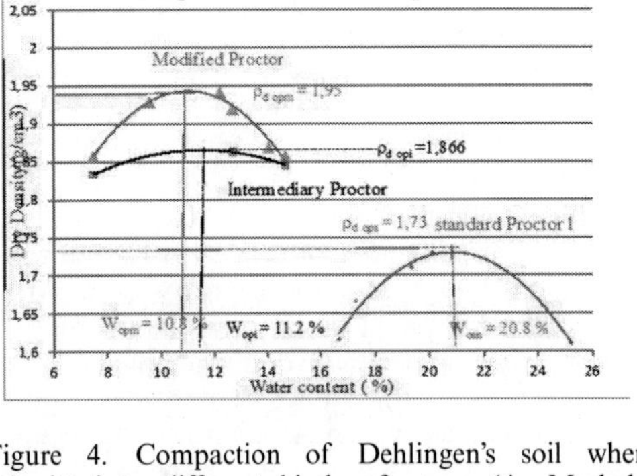

Figure 4. Compaction of Dehlingen's soil when submitted to different kinds of stress (A. Mesbah, ENTPE).

	Comp strength (Mpa)		Tens strength (Mpa)
Treatment	Dry situation	Wet situation	Dry situation
Cement (CEM I) 6%	2,4	0,859	0,2
Lime 2% + Cement 2%	2,45	0,574	0,23

Figure 5. Mechanical characteristics of Dehlingen's soil with binders (A. Mesbah, ENTPE).

2.3 *Moisture migration*

Will moisture migration through a complex "triple-walled skin" cause any accumulation of water? If so, can it cause mechanical or thermal disorders?

The study [Frey, N 2011] was focused on the determination of the thermal and hygrothermal characteristics of rammed-earth walls and on the simulation of their behavior in response to moisture migration.

The weather data file used comes from TRIER University in Germany (100 km away from Dehlingen). The climate there is harsher in temperature and precipitations than in Dehlingen. The simulations were performed for a wall oriented towards rainy winds. Our building, being located in the center of the town, is protected by roof overhangs and adjacent buildings.

At the same time, laboratory tests [Serreau, L & Surma, F 2011] have allowed to define the necessary parameters for simulations with WUFI (software). (Fig. 6).

Firstly, studies were focused on Glazer's dynamic method allowing studying vapor diffusion exclusively. The simulation first observed the water content of the internal wall, the insulator, and the exterior wall, as well as the condensation in the thickness of the insulating wall.

Next, the study focused on the integration of the concept of liquid-phase transfer, incorporating

Paramètres	Unit	Estimated Value (for the 1st approach)	Measured Value (earth+cement)	Measured Value (earth+lime+cement)
Dry density	$[kg.m^{-3}]$	1860	1740	1750
Porosity	[%]	29	34	34
Thermal condctivity	$[W.m^{-1}.K^{-1}]$	0,93	0,9	1,17
Specific heat	$[J.kg^{-1}.K^{-1}]$	830	980	1044
Résistance to water vapeur d'eau	[-]	10	11,2	11,1

Figure 6. Results of tests to determine the necessary values of the coefficient Dw (nunc architectes).

the concepts of sorption and suction of materials. WUFI simulates the behavior of a material with a liquid conductivity coefficient Dw. The study tested this behavior with two Dw opposite values to understand the range of influence of this parameter. The results of the tests on the samples showed that the used soil is well within this spectrum (Fig. 7).

The results of these simulations have highlighted several points:

- Preponderance of moisture coming from the rammed-earth wall in the behavior of the wall in the early years of the structure;
- Preponderance of liquid-phase transfers in total mass transfers (which makes the results more favorable than with vapor alone);
- The heat transfer coefficient U [w.m–2.K-1] confirmed a value close to the objectives of the thermal study 0.17 (fig.7);
- The impact of the operating constraints (humidity generated by the occupants of the museum) is very small, even insignificant on the inside wall;
- The exterior wall is more strongly subject to changes due to weather and seasons;
- The risks of condensation are almost nonexistent and exclusively located on the inner face of the outer wall.

2.4 *Decisions*

At the end of these specific studies, important decisions were taken:

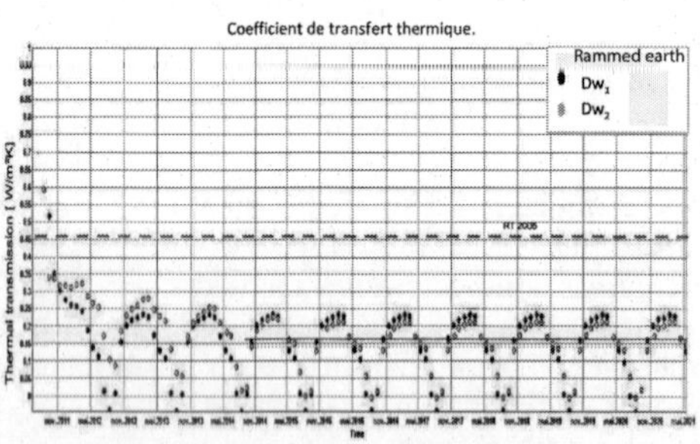

Figure 7. Results of the tests (N. Frey, nunc, INSA).

Figure 8. Prototype (Nunc architectes).

- The construction of the complex had to be divided into two distinct phases in order to allow the walls to dry for at least 6 months before assembling them.
- In that perspective, the exterior skin was made up of prefabricated modules. These modules were manufactured at the same time as the monolithic inside wall.
- The necessary use of cork insulation to prevent saturation during the drying phase of the inner wall was confirmed.
- Different binder dosages were definitely necessary in the composition of the exterior wall.
- Maintaining openings in the lower part of the wall had to be planned in order to control the moisture content of the insulator periodically.

2.5 *Prototype*

Finally, a prototype was completed. This 2-m long, 2-m high, 0.5 m-wide wall portion integrating 4 facing blocks was used as a reference in the treatment of both internal (brushing and fixative) and external surfaces (facing blocks' joint treatment), and to test the attachment system of prefabricated blocks with angle brackets and kingpins. It was also used for pull-off tests on the bearing wall's prefabricated-block fasteners.

Wind-induced tearing stress was calculated at 50 daN per angle bracket. Pull-out tests showed an average resistance of the angle bracket of 220 daN.

3 STRUCTURAL BEHAVIOUR

The wall consists of three parts. The structural wall inside the building is made up of successive layers of 60 cm in width on the ground floor and 40 cm on the first floor. On the other hand, the outside wall is constructed with prefabricated rammed-earth blocks whose dimensions are 50 cm × 125 cm by 30 cm in thickness. The space between them is filled with 20 cm of granulated cork.

The floors are made of solid wood keyed with a thickness of 24.3 cm, with a capacity for important crossing. Each panel is 2.50 m wide. The finishing on the underside of the slab is rough wood.

The shape of the timber frame is based on the principle of the typical Alsatian house (no upper purlin), with 1.25 m between each roof truss. It allows for large storage space in the attic. The cover follows the shape of the old house and offers a lateral overhang to protect the rammed earth walls from possible runoff during heavy rains. The roof is insulated by 28 cm of cellulose wadding and 8 cm of rigid fibre board wood.

The structure of the timber frame has been designed so as to apply rammed earth exclusively in compression. Horizontal forces generated are taken

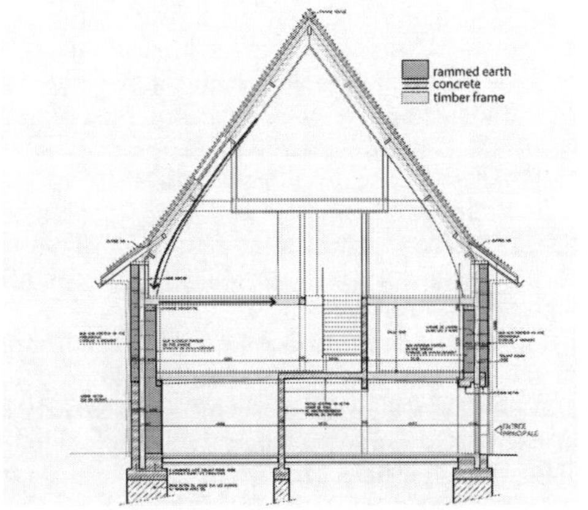

Figure 9. Building section (Nunc architectes).

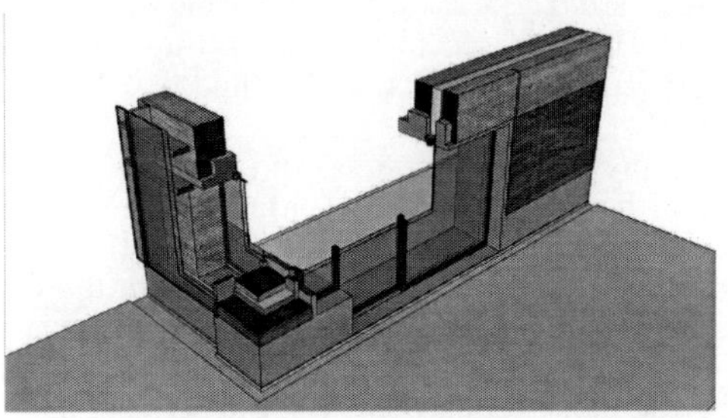

Figure 10. Axonometric detail of wall complex (Nunc architectes).

up by the wooden floor which is used as anchoring. This avoids submitting the wall to bending.

In order to avoid localized loads, trusses are laid on the wooden floor. At each level, the wooden floor is placed on a reinforced concrete chaining which has no structural role. It allows for a better load distribution on the wall, and also provides a better contact area between wood and earth.

The set is based on strip footing foundations to minimize differential settlement.

To relieve rammed-earth walls from potential torsional stress in the timber frame, the bracing is ensured by a concrete core which distributes the stress generated by lateral thrust or by the twisting of the timber frame through the floor.

Wind-induced horizontal stress on the south glazed façade is transmitted through the reinforced concrete chaining to the wooden floor which acts as a diaphragm. It is finally taken back to the concrete core.

3.1.1 *Protection of the top and base of the walls*

A 50 cm high (minimum) watertight concrete beam has been placed at the bottom of each wall in order to protect it from rain splashing and moisture infiltrations. The concrete ground beams are placed below the supporting element of each adobe wall so as to generate a hollow section under the insulation layer. This way, any condensation in the insulation and excessive moisture can be evacuated below the walls' base. On the Southern façade, the rammed-earth wall is entirely covered with a glazed skin.

3.2 *Prefabrication*

The building site being located in the heart of the village, little space was available to manipulate prefabricated modules. It was therefore decided to limit their size (125 cm × 70 cm high and 30 cm thick) to ensure a quality installation. We developed a system to grip and transport the blocks to the storage area.

Vertical joints were integrated in the façade to anticipate cracking between blocks and near the lintels. Horizontal joints were replaced with a bed of clay mortar in the same color as the wall. Holes associated with prefabrication (for handling) were closed, except on the first row to allow the regular monitoring of the inside insulation moisture rates.

4 WORK EXECUTION

4.1 *Mixture preparation for rammed earth*

The soil was mixed in a 300l planetary mixer. The dosage in the mixer was performed using buckets for measuring the precise quantity of hydraulic stabilizer incorporated into the mixture. The mixing time was at least 3 minutes to 5 minutes maximum. 10–12% of water was added during mixing. A homogeneous mixture was obtained; it had to be applied in half a day.

4.2 *In-situ rammed earth*

Rammed-earth walls were erected immediately after the completion of structural work (basement slab and foundations). The construction period began on July 2nd and ended in early October, to avoid temperatures any lower than 5 degrees.

The structural rammed-earth wall was constructed in two phases:

- phase 1: first-floor walls (3 m high max). Duration of the treatment: 2 weeks, under plastic sheeting; 1 additional month for drying
- phase 2: second-floor walls. Duration of the treatment: 2 weeks, under plastic sheeting

The in-situ execution of the walls was 25 days total (with 8 people)–10 days for the base and 15 days for the top.

Figure 11. First-floor wall in progress (Nunc architectes).

Figure 12. Prefabricated rammed earth block (Y. Meier).

Figure 13. Façade with prefabricated rammed earth blocks (L. Boegly).

The erection of the roof timber structure took place on December 17th, i.e. 60 days after the end of the curing.

The construction of the rammed-earth wall was achieved using continuous formwork in aluminium. Each side of the wall received complete horizontal formwork in order to obtain consistently homogeneous layers.

The mud was applied using two pneumatic rammers. The angles of the rammed-earth walls were chamfered using wooden chopsticks to increase their resistance to abrasion and weathering. The layers of earth were compacted from 15 cm to 10 cm. The regularity of the layers was checked using a gauge throughout the filling of the earth in formwork, to ensure perfect horizontality. The refilling of formwork holes and small cutaways was performed at the end. The same type of earth was used; only this time, it was applied manually, with mallets (vertical rammed earth).

The reinforced concrete chaining was executed in situ (dry mortar was deposited along the formwork to avoid drips on the wall) two weeks after the wall had been erected.

Temporary protection of rammed-earth structures was provided daily by installing wood panels on top of the wall with overhangs of at least 10 cm on each side. These panels were slightly tilted to allow rainwater drainage.

4.3 *Rammed earth prefabricated blocks*

Block size (1.25 m long × 0.7 m high and 30 cm thick) was defined to facilitate lifting and transport. Two reservations on the mold base were put in place for lifting straps.

A prefabrication area was provided on site with a sill height of 10 cm to ensure a perfectly flat bottom of the mold. The blocks and the in-situ rammed-earth walls were produced with the same method and at the same period.

Vertical grooves were provided on each side of the block to allow the pouring of clay mortar between each block. The earth was scraped off to level the blocks before stripping. They were then stored on pallets under breathable protection for six months. The installation was carried out in March. 2 cm of earth mortar lime were applied onto each row of blocks. Particular attention was paid to the formulation of the mortar to obtain the same strength as with rammed earth.

Brackets and stainless connectors were installed between the precast blocks and the bearing wall in order to adjust their positions. The reservations left for transport under the blocks were then used again to absorb the thickness of stainless connectors into the lower blocks.

REFERENCES

Frey, N 2011, *Determination of the thermal and hygrothermal characteristics of rammed earth through digital simulations assessing its behavior in case of moisture migration*–INSA de Strasbourg.

Mesbah, A 2011, *Study of the influence of concrete and/or concrete-lime additions onto Lorentzen material for the construction of a rammed-earth wall in Dehlingen*-ENTPE.

Serreau, L & Surma, F 2011, Characterization of a mud sample—report reference—*CRITT Matériaux Alsace Type B ATEx n°1871–2, 25 June 2012*—CSTB.

Earthen Architecture: Past, Present and Future – Mileto, Vegas, García Soriano & Cristini (Eds)
© 2015 Taylor & Francis Group, London, ISBN 978-1-138-02711-4

The other architecture: From natural caves to excavated housing

B. Piedecausa García & J.C. Pérez Sánchez
Department of Building and Urbanism, University of Alicante, Alicante, Spain

ABSTRACT: Excavated towns are an attractive example of underground urbanism that managed to solve, in a very interesting way, thermal problems thanks to natural ground inertia or transition spaces. The aim of this paper is to show the typological evolution of excavated dwellings worldwide, as an architectural proposal and urban solution. The proposed methodology provides an analysis of underground architectures from natural caves to excavated housing, focusing on the study of global constructive solutions to specific problems. Thus, architectures as a natural geography correction (horizontal excavation), buried underground architectures (vertical excavation), subtractive architectures (shallow excavation) and combined architectures (mixed excavation) are studied. In conclusion, there are many examples of typological combinations since troglodyte architects tried to adapt the most elementary constructive rules to get greatly enriched results. These proposals of different underground structures deal with each territory and its geographical features, and obtain urban and architectural solutions transferable to current configurations.

1 FROM CAVES TO EGYPTIAN MONUMENTALITY

For many years, both historians and architects have shared the view that the origins of architectural production mainly dated back to the Egyptian period, over 5,000 years ago. However, this vision put aside popular architecture that, prior to this great civilization, had sheltered man many centuries ago and forgot those constructions that endured and evolved over time, even to the present. Because of this, we are convinced that it is necessary to consider also this other kind of primitive architecture as, despite not being able to compete with Egyptian monumentality, it is also part of the collective experience of human habitat.

Thus, the starting point of this work is based on the study of the first permanent habitat used by early man: the cave. This natural element protected from rain or cold in winter, and from sun or heat in summer, serving even to protect against wild animals or other tribes attacks. Probably, for many years man made no alteration of these spaces but, over time, learned to modify them as convenience: smooth floors, enlarged natural cavities or dug new spaces linked to others; that was the precise moment when man really began to create architecture.

Therefore, these natural caves, used for a living purpose and sheltering from bad weather, improved slowly the initial spatial conditions and generated new constructive proposals with simple architectural interventions. These primitive actions established the basis for later apparition of additional excavations and small structures outside the constructions, improving bioclimatic conditions in the built environment.

2 URBAN TROGLODYTISM: THE TERRITORY AS A SCULPTED MATERIAL

The development of early planning is based, fundamentally, in the proper growth of an isolated unit of housing. This minimal unit was primarily developed as a shelter for man, first in a single simple space and afterwards, as life became more complex, emerging additional units from the basic volume and creating complex and larger structures.

From the beginning, excavations evolved in space and size demands, with new shapes reflecting specific functions, being also externally influenced. It was very advantageous for troglodyte groups to join each other for collective defence or common work and, in most cases, they formed isolated communities in the territory, trying to survive as independently as possible organized in rows or in clusters of houses.

This new two-way social relationship among members of the group limited the expansion of the unit due to the need of protection, but also brought greater benefits to the community with the purpose of isolation, creating and sharing materials and common tools. This community life drifted to a house with new functions to facilitate greater privacy but not because of a protection need from the environment but from other components of the group where they belong.

The need for security of those citizens conditioned a tendency to build clusters in locations with difficult access but, at the same time, this decision was undoubtedly even more related with the need to define a recognizable perimeter delimited by an own community. Therefore, it can be supposed that most of the origins of troglodyte villages were arisen from protection against enemy attacks.

Then, the beginning of community systems was developed as a result of social needs, family demands, community organizations or livelihood of the inhabitants; but afterwards, also these clusters are a result of local or regional characteristics as climate, topography and available materials.

In conclusion, the present work analyses how some different examples of underground architecture are included within a larger system, developing clusters and generating specific urban interventions.

2.1 *Horizontality as a group: A steep site*

In this group we can consider clusters of horizontally excavated architecture where excavation runs forward from the front of a mountain or cliff, building living spaces deepening into their vertical walls. Depending on the natural settings of the cliff, we can distinguish between linear or amphitheatre organizations, described in the following paragraphs.

2.1.1 *Linear groups: Cliff towns*

This type of group is an extreme example of hillside configuration in areas with pronounced, almost vertical, slope which can be considered as part of a cliff wall. Unlike other places, in this case not only the rock is important but also very specific locations are needed, as natural cliffs, resulting tiny villages fully sited in steep ledges.

Cliff caves had an essentially defensive significance, however, when the site was selected it wasn´t forgotten to choose the best orientation to shelter from the prevailing winds, benefiting with a particular microclimate. The best known examples of these groups are found in the south of the United States (Colorado, Utah, Arizona and New Mexico) but also there are other interesting groups in the valley of Zelva (Turkey), in the valley of Loire (France), in Vienna or in India. In the Iberian Peninsula, we can find natural cornices in Cádiz (Setenil), Huesca (Monasterio de San Juan de la Peña) or Portugal (Monsanto). It is notably known the case of Almanzora caves in Andalusia, where there are parallel rows of excavations in the rock forming even three stories of housing around large public squares, depending on the ground conditions.

Focusing on the most famous "cliff towns", there are spectacular troglodyte crowds at Cliff Palace, located in Mesa Verde, a region in Colorado (USA). It is an interesting group carved into the rock facing southwest, with a great cavity of 100 m length and 20 m width, hosting more than 220 spaces and a population of about 250 inhabitants. The steepness of the slope in the urbanized part is substantial, with a system of outer circulation and distribution of houses, located on a site providing protection from 40 degrees in the shade summer sun and from the low temperatures in winter with heavy snowfalls.

2.1.2 *Circular groups: Amphitheatre towns*

This group is another example of slope configurations arisen when settlements are located in nonlinear cliffs, but describing a curve. In these cases, the specific morphology of the site allows a more focused organization, reinforcing the isolation and autonomy of the group. Contrary to what happened in the previous case, where the main element is based on the linearity of the placement, this setting is determined by a natural amphitheatre, leading to a convergent central space. Although there are groups in other places like Chenini and Ghomrassen, both in Tunisia, the most characteristic examples are the amphitheatres of Göreme and Cavusin (Turkey).

In the Göreme Circus (Cappadocia) there is a circular shape with a large number of churches, monastic establishments and cells, by organizing in a large community different housing, services and refectories of religious communities.

2.2 *Verticality as a group: An excavated plain*

In this group we find small communities of underground dwellings built under the ground in a completely vertical excavation, resulting one of the first cave dwellings found, and thus, the most rudimentary. Because of this, there are very few examples of this early architecture as this type of construction soon suffered a great evolution due to the improvement of tools, giving way to mixed excavations and constituting more complex interventions.

These vertical excavations were very elementary, primary and ineffective, creating small size habitats. At that time, they needed additional buildings as they were too exposed to the elements, and it was important to protect the main hole with a simple cover in the surface.

Thus, a straight vertically executed cave responds to a primitive habitat, not providing effective performance. It was built to give a minimum shelter from the wind or keep food and belongings. It was an initial shelter that could be used as a stable habitat if an outdoor deck and a basic access were added.

In contrast to additive or built construction, where the traces of previous buildings at the site are part of new constructions either physically or conceptually, excavated architecture cannot be archaeologically studied such as it is understood today, since the construction is not supported by earlier remains or added layers. These structures are composed by slowly removed materials, sculpting and carving the remaining traces of previous constructions. Thus, there are not previous architectural grounds to preserve since the spaces are widen by destroying their original settings. In this architecture, its physical memory can be deleted but not its essence, leaving free imagination to any prior spatial assumptions.

2.3 *Weightlessness as a group: A defensive landscape*

To this group belong those architectures that developed shallow excavations in rocky elements, such as the troglodyte villages in the cones of Cappadocia (Turkey) that used pre-existing natural mounds in the environment to locate their homes by selecting strategic points in a global structure. This is a subtractive action digging on a natural configuration to a surface urbanism above the ground line.

Therefore, it is in the region of Anatolia where some of the oldest civilizations in the world were born since prehistoric times: natural caves in Cappadocia refuged people running away from successive waves of invaders, as it is testified by different archaeological locations like Catalkáyük, approximately founded in 6500 BC.

The formal aspect characteristic of a urban set is the result of the material surroundings, a tuff of volcanic origin that, due to wind erosion, offers unique natural forms. These special forms in the rock also welcomed different civilizations over time, serving as a refuge for such singular societies as Hittite, Greek, Roman, Christian and Ottoman. Their abrupt geography allow the generation of a unique defence system as the occupation of these natural "watchtowers" is an interesting way of controlling the territory, making it a suitable place for a great excavated development in height.

In this community, there are lots of dug housing but also lots of churches or chapels occupying different cones. Housing with rectangular, compact and closed spaces usually have two or more levels connected internally or externally. On the entrance floor there is usually a bedroom and the visit room; the rest are private rooms, all of them with three meters height. Because the water is scarce in the area, there are no bathrooms inside the house. The only decorated room is the one for visitors, the rest retain the natural colour and the finish of the tuff.

2.4 *Complexity as a group: An underground metropolis*

Several examples can be included within these troglodyte groups, where the two most important sites are the underground cities of Kaymakli and Derinkuyu, in the plateau of Cappadocia in central Anatolia (Turkey). According to different studies, there are over 30 cities yet to be discovered in the area, many of them interconnected. In these clusters, located in a large, irregular and arid region, the defensive demands led to a much more extreme solution, leading housing to a total burial, resulting underground cities with an important defensive function.

In these regions, the soil is composed of tuff eruptions from the surrounding volcanoes that were active three million years ago. A pretty soft and manageable material as a mixture of lava, ash and mud, provokes landscapes of rocky cones eroded by wind which can be very easy to work with few tools. This rock hardens in contact with air, providing sufficient strength to be excavated and to ensure its hardening. It is so a fairly slow process that, according to different authors, the excavation of a local in these conditions could take about three years time.

The underground city of Kaymakli was founded in the seventh century and it is located in a volcanic soil. It is a group that started to be built towards the first century in an area with particularly favourable natural conditions and was aimed to be established as a fortified city for defence against political/religious persecution or neighbours invasions: Persians, Macedonians, Romans and subsequently, the Islamic oppression in the seventh century.

This excavated city, holding about 20,000 families and occupying up to ten levels below ground, was extended through several kilometres developing a defensive underground urbanism. The different spaces, with extremely varied dimensions according to their functions, were linked by a network of interconnected galleries that, if necessary, could be blocked rolling a heavy rock that isolated populations of a potential aggressor. Exterior communication was only carried through some very localized and secure entries.

In this type of underground cities the organization was very complex having plazas, warehouses, mills, presses, stocks of provisions, churches and cemeteries put through a network of vent ducts. This organization served as a defensive system while allowed productive development assuring people a long period of autonomy.

Moreover, the excavated city of Derinkuyu was built in eight underground levels and its lower zone sits about eighty meters below the surface. It was composed of all spaces required for life in total isolation thanks to the construction of mills, tanks or silos for storage of provisions. There were also relational spaces such as streets, plazas, churches or markets where economy was developed while its inhabitants were protected for long periods. It is estimated that the city accommodated about ten thousand inhabitants in a great structure similar to an ant complex. In many points stays were intersected to meet each other, organizing a great urban complex, with thousands of inclined passageways and small sized corridors joining family rooms or common areas as a tight defensive system.

It is also interesting to consider that underground architecture in areas of extreme temperatures favours regulation of climatic conditions in housing, providing a high thermal stability. These buildings show a large thermal inertia because of their completely burial, thus, the rooms get a perfect temperature for both to live and to be used as a good food storage, which is also essential for self-sufficiency for long time periods.

3 CONCLUSION

In all, regarding excavated architecture it is true that there are many examples that respond to different typological combinations since troglodyte architects tried to adapt elementary rules enhancing the result greatly. This game of horizontal and vertical excavation, which initially could be considered as a part of an hazardous situation, may respond to climatic reasons requiring greater protection. In short, excavated towns are an attractive example of underground urbanism solved in a very interesting way, opening its constructions to the exterior and achieving a climate benefit through the thermal inertia of large ground walls. These spaces distribute and organize rooms in the house, ensuring the indirect penetration of light and heat in a very comfortable way.

Within these excavation types you can find different subtypes, some of which only focus on the excavation of existing soils, such as those studied in this work, while others rely on the contribution of certain materials to the excavation. Therefore, it should be noted that, contrary to what happened in previous typologies in which the excavation was the star item, there are other excavated mixed architectures in which not only dominates the material removal but also the combination with a contribution of other elements, usually earth, to modify the existing environment. The excavated buildings are combined with other additives to generate more complex structures and to create strong relationships between spaces with a different conceptual creation.

On the other hand, concluding about the troglodyte urbanism, and although throughout human evolution there have been a wide variety of social organizations and cities, certain basic concepts can be considered common to all vernacular towns. The wide variety of these groups is a testimony to the troglodytic architectural adaptability to the conditions of the soil, topography and climate, considering that the potential of the site determine the existence of certain kind of buildings such as excavated housing, defensive items or places for worship.

Finally, it is important to say that one of the main characteristics identifying these towns is their relationship with the natural environment, as a necessary link, reaching a balance with nature either by moulding the human structure to a complex territory in a submissive way, as in the case of Matmata (Turkey) or imposing artificiality to a territory with no accidents, in the case of Luoyang (China).

In conclusion, the present work is the first phase of a larger project which seeks to explore and evaluate the significance of the troglodytic architecture in an international, national and regional study, and how certain solutions can be transferable to today´s architecture and urbanism. Following the study shown here, a detailed research of different solutions in Spain and their practical implications on regional geography is being carried out.

REFERENCES

AA.VV. 1989. *Arquitectura subterránea. Tomos I y II.* Andalucía: Consejería de Obras Públicas y Transportes.

AA.VV. 1993. *Cobijo*. Madrid: Tursen Hermann Blume Ediciones.

Algarín, M. 2006. *Arquitecturas excavadas. El proyecto frente a la construcción de espacio*. Colección Arquíthesis, num.21. Barcelona: Fundación Caja de Arquitectos.

Deffontaines, P. 1980. *L'homme et sa maison*. Collection Géographie Humaine. Mayenne: Éditions Gallimard.

Goldfinger, M. 1993. *Arquitectura popular mediterránea*. Barcelona: Gustavo Gili.

Loubes, J.P. 1985. *Arquitectura subterránea. Aproximación a un hábitat natural*. Colección Tecnología y Arquitectura. Barcelona: Gustavo Gili.

Neila, F.J. 2004. *Arquitectura bioclimática en un entorno sostenible.* Colección Arquitectura y Tecnología. Madrid: Munilla-Lería.

Oliver, P. 2007. *Dwellings: the vernacular house worldwide*. Hong Kong: Phaidon.

Omedes, A.; Piqué, J. 2003. *Los otros arquitectos.* Catálogo de exposición. Museu de les Ciències Naturals de la Ciutadella. Barcelona: Gustavo Gili.

Rudofsky, B. 2002. *Architecture without* architects. *A short introduction to non-pedigreed architecture.* Albuquerque: University of New Mexico.

Seijo, F.G. 1979. *La vivienda popular rural alicantina: tomos I y II.* Alicante: Ediciones Alicante.

Shoenauer, N. 1984. *6.000 años de hábitat. De los poblados primitivos a la vivienda urbana en las culturas de oriente y occidente.* Colección Arquitectura Perspectivas. Barcelona: Gustavo Gili.

Soriano, V. 2006. *Arquitectura de tierra en el sur de Marruecos. El oasis de Skoura.* Colección Arquíthesis. Barcelona: Caja de Arquitectos.

Earthen Architecture: Past, Present and Future – Mileto, Vegas, García Soriano & Cristini (Eds)
© 2015 Taylor & Francis Group, London, ISBN 978-1-138-02711-4

First considerations for an adobes mensio-chronology

D. Pittaluga
Scuola di Specializzazione in Beni Architettonici e del Paesaggio, Università di Genova, Genoa, Italy

R. Pagella
ISCUM, Istituto di Storia della Cultura Materiale, Genoa, Italy

ABSTRACT: This research studies adobe in the province of Alessandria, with the purpose to expand the knowledge about these buildings and to establish new mud-brick chronology research tools for direct dating. One of the aspects analyzed, the object of this contribution, has been to a possible dating of the elements of direct sources. "Direct sources" are those sources that draw information from the building itself. This is in line with the archaeological method, and these tools fall into that discipline called "Archaeology Architecture", of which was Tiziano Mannoni precursor. Increasing the tools of knowledge structures in clay and mud-brick is helpful especially in cases of "minor building" where it is not possible to have any other form of information (e.g. archives, etc.). More specifically, for the above-mentioned area (Alessandria Province, Italy), among other tools, we assessed the possibility of applying the mud-brick mensio-chronology technique.

1 THE NEED FOR A MENSIO-CHRONOLOGICAL ANALYSIS

1.1 *The international debate*

Several recent studies have shown a widespread use of earth as a building material in certain contexts, and specifically the mud-brick technique used in "minor" buildings. Such buildings strongly characterize some areas and therefore the possible loss of this architectural heritage should be avoided. These are often situations that lack written documents on their history and evolution. A way to overcome this problem is to develop and use direct tools to date these artifacts, and mensio-chronology combined with stratigraphical-analysis represents a good possible solution.

Much has been debated on dating techniques and their importance in restoration; this is not the place to address this topic. However, it must be noticed that the ability to specify chronologically the interventions on a building can be useful if there are several different phases, in order to understand their possible temporal order or to evaluate the effectiveness of any prior repair (the awareness that a two hundred years old repair is still in perfect condition may be important for a positive assessment of the adopted solution, conversely observed degradation of a recent intervention might lead to rethink that type of solutions).

1.2 *Reasons for a possible mensio-chronological analysis of mudbricks*

Mensio-chronology of baked bricks has been confirmed as a useful dating technique by several national and international studies. This method can be applied in regions and periods where the use of fired bricks is homogeneous in terms of socio-economic and trading-monopoly conditions. One of the basic considerations of the study presented here is related to the ability to purchase adobes and bricks from the same furnace, especially in those cases where the two types are both used in the same building. In these cases, it is therefore possible that the same box cases are used for the two types, for practical convenience, and the same production rules were applied, and the only difference could be at the end of the cycle, when a part of bricks was fired while the other skipped this final step and could already be sold.

1.3 *Differences with "baked bricks" mensio-chronology.*

There are still differences between the two productions.

In mud-brick there is less variability in measurements between a brick and another. In fired bricks, indeed, depending on the firing method and the more or less heating uniformity inside the furnace itself, thickness variations can be several millimeters (Pittaluga 2009b).

However, adobe is more vulnerable to degradation, especially when it is used in masonry exposed to weather elements, not adequately protected from protrusions, cornices or plaster layers and it can easily lose some of the material and in more severe cases even the shape itself may be corrupted.

2 THE BUILDING MATERIAL

Both worldwide most common earth-building techniques can be found in Alessandria province: rammed earth, also known as "pisè", and the adobe. This paper refers to this second technique, adobe, that still retains many features of clay, such as energy saving, physical well-being, static resistance, but here these are expressed to a lesser extent because adobes happen to be used only in certain parts of the building, that is completed with bricks or stones.

The type of earth material used for adobe walls is different from the one used for the realization of clay, in fact it must have a negligible amount of aggregates inside. The earth taken from the quarry must be sifted in a very fine way, and this would be a problem if large aggregates were inside. Then the material is properly humidified and it is inserted into moulds. Finally the blocks are left drying in the sun and put in place with mortar of earth or lime.

Adobe structures can be found in several areas in Alessandria Province, for example in Fraschetta, Bassignana, Montecastello, Rivarone, Mugarone, Isola S. Antonio, Sale, Castelnuovo Scrivia, Castelnuovo Bormida, Castellazzo, Cassine, Castelspina, Sezzadio, Predosa, Basaluzzo, Novi Ligure, Fresonara, Bosco Marengo, Viguzzolo (Pagella Poggio 1992).

In these areas adobes are put in place over basements in brick or stone, sometimes they are bonded with these other materials weaving multicolor structures. A popular say is "in mou, dui trou e trèi prijou", that means "one brick, two adobes and three big stones" (Cermielli & Barco 2008). Otherwise they may be put beside rammed earth walls, mostly in the upper parts of the buildings, because of a better ease of use and lightness. Adobe bearing walls come up with good resistance to compression and atmospheric agents, provided that they are defended by the projection of the eaves and placed on high pedestals.

3 GEOGRAPHICAL NOTES

In municipalities of Castelcermelli and Castellazzo Bormida, in Alessandria, Italy, there are several adobe buildings. Most of them date back to the 19th and 20th centuries.

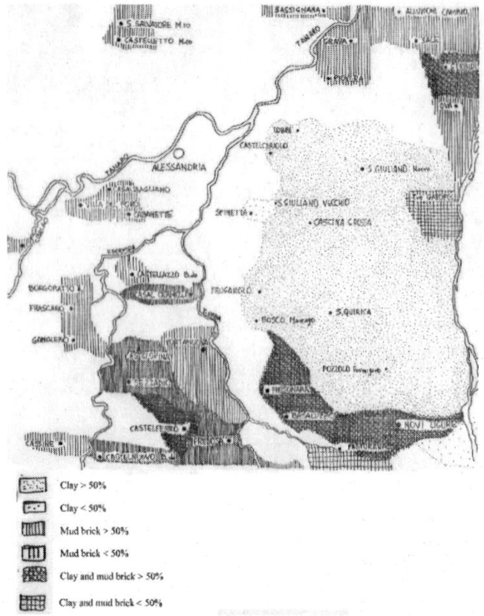

Figure 1. Map of the region (adapted from Pagella, 1987).

3.1 *Casalcermelli*

Casalcermelli is a municipality in the Alessandria Province, a flat area on the left bank of Orba river; its Fontanasse and Rossina hamlets are typical agricultural 19th century villages.

In 1280 A.D the Cermelli family founded the village, giving it its name. The main Casalcermelli parish church is dedicated to the Blessed Virgin of the Assumption, built in 1773 with bricks prepared in Cascina Campagna and donated by the family Trotti. The oratory of the Confraternity of St John dates back to 1614, while the original building dated back to 1420 was built of brick and raw wood. The parish church of San Carlo was started in 1773 and it was consecrated in 1819. The chapel of Campagnotta dates back to 1733, it is located near the road to the Cascina Rosa.

The first asylum for poor girls and elderly people was realized in 1888 in Via Gerbidi, funded by a bequest in favor of the Cottolengo monastic and assistance organization; a piece of the wall in mud bricks is still in place today. Then it became a nursery.

Cascina Campagna is a kind of barn-fortress of medieval origin, it retains a 15th century hall from its primitive nucleus.

The "Cascinotti" are typical buildings dating back to the 19th century, that have a residential ground floor used also as a stable, and a barn above.

3.2 *Castellazzo Bormida*

This municipality is on Bormida and Orba rivers, in Alessandria Province. From 1707 the village passed to the Savoy, who nominated seignor Ottavio Pallavicini, from Genoa, whose family went out in 1778. In 1863 the village took the current name of Castellazzo.

19th and mid-20th century saw the emergence of craft activities related to silk industry, serum and construction, including Hoffman system kilns for bricks production.

At the time of the French Revolution, Castellazzo already had three mills in operation; in 1850 there were eleven; in 1880 the three companies were consolidated, Boidi, Pistarini and Ricagni. The local silk production was particularly valuable for its shine and quality (Reposi 2008). Currently, some of the buildings of the mills and furnaces have been refurbished as residential buildings.

4 THE RESEARCH METHOD ADOPTED

The first phase of the field investigation was the identification of several building units in Alessandria Province area in which both mud-bricks and baked bricks are present together.

Figure 2. Adobe wall of Casalcermelli ex-Nursery (D. Pittaluga).

A quick stratigraphic analysis was made on these units, to verify that the part of the wall containing crude bricks and baked bricks could be interpreted as a single stratigraphic unit (i.e. it was originated by a single construction action and not by changes at different times).

Then bricks measures were taken, with millimeter precision, over twenty brick and twenty adobe units (where possible). The survey registration form collected also notes on the conservation status of the measured elements and on the color of the bricks.

These remarks can be very important because different colors can be related to different firing levels, and this can originate at difference in the final element dimensions (Pittaluga 2009b).

Statistical analyses were made on sample data, as a base for mensiochronology:

- Average thickness, length, width and standard deviation
- Average thickness by length, thickness by width, width by length products and standard deviation
- Average volumes (thickness by width by length) and standard deviation.

5 DATA INTERPRETATION

An attempt was made at this early stage to rely on cases where the date of the stratigraphic unit was identified by means of other archaeological sources, in order to associate the material datum and its characteristics to a specific period of time. In doing this it was necessary to proceed to an archaeological reading on these walls (Pittaluga 2009a, b), to determine whether the chronological attribution could be more or less compatible with the sample studied.

In second phase only mud bricks were examined.

In third phase we will work more raw material analysis and their origins, using both direct

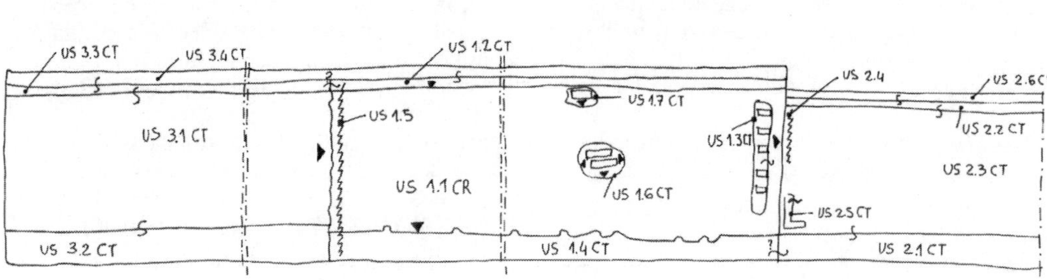

Figure 3. Stratigraphic units over the wall.

USM/US	Thickness statistical distribution	Thickness (std. dev.) [cm]	Analysis notes	Chrono
US_1.1_CR	two phases		need to be split, see two rows below	post 1888
US_1.1_CR_A	multimodal gaussian	5.64 (0.36) Pavement brick ?	less thick than bricks	post 1888
US_1.1_CR_B	unimodal simmetric gaussian	6.01 (0.13)	like bricks	post 1888
US_1.2_CT	unimodal asimmetric right-handed gaussian	5.9 (0.36)	Like adobes	post 1888
US_1.3_CT	unimodal asimmetric right-handed gaussian	6.06 (0.14)	Like adobes	post 1888
US_2.3_CT	unimodal simmetric gaussian	5.95 (0.54)	Like adobes	1888

Figure 4. Sample of measure values for stratigraphic units.

analysis, including diffractometrics, and indirect sources. At this stage, research will be done on production furnaces.

6 FIRST RESULTS OF THE ANALYSIS

The following cases of masonry techniques have been identified:

1. Mixed masonry, with bricks and adobes in alternating rows, with a clear separation of two brick types. Typically there is a block of more subsequent courses of adobes followed by a blocks of multiple courses of bricks. Sometimes the block height (in number of courses) decreases from bottom to top of the wall. The number of subsequent courses of bricks is generally comparable to the number of adobe courses, or a slightly smaller, e.g. a block of 4 courses of mud bricks is overlapped by a block of 3 courses of bricks.
2. Mixed masonry, with bricks and adobes in the same course. This usually means that the large majority of elements are baked, More rare cases where adobe bricks prevail in number
3. Alternation of adobe and brick along wall thickness. For example a masonry made by two walls tied together side by side, the one towards the outside of the dwelling uses only bricks, the internal one uses adobe.
4. Masonry entirely of adobe, except for baked bricks corner cantonal
5. Masonry entirely of adobe
6. Masonry entirely of brick

Each of the cases above has a special influence on dimensions of adobe:

1. Where brick and adobes lay on different courses, there is a greater freedom, i.e. the two types of bricks must not have the same thickness. However a good practice for a correct balance is to have in the whole masonry all brick and adobe elements with similar volumes. In the future more, when more cases will be available, we plan to better analyze and evaluate the variability in number of overlapping courses of each block
2. In this case, there is the absolute need to have the same thickness for every unit, both the bricks and the adobes.
3. In this case there is no dimensional constraint for the bricks (both adobe and baked)
4. In this case adobe and bricks must have the same thickness
5. No constraint in this case
6. No constraint in this case

7 RESULTS FROM CASALCERMELLI

The stratigraphic analysis of the structures is essential to make mensio-chronological analysis of these structures. It is necessary to compare the measures of the units (both, adobe and bricks) with the dates obtained from the documents. These dates could concern the first phase building or later interventions. The stratigraphic analysis is the key to establish this sequence accurately.

The wall of Casalcermelli Nursery has been investigated. This is a boundary wall of a large estate that has been used as a kindergarten since 1888.

The analysis highlights three major archaeological interventions that characterize three distinct parts of the wall; these three parts have different materials and textures. The three parts are named USM_1, USM_2 and USM_3 (USM = *Unita' Stratigrafica Muratura*, i.e. stratigraphic wall unit).

USM_1: almost entirely of adobe wall, the only bricks are at the base of the wall (US_1.4_CT) and in the crowning and top area immediately below

USM	Description	Chrono (source)	US	Description	Stratigraphic notes
1	Nursery enclosure wall, adobes, few bricks	post 1888 (stratigraphic analysis)	1.1_CR	positive, wall adobes	post 1.4_CT coeval 1.3_CT ante 1.2_CT ante 1.5 ante 1.6_CT ante 1.7_CT post 2.5_CT
			1.2_CT	positive, crowning, bricks	post 1.1_CR coeval ? 3.3_CT coeval ? 3.4_CT
			1.3_CT	positive, cantonal, bricks	post 2.3
			1.4_CT	positive, base, bricks	coeval ? 2.1_CT ante 1.1_CR ante 1.5
			1.5	negative, break along boundary with USM_3	post 1.1_CR post 1.4_CT
			1.6CT	positive, break repair, bricks	post 1.1_CR like 1.7_CT
			1.7CT	positive, break repair, bricks	like 1.6_CT coeval ? 1.2__CT post 1.2_CR
2	Nursery enclosure wall, stones and bricks	1888 (bibliographic analysis)	2.1_CT	positive, base, bricks	coeval 2.3_CT coeval ? 1.4_CT
			2.2_CT	positive, crowning, bricks	coeval 2.6_CT coeval 2.3_CT
			2.3_CT	positive, wall in stones and bricks, bricks	ante 2.4 coeval 2.1_CT coeval 2.2_CT
			2.4	negative, break in the wall	post 2.3_CT
			2.5_CT	positive, cantonal, bricks	coeval 2.3_CT ante 1.1_CR
			2.6_CT	positive, cover in pantiles	coeval 2.2_CT ante 3.4_CT
3	Nursery enclosure wall, bricks	post 1888 (stratigraphic analysis)	3.1_CT	positive: wall in bricks	post 1.5 coeval 3.2_CT coeval ? 3.3_CT
			3.2_CT	positive, base, bricks	coeval 3.1_CT post 1.5
			3.3_CT	positive, bricks crowning	coeval ? 3.1_CT coeval 3.4_CT
			3.4_CT	positive, cover in pantiles, bricks	coeval 3.3_CT

Figure 5. Meaningful data for mensio-chronological analysis.

the roof (US_1.2_CT), and in the corner (US_1.3_CT, adjacent to USM_2).

USM_2: cobblestone and bricks wall, the texture is "edged" with alternating 50 cm high blocks of pebbles courses and blocks of two bricks courses, arranged in header and stretcher bond.

USM_3: wall entirely made of bricks, arranged mainly in stretcher bond, and only few ones are in header bond.

Other stratigraphic units have been defined for other parts of the w—all.

8 STRATIGRAPHIC RELATIONSHIP AMONG THE USM OF CASARCERMELLI NURSERY

Here is the stratigraphic analysys of Casalcermelli nursery.

USM_3 is the most recent wall, USM_2 is the oldest one. These conclusions are drawn from the analysis of the border areas between these units; looking at how a unit leans over the other one, it is possible to establish their chronological relationship (before, contemporary, after).

Archaeological analysis states that USM_1 is more ancient than USM_2 but more recent than USM_3. In USM_1 most of the masonry is in adobes (US_1.1_CR), and it is possible to locate other transformations inside it. Just above the bricks base there are sections of bricks wall (broken pillars? US_1.4_CT); this suggests that the original brick solution was replaced by an adobe wall later on. At a later stage of the research mensio-chronological bricks analysis will be applied to both this basement and the pillars to try to determine whether or not this arrangement is connected to the first phase (USM_2).

The top course crowning USM 1.2 wall, named US_1.2_CT, is made of bricks arranged in header bond, the mortar is made of earth, and from the stratigraphic point of view this layer is different from the one in adobes. In the next phase of the research the measures of these bricks will be compared with those from US_3.1_CT (common coverage to USM_3 and USM_1 walls). Bricks on the extreme right edge of Casalcermelli Nursery external wall, identified as US_1.2_CT, are used for the corner, and are adjacent to USM_2. The mortar analysis determined that these bricks and those of USM_1 are contemporary.

USM_2 is older than USM_1, this can be determined examining the border area between the two units. In the upper part of USM_2 there is a break over which USM_1 leans. In the lower part there seems to be a continuity between USM_2 and USM_1, however, a more detailed analysis of this parts will be made at a later stage of the research.

9 CONCLUSIONS

The situation shown in the example above (Casalcermelli Nursery wall) is also present in several other cases in the same town and in the nearby town Castellazzo Bormida.

In total 23 different situations with mixed walls have been analyzed, in comparison with documents, archive dating and descriptions taken from other studies (Pagella 1987). This first phase of the research has found this result: when adobes and bricks coexist in the same mixed wall, all units have similar measures, and this means probably that these bricks were made from the same mould. In detail, measures of adobes are on average a few millimeters over the backed ones, and this makes sense, as the baking process includes a little shrinkage.

As a consequence the bricks mensio chronology technique seems to be applicable not only to baked bricks, but also to adobe bricks, using the same curves, with minimum adjustments.

But the analysis of 23 mixed wall situations is not enough to promote this result to a general rule. The following step in the research will be to extend the measurement and analysis to more buildings, obtaining a larger data base. For this purpose 57 situations have already been identified as targets in Alessandria province. These have been selected from a 169 elements list (Pagella 1987). It will also be important to collect updates on conservation states of these buildings after thirty years.

A third phase will analyze building made of adobes only, with further comparison to archive dates and documents.

If the results of these phases will confirm the preliminary results already obtained now, a mensiochronological analysis of adobes could become a reality.

NOTE

The research is a joint effort of both authors, however Rosa Pagella wrote chapters 2 and 3 of this paper while Daniela Pittaluga wrote chapters 1, 4, 5, 6, 7, 8, 9.

REFERENCES

Buzzi G. 1878, "*Storia di Gamundio antica o Castellazzo di Alessandria*", voll. 4, Alessandria 1863–64; in Boidi G.A., "*Documenti sulla storia di Castellazzo Bormida*", Torino.

Cermelli D., Barco G. 2008, "*U spapiari dir Casà. Storia, tradizioni, leggende, proverbi e dialetto di un lembo di terra specchiato sull'Orba*", Alessandria: i Grafismi Boccassi.

Mannoni T., Milanese 1987: "Mensiocronologia", Firenze: ed. All'insegna del Giglio.

Moretti Cristoforo, 2001, "*Catalogo di Edilizia Ecclesiastica nel territorio di Castellazzo Bormida*", Alessandria: Boccassi Editore.

Pagella, Rosa 1987. "*Edilizia in terra cruda nella provincia di Alessandria. Tipologia, cronologia, materiali, tecniche costruttive*", tesi di laurea a.a. 1986–1987, Facoltà di Architettura di Genova, relatore prof. T. Mannoni, correlatore prof. Andrea Buti.

Pagella Poggio, Rosa 1992. *Costruzioni in terra cruda nella Provincia di Alessandria, Tecniche costruttive, Materiali, Forme architettoniche e tipologia*, Castelnuovo Scrivia, Maxmi.

Pipino G., 2005, "*Liguria mineraria. Miscellanea di giacimentologia e storia estrattiva*", Ovara: Tipografia Pesce.

Pittaluga Daniela 2009a, "*Questioni di archeologia dell'architettura e restauro*", Genova: ECIG.

Pittaluga Daniela 2009b, "*La mensiocronologia dei mattoni*", Genova: ECIG.

Pittaluga Daniela, Ghislanzoni Paola 1991, "*Informazioni storiche e tecniche leggibili sulle superfici in laterizio", in* in *"in Atti dell'VIII convegno di studi Scienza e Beni Culturali "Le superfici dell'architettura: il cotto. Caratterizzazione e trattamenti", ed.Libreria Progetto Padova, pp. 11–22.*

Pittaluga Daniela, Ghislanzoni Paola 1991., "*La mensiocronologia dei mattoni, la statstica applicata all'analisi*", in Archeologia Medievale.

Prati Mario, 2008, "*Tesi di Laurea In Economia e Commercio*", Università di Torino.

Regione Piemonte, "*Caratterizzazione dell'area idrografica e relativi corpi idrici. Caratteristiche generali.*"

Reposi Milena. 2008, "*La seta a Castellazzo Bormida*", Alessandria: Edizioni dellOrso.

Earthen Architecture: Past, Present and Future – Mileto, Vegas, García Soriano & Cristini (Eds)
 ISBN 978-1-138-02711-4

T.E.R.R.E.—Technological Experiments and Research on Earth buildings

S. Pollak & A. Fasoli
AK0—Architettura a Kilometro Zero, Rome, Italy

ABSTRACT: The challenge of a "real" construction in education goes along with values such as technical dexterity, team work, familiarity with schedules and layouts and similar concepts that are difficult to transmit sitting at a desk. Moreover, it is acknowledged that a factual goal fuels enthusiasm, commitment and self-estimation among the participants. Some of these aspects are of interest for the changing role of architects nowadays. Appropriate solutions can emerge from working methods that blur the borders between professional practice, education and research. Construction techniques that use easily available natural materials enhance the participation of non-experts such as students or community members. In this framework, building techniques based on raw earth play a central role. They are acknowledged to be one viable method of providing buildings for a rising population with limited resources. Along with its economic and ecological benefits, building with earth allows for a easy and safe experimental approach.

1 BACKGROUND

For long years architecture practice and architectural debate believed that "design has become the most powerful tool with which man shapes his tools and environments (and, by extension, society and himself)" (Papanek 1971).

In nowadays spatial transformation processes we can observe two main aspects: architects are involved in less than 5% of all building and transformation processes and when they are, they often uncritically reiterate consolidated spatial and material solutions. Technological, economical and cultural constraints apparently condition the design results more than the peculiar situation to act on.

So what happens if we pass from asking "What designs or plans shall we use to improve the world?" to wondering how possible improvements can act on already ongoing processes of spatial transformation? Which already available resources can be activated? In how far can new approaches for a sound environmental transformation emerge from a direct interaction of involved actors?

All these questions affect professional design activity as well as education; not only because education is the preparation of new professionals but, as shown later, students can have an active part in getting out of the described impasse.

2 LEARNING

Learning by doing methods have been introduced in architecture schools more then 30 years ago. Nevertheless, in most schools the educational patterns are mainly theoretical.

Experimental activity is relegated to sketch books or computer screens and often enough architecture students complete their carrier without seeing a "real" building site. This lack of practice makes it difficult to interact with enterprises, craftsmen or other persons in charge with the physical production of the designs.

The awareness on these limitation is rising and young professionals demand for educational opportunities with a strong practical approach.

The association AK0 conduces from three to six building workshops each year. Each of them integrate theory and practice, however the range of formats for the events is rather large. Some initiatives aim at providing a general introduction on specific building techniques, mainly in the field of natural materials. In other cases the educational activity is integrated in a larger building process commissioned by public or private clients. In some cases small experimental architectures have been designed with the specific purpose to get familiar with peculiar techniques or materials.

A joint feature of all workshops is the direct contact the participants get with building materials and constructive processes; they actually have to get their hands dirty. Especially for participants without former experience on building sites the manipulation of matter activates a learning process that makes the absorption of abstract information—generally provided in a second step—easier. Moreover, the more craft oriented working steps of a construction simply cannot be taught theoretically.

The use of raw earth in this framework plays a crucial role. Besides its benefits with respect to its large availability and its environmental performance, two are the features that make it an appropriate material for workshop activities:

On one hand it is nontoxic, which makes its manipulation, even with nude hands (or feet), easy. On the other, this same manipulation proved to be an emotionally involving activity that makes the act of knowledge acquiring literally "feel" (Fig. 1). Getting one's hands dirty activates the limbic part of the brain, which is capable of processing a much higher amount of information compared to a rational elaboration of new knowledge.

Earth as a natural material happens to have different qualities according to the place of excavation. Some of these features can only be assessed empirically, by handling the matter. Traditionally these differences in granulometry or clay content led to the definition of different building techniques, which strongly contribute to the identity of the respective built environment.

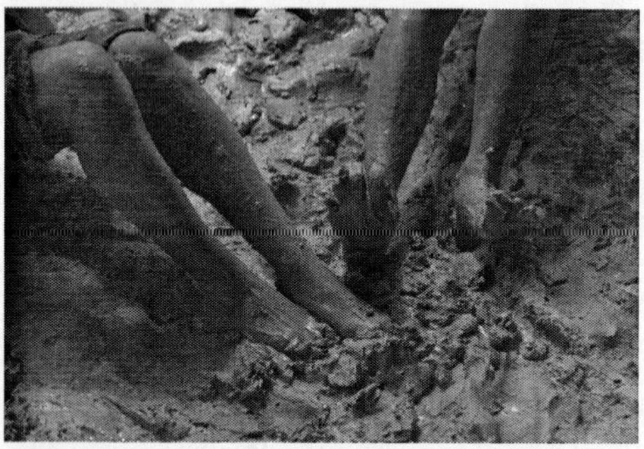

Figure 1. Direct interaction with building material (M. Saccone).

Figure 2. Multiple reiteration makes new knowledge emerge.

The educational program TERRA by AK0 offers short introduction courses on some of the most used techniques. Besides a basic course which provides a general introduction on building with earth, participants have the possibility to attend workshops on monolithic techniques like rammed earth or cob, masonry including adobe, earthbags or *cob*, and "structures", which includes all those techniques such as wattle-and-daub where raw earth is applied on a bearing frame made of other materials. An additional course is offered on rendering solutions made of natural materials including earth, clay and vegetable additives.

All construction issues are practised in the field and the participants are invited to skip constantly between abstract learning and practical application. Teamwork is requisite to carry the rather tiring work out; this activates additional knowledge exchanges among the participants, who often enough find solutions for unexpected situations and combine them to design.

Doing allows for an immediate understanding of operational sequences: What has to be done first? What can wait? What can be done at all? What can only be done with the help of someone more experienced than me?

This teambuilding process is an educational issue on itself. Participants get to know what they are good in or how own weaknesses can be compensated through collaboration. This awareness is particularly valuable in a framework of activities that can be compared to contract work as i.e. in case of building workshops for small experimental architectures or within the sequence of a wider building site.

3 BUILDING

When possible AK0 encourages the integration of education with professional practice inviting students and young professionals in building workshops with "real" briefs. For the participants the factual goal is a strong motivation to measure their skills. This usually fuels enthusiasm, commitment and self-estimation among them and can help in reaching project goals that were unreachable otherwise. Some of the more labor-intesive building solutions, especially in the realm of earth construction, were not viable in more industrialized contexts if not thanks to the combination of education and building.

However the value of involving a group of non fully trained operators on a building site is not only economic. In certain situations they can act as vectors for technological and cultural transfer. This is particularly true for those projects where entire communities are concerned as in the C.A.S.A.-project (Fig. 3), which led to the construction of a small trade school in a rural settlement in Guatemala.

At a first glance, the presence of a large group of non-experts on a building site may have the taste of a disturbance or a loss of time. Procedures have to be explained many times and, at least in the first attempts, the quality level is low. Especially for those techniques that belong to the realm of conventional building solutions, expert builders can easily get the impression that doing the job alone instead of instructing a bunch of laymen would be certainly better.

In the Guatemalan experience the perception changed radically since the project introduced less standardized technologies. Here local construction workers and European workshop participants needed each other. In order to bear the challenge of reintroducing rammed earth structures, which used to exist in the Mayan culture but had been

Figure 3. The building site as an exchange hub (T. Chiaravellotti).

Figure 4. Local workers apply their skills to unconventional techniques and make them viable (T. Chiaravellotti).

abandoned for the benefit of apparently more "modern" concrete and brick constructions, it was necessary to join the cultural enthusiasm of the young volunteers from abroad with the operative know-how of skilled construction workers, even if they had no experience on the specific technique.

The presence of a large group of volunteers helps in facing the sense of uncertainty that goes along with the use of non-standard solutions. In such a framework participants with an academic background benefit from the fact that they are used to trial and error approaches. Even a nonchalant attitude can bring fruitful results in those situations where small inventions are needed to solve construction issues.

Large numbers are all but a problem if certain conditions are given:

- communication within the group must be granted;
- achieved results must be reported and others must have the opportunity to benefit from them;
- participants with more experience such as instructors or experts for specific working steps must accept to teach and execute at the same time. Working steps don't have to be only carried out but explained and made replicable;
- the building site must allow for enough space and safe working conditions.

Using manual interaction as an investigative design tool brings the design process nearer to *making from the inside* rather than *looking from outside* and the building site becomes a laboratory and the meeting point to share competences and experiences.

4 RESEARCH

The above mentioned working method, based on collaboration and direct manipulation of building materials also provides fertile ground for experimentation and research activities.

The sequence of tests and buildings contributes to the definition of a contemporary interpretation in the use of millenary materials like raw earth or other natural resources. This long term investigation includes more specific and circumscribed assessments with respect to structural behaviour, construction progress, climatic performance and similar technical aspects of building.

While getting familiar with earth as a building material, workshop participants actually assess the resource's properties. This is generally done with the earth-matrix, a simple exercise that allows to understand which techniques suit the available earth better than others. Since the workshops are

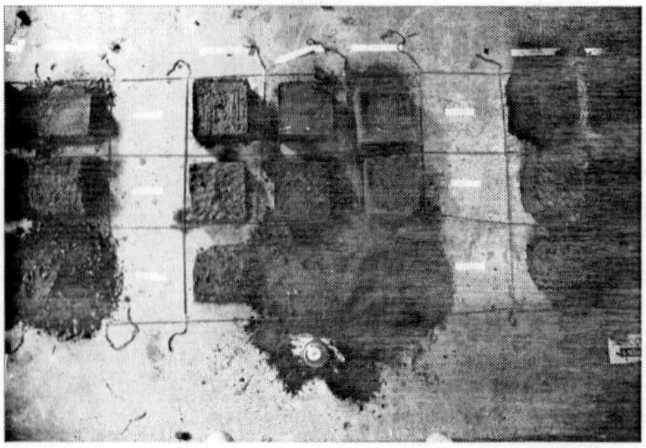

Figure 5. Simple assessment methods for earth properties.

held in different places, the findings can be crossed and operative conclusions can be drawn from them.

On one hand the standardization of procedures is requisite to make the findings comparable, on the other it proved to be extremely valuable to have moments of questioning these routines. Within the workshop framework this generally happens when persons with different cultural background and skills come together. Generally participants in experimental workshops are happy to share their knowledge. Approaching specific items from the viewpoint of different competences may be initially demanding but ends up in letting unexpected results emerge.

Among the specific issues AK0 investigated in the field of earth-construction, the focus is posed on how this millenary material can be integrated in contemporary building processes. More in detail this includes questions like:

- How must a wattle-and-daub wall made of timber, reed and earth be built in order to suit with the climatic and legal requisites of the building sector of industrialized countries?
- Is it possible to separate the single steps of a construction an introduce some aspects of prefabrication in order to improve the control of time and costs?
- What expressive possibilities can come from a combination of prefabrication and self-construction and how can a building with these features be organised?

In one of the first experiments the wattle-and-daub system from the Peruvian Andes has been adapted to the Mediterranean climate of central Italy by using a double layer of panels with an insulating infill made of compressed wood fibre or a straw-earth mixture in between. A SHELLter-prototype has been built in a one week workshop by a team of students, professionals and inhabitants (Fig. 6). A second week has been dedicated to the interior and exterior finishing applying mixtures of clay and lime plasters with different natural additives.

While in the SHELLter project all steps have been carried out on site, the wattle-and-daub system appeared to be suitable for an approach in terms of prefabrication. The SPORA project (*Self-build Prefabricated Organic Renewable Architecture*) provided the opportunity to define a complete scenario where small but technologically well equipped enterprises could care about the prefabrication of timber frames, even with unconventional shapes, i.e. irregular polyhedrons (Fig. 7).

In a design proposal for a public building in the Peruvian town of Qotowincho, thought to be erected with the direct contribution of the local community, the *quincha* system (wattle-and-daub) has been reinterpreted as a multilayered climatically performing wall: while the interior panels follow the traditional rectangular shape and act as support for an insulating layer, the external panels have been redesigned to form a ruled surface. The assembling of all the panels generates a sort of waving façade, with reed-screens as external layer (Fig. 8).

Figure 6. Assembling of the prototype SHELLter.

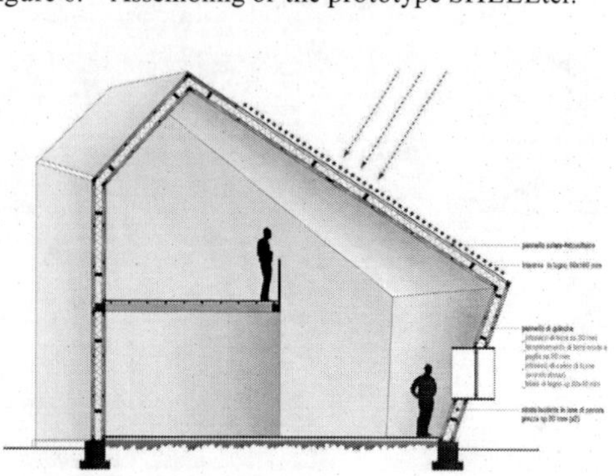

Figure 7. Polyhedrally shaped structure of prefabricated wattle-and-daub panels to be rendered in self-construction.

Figure 8. Ruled surfaces (G. Puri).

Figure 9. Wattle-and-daub panels allow for pre-fabrication and later completion in self-construction (F. Liberati).

Figure 10. Students test the feasibility of prefabricated rammed earth elements.

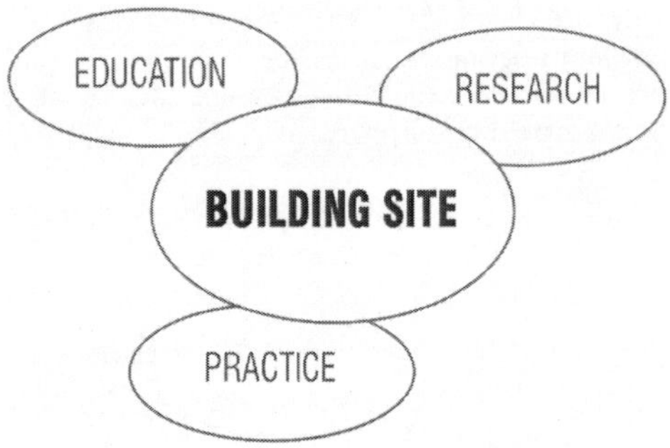

Figure 11. The building site as exchange hub (V. Gori).

The professionally manufactured pieces grant structural reliability, control of the assembly sequence with precise lose-fit connections and short construction times as a consequence.

Once the timber skeleton is produced, it can be filled in with reed and raw earth by persons without specific skill in the field of construction. This allows for considerable reduction of expenses and provides the opportunity to really get involved in the realisation of the building. Especially for small residential or public buildings this endows the architecture with a strong identity.

A building workshop within the postgraduate course in housing (Master Housing) at Roma Tre University offered the opportunity to prototype a small structure with non stereo-metric shape (Fig. 9).

Another student workshop gave the possibility to test the concepts of prefabrication applied to rammed earth. Small elements were manufactured by hand in a way that allowed their displacement within the same building site. This technique is not new, however it is not fathomed yet with respect to self-construction, which can open new ways to design rammed earth structures, especially in less accessible urban spaces.

5 CONCLUSIONS

All the illustrated cases take advantage from the collaboration of experienced workers with less skilled students or young professionals. Their inherent curiosity and their disposition to explore non familiar issues is a valuable source of innovation.

The organisational asset that goes along with large groups of actors allows for investigation in several directions. In those cases where such an apparently random investigation meets well working communication fluxes within the team elements of knowledge can emerge that are not explainable as additions of single inputs but rather have to be considered of a higher order. In order to get this, the building site itself is looked at as an exchange hub (Fig. 11) and those techniques are favoured that enhance direct manipulation, and thus a limbic elaboration of knowledge. This goes along with a constant readiness of all actors to question the achieved findings and define new goals.

While design activity is often constrained by best practices or unquestioned conventions, a direct involvement of students and other non-experts can bring valid inputs to construction processes. This opens new opportunities for the spreading of architectural know-how in contexts that are generally difficult to attain with innovative approaches. If well managed, this can be a starting point for a greater performance of the built environment and hence for a more sustainable design.

The workshop approach described above can be seen as complementary for more conventionally structured educational carriers as well as for the realm of on-the-job training. Cooperation experiences with universities and other professionally oriented associations proved this integration to be fruitful.

Involving the future generations of architects from the beginning in more globally oriented issues not only helps them to become aware of current and upcoming professional challenges, but in certain contexts, their involvement can be the key to making things happen.

REFERENCES

Barriola, N. (ed.) 2000. *Diseño y construcción en Quincha.*, Lima: Ediciones PUCP.

Barrionuevo de Machicao, R., 2005. *Centro Virtual de Salud, Qotowincho*, in Waka XXI, n°3, Lima: FAUA UNI.

Houben, H. & Guillaud, H. 1994. *Earth Construction—a comprehensive guide.* Rugby (GB): ITDG Publishing.

Hudson, D. 2013. *New trade school in Guatemala empowers its citizens.* In Designboom—digital architecture and design magazine. Milan: Designboom.

Minke, G. 2000. *Earth Construction Handbook—The Building Material Earth in Modern Architecture.* Southhampton: WIT Press.

Minke, G. 2006. *Building with Earth—Design and Technology for a Sustainable Architecture.* Basel/Berlin: Birkhäuser.

Papanek, V. 1971. *Design for the Real World: Human Ecology and Social Change.* New York: Pantheon Books.

Pollak, S. 2007. *Emerging Knowledge, Students as Intermediate Agents in Construction and Transformation Processes.* In Transactions, the online journal of the Centre for Education in the Built Environment. Cardiff: JEBE.

Earthen Architecture: Past, Present and Future – Mileto, Vegas, García Soriano & Cristini (Eds)
© 2015 Taylor & Francis Group, London, ISBN 978-1-138-02711-4

An itinerary through Sardinian earth built villages

E. Pusceddu
Associazione Nazionale Città della Terra Cruda, Samassi (CA), Italy

ABSTRACT: The pilot project '*Itinerary through earthen built villages*' has been developed in the past two years in Sardinia, and it is now refining its activities. The main goal of the project is to offer a new and fully sustainable local development based on the conservation and reuse of the local traditional earthen building culture. It focuses on the consistent vernacular heritage still in use in Sardinia, and on the intangible patrimony of 'good lessons' coming from the past. For this reason, this project considers the Sardinian earthen culture as the main strategy to pair the conservation of the regional heritage with a new development policy, able to assess the community's traditional culture and foster a new local '*green*' economy.

1 INTRODUCTION

During the last fifteen years, local administrations, Universities and small-medium enterprises (SMEs) shared common important efforts on local earthen heritage, recognizing in it both the historical-traditional identity of the Campidano flatland and an economical source located in the Middle-South area of Sardinia and is characterized by the adobe building technique. The development of this building culture is due to the soil properties, constituted mainly of Quaternary sediments such as gravel, silt and clay, which have caused a predominantly flat, agricultural area (Sanna in Achenza, 2004). A decisive reconnaissance of the building material earth and a strong solicitation run by public and private interlocutors has certainly favoured, if not caused, some recent events of absolute importance, among those are:

- The foundation of the National Association of the Earth Towns in 2002, which includes over 30 Sardinian local administrations and about ten others in representation of four different Italian regions (Abruzzo, Marche, Piemonte and Basilicata). The corporate charter of the National Association of the Earth Towns has been signed in Samassi (Sardinia) in February 2002. Founders of the Association are the Municipalities of Samassi, Villamassargia, Serramanna, Vallermosa, San Sperate, Musei, Samatzai, Segariu, Serrenti, Nuraminis, Selargius (all in the province of Cagliari), Montegranaro (pr. Ascoli Piceno), Casalincontrada and Roccamontepiano (pr. Chieti), Treia (pr. Macerata). (http://www.terracruda.org);
- The publication of the Manuals for the Rehabilitation of Sardinian Historical City Centres (Sanna et al., 2009), which are at the same time a vehicle of knowledge on Sardinian earthen building cultures and techniques, and a guidebook for local administrations confronted with recovery actions and city plan upgrades;

Figure 1. Typical earthen house entrance, San Gavino Monreale, Sardinia—Italy (credit: Associazione Nazionale Città della Terra Cruda).

- The establishment of the Labterra, the Centre of Studies and Research on Earthen Architecture, at the Department of Architecture of the University of Cagliari in 1997, coordinated by architect M. Achenza (http://www.terracruda.org/content/lab-terra);
- The institution of the Earthen Building Cultures course into the Architecture BSc at the University of Cagliari. The course, held by the architect M. Achenza, is studied as part of the UNESCO Chair "Earthen Architecture. Building cultures and sustainable development". The UNESCO Chairs on Earthen Architecture number 31 in the world and only 4 in Europe. The University of Cagliari gained the Chair in 2001 and from then, it fostered several teaching and awareness activities. (http://www.unesco.org/fr/university-twinning-and-networking/access-by-region/europeand-north-america/italy/).

In these years, the close collaboration between the Labterra and the local governments has led to recovering a remarkable part of both the tangible and intangible earthen heritage of the Campidano area, fostering the growth of a new local adobe producing SMEs. Although these are mostly related to the buildings' rehabilitation and preservation, the earthen products market highlights an upward trend also in relation to the new construction and correlated activities. Getting the master craftsmen started on reassessing the local construction techniques and reactivating the local manufacturing—maintenance system, showing clearly that the reinstatement actions on the building heritage would be able to bring about the recovery of the whole area. This chain reaction, strictly related to the place's potentiality, prepares the ground for that sustainable developing model, pointed out by scholars starting in the 1970s, who identified the local sources as the driving force for the economy and the development of the whole territory. The current limits to growth, resulting from the past consumistic-materialistic ideal based both on profit and the misleading belief of a limitless progress, demands an alternative sustainable modelable to replace the Market idol with the quality of human life, in balance with the surrounding environment (Mumford, 1970; Barbier, 1987; WCED, 1987).

A radical change well-expressed through the Mumford quotation: "The goal is not «more goods for more consumers», but «more life's possibility for more men»" (1970, p. 455).

During the last forty years, the advances in the sustainability issues overcame the building size in favour of the environment (Garofalo, 2003; Droge, 2006, Lehmann, 2008). In this broader scenario, the city is considered as a small part of a breakable ecosystem through which it feeds itself and develops (Rees, 1997; Scandurra, 1995).

This one-to-one relationship between the human environment and the local supplies, of which the history of the city highlights several significant

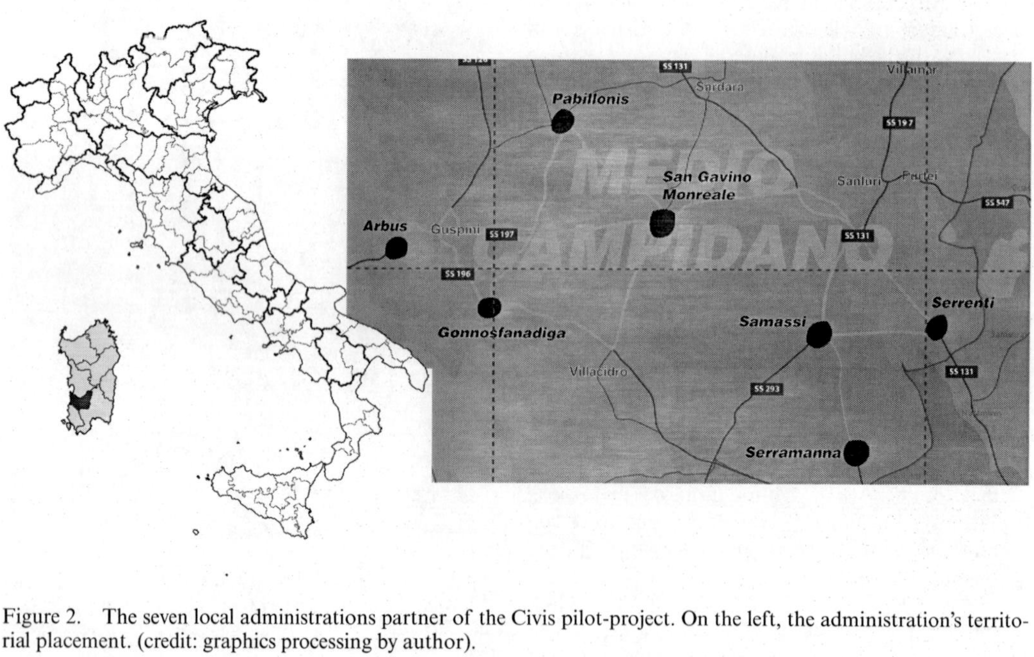

Figure 2. The seven local administrations partner of the Civis pilot-project. On the left, the administration's territorial placement. (credit: graphics processing by author).

examples (Mumford, 1967; Smil, 2000; Benevolo, 2006; Gallo, 1996), has led the Sardinian case study to identify in local sources the necessary challenge for its own economic and productive relaunch. In this light, the CIVIS pilot-project 'The 'earthen cities' itinerary' induces the Earth Towns network to look on the Campidano earth as the main strategy to combine the conservation and rehabilitation of its heritage with a new development policy, able to appraise the communities' traditional culture, and foster a new local 'green' economy.

2 CIVIS PILOT-PROJECT

The project CIVIS proposes a touristic itinerary through seven villages in the Campidano plain (central-south Sardinia). This network of villages, which is member of the National Earthen Cities Association since 2001, collaborates in the achievement of three specific goals:

- promoting the urban regeneration and the local sustainable development supporting the 'good building practices' through the rehabilitation and/or the new construction of earthen architectures underlining the 'sustainable' characteristics of earth and its deep relationship with the current need of sustainability in architecture.
- promoting a sustainable regional touristic development through the creation of cultural participated itineraries. This is joined to laboratories offered by local traditional craftsmen open to visitors.
- supporting a better understanding of technical and administrative cultures and promoting a higher social awareness about the use of earth.

The paper illustrates this pilot project as an example of how communities can find in their local cultural heritage the ideal tool through which they can build an innovative sustainable development model based just on local resources.

Difficulties of the Campidano area, mainly concerning the town's ongoing depopulation, had led seven local administrations to draw up the CIVIS pilot project through which a new future development centred on the earth could be built.

The project, started in 2006 but is still in progress, has become the load-bearing structure for the social, cultural and economic recovery of

Figure 3. Typical earthen house, Samassi, Sardinia—Italy, (credit: Associazione Nazionale Città della Terra Cruda).

Figure 4. Loggia of a typical earthen house, Pabillonis, Sardinia, Italy, (credit: Associazione Nazionale Città della Terra Cruda).

the territory, starting with the rehabilitation of the existing heritage. It organises itself around the achievement of three main specific goals which require the city's network cooperation. Those are:

- Promoting both the urban regeneration and the local sustainable development supporting the "building good practices" into the rehabilitation, and new construction of earthen architectures and public spaces. It gives specific prominence to the application of local, natural and traditional materials. This goal involves both the environmental-architectural quality of interventions and the "sustainable" features of the earth (healthiness, recycling, environmental friendly and energy-saving property, etc.). The latter in particular also responds to the current need of "sustainability", widely supported by the European policies during the last few years, with the aim to reduce the high energy and material consumptions conferred on the construction sector (Pèrez-Lombard et.al., 2008). In fact, the material and architectural features of earth respond promptly to the 3R policy-Reduce, Reuse and Recycle originated from the Kobe G8 Conference in 2008, through which this ancient material regains new values as up-to-date construction material, able to guarantee high levels of thermal comfort and sustainability.

This is not restricted to its physical characteristics, but involves the whole construction, maintenance and decommissioning building processes and culture. These "sustainable" issues, shared among the CIVIS's administrations, do not restrict themselves to the construction sector, but broaden to more natural and healthy living standards, well-integrated into the 'earthen' environment which involves all the local productive processes and techniques. In particular, among the project's practical goals, it aims to combine the earthen heritage with the handcrafted and agricultural-and-food chain in the territory, using the adobe buildings (existent or new) to exhibit and trade in branded local products.

- Promoting the development of sustainable tourism, through the creation of cultural-tourist itineraries. The latter are joined with urban and architectural requalification actions with the purpose of incentivizing and supporting the 'horizontal' or 'scattered' hotels on the territory. Specifically, the project achieves high-quality scattered hotels in these seven different towns, recovering the existing earthen structures or building new ones, in both cases according to the best sustainable standards as introduced previously.

Figure 5. Interior of a traditional earthen house, Gonnosfanadiga, Sardinia—Italy, (credit: Associazione Nazionale Città della Terra Cruda).

Figure 6. The oven for the bread, Serramanna, Sardinia—Italy, (credit: Associazione Nazionale Città della Terra Cruda).

Some representative traditional buildings will be recovered and used as domus amigas: these are "stations" and informational focal points of the earthen path.

These "containers" function as a territorial network of info points, laboratories and launch windows, where showing and selling handcrafted and food farming products. Furthermore, in a similar manner to the previous goal, the itinerary interlaces the town's earthen buildings (belonging to the traditional heritage or new ones) with the cultural assets located in the territory, such as local customs, traditional knowledge, wine-and-food traditions, archaeological sites and monuments, countryside festivals, etc. In this light, the former are responsible for the task of developing and enhancing the hospitality industry in support of local events and attractions. Both the hotel industry and the cultural assets have being mapped and catalogued in a database shared among the administrations. This mapping system is applied to each earthen building and collects, through a data-sheet, the following main information:

- the building's technical features and integrity;
- the building's history and peculiarities;
- the building owners' tendency, both in sticking to the itinerary and research activities;
- a final evaluation by the technical experts on the possible admission to the itinerary.

Finally, of the practical actions still in progress, it is planned to support tourism through physical and virtual products, as itinerary web site and guides, useful to the dissemination of the touristic and cultural information.

- Promoting both technical-administrative cultures and social perception links to the earth. The aim is the creation of real technical and administrative local cultures able to manage the earthen heritage and 'rehabilitate' people's traditional building culture, abandoned since the 1950s and 1960s and at present belonging to few. This goal is reached through the dissemination of knowledge as professional training courses, related to both the heritage conservation and the sustainable design.

Concerning these topics, the Labterra plays a referential role, having significant experiences both in educational and professional courses. Regarding the educational ones, the Labterra has carried out important measures to integrate and promote the earthen culture and techniques, at different levels of educational modules (such as university educational courses, national and international workshops, summer schools, lectures at secondary schools on construction, etc.), whereas in relation to the professionals, the formation course 'for experts in adobe production and

Figure 7. Typical blacksmith's workshop, Arbus, Sardinia—Italy, (credit: Associazione Nazionale Città della Terra Cruda).

Figure 8. The loggia onto the courtyard, Samassi, Sardinia—Italy, (credit: Associazione Nazionale Città della Terra Cruda).

construction with earthen elements' represents an important example of the Labterra's active involvement in the training of specialized professionals (Achenza, 2008).

3 CONCLUSION

In 2006, seven Sardinian local administrations presented the CIVIS pilot project, and won the regional selection, through which they envisaged a different territorial model centred on tangible and intangible local earthen heritage.

The main idea is based on the rehabilitation of the traditional earthen architectures and the earthen building culture, which has become the driving force to regenerate the whole area. This force does not restrict itself to the construction sector, but more broadly to more natural and healthy living standards, involving all the local productive process and techniques.

In fact, the "sustainable" features of the earth, both as construction material and building culture strictly related to the territorial sources, combine well with the natural, biological and sustainable local handcrafted and agricultural and-food productive chain, each becoming the launch window of the other. In this way, the project promotes, through the earthen heritage's recovery, a different developing model for the whole territory centred on sustainability (sustainable architecture, sustainable construction materials, sustainable productive and construction cultures, sustainable tourism, sustainable hotel industry, etc.) and the dissemination of knowledge. Concerning the latter, the project particularly endorses the local cultural assets, promoting touristic-cultural itineraries and activating different levels of educational-training courses, related both to heritage conservation and sustainable design.

Thanks to the close collaboration with the University of Cagliari (Labterra Research Centre), the local administrations and the SMEs, these initiatives will stimulate a new territorial economy.

REFERENCES

Achenza M., 2003, *Recent developments in building production and technology: the new season of a sustainable material*, The 9th Conference on the Study and Conservation of Earthen Architecture—Terra 2003, Iran.

Achenza M., edited by, 2004, *Terra 2000*, Atti del convegno—Serrenti 13–14 Dicembre 2002. Una normativa per la terra cruda—architettura sostenibile, EdicomEdizioni, Gorizia.

Barbier E.B., 1987, *The concept of sustainable economic development*, «Environmental Conservation», vol. 14, n.02, pp. 101–110.

Benevolo, 2006, *Storia della città*, Laterza, vol. I–IV, Roma.

Droge P., 2006, *The renewable city—A comprehensive guide to an urban revolution*, J. Wiley & Sons, England. Trad. It. Bossi G., Pozzi M.G., 2008, *La città rinnovabile. Guida completa ad una rivoluzione urbana*, Edizioni Ambiente, Milano.

Gallo C., 1996, *Passive cooling as design methodology: some examples from the past to the present, special issue on World Renewable Energy Congress Renewable Energy, Energy Efficiency and the Environment*, «Renewable Energy», vol. 8, issues 1—4, May–August, pp. 309–314.

Garofano I., 2003, *Le città per un costruire sostenibile.* Catalogo della mostra 2003, EdicoEdizioni, Gorizia.

Lehmann S., 2008, *Sustainability on urban scale: Green Urbanism—New models for urban growth and neighborhoods*, in Droge P., 2008, *Urban energy transition. From Fossil Fuels to Renewable Power*, Elsevier Science.

Lerner J., 2009, *Le città non sono problemi, sono soluzioni*, CityFutures. Architettura Design Tecnologia per il futuro della città, International Conference Proceedings, 4th–5th February, Milan.

Mumford L., 1961, *The City in History. Its Origins, its Transformations and its prospects*, New York. Trad. It. 1967, La città nella storia, Etas Kompass, Milano.

Mumford L., 1970, *The culture of cities*, Harcourt Brace & Companies, London.

Pèrez-Lombard L., Ortiz J., Pout C., 2008, *A review on building energy consumption information*, «Energy and Building», vol. 40, n. 3, pp. 394–398.

Sanna A., edited by, 2009, *Manuali del Recupero dei centri storici della Sardegna*, Vol I–V, Roma, Dei.

Smil V., 1994, *Energy in World History. Westview Press,* Boulder. Trad. it., 2000, Storia dell'energia, Il Mulino, Bologna.

Earthen Architecture: Past, Present and Future – Mileto, Vegas, García Soriano & Cristini (Eds)
© 2015 Taylor & Francis Group, London, ISBN 978-1-138-02711-4

Light Earth Designs: Natural material, natural structure

M.H. Ramage
Light Earth Designs, Cambridge, UK
Department of Architecture, Cambridge University, UK

T. Hall & P. Rich
Light Earth Designs, Cambridge, UK

ABSTRACT: Light Earth Designs is a partnership founded by Peter Rich, Tim Hall, and Michael Ramage. The practice brings together complementary backgrounds that each of them has developed in practice, research and academia to design locally sourced, community-based projects in Africa and around the world. Their built work can be found in Africa, Europe and North America, and has received numerous architectural and structural awards, including the 2009 World Building of the Year and a short-list nomination for the 2013 Aga Khan Award for Architecture. The present text explains its approach to sustainability in architecture with several built examples.

1 INTRODUCTION

Imagine arriving at a future construction site, surveying the materials available, and designing with what's at hand. This is the way we work. Wherever possible, we make use of local materials in construction and we use local knowledge, history and culture to inform our designs. Our designs and construction methods are often rooted in tradition, but updated for contemporary demands, so we work with craftsmen and craftswomen from each region where we build to realize our architectural designs. The process is always exciting, challenging, and rewarding; we learn a tremendous amount, and at the same time, teach new techniques to our construction teams.

2 SUSTAINABILITY

Sustainability is easy to say but much harder to achieve. The most sustainable designs are those that people cherish, and therefore look after, so sustainability, for our practice, is rooted in doing the best work possible, rather than achieving a specific metric or performance goal. Naturally, we strive to use materials with low embodied energy, and we make buildings that are designed to consume as little energy in their operation as is feasible. But even the most efficient building, in quantitative terms, can be inefficient overall if it is replaced, due to neglect, fashion change, or some other factor before the lifetime of its materials is up.

Sustainability is about much more than the energy performance of a building, or its economic costs. The lives of the people around the building, and those who make it and use it, are deeply embedded in our concept of a building that can sustain a community. As such, we work closely with builders, not just on the quality of their work, but also on the underlying skills and training. This approach is valuable in the long run, as we continue to return to the same group of builders, but we also see how they have trained others to work on new projects.

An example of the value of training and its ongoing input to our work starts with the Pines Calyx (2006) (Fig. 1), a conference centre in Dover, England, where the twin roof domes were designed and built by Michael Ramage in collaboration with John Ochsendorf and Wanda Lau of MIT. The overall design of the building, two intersecting circles in plan surmounted by 12 m span shallow masonry domes (Fig. 2), is by South African Architect Issy Benjamin and engineered by UK-based Philip Cooper. Our work was to design and build the masonry domes using the Mediterranean construction system of thin tile vaults (*bóvedas tabicadas*), an art of building that had flourished in Spain and the U.S. in the 19th and early 20th centuries, largely due to the influence and output of the construction firm of Rafael Guastavino. The construction system, also called "timbrel vaulting," "Catalan vaulting," and "Guastavino vaulting," is not in common use today, and had never been used in the UK, so we needed to teach

Figure 1. Pines Calyx, conference centre, Dover, England (M. Ramage).

Figure 2. Pines Calyx, interior, masonry dome (M. Ramage).

Figure 3. Pines Calyx, construction process layers of thin tiles (M. Ramage).

a group of bricklayers how to do it. Ramage had taught himself, so he knew enough to make a trial dome of 3 m span that convinced the contractor that the building was feasible. We also worked with Maximo Portal, a Spanish expert from Extremadura, to bring long-standing experience to the construction. Over the course of two weeks Maximo and his team taught a group of English bricklayers how the technique worked, using rapid-setting gypsum mortar and layers of thin tiles (Fig. 3), and completed the first of two domes (each was 100 m^2). We have subsequently worked with all members of that construction team on four more projects, and used them to train colleagues in London, Chicago and the remote South African bush. Our work ever since has strongly featured, or been informed by, the structural efficiency and construction systems inherent in thin-tile vaulting. Our projects each have a deep commitment to green design, qualitative and quantitative, which is integral to, rather than in addition to, the overall architectural design.

3 HISTORY

We draw on history not just for inspiration, but also for the knowledge of structure, construction, and design embedded in all of the great historic buildings, and many lesser-known ones. In architecture, constraints drive good design, and our forebears faced many more technological and material constraints than we have today. At King's College Chapel, Cambridge (Fig. 4), the 16th Century stonemasons and master craftsmen didn't have a theoretical way to understand their buildings, but their generations of tradition gave them an empirical knowledge unsurpassed even today. They knew that the forces of the structural fan vaults needed to stay within the geometry, and thus their form and the required structure are the same. This allowed them to build vaults with the proportions of an eggshell, which only a few hundred years later did we begin to understand theoretically.

Historical structural engineering based in good design and understanding, rather than electronic computation, resulted in efficient and expressive structures that we still hold up as icons today, such as Robert Maillart's Salginatobel Bridge in Switzerland (Fig. 5). Our practice relies on many of the same historical techniques used by Maillart, primarily analysis of structural forces with graphic statics.

Light Earth makes buildings that have great respect for the people of the world and our environment, drawing from the best lessons of the past to learn for the future. Through our reading of history and culture we come to an approach that acknowledges that simplicity is more difficult than complexity, and more valuable.

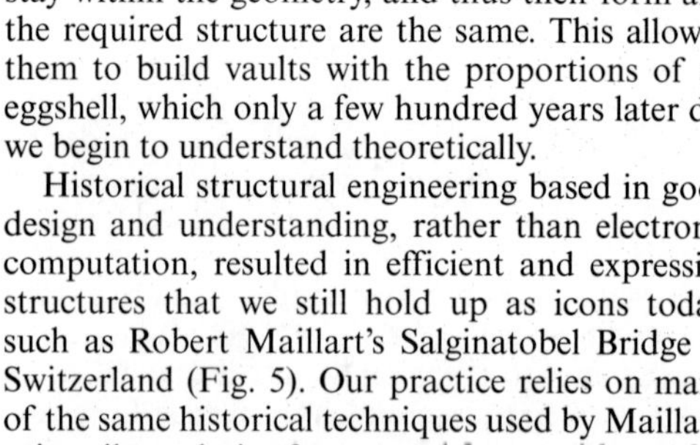

Figure 4. King's College Chapel, Cambridge (M. Ramage).

Figure 5. Robert Maillart's Salginatobel Bridge, Switzerland (M. Ramage).

Figure 6. Bridge at Cow Hollow, UK (M. Ramage).

Figure 7. Bronze Age willow fish trap excavated near the construction site (Must Farm).

Figure 8. Coppiced willow, fresh growth in spring (M. Ramage).

4 THE LIVING LINK

How do you stop a timber footbridge from decaying? At Cow Hollow we have planted a bridge by keeping it alive we increase its strength over time, and we prevent it from rotting (Fig. 6). Using coppiced willow from the surrounding plantation we have woven a willow arch and deck that has been planted into the ditch bank. The Living Link was designed and engineered by Michael Ramage, Smith and Wallwork Engineers and Patrick Fleming for the Woodland Trust, a UK charity. The region around Cambridge has a long history of using woven willow as a material, but this is the first living willow bridge that we know of. We drew inspiration from the living bridges of northeast India, and learned the weaving craft from local Cambridgeshire builders who can trace the tradition back thousands of years. Recently a Bronze Age willow fish trap was excavated near the construction site (Fig. 7).

Our approach to sustainability is straightforward: we've used only material from the site for the construction, and we've let nature provide the foundations, the preservation and the structure. There is no metal in the bridge, and the only materials imported are the smooth boards for the walking surface and the natural twine that ties the handrail. So the entire structure has essentially zero embodied energy, and over time it will even sequester CO_2. It's a small bridge, so that will be a small amount, but we're demonstrating that a little bit of lateral thinking with tradition and material can go a long way.

But more importantly, we've designed the bridge specifically tailored to the constraints of the material. We don't see the constraints as limitations, but rather opportunities to apply rigorous design and engineering to maximise the potential of the materials we are working with. Coppiced willow (Fig. 8) is not typically a structural material, as it is weak and rots quickly. But as with many of the materials we work with, when used thoughtfully, it is strong enough. The inherent weakness is a constraint that guides us toward an enduring design solution. Willow also has an amazing regenerative capacity, so a bare branch cut from a tree and stuck in the ground in winter is likely to be growing and green in the spring (Fig. 8).

The arch curves across the stream with members that are 45 mm in diameter, and also in the perpendicular direction across the bank. This double curvature, imported from our experience in shells, gives the bridge structure stiffness not otherwise inherent in thin willow (Fig. 9). Each pole is embedded a meter into the bank, ensuring that once it

Figure 9. Construction of the bridge structure (M. Ramage).

takes root, there will be sufficient root contact with the earth for the structure to take increasing load as it grows older. The main structural poles are then woven, like the basket, to hold them together and give added structure. The finished bridge has a handrail made of braided willow, with three strands planted at each end and intertwined in the middle. Over time, the fresh growth will be trained to strengthen and stiffen the handrail.

5 BUILDING LIGHT AND STRONG

The same principles of exploiting constraints for design and engineering advantage drive our work with thin-tile masonry and earth construction. A number of successful projects combine the two, first used in the Mapungubwe Interpretive Centre in South Africa, and subsequently for projects in London and Chicago.

5.1 *Mapungubwe*

The Mapungubwe Interpretive Centre (Figs. 10–11) in Limpopo Province, South Africa, designed by Peter Rich, Michael Ramage and John Ochsendorf, embodies our thinking on sustainability. We derive its design from cultural cues taken from the Venda culture, on which the equilateral triangular plan organization is based, the cairn route markers of southern African cultures, and stone masonry of nearby Great Zimbabwe, to name a few. Because of contested histories, these references are subtly intertwined and never overt, but guided the development of the architectural design.

Mapungubwe National Park is in far northern South Africa, at the confluence of the Shase and the Limpopo Rivers that define the borders with Botswana and Zimbabwe. It is remote and surrounded by poverty, and part of a UNESCO World Heritage Cultural Landscape. The South African

Figure 10. Mapungubwe Interpretive Centre, Limpopo Province, South Africa (P. Rich).

Figure 11. Construction of the Interpretive Centre (J. Bellamy).

Figure 12. Locally made tiles visible in the interior (I. Baan).

government, as part of the contract, included significant funds for poverty relief, but only if local unemployed people could be put to work. As such, we put together a design program that not only used a lot of labour in construction, but also used labour in the production of the building material from the earth of the site. In this way we put over two dozen people to work for two years making the 300,000 tiles needed for construction (Fig. 12). Because the tiles were made using a hand press, we simultaneously addressed community and environmental elements of sustainability.

The building took over a year to construct, with the vaults taking eight months themselves. During that time we trained over 80 bricklayers in the art of thin-tile construction (Fig. 13). This, in fact, was the most rewarding aspect of the project—teaching people to take pride in their work, and giving them a new skill. Because of the legacy of apartheid in South Africa, there was no history of craftsmanship among the black labourers; they were accustomed to completing a job without any expectation that they would be taught how to improve, and move up to greater skills and higher wages. But thin-tile vaults can't be built without good judgement, so we had to teach that judgement, which is part of becoming skilful.

The finished building uses local soil in the primary structure, and local stones to clad and protect the vaults, and to give a structural surcharge so they weigh more, which makes them more stable. Its materials fit the place exactly, helping the building to merge with its surrounding landscape. The vaults, with curvature in both directions, gives them stiffness and a structure that works in pure compression. This means that the geometry is optimal for the structure, just as in the vaults of Gothic Kings College Chapel. Because we designed the vaults to work in pure compression, they need no reinforcing steel to act against bending. This material savings adds tremendously to the environmental credentials of the building, and to the economic savings. During the construction of the building, at a time of construction boom in South Africa due to FIFA World Cup stadiums being built, and global demand for steel, the cost of steel rose over 200%. Mapungubwe was immune to that external factor, as the only steel is limited to the reinforced concrete supporting buttresses. If the vaults had had steel in them, the increased costs would have crippled the project.

5.2 *The Earth Pavilion*

The lack of reinforcement, however, means that the vaults aren't suitable for seismic areas, or where asymmetrical loads are significant, such as hurricane or cyclone areas. Unfortunately, much of the developing world coincides with one or both of those natural constraints, so we have been working to develop the building technique further for dynamic loads without using steel. We've turned to a polymer geogrid that is typically used in road construction, but in our experimental research performs admirably. We first used it with earth masonry on a built project for HRH Prince Charles's START Festival 2010, a celebration of sustainable living. Peter Rich, Tim Hall and Michael Ramage designed The Earth Pavilion, (Fig.14) our architecture submission, using the underlying principles and techniques from Mapungubwe, but updated from our experience there, and for the UK context.

The use of a geogrid allows us to use only two layers, further reducing material, for a total thickness of 50 mm. The geogrid gives the structure some ductility and bending capacity, without making it susceptible to weathering and rusting the way steel reinforcing would. Furthermore, the geogrid is light and relatively inexpensive; two men can carry a 300 m^2 roll of it. (The same would not be possible with steel.)

In London, we built with soil from a construction site in Barking that otherwise would have been waste. With the British history of exquisite brick masonry, we decided to make the inside and outside surfaces special, and worked with stonemason and brick layer Sarah Pennal, who had initially trained on thin-tile vault construction on the Pines Calyx five years earlier. We also had access to more sophisticated building tools, including computer numberical control (CNC) machines for making the formwork, which let us be more exuberant with the form (Fig. 15).

This project for Prince Charles incorporates our final element of sustainable thinking: aspiration. There is no benefit in building energy efficient, green, sustainable, bioclimatic buildings if no one wants them. Our earth buildings are made

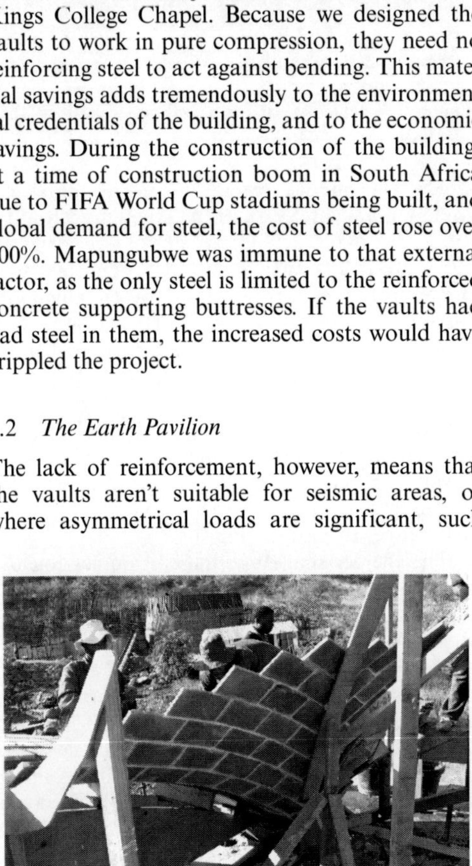

Figure 13. Training bricklayers (J. Bellamy).

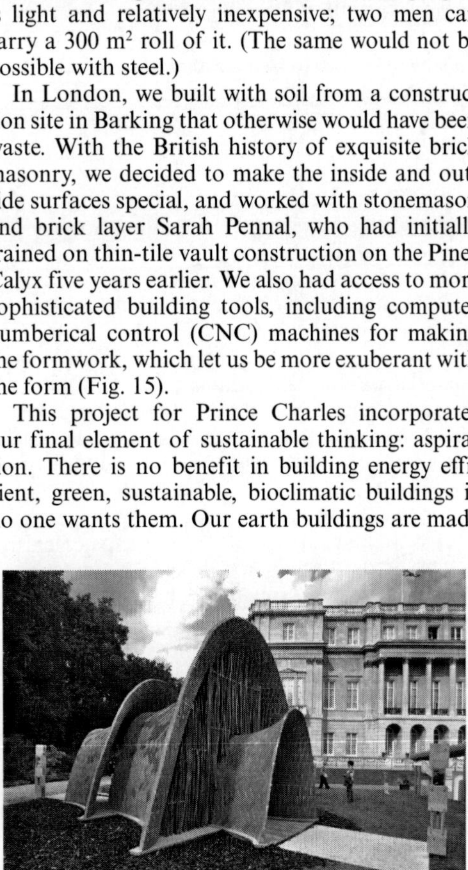

Figure 14. The Earth Pavilion, London (M. Ramage).

Figure 15. Formwork visible while building the Pavilion (M. Ramage).

from the ground, for the earth. But an uncharitable view could classify them as mud huts, something nobody wants, and everybody's ancestors at some point probably lived in. Thus we need to make people aspire to living in an earthen, efficient building, and that will happen only if they are seen as a luxury, something to aspire to. This doesn't mean they have to be expensive, or unreachable: it simply means they have to be special, and to be celebrated. That is what we aspire to.

5.3 *Chicago*

Light Earth Designs' most recent project is in Chicago, USA, where we have designed the interior of an office building for one of our clients. Our brief was very simple: the client wanted the architecture of their offices to highlight that they think and work differently (Fig. 16). As much as possible, we were to draw on nature. We limited the material palette, and primarily worked with timber, earth masonry, and steel. The timber is beetle-damaged wood that is otherwise a waste product (or source of energy, which isn't a good use of virgin wood), while the earth comes from the client's farm a few miles away.

The architecture creates a landscape of earthen vaults in the ceiling, which spread out over the floor and furniture made of beetle-kill pine. All the materials are bright, enhancing the natural light and limiting the amount of artificial lighting needed. This is further enhanced by a reflecting pool that increases the perceived height of the space as well as the light.

In order to build the 200 units of the ceiling with precision and minimal waste, we had to bring together a number of construction elements from previous projects, and to develop new ways of working. We've managed to maintain high craft with digital data underlying it all. The ceiling vaults we designed and printed as surfaces at 1:10 in 3D, from which we made a model at 1:3 in

Figure 16. Office building, Chicago (T. Hall).

Figure 17. Detail of the ceiling, repetitive units (T. Hall).

stone. This model, by Sarah Pennal, allowed us to define and understand the coursing. From this model we made a CAD model of every tile and every grout line, which we sent to Chicago to be milled in timber using a CNC router. This model served as positive formwork out of timber, on which we then cast multiple negatives in silicone. The silicone molds held individual tiles as we built the vaults. The tiles themselves vary slightly, as we mixed all the batches. This, plus the replication of craft construction in digital form, lends the vaults a hand crafted look while taking advantage of a manufacturing process that allowed us to make many repetitive units. (Fig. 17)

The themes that draw these examples together, and are present throughout our work, are efficient and intelligent use of materials, expressive structure, and construction based in historic techniques adapted for the 21st century. At Light Earth Designs we see sustainability rooted in good architectural and engineering design. Buildings which communities hold dear tend to have long lifespans, and where possible are sympathetically upgraded throughout their lives. We strive for simplicity and efficiency in all aspects of our work: architectural, structural, environmental and material. But our greatest challenge, and our fiercest goal, is to put people at the heart of our architecture, and to engage communities as the basis for a sustainable future.

Earthen Architecture: Past, Present and Future – Mileto, Vegas, García Soriano & Cristini (Eds)
 ISBN 978-1-138-02711-4

Experimental rammed earth prototypes in Mediterranean climate

L. Rincón, S. Serrano & L.F. Cabeza
GREA Innovació Concurrent, Universitat de Lleida, Lleida, Spain

B. González, A. Navarro & M. Bosch
GICITED, Universitat Politècnica de Catalunya, Barcelona, Spain

ABSTRACT: This paper presents two different rammed earth building prototypes, which have been constructed and monitored in Barcelona and Lleida (Spain). The experimental set-up located in Barcelona consists of one cubicle (2.48 × 2.15 × 2.50 m), with no load-bearing rammed earth walls (50 cm thickness) that has been manually rammed, comparing two types of stabilizers. It has a wooden load structure and two different wooden green roofs. The experimental set-up located in Lleida consists of two cubicles (2.4 × 2.4 × 2.4 m), with load-bearing rammed earth walls (29 cm thickness) that have been mechanically rammed, and wooden green roof. Both prototypes are used to show the construction possibilities of rammed earth. Furthermore, the soil characterization and the compressive strength of the rammed earth are tested at laboratory scale. Indoor and outdoor temperatures, humidity, and inner and outer surface wall temperatures were measured in order to study the thermal behavior of rammed earth walls.

1 INTRODUCTION

1.1 *Justification*

The building sector accounts for more than 40% of CO_2 emissions in Europe (Pérez-Lombard et al. 2008). The energy consumption for thermal comfort in buildings has grown recently because the requirements of the user's thermal comfort are stricter.

Today, the industrialized countries use construction materials with high embodied energy. Conventional architecture implies an important energy cost in all phases of the material life cycle, from manufacture, transportation, construction and disposal.

Historically, earth building has been an answer to the housing demand of population from all over the world. Today, it is still widely used in non-industrialized countries. The use of earth as a building material decreased globally due to the extensive use of concrete and masonry and its cultural linkage to poverty. Earth, as a building material, is locally available and abundant. Furthermore, an earth wall can contribute to the indoor building climate comfort providing high thermal inertia to deal with the day-night temperature changes in moderate, warm and even cold climates (Heathcote 2011).

New policies have been promoted all over the world to construct sustainable buildings and hence to reduce CO_2 emissions. In Spain, earth as a building material is not included in the current building regulations (Código Técnico de la Edificación), what hinders its recovery. On the other hand, the reduction of the building sector carbon emissions is mandatory in the European Union (European Commission 2002; IPCC 2007). Rammed earth is considered a very sustainable solution due to its low embodied energy, small materials treatment process, long service life, and high recyclability. For these reasons, rammed earth follows the European requirements. This fact increases the scientific interest of its recovery.

1.2 *Objectives*

The objective of this paper is to characterize the earth materials used in the construction of two different rammed earth prototypes using particle size and compressive strength test at laboratory scale. Furthermore, walls thermal performances of both cubicles are presented in order to compare the behavior of the different rammed earth wall compositions. The cubicles have been constructed and monitored in Barcelona and Lleida (Spain).

2 MATERIALS AND METHODOLOGY

2.1 *Rammed earth characterization*

The earth characterization is carried out with a particle size test and a compressive strength test to determine the properties of the earth used in the construction of both cubicles. Furthermore, the variability of these properties caused by the addition of additives is also studied.

2.1.1 *Materials*
The cubicle located in Barcelona was constructed with three different types of rammed earth (Figs. 1, 2): north-western and south-western walls have no additives, the northern wall is made of earth and 40% (in vol.) of expanded clay in order to improve the thermal behaviour, and the southern wall is made of earth with 3% (in vol.) of cement to improve the compressive strength of the wall. The composition of the earth is (in vol.): 71% of clay and 29% of sand, according to the earth available in the site excavation.

The cubicle located in Lleida was constructed with one type of rammed earth and it is composed by earth and 10% (in vol.) of straw. The earth composition is (in vol.): 38% of clay, 52% of sand, according to literature (Jiménez Delgado et al. 2007).

2.1.2 *Particle size test*
The Unified System of Soil Classification (USSC) developed by A. Casagrande is used to determine the particle size composition of the earth used in the construction of both cubicles following the standard UNE 103101:1995. In this test, the earth of both cubicles without additives is analysed. Moreover, both the earth with expanded clay and cement, and the earth with straw are analysed in order to evaluate the variability of the granulometric compositions of the earth.

2.1.3 *Compressive strength test*
The compressive strength of rammed earth was tested at laboratory scale with samples of 25 × 30 × 30 cm (four samples per type) and 30 × 30 × 30 cm (two samples per type) at Barcelona and Lleida, respectively. The compositions studied in Barcelona were: earth without stabilizer and earth with 40% of expanded clay (manually rammed). In Lleida, the compositions studied were: earth without stabiliser and earth with 10% of straw (mechanically rammed).

2.2 *Experimental set-up*

2.2.1 *Experimental set-up in Barcelona*
The experimental set-up located in Barcelona, Spain, is located at N 41°23'1,39", E 2°6'48,82", with an elevation of 9 m above sea level, and a Mediterranean central coast climate. The area corresponds to the climate Csa according to the Geiger climate classification (Kotterk et al. 2006). The annual number of heating degree days in Barcelona is 573 and the number of cooling degree days is 354 (Margarit i Roset 2003). The average summer and winter temperatures are 23.8°C and 6.1 °C, respectively, and the annual precipitation is 600 mm, according to the Catalonia Weather Forecast Service.

The dimensions of the cubicle are 2.48 × 2.15 × 2.50 m. It consists of no load-bearing rammed earth walls of 50 cm (manually rammed), a wooden load structure, and two different wooden green roofs (Fig. 1). The orientation of the cubicle is North-74°.

The set-up at Barcelona shows the temperature profile of three types of rammed earth walls: one wall without additives and two walls with two different types of additives (Figs. 2, 3). Cement as additive is added in order to test the improvement of the mechanical strength (3% in vol., placed in the south façade), and expanded clay in order to test the thermal behaviour (expanded clay

Figure 1. Experimental set-up in Barcelona, Spain (authors).

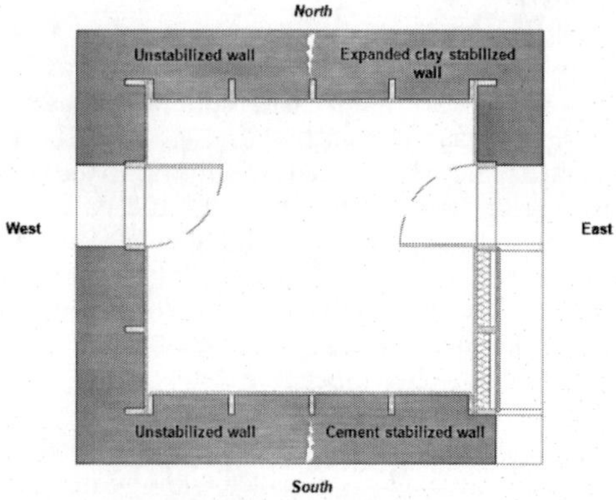

Figure 2. Plan of the cubicle in Barcelona (not to scale) (authors).

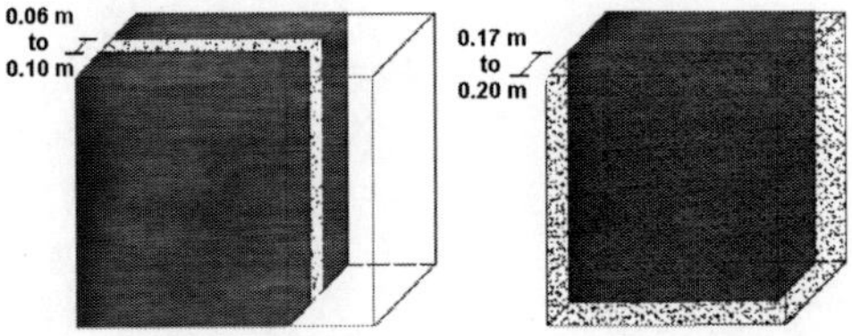

Figure 3. Scheme of rammed earth with expanded clay (left) and rammed earth with cement (right) (authors).

of 3–10 mm diameter, 40% in vol., placed in the northern facade).

The set up at Barcelona allows measuring wall temperatures using thermocouples type K located on the:

- Interior surface (north, south)
- Inside the wall (north, south)
- Exterior surface (north, south)

2.2.2 *Experimental set-up in Lleida*

The experimental set-up is located in Puigverd de Lleida, Spain, located at N 41° 32', E 0° 44', with an elevation of 219 m above sea level, and Mediterranean continental climate (Fig. 4). The area corresponds to the climate Csa/Cfa according to the Geiger climate classification (Kottek et al. 2006). The annual number of heating degree days in Puigverd de Lleida is 1,230 and the cooling degree days is 423 (Margarit i Roset 2003). The average summer and winter temperatures are 22.6°C and 8 °C, respectively. The annual precipitation is 456 mm.

The experimental set-up consists of two rammed earth wall cubicles (2.4 × 2.4 × 2.4 m), orientation North 0º, with 29 cm of load-bearing rammed earth walls, which has been mechanically rammed, and a wooden green roof (Fig. 5).

The set-up at Lleida allows measuring the thermal performance of the rammed earth cubicle. In order to evaluate the thermal performance of the cubicle the following data were registered:

- Interior surface wall temperature (east, west, north, south, ceiling, and floor)
- Inside wall temperature (3 equidistant points in the south wall)
- Exterior surface wall temperature (south)
- Indoor temperature and air humidity
- Solar radiation
- Outdoor temperature
- Wind speed

The constructive solutions were analyzed for summer and winter conditions, measuring free floating indoor temperature. All temperatures were measured using Pt-100 DIN B sensors, calibrated with a maximum error of ±0.3°C.

Figure 4. Experimental set-up in Puigverd de Lleida, Spain (authors).

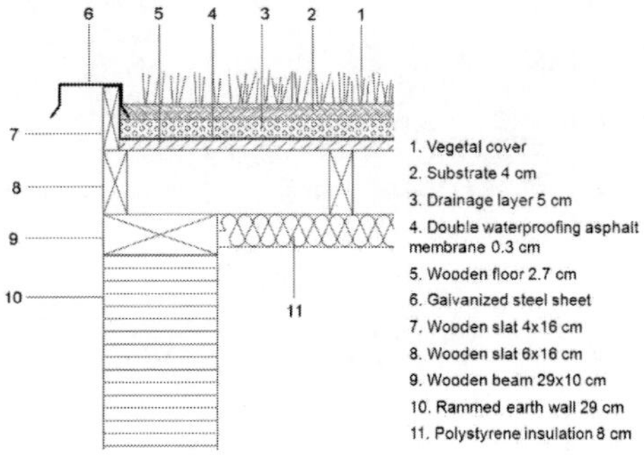

Figure 5. Constructive section facade-roof of the cubicle in Lleida (not to scale) (authors).

3 RESULTS

3.1 *Rammed earth characterization*

The granulometry of the soils used to build both cubicles is determined following the standard UNE 103101 (Figs. 6–8).

The earth of the cubicle located in Barcelona has been studied without stabiliser, with 3% of cement, and with 40% of expanded clay. The earth of the cubicle located in Lleida has been studied without stabilisers, the straw separately, and the earth stabilized with 10% straw.

According to the Unified System of Soil Classification Casagrande, the earth of the cubicle located in Barcelona corresponds to a cohesive soil of clay with medium plasticity. However, the earth of the cubicle located in Lleida has a granular soil of sand properly mixed with 6% clay. The difference in the properties of the earth is due to the difference between the origins of the soil, since the earth used in Barcelona came from the cubicle site and the earth used in Lleida was bought and mixed properly.

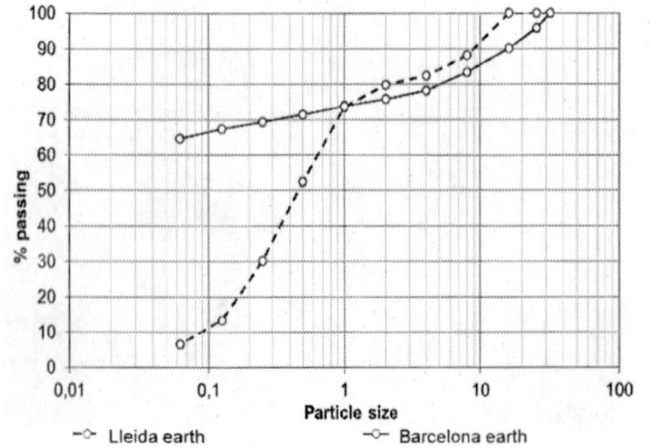

Figure 6. Granulometry of the soils used in Lleida and in Barcelona (authors).

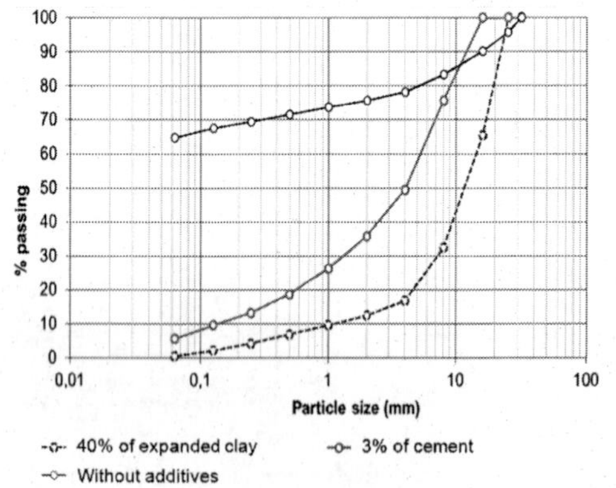

Figure 7. Granulometry of the soils used in Barcelona: earth without additives, with 40% of clay and with 3% of cement (authors).

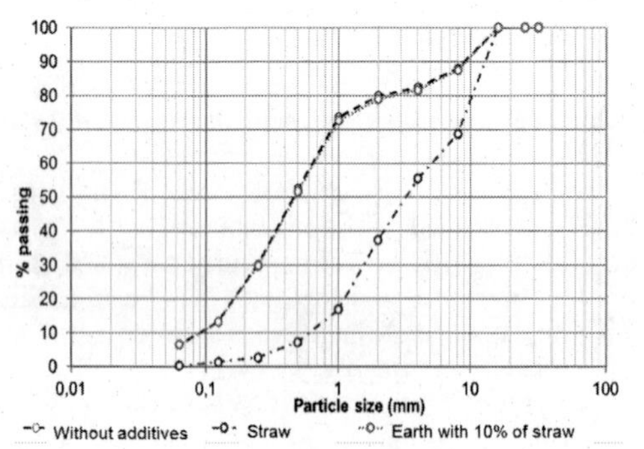

Figure 8. Granulometry of the soils used in Lleida: earth, straw, and earth with 10% of straw (authors).

The results obtained of compressive strength testing are shown in Table 1. The addition of 10% straw increases the compressive strength around 2 times while with the addition of 40% of expanded clay the compressive strength remains unchanged.

Table 1. Compressive strength results.

	Type of sample according to the stabilizer		Compressive Strength (N/mm²)* Manual Compaction	Mechanical Compaction
Lleida	Non-stabilised	Reference	–	0.944
	Physical stabilizer	Straw (10%)	–	1.794
Barcelona	Non-stabilised	Reference	1.076	–
	Physical stabilizer	Expanded clay (18.52%)	1.065	–

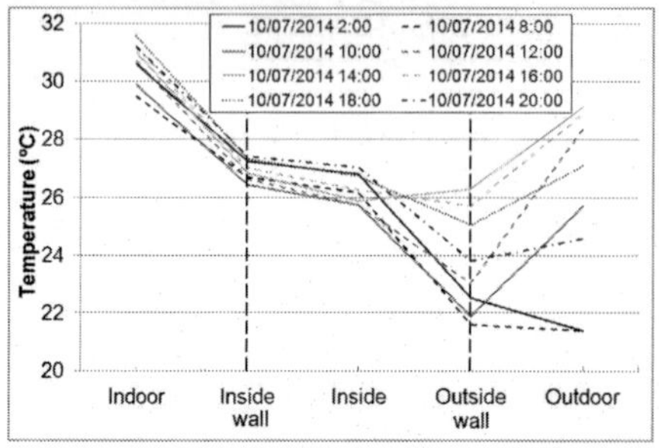

Figure 9. Temperatures of the south wall in summer conditions (July 10th, 2013) (authors).

3.2 *Temperature profiles of Barcelona rammed earth walls*

The temperature profile in free floating through the south rammed earth wall is shown for one representative day in winter and summer in Figure 9 and Figure 10, respectively.

The behavior of the inner surface is very similar in both summer and winter. The tendency is to unify the outside thermal amplitude, transmitting a constant temperature inside the prototype. Both summer and winter, large temperature variations are not seen in the inner surface of the wall. However, throughout the day the outer surface of the wall denotes a differential of 5°C in summer and 1°C in winter. In winter, the interior surface wall temperature remains between 11–16°C, while in summer between 21–28°C.

3.3 *Thermal behavior of Lleida rammed earth cubicle*

The indoor and outdoor temperature profiles of figures 11 and 12 show that a thermal stability inside the rammed earth cubicle is achieved both in winter and in summer conditions. The indoor temperature fluctuation was about 2–3°C while the outdoor could reach an amplitude of 18°C.

This indoor thermal stability is also visible in the temperature profile through the south rammed

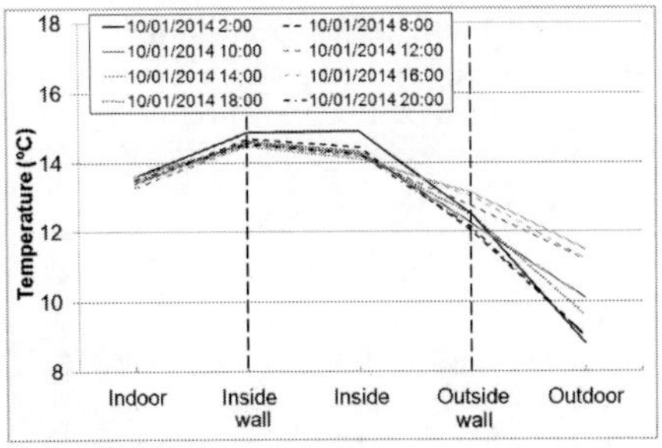

Figure 10. Temperatures of the south wall in winter conditions (January 10th, 2014) (authors).

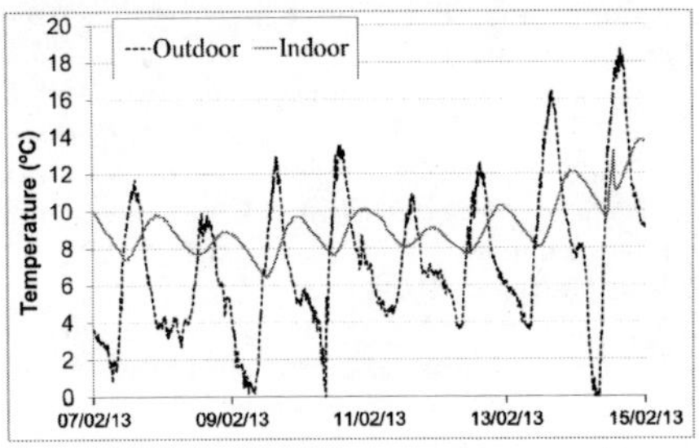

Figure 11. Temperature profile from February 7th to 15th, 2013 (winter conditions) (authors).

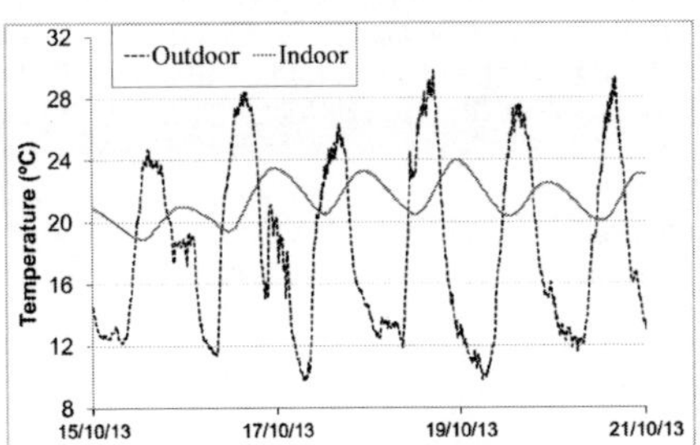

Figure 12. Temperature profile from October 15th to 21th, 2013 (summer conditions) (authors).

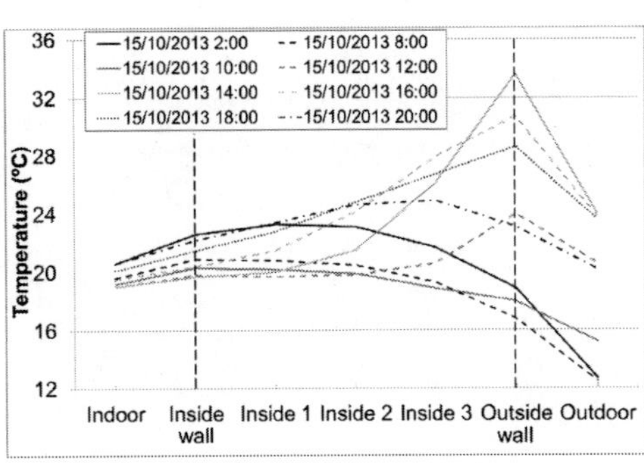

Figure 13. Temperatures of the south wall in summer conditions (October 15th, 2013) (authors).

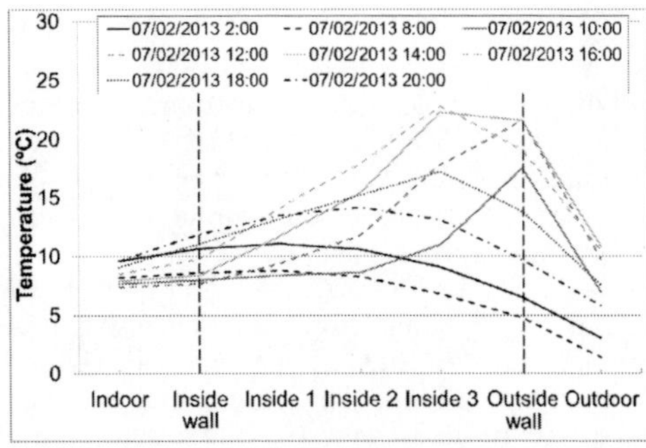

Figure 14. Temperatures of the south wall in winter conditions (February 7th, 2013) (authors).

earth wall for one representative day in winter (Fig. 13) and summer (Fig. 14). Both profiles show the decrease in the thermal amplitude from outdoors to indoor through the wall. Nevertheless, the effect of the roof and floor in the indoor temperature fluctuation cannot be neglected.

Whereas in summer the indoor temperature was within the range comfort, in winter the severe low temperatures did not allowed reaching a comfort indoor temperature with no heating system. The results show that the average indoor temperature could be higher in sever summer conditions.

4 CONCLUSIONS

The results of compression strength test and temperature profiles are comparable even though two different types of rammed earth have been used in Barcelona and Lleida. Both earths were characterized according to the Unified System of Soil Classification (USSC) developed by A. Casagrande showing differences in their properties due to its different origin.

Compressive strength test have shown that, despite the use of different materials, the behavior is similar. Compressive strength depends mostly on the dosage of stabilizers and the construction of the wall.

It has been shown that the thermal performance of rammed earth walls depends on different factors. When both prototypes are compared, it can be observed that the thickness of the wall is critical. As expected, the benefits of rammed earth thermal mass are visible by using 50 cm of thickness showing constant temperature profiles inside the wall. However, the benefits of rammed earth thermal mass are less evident with 29 cm thickness. Since modern earth buildings demand thinner walls than traditional earth buildings, passive design is needed. The incorporation of an insulation material attached to the outer side of the wall, and the solar gains due to the opening of a window in the south wall would provide higher indoor temperatures in winter conditions. The night cooling and solar protection elements would provide lower indoor temperatures in summer conditions.

Furthermore, the thermal amplitude between day and night outer temperatures is also a determining factor of the thermal mass behavior of rammed earth walls. The continental climate in Lleida has more extreme thermal amplitude than the Mediterranean climate in Barcelona.

Theoretical transmittance, decrement factor and thermal lag are required to obtain a proper thermal characterization of the rammed earth wall. Moreover, other parameters should be considered, such as absorptance of the outer facade and roof against direct solar radiation, especially in summer conditions, and the natural cooling capacity of the outer surfaces of roofs and facades during the summer night.

ACKNOWLEDGMENTS

The work was partially funded by the Spanish government (ENE2011-28269-C03-02 and ENE2011-28269-C03-03), in collaboration with the City Hall of Puigverd de Lleida. The authors would like to thank the Catalan Government for the quality accreditation given to their research group (2014 SGR 123). The research leading to these results has received funding from the European Union's Seventh Framework Programme (FP7/2007-2013) under grant agreement n° PIRSES-GA-2013-610692 (INNOSTORAGE). Barcelona cubicle has been conducted under the direction of the company Casa S-Low, with the collaboration of Luis Allepuz y Cristian Poza in their final degree project at EPSEB. UPC.

REFERENCES

European Commission 2002. *Directive 2002/91/EC of the European Parliament and of the Council of 16th December 2002 on the energy performance of buildings, OJEC, L1/65, 04/01/2003.*

Heathcote, K. 2011. El comportamiento térmico de los edificios de tierra. *Informes de la Construcción* 63(523): 117–126.

Intergovernmental Panel on Climate Change (IPCC) 2007. Summary for policymakers in climate change, Mitigation. In: Metz B, Davidson OR, Bosch PR, Dave R, Meyer. *Contribution of Working Group III to the Fourth Assessment Report of the Intergovernmental Panel on Climate Change* LA, (ed), Cambridge: United Kingdom and New York, NY (USA): Cambridge University Press.

Jiménez Delgado, MC. et al. 2007. The selection of soils for unstabilised earth building: A normative review. *Construction and Building Materials* 21:237–251.

Kottek, M. et al. 2006. World Map of the Köppen-Geiger climate classification updated. *Meteorologische Zeitschrift* 15(3): 259–263.

Margarit i Roset, J. 2003. *Els graus-dia de calefacció i refrigeració de Catalunya: resultats a nivell municipal. no14*. Barcelona: Generalitat de Catalunya–ICAEN.

McDonough, W., Braungart, M. 2002. *Remaking the way we make things: Cradle to cradle*. New York: North Point Press.

Pérez-Lombard et al. 2008. A review on buildings energy consumption information. *Energy and Buildings* 40: 394–8.

Earthen Architecture: Past, Present and Future – Mileto, Vegas, García Soriano & Cristini (Eds)
© 2015 Taylor & Francis Group, London, ISBN 978-1-138-02711-4

Inventory of Outat valley *ksour* in Morocco

P. Rodríguez-Navarro & T. Gil Piqueras
Universitat Politècnica de València, Valencia, Spain

ABSTRACT: This writing presents a preview of the results of the I+D Project (20110499 Cód. ADSIDEO-2011 y NR 1926 PAID-06-11-2011—Universitat Politècnica de València) carried out in the Midelt province in Morocco between 2011-2013. Research began with the search for graphic sources and documentation as references to earthen cities (*ksar* sing., *ksour* pl.) in the valley. We were able to verify that there were no documents that gathered—as an inventory—all the *ksour*, apart from the records made by Charles de Foucauld after his trip to Morocco, which was reflected in his work *Reconnoissance au Maroc* (1883-84). This writing shows the results of the inventory of *ksour* in the Outat valley, where for the first time all their names are collected, as well as their geographical, architectural and urban information, which will serve as a basis for further detailed analysis of each one of them.

1 INTRODUCTION

Throughout this research project, whose partial results we present here, a constant work of recording and inventory was carried out, of all the cities in the land of the Outat valley. The first references to these constructions were provided by Foucauld (Foucauld 1888) in the first edition of his book *Reconnaissance au Maroc* 1883–1884, but unfortunately in the following editions this chapter disappears, plunging this architecture into profound oblivion and beyond any study, until the development of this research project. This project started in 2008. In our first expedition to the High Atlas we located an important group of *ksour* in the Outat valley, in the Midelt province. After exhaustive bibliographical search we found some historical and geographical references to this place (Raynal 1961) (Peyron 1976 and 1984) (Aouchar 1988–89), due to the fact that this valley had been the refuge of the Ait Izdeg, one of the most politically relevant tribes, as well as one of the most belligerent, in the Oriental High Atlas throughout the nineteenth century. Similarly, we also observed that there were no comprehensive studies about its architecture: only Foucauld, and to a lesser extent Seconzac (Seconzac 1903), had come across its existence, always viewing it according to the geographical and military interests of the French people aiming to control Morocco. The references taken into account in our research served as a basis of study, although through our own fieldwork we found that not all the data gathered in them ended up being reliable or even complete. Despite this, we do recognize their work to be praiseworthy as the means with which those challenges were faced were not ideal: travelling was done on horseback or on foot, mapping was non-existent and general means were minimal. These works indicate that the route taken was the river, following the course of the water, meaning that all the *ksour* outside this route remained unrecorded. However, Seconzac's study collects data on population, the river's course and the name of two *ksour*, Berrum and Ait Ouafella; de Foucauld's study, much broader registers the name of 12 *ksour* (although it lists 16 *ksour*), hence it being the set of records we chose as the main reference for this research. The fact that there were not many references to the Outat valley and the fact that it lies in a relatively inaccessible place, as it goes into the mountains along a narrow channel until it reaches the bottom of Mount Ayachi (3,747 m), are perhaps the reasons why it had remained unnoticed; even today and despite Midelt's geographical location within Morocco makes it an obligatory stopping point for tourists heading south through the country's inland regions, it is hard to see them because access is achieved via dirt roads far away from conventional highways. Right from our first expedition, and throughout the subsequent ones, our intention was to locate, visit, record and study all these *ksour* with the goal of drafting a complete inventory of the valley, in order to give this architecture the importance it deserves within the national map or at least to document it due to its state of deterioration.

2 FIRST SETTLEMENTS IN THE OUTAT VALLEY (19TH–20TH CENTURIES)

The dangers posed, even before the sixteenth century, to the Tizi n'Talghoumt passage, along which the important trade route linking Fez with Sijil-

Figure 1. Ksar Tabenaâtout. Outat valley, Midelt. (P. Rodríguez-Navarro).

Table 1. Right side of the Outat river:

KSAR'S name (sic)	Number of rifles
Tatteouin	60 rifles
Afelilou (2 *ksour*)	150 rifles
Tissouit	20 rifles
Asselim (2 *ksour*)	150 rifles
Ait ou Afella (3 *ksour*)	100 rifles
Ikher Imzioun	60 rifles

Table 2. Left side of the Outat river:

KSAR'S name (sic)	Number of rifles
Berrom	100 rifles
Tabnattout	50 rifles
Semmoura	60 rifles
Bou Zmella	60 rifles
Ai Otman ou Moussa	150 rifles
Teççaouit	100 rifles

massa ran, makes the Tizi n'Tssardount passage, further west, acquire great importance among local tribes, who seeking refuge in the mountains, move across it in search for new opportunities.

Thus, from 1810 and after the battle of Dar Eddel against the Ait Idrassen, the Ait Izdeg arrive in Outat seizing power in the region.

In the early nineteenth century sedentarization starts in the Outat valley, becoming inhabited by two major tribes, the Ait Izdeg and the Ait Ouafella, who will gradually be joined by others like the Ait Hdiddou, the Iguerrouane, the Ait Seghrouchen, the Ait Merghad and some Jewish Berber groups.

3 FIRST RECORDING ATTEMPTS

As mentioned earlier, the first *ksour* records were made by the priest Charles de Foucauld between 1882–1886 throughout his journey around Morocco.

In these records, in addition to citing the name of *ksour* according to the riverside where they were found, he referred to their demographics, taking the number of rifles as a unit of measure.

4 OUTAT VALLEY INVENTORY SUMMARY

Below is a summary of the inventory of *ksour* in the Outat valley:

1 *Taddamout*. 32°42′15,51″ N; 4°46′47,83″ W
Located within the fertile region, this *ksar* probably had, originally, a square shaped floor plan, although it currently only preserves half of its structure, sustaining two towers that face west. Its entrance lies to the north. This *ksar's* usable area is 1.109 ². Today it is inhabited and in a state of semi-ruin.

2 *Tajilalit*. 32°41′54,58″ N; 4°46′47,97″ W
With an area of 690 m², this *ksar* has a square-shaped floor plan, topped by four towers in its corners and has its access situated on the eastern side. The inner layout consists of a central street from which access is given to the housing. It currently is in a state of ruin.

3 *Ait Ali Oulhsan*. 32°42′06,35″ N; 4°46′45,66″ W
This *ksar* covers an area of 1.308 m² and has a square shaped floor plan and towers in each corner. Its entrance is situated on the east side, leading to the main street from which access is provided to the housing. Its location corresponds to the outer limit of the fertile area.

4 *Tachaouit*. 32°41′46″ N; 4°44′49″ W
This ksar has an organic floor plan, following an almond shaped layout. Its entrance is located on the north side, where we also find the orchard. It has a surface area of 6.200 m². Its location situates it within the fertile area. Today it is restored and inhabited.

5 *Iguerrouane*. 32°41′11,74″ N; 4°46′34,23″ W
The Iguerrouane *ksar* has a square floor plan, accessible through an angled entrance on its eastern side. A new population center has developed next to its southern facade. This *ksar*, with its ground floor and two floors at its highest point, is limited by towers in its corners, which heighten the set one more level. Many of them still retain the decoration in their upper part. Its total area

is 5.400 m². It is partially in ruins and semi-inhabited.

6 *Ait Sidi L'Hbib*. 32°41′07,14″ N; 4°46′37,89″ W
It currently only preserves some of its outer walls, corner towers and a few inner walls. The entrance seems to coincide with a tower situated in the central area of its western wall. Its total area is 1.990 m². Today it does not preserve any flooring or covering. It is in a very advanced state of ruin.

7 *Ikhramjiouine* (Midelt)
Currently this *ksar* has disappeared and there are no visible remains to verify its original location.

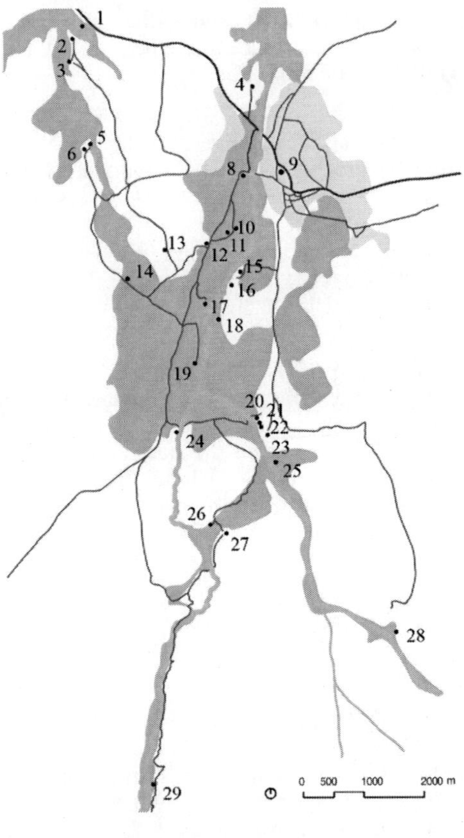

Figure 2. Location map of the *Ksour* (Outat valley). 1 Taddamout. 2 Tajilalit. 3 Ait Ali Oulhsan. 4 Tachaouit. 5 Iguerrouane. 6 Ait Sidi L'Hbib. 7 Ikhramjiouine (desaparecido). 8 Otmane ou Moussa. 9 Tachiouine. 10 Taâkit. 11 Tamoussa ou Ali. 12 Bouzmellah. 13 Al Zaouiat. 14 Assaka. 15 Ait Ouafella. 16 Ait L'Kaid. 17 Smoura. 18 Ugerja. 19 Tabenaâtout. 20 Asselim Ait Echo. 21 Asselim Ait Said. 22 Asselim Ait Amo. 23 Ibzazn. 24 Berrum. 25 Tissouit Sidi Hamza. 26 Flililou Tahinoust. 27 Flililou Taffraout. 28 Tissouit Ait Seghrouchen. 29 Tatiouine. (T. Gil Piqueras & P. Rodríguez-Navarro).

However and according to oral sources, its location was supposed to be the Midelt's municipal market lies today.

8 *Otmane ou Moussa*. 32°40′53,73″N; 4°44′56,6″ W
With its 16.811 m² this is the largest *ksar* in all of Outat. Its floor plan has an organic circular shape and its interior has a labyrinthine structure with narrow streets that sometimes run for a long distance, barely letting external light in. At its highest point we can observe a ground floor and two upper floors. It has no towers. Following the intervention carried out between 2008 and 2010 some small patches of swelling appear in its layout, as well as new entrances. Today it is in a good state of preservation, being inhabited almost in its entirety.

9 *Tachiouine*. 32°40′17,32″ N; 4°44′02,53″ W
Today it is very difficult to recognize this *ksar* as new constructions have absorbed it completely. According to what we can see, its area is approximately 924 m², presenting a square, shaped floor plan with towers in its corners and only one entrance to the north.

10 *Taâkit*. 32°40′24,57″ N; 4°45′05,96″ W
Currently, its geometrical floor plan has been greatly altered due to the annexation of new housing. This *ksar*, with an area of 3.451 m², features two main streets which start from the original entrance located to the southeast, dividing its inner structure into two neighbourhoods. Today the towers on its walls are not visible.

11 *Tamoussa ou Ali*. 32°40′26,80″ N; 4°44′59,89″ W
The Tamoussa ou Ali *ksar* occupies an area of 3.425 m² and exhibits a geometrical floor plan. Its proximity to the city of Midelt means that this *ksar* has been altered by the annexation of new housing, making it difficult to study its original morphology. Its entrance is of a direct kind and is located to the east.

12 *Bouzmellah*. 32°40′18,72″ N; 4°45′19,86″ W
This *ksar*, of Jewish origin, is located to the south of Midelt and has an approximate area of 7.446 m². Its geometrical floor plan is rectangular and its inner layout is orthogonal. It is limited by slender towers, or what remains of them, which are repeated midway through the outer walls. Originally, there were two towers flanking the main entrance located at the center of the eastern façade. Today these towers have disappeared and those that remain have lost at least one floor. There are also three new entrances, apart from the original one. It is largely inhabited, although it is in a state of semi-ruin. A mosque has recently been built in its eastern facade, breaking its continuity, thus altering its appearance.

13 *Al Zaouiat*. 32°40′14,91″ N; 4°45′46,63″ W
This *ksar* exhibits a geometrical floor plan that is square shaped and displays a regular urban layout. Its area is 2.471 m², although its rear part located to the west, has a similar sized walled enclosure, which illustrates a possible extension attempt of the *ksar*. Its entrance is located on the east side, leading to the main street from which new perpendicular, narrower, streets emerge.

14 *Assaka*. 32°40′00″ N; 4°46′07″ W
The Assaka *ksar* occupies an area of 3.590 m² and possesses a regular structure which is accessed by a door in its eastern facade. The height of the current *ksar* comprises a ground floor and one upper floor, showing in places signs of an extra upper floor. Inside, the layout of its streets is still clearly visible today. Of all these streets one that stands out actually emerges from the very entrance, independently from the rest, possibly made for the private use of a family of higher social status.

The outer wall is limited in its corners by towers made with earthen walls and topped with adobe.

Lowly populated, it only preserves the houses that are closest to the entrance area. The rest of the population can be found in the surrounding areas.

15 *Ait Ouafella*. 32°40′03,45″ N; 4°44′57,86″ W
Covering an area of 8.707 m², the Air Ouafella *ksar* displays an organic almond shaped floor plan, set on a raised level that runs into the fertile area south of Midelt. With the exception of one that reaches two upper floors, the *ksar's* housing generally consists of a ground floor and one upper floor. The inner layout stretches along a central longitudinal axis from which side streets emerge, not facing each other.

Presently, all the streets are covered and it is unknown if this was the case originally, as there is no distinctive evidence. The angled main entrance is located on the north side, although another access has been opened to the south. It is in a state of advanced ruin.

16 *Ait L'Kaid*. 32°39′57,52″ N; 4°45′03,53″ W
The geometric floor plan of this *ksar* is rectangular, due to the enlargement of the square shaped primal *ksar* towards the north. Its total area is 3.124 m². It has two angled, north facing entrances. The first leads to the main street, up to where this street bends and a door there leads to the second entrance, also angled, which allows access to the old *ksar*, which in turn is completely ruined.

17 *Smoura*. 32°39′46,86″ N; 4°45′20,57″ W
This ksar occupies 1.184 m² and its floor plan is square shaped with towers in the corners. It is located on a hill bordered by the banks of the Outat River to the east and by the extensive network of canals that supply water to the western side of the valley. Its entrance, topped by two towers, is situated on the eastern façade. To the north there are remains of what could have been a barn. It has a mill next to the river at the bottom of the entrance. The cemetery is located to the northwest. The interior lies in ruins. Its layout is organized originating from a central street that gives way to housing and some small streets.

The *ksar* was abandoned after French occupation due to the death of its owner. It was closed in 2009. Currently, it can be accessed via the southeastern tower, which has collapsed, and through a hole that has been opened by smashing in its northern facade.

18 *Ugerja*. 32°39′35,02″ N; 4°45′14,39″ W
At present this *ksar* has undergone major transformation as various houses have been added, invading its 1.389 m² area.

19 *Tabenaâtout*. 32°39′15,66″ N; 4°45′27,65″ W
The Tabenaâtout *ksar's* rectangular floor plan is accessible from its angled eastern side entrance. Its interior is formed by a set of streets that are covered, except at junctions, from which as from a sort of well, the *ksar* gets external light and ventilation.

It covers a 3.024 m² area and has an ample space intended for the orchard in its western part.

20 *Asselim Ait Echo*. 32°38′46,92″ N; 4°44′48,97″ W
Originally, this *ksar* is supposed to have had a geometrically shaped floor plan, although today that is difficult to see given the state of ruin it is in. It occupies an area of 1.636 m². The entrance is located to the north and its inner urban layout is organized around a main road that starts right at the entrance.

21 *Asselim Ait Said*. 32°38′43,93″ N; 4°44′46,52″ W
This *ksar* has a surface area of 1.170 m². Access is provided through a door located to the north. It is in a state of semi-ruin.

22 *Asselim Ait Amo*. 32°38′42,12″ N; 4°44′45,98″ W
This 4.288 m² structure has two entrance doors at both ends of the main street. This street divides the *ksar* into two clearly different parts. On both sides of the street we see doors that lead to new, narrower and covered streets, from which access is granted to the housing.

23 *Ibzazn*. 32°38′37,72″ N; 4°44′41,82″ W
Covering an area of 1.207 m², this *ksar* has a square shaped floor plan with towers in the corners and to both sides of the entrance door. The inner layout is based around one single street, from which access is given to the housing. It lies abandoned and in a state of semi-ruin.

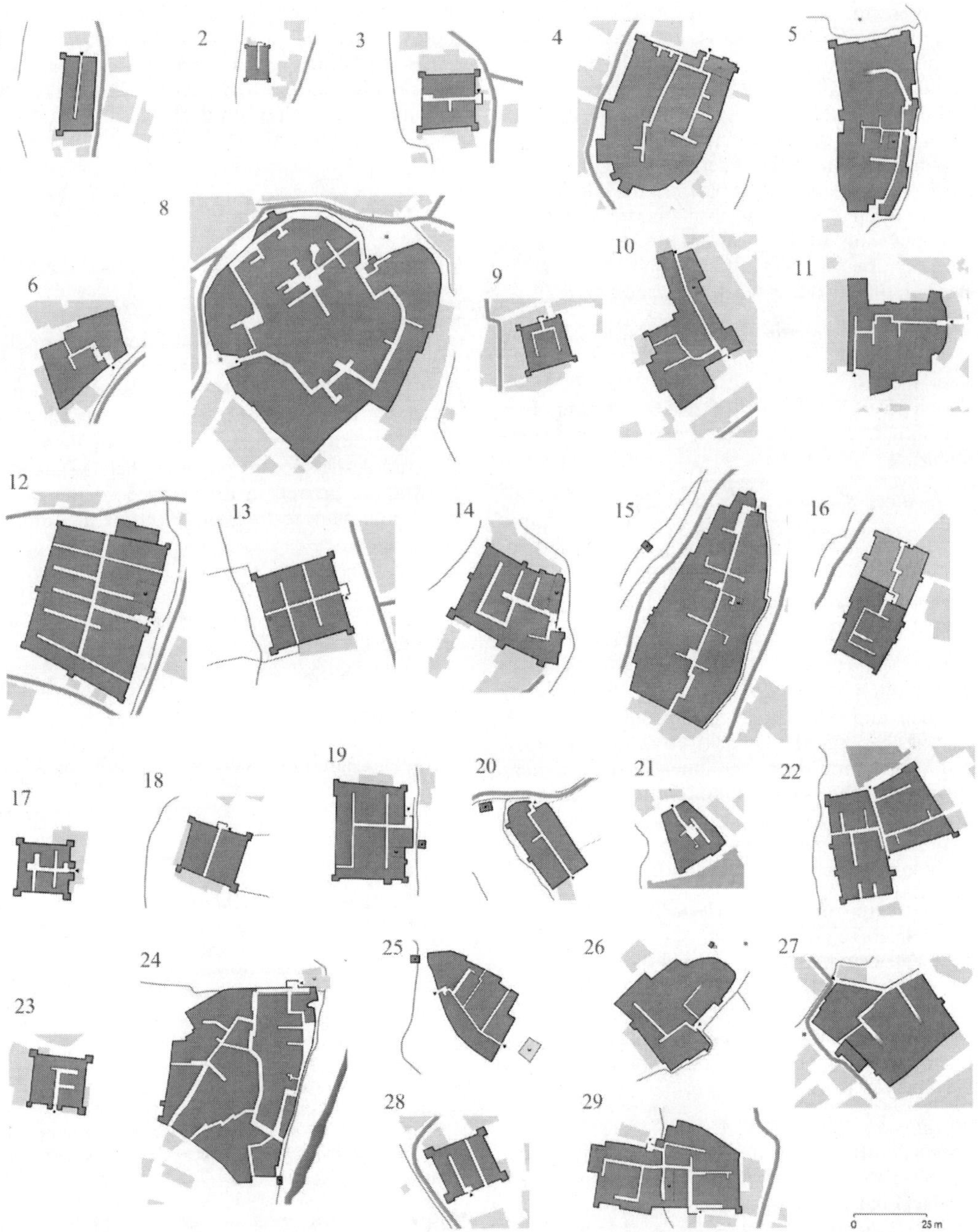

Figure 3. Plants *Ksour* Outat valley. 1 Taddamout. 2 Tajilalit. 3 Ait Ali Oulhsan. 4 Tachaouit. 5 Iguerrouane. 6 Ait Sidi L'Hbib. 7 Ikhramjiouine (desaparecido). 8 Otmane ou Moussa. 9 Tachiouine. 10 Taâkit. 11 Tamoussa ou Ali. 12 Bouzmellah. 13 Al Zaouiat. 14 Assaka. 15 Ait Ouafella. 16 Ait L'Kaid. 17 Smoura. 18 Ugerja. 19 Tabenaâtout. 20 Asselim Ait Echo. 21 Asselim Ait Said. 22 Asselim Ait Amo. 23 Ibzazn. 24 Berrum. 25 Tissouit Sidi Hamza. 26 Flililou Tahinoust. 27 Flililou Taffraout. 28 Tissouit Ait Seghrouchen. 29 Tatiouine. (P. Rodríguez-Navarro & T. Gil Piqueras).

24 *Berrum*. 32°38′38,19″ N; 4°45′40,72″ W
The Berrum *ksar* occupies an area of 9.593 m² and has an irregular structure. Its remarkable geographical location guarantees it a high level of protection, especially for the side facing the river.

It has various entrances, all of them direct, but its main entrance door is situated to the north. It has several mills. The ditch bordering the *ksar* to the east turns north, becoming visible from the interior.

Various paths emerge from this *ksar*, like the one that leads to the great limestone gorge where caves can be seen, caves that usually house shepherds and their flocks when the weather is cold.

25 *Tissouit Sidi Hamza*. 32°38′23,42″ N; 4°44′36,9″ W
Organically shaped *ksar* with a 2.471 m² surface area, located in the valley, right beside the Mimachoun river, a tributary of the Outat river. It is currently in ruins, being used only as a barn and stable.

26 *Flililou Tahinoust*. 32°37′50,41″ N; 4°45′18,1″ W
This *ksar* has a surface area of 3.328 m² and is completely abandoned. The state of ruin it is in makes it impossible to analyse its inner layout. Access is gained through an angled entrance to the east of the *ksar*.

27 *Flililou Taffraout*. 32°37′45,94″ N; 4°45′08,60″ W
The Flililou Taffraout *ksar* covers an area of 4.430 m² and is located at the bottom of the valley that lies between the Outat river and its tributary, the Mimachoun river. The inner layout is similar to that of other *ksour*, although its structure has collapsed and changed to such an extent that recognizing its original layout is no longer possible. Only some towers can be seen in the corners.

28 *Tissouit Ait Seghrouchen*. 32°36′54,29″ N; 4°43′20,76″ W
This *ksar* occupies a surface area of 1.330 m² and has a square shaped floor plan, marked by towers at the corners. The inner layout is comprised of two streets that are accessible from the exterior through two doors situated in the southern wall. It is used as a barn and stable.

29 *Tatiouine*. 32°35′35,10″ N; 4°45′56,32″ W
Originally, the Tatiouine *ksar* was formed by a square shaped floor plan located in the upper part of the Outat river, right by the riverbank. Eventually, another *ksar* was annexed to its highest section, forming between them a central square where a mosque and a *hammam* were built. The entrances can be deduced from the remains and are located on the northern and southern facades, both angled. According to current oral sources, streets were mostly covered and families were distributed along them. An irrigation ditch runs across the *ksar* and there is also a well inside, near the river.

Currently the *ksar* has an area of 4.575 m² and is in an advanced state of ruin: the western part as a result of river floods and the eastern part as a result of emigration.

5 CONCLUSIONS

We now have for the first time a complete inventory of the *ksour* in the Outat valley, which will facilitate gathering information about the research carried out on a unique architecture that presents singularities due to the adaptation to its geophysical environment, which in turn is not found elsewhere.

In addition although the process of deterioration and destruction of many of them seems irreversible, we arrived in time to record the urban layout and main features of 28 out of the total 29 that existed.

Finally, as well as providing a map of the valley detailing the state of the various *ksour* and their geographical coordinates, we here describe their exact localization.

REFERENCES

Foucauld, Charles de. 1888. *Reconnaissance au Maroc 1883–84. 1ª Ed.* Challamel. Paris.

Ihrai-Aouchar, Amina. 1988–1989. Communautés rurales de la Haute Moulouya du XVIIème siècle à nos jours. Administration locale et pouvoir central. *Hespéris Tamuda*. Vol. 26–27. Nº 1. Editions Techniques Nord-Africaines, Marruecos.

Peyron, M. 1984. Contribution à l' historie du Haut-Atlas Oriental: les Ayt Yafelman. *Reveu de l' Occident musulman et de la Méditerranée*. nº 38.

Raynal, René. 1961. La terre et l'homme en haute Moulouya. *Bulletin economique et social du Maroc*. Société d'Études Economiques, Sociales et Stadistiques. Rabat.

Rodríguez-Navarro, Pablo y Fantini, Filippo. 2011. The interpretation of archaeological persistence to generate digital 3D architectural typologies: the case of *Ksar* Tatiouine in the Moroccan High Atlas. *International Conference on Cultural Heritage and New Technologies*. Museen der Stadt Wien. Viena.

Rodríguez-Navarro, Pablo y Gil Piqueras, Maria Teresa. 2012. Integrated Methodology for Urban Survey and Representation of the Morocco's High Atlas. *18th International Conference on Virtual Systems and Multimedia (VSMM 2012)*. IEEE. Milán.

Seconzac, Mis. Voyages au Maroc (1899–1091). 1903. Librairie Armand Colin. Paris.

Earthen Architecture: Past, Present and Future – Mileto, Vegas, García Soriano & Cristini (Eds)
© 2015 Taylor & Francis Group, London, ISBN 978-1-138-02711-4

Knowledge exchange for innovation in the vernacular rammed earth technique, Peru

S. Rodríguez-Larraín & J. Vargas-Neumann
Departamento de Arquitectura, Centro Tierra, INTE, Pontificia Universidad Católica del Perú, Lima, Perú

S. Onnis & R. Correa
Centro Tierra, INTE, Pontificia Universidad Católica del Perú, Lima, Perú

ABSTRACT: Vernacular earthen construction in seismic areas of the Andes, the vulnerability of which is compounded by the effects of climate changes, requires the intervention of Academia to develop and transmit innovative technologies that ensure its sustainability. The Centro Tierra research group at Pontificia Universidad Católica del Perú, has developed a project for improvement of the infrastructure of a Colonial hacienda manor currently used as a hotel, of adobe and rammed earth in Tarma and the training of local personnel for strengthening existing buildings and new construction works in rammed earth. Mitigating actions to prevent risks by adapting vernacular rammed earth techniques to innovative anti-seismic and rain-proof aspects are proposed. Findings of the project are reached on the following areas: vernacular and academic knowledge exchange for present-day rammed-earth building, understanding the difficulties in the transmission of new technologies, application of techniques validated in the economic and practical aspects.

1 INTRODUCTION

At present, the vulnerability of vernacular earthen constructions in seismic areas of the intertropical Andean region is compounded by climate changes that are having an effect in the form of rainfall variations and increases, one of the main causes of deterioration of earthen constructions.

Within the context of mitigation processes, the academia emphasizes development and transmission of innovative technology research applied to the safety and sustainability of rural vernacular and heritage earthen housing.

The Peruvian government, through competitive awards of Funds for Innovation, Science and Technology (Fondos para la Innovación, Ciencia y Tecnología), Fincyt, promotes financing projects with the involvement of private enterprises as beneficiaries of research results and participants in the transfer of applied knowledge.

In this context, *Centro Tierra*, jointly with Axis Arte, research groups working at Pontificia Universidad Católica del Perú, have developed and started implementation of a project for improving the infrastructure of the Hacienda Santa María Colonial manor, currently used as a hotel, located in the rural outskirts of the city of Tarma, Department of Junín, Peru (Fig. 1).

Figure 1. Hacienda Santa Maria, view from the patio (Centro Tierra).

For the company, the goal is to improve its financial returns as a result of the renovations to the historic manor and the expansion of its lodging capacity. Innovations are to be applied to earthen construction technology and knowledge exchange and transfer, and seek to train local workmen in structural consolidation and improving the response to moisture of existing constructions, both heritage and otherwise, and in new earthen constructions.

The project leverages this partnership between government, academia and private enterprise to generate investment of public funds in heritage properties and to transmit and use the results of interventions for conservation of local vernacular architecture, thereby benefitting society as a whole.

The experience will help to test the transfer methodology from the standpoint of repeatability of the proposed innovations.

2 RESEARCH CONTEXT AND FOCUS

The focus of this pilot intervention at the Colonial Hacienda Santa María is specific: it must initially observe current construction techniques used in earthen vernacular architecture in order to meet one of the study's main purposes, which is the development of innovative solutions based on vernacular knowledge.

How can a construction with these characteristics be designed and built taking academic knowledge and experiences as a starting point? The technical team's first approach to local techniques has been through a study of earthen constructions in Tarma and its surrounding area. In general, it was observed that Colonial constructions were made of adobe, while present constructions use the *tapial*, or rammed-earth, method.

In addition to site visits, local master builders were interviewed about existing building techniques.

The use of tapial technology is common in rural areas adjoining the city of Tarma, where the "self-help housing" approach is used. In Peru, self-help housing, referred to as autoconstrucción, is understood as construction by an artisanal builder specializing in the technique to be used in the dwelling—in this case, the tapial method—, who proposes the design of the home. Construction quality will depend on the experience and technical expertise of the builder. The layout and volumes are typical. No professional engineers or architects support the builder in the design.

Constructions, usually homes or homes with commercial spaces, are rectangular and have one or two floors, exceptionally three. The relationship with the topography is essential for the home's orientation which, to avoid steep slopes or land fillings, is longitudinally aligned with contour lines (Fig. 2).

A study of traditional construction techniques allows for validating solutions and, at the same time, detecting a series of omissions and limitations that cause failures and defects in existing constructions. A lack of foundations and stem walls was observed, as well as a lack of collar beams, a construction element that is unknown in the area, among others.

The proposal for the pilot project takes observations into account in order to try to solve and share solutions before onsite validation and validation by users. Note that innovations involve introducing new elements made with the materials used in the project (wood, stones) and that the introduction of new materials is limited to easily accessible and inexpensive materials.

Figure 2. Traditional rammed earth house (Centro Tierra).

No alterations are being made to the *tapial* construction method nor to the basic dimensions of the formwork or mould—the basis for their modular building scale—nor to the earth mixture, which gives solidity to the walls.

3 TECHNOLOGICAL INNOVATION FOR RISK MITIGATION

The city of Tarma is located in the central Andes of Peru, 240 km (150 miles) east of the Peruvian capital, on the eastern slope of the mountain range. It was founded by Spaniards in 1534 in the lower valley, where four ravines meet at 3050 meters (10,000 feet) above sea level. Pre-Hispanic, Colonial and present rural developments are located in the surroundings of the city at higher elevations on steeper slopes. The valley is characterized by agricultural production, particularly floriculture. Due to the layout of the terrain, rural constructions on the slopes receive runoff water, rainwater or irrigation water in the dry season. The city center, located in the depression, receives water both from the surrounding slopes and from the water table which, because of its proximity to the surface, creates serious moisture problems in walls, which have caused the collapse of many historical downtown constructions.

Located in Seismic Zone 2, the area is exposed to local earthquakes. However, since recently there has been little seismic activity, the population does not feel threatened by this phenomenon. How much this factor may prevent transmission of structural reinforcement technologies for buildings remains to be seen.

Hacienda Santa María is the result of at least four construction stages. The initial construction was built in Colonial times (circa 1724), later expanded in 1840 and then in 1900, and the last intervention has been the 2013 pilot project.

It is an adobe house with a gallery consisting of wood columns and balconies organized around two patios connected through a hallway.

The more important spaces like chapel and drawing rooms face the main patio. The manor is built on a slope and its volumes are distributed in one or two levels. The volumes on higher ground are older and are built on terraces made of stone walls and fillers.

The pilot intervention in the section called Los Domingos led to knowledge transfer and exchange. The project consists of building a new volume replacing some storerooms with no historical value and near collapse due to the high humidity of the soil in this area adjacent to the orchard in the lower part of the manor.

Existing volumes were replaced by a new building made using the tapial method at the same location, and maintaining the characteristics of the heritage construction without altering the overall appearance of the home. This decision was made pursuant to the Charter on the Built Vernacular Heritage, ICOMOS, 1999. The new construction also increased lodging capacity by three rooms.

The "Los Domingos" pilot project consisted of an improved tapial construction, innovating the vernacular tapial method with the elements described in Adobe Standard E080 of the National Building Regulations, which has not been disseminated in rural areas and is unknown even to professionals.

Elements new to the Tarma construction context were added, such as collar beams and the connections between the walls and the roof through this beam, and rope ties between collar beams (Figs. 3, 4). The concept of partitions or light structures for non-load bearing divisions in indoor spaces was introduced.

Foundations were reinforced to counter the thrust of the soil and stones were set with a lime mortar that is more resistant to moisture (Fig. 5).

For insulation from soil moisture, indoor floors were built on a filler made up of layers of stones of different sizes. Sidewalks with drains were also built to protect the base of the walls.

The roof structure was calculated in order to determine the center to center distance between beams according to the type and strength of the wood being used. For the roof cover, various layers were carefully installed to ensure insulation and final sealing with a mixture made up of a layer of mud and straw mixture, plastic and properly laid roof tiling.

The sum of new and improved elements enhance the overall performance of the structure against earthquakes, settling and moisture, thereby limiting damages. The above mentioned construction details were developed based on vernacular construction tradition, the knowledge of the master builder, the standard, available materials considerations, and cost.

Figure 3. Introducing collar beams (Centro Tierra).

Figure 4. Connected collar beams (Centro Tierra).

Figure 5. Reinforced foundations with lime mortar (Centro Tierra).

The presence on site of the resident architect provided for instances of discussion and exchange: some construction solutions proposed were rejected and replaced by others. Such was the case of the quincha (mud-covered reed frame) tympanum suggested by the project. This technique is not known in Tarma and the mason suggested building it with the traditional tapial. The need to reinforce the higher portion of the wall led to installing a slanted collar beam connected by ropes to the horizontal collar beam. This solution is the result of knowledge-sharing between the technical team and the builders.

Reconstruction of the small building provided the possibility of organizing the project in a way that ensured continuous exchanges with local workmen (masons, carpenters) and technological transfer and sharing of results with workmen, local residents, students, professionals and members of the Municipality through a workshop organized at an intermediate stage of the project.

4 METHODOLOGY FOR TECHNOLOGICAL TRANSFER

Repeatability of the proposed solutions is the objective of knowledge transfer during the process.

The choice of workmen (mason, carpenter) is paramount in this type of project. Their knowledge and experience are necessary to contribute and validate construction solutions on site in a continuous exchange with the resident professional. The design was proposed and approved in terms of feasibility and repeatability, while the professional team members, who were ultimately responsible for the project's execution, approached local knowledge with respect and from a research standpoint.

Communication of technical solutions was adapted for better understanding by the workmen and a relationship of trust developed due to the interest in vernacular techniques shown by the professional team members. As a result, tests and prototypes of new elements, such as straw-mud blocks, were made possible.

A workshop for sharing construction experiences was organized at Hacienda Santa María, with the attendance of workmen, interested citizens, students of Pontificia Universidad Católica del Perú (PUCP) and Universidad Católica Sedes Sapientiae—Tarma, municipal officials and local professionals. The workshop had a theoretical component, as well as discussion panels and a demonstration of field tests.

During the sessions, the resident architect/master builder team presented the project. The resident architect explained the technical aspect, while the master builder spoke of the difficulty of executing the new elements, overcoming initial rejection of some elements and his perception of the importance of the improvements. Later, the same information was repeated during a guided visit of the construction. An exhibit was prepared showing traditional and proposed materials, including the historic adobe of the hacienda, adobe forms, good and defective artisanal roof tiles, various types of fibers, and lime and its possible applications.

The exchange portion was the most successful. Each discussion panel included students, workmen, architects, engineers, municipal officials and was directed by one of PUCP's volunteer students, trained as panel coordinators by the Axis Arte architects.

As to information dissemination, a manual in two versions, in booklet and in poster form, has been developed and is currently under review. Their effectiveness will have to be assessed over time.

5 RESULTS AND CONCLUSIONS

Self-esteem is strengthened when theoretical knowledge is validated by professionals and shared on equal terms. One of the insights gained by the technical team as a result of the workshop was the growing interest shown by the population as they valued their own environment and vernacular heritage. However, this initial awareness-raising requires ongoing and sustained follow-up over time.

Among the elements introduced as technical improvements, the collar beam has gained the most acceptance, so much so that the master mason has systematically incorporated it in his subsequent designs. In view of objections from clients who consider this innovation an unnecessary additional expense, the transfer was achieved through an adaptation that he proposed: a half collar beam to adjust to financial requirements.

Results need to be monitored by following up on the activities of the trained master workmen in their future projects. Likewise, in order to aid in building a safe habitat, the academia must assume and meet a commitment to cooperate and provide advice to builders on an ongoing basis.

REFERENCES

Corporación de Desarrollo Tecnológico, 2012. *Evaluación de daños y soluciones para construcciones en tierra cruda Manual de terreno*. ed. Grafica Andes, Chile.

FINCyT-FIDECOM-PIMEN–2012. Mejorar las aéreas de hospedaje de la hacienda Santa María en Tarma mediante la adaptación de la técnica contemporánea de tapial reforzado para la construcción con tierra sismoresistentes, manteniendo sus características arquitectónicas patrimoniales (tipología de casa/patio con galería de columnas y balconera).

Fincyt project documents—study on techniques, validated technique manual.

ICOMOS, 1999. *Charter on the Built Vernacular Heritage*.

Norma E-080, Reglamento Nacional de Edificaciones.

Puente Zambrano, F. *Historical study by historian*, unpublished, developed for the Fincyt project.

Earthen Architecture: Past, Present and Future – Mileto, Vegas, García Soriano & Cristini (Eds)
© 2015 Taylor & Francis Group, London, ISBN 978-1-138-02711-4

The Jiayuguan Fortress: Structural consistency and safeguarding programs

L. Rovero & U. Tonietti
Dipartimento di Architettura, Università degli Studi di Firenze, Florence, Italy

ABSTRACT: The Jiayuguan Fortress (Ming period, 14th century), stands at the end of the Great Wall of China, facing the Gobi desert. The building consists of a large square-shaped enclosure wall, enhanced by two entrance courtyards, to which a further external wall was added in the 15th century. The building is made by rammed earth at the bottom and by adobe bricks at the top; but a brick "coat" was added on some parts of the original artefact causing an obstruction in the "breathing" movements of the massive inner core. This masterpiece represents an important testimony of the old defense techniques based on earthen walls or embankments. After a survey on the principal techniques used for the construction of defensive earthen walls, the paper investigate the main structural weakness, analyzing the connection among employed building techniques, cracks pattern, intervention phases and exposure to meteoric actions, suggesting guidelines for a safeguarding strategy.

1 INTRODUCTION

1.1 *The Great Wall, the origin and spread of rammed earth techniques for defensive ramparts*

The Great Wall history is very complex; it begins 4000 years from the present time. So the Wall is like a palimpsest where different materials and building techniques are shown, declined and incorporated over the centuries. Emperor Qin (260–210 BC) ordered "to build according to topographical conditions" and the consequence was that stone, clay, earth, reeds, wooden bars, coexist, depending on the position in the mountains or in the plains or in the desert.

The issue of fortification and defense is shared throughout the ancient world and, at the same time, the massive use of earth as the main building solution. The use of earth in defensive ramparts may present different solutions, consisting of various forms of rammed earth, characterized by the use or not of formwork and by the presence or absence of added timber elements. The oldest examples of rammed earth walls can be found in China and Central Asia (Warren1993, Jaquin et al. 2008) starting from Lung-shan era (about 2000 BC) in the primitive form of mounds obtained compacting soil. Then the technique developed by the use of "vertical timber walls with soil heaped against each one", and finally the use of formwork established (Jaquin et al. 2008). The Great Wall shows the superposition of many version of rammed earth, since Han dynasty (200 BC–200 AD) to the Three Kingdoms Period (220–580 AD), considered by Houben and Guillaud the first example of "true" rammed earth technique (Houben & Guillaud, 1994). Middle East and Central Asia show the presence both of rammed earth and cob (this second technique prevailing in different versions, being known as *chineh* in Persian area or *pakhsa* in Uzbek and spread in the countries crossed by the Silk Route) (Fodde 2009, Jaquin et al. 2008).

In the Mediterranean area, after an initial use of rough rammed earth dating back to the Phoenician settlements, then spread in North African countries, perfectly executed earthen walls became usual in Spain and Morocco as a result of Muslim expansion (Gamrani et al. 2011, Baglioni et al. 2012).

It's probably less well-known the description, of the construction techniques employed in the fortifications for defense purposes, starting from Vitruvius, J. Caesar and then picked up by many treatises in the Italian Renaissance period. In this regard there is an extensive literature dedicated to building of walls and embankments on earth with a detailed illustration of the different techniques. From these documents it can be seen that the construction of defensive walls present a range of techniques quite shared by people and cultures around the world.

We can firstly refer to Vitruvius' *De Architectura*, Book I, chapter V (and Book VI, c. XI) where foundation and elevation of embankments are described, introducing the use of *lateres* (adobe) and *terrae* (mounds), the use of *fibulae* (wooden chain) (or *tabulae ustilatae* as smoked poles with the role of *diatoni*) in order to well connect the large section of the ramparts (Vitruvius 2002). Then we owe to the important chronicles of Caesar' *De bello gallico'* the description of the *murus gallicus*, based on the association of rammed earth and wooden poles well connected (".. *Trabes derectae perpetuae in longitudinem paribus intervallis.. (..) in solo collocantur*...). Caesar declares to appreciate this new technique and commends its effectiveness and attractiveness (Caesar 2001). We can easily recognize a primary declination of a technique, then spread around the world, where earth and wood are pushed together in big mounds. Lastly we cannot forget the important G.B. Belluzzi' *Trattato delle fortificazioni di terra*, a treaty (16th century), almost unknown, totally dedicated to earth platforms and ramparts, where a detailed description, accompanied by drawings, is carried out—through the watchful eye of an experienced in the Renaissance period in Italy—as regards the common used techniques (Lamberini 2007, Magini 2013).

1.2 *The Jiayuguan Fortress*

The Jiayuguan Fortress is located at the "end" of the Great Wall in the narrowest point of the western section of the Hexi corridor, few kilometers from the town of Jiayuguan in Gansu Province in China. The Fortress, known also as Jiayu Pass because of the meaning of "guan" in Chinese, is the most intact surviving ancient military building among the "passes" on the Great Wall. It belongs to the list of the UNESCO World Heritage Sites since 1987. The fortification was built in the early Ming period, during 14th century, facing the Gobi desert, as a bulwark against the Mongol (and Timurid) invasions. The walls are completely built by earth, part through rammed technique, part through adobe brickwork, and only some external delimited surfaces were later covered by a baked bricks covering. So the Jiayuguan Citadel is an important example both of the fortification techniques and of the earthen building techniques at the beginning of Ming Dinasty. The paper exposes a direct investigation on the structural consistency of the Fortress, aimed to suggest guidelines for a safeguarding strategy.

2 THE BUILDING SYSTEM

2.1 *Historical overview of construction phases*

The Fortress appears as the result of a series of interventions and a stratification of materials and techniques, carried out throughout different construction phases. The building consists of a large square-shaped enclosure wall, enhanced by two entrance courtyards facing the eastern and western gates respectively. The cross-section features a typical trapezoidal shape: the lower part, about 6 meters high, is built in rammed earth, while the upper part, about 4 meters high, is made of earthen bricks finished with an earthen plaster. In a later period, an additional imposing earthen wall was built outside the western side of the enclosure wall. This later wall is entirely covered with baked bricks and is embellished by a monumental tower building in the middle, the Jiayuguan Tower (dating back to the end of 15th century). A few years later (beginning of 16th century), two similar towers (Rouyuan and Gonghua towers) covered with an outer brick facework, were added in the middle of the east and west sides of the older square-shaped enclosure (Fig. 1).

Considerable renovation works were carried out at the end of the 18th century. These works included: a) the protection of the tops of the outer side of the enclosure walls by building a crenellated battlements in brickwork; b) the contextual reinforcement of the four watchtowers at the corners and, a few years later; c) the erection of two new Enemy Towers in the middle of south and north sides, all of them covered with a layer of brickwork.

The fortress underwent other later interventions, many of which were just maintenance works. The most important of these concerned the rebuild-

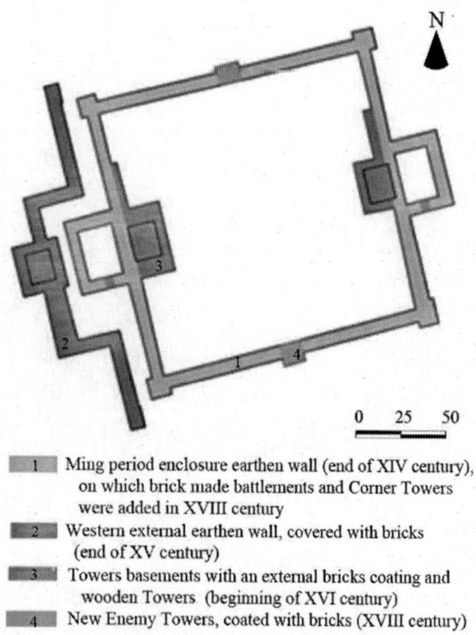

Figure 1. The Jiayuguan Fortress: plan (scale is in meters) and construction phases (Rovero & Tonietti).

Figure 2. a. The corner solution in the external wall of Jiayuguan Fortress. b. The Great Wall Museum sketch showing the rough wooden "formwork". c. The rammed earth enclosure wall: evidence of consecutive construction phases (Rovero & Tonietti).

Figure 3. Fractures on the walls corners, expression of the building technique and phasing (Rovero & Tonietti).

ing of the battlements on top of the western wall (19th-20th centuries).

2.2 *Investigation into the building techniques*

The original square-shaped enclosure wall features two superimposed building techniques. The lower part of the wall consists of rammed earth laid on a base of stones (having a drainage function); the upper part is made of adobe, coated by an earthen plaster. The lower part of the wall is built by using formworks, made of wooden boards overlapped and connected each other by ropes. The use of such technique leaves, as shown by evidence, an imprint on the external surface of the wall (Fig. 2). The large thickness of the wall called for a phased construction of the rammed earth layers, determining a series of portions of "construction sites". This construction method and phasing creates some discontinuities or 'construction joints' inside the wall between any given construction phase and the subsequent one.

Another series of discontinuities is created each time the rough formwork needs to be moved in the longitudinal direction, to build the length of the wall portion after portion. These discontinuities are visible on the external surface all along the wall, with a span of around two-three meters, depending on the length of the wooden boards that were used (Fig.2).

But the crucial point regards the constructive solution adopted for the wall's corners. When a section of rammed earthen reaches the corner, it cannot be toothed with the portion coming from an orthogonal direction. Hence, the latter portion is stopped about one meter before, as it clearly appears at all the corners of the wall. (Figs 2–3). The constructional solution of this key point of the wall is more complex also due to the inclination of the wall surface. The upper part of the wall is made of adobe most probably due to the difficulty of setting up the formwork at this height. A layer of wooden boards is inserted between the two parts, in order to create a smooth surface and to achieve a good transition. The insertion of wooden elements is also useful for the creation of scaffolding.

2.3 *Protection from water infiltration*

The ancient masons working at the Fortress faced and solved this problem by adopting a common system, which consists in building a stone plinth with draining function at the base of the wall. When, later on, they had to cover some parts of the earthen walls with extended brickwork, they inserted at the bottom of the wall a well-made stone base (Fig. 4). Similarly, they also paid attention to the protection of the upper part of the walkways on top of the walls by leaving drain holes in the parapet masonry, placed at regular intervals and close distance, to channel the water collected on the floor towards the exterior (nowadays only the drainage system created during the renovation works conducted in the 16th–18th century is visible). An earthen plaster coat was then applied to the outer wall surface, to effectively protect the upper part of the walls, made of adobe. Later (19th century), four special and useful drainage holes and channels were created on the walkways and down the internal façade of the later western wall (three of them are now closed). Moreover, in the 20th century several drain holes (still visible on the walkway floor) and pipes in iron were placed inside the earthen structure of the enclosure wall, in order to collect and channel down the water from the walkway. Afterwards, these drain holes were closed due their difficult maintenance.

Figure 4. Stone plinth in the wall bottom (Rovero & Tonietti).

3 IDENTIFICATION OF STRUCTURAL WEAKNESS

A preliminary evaluation of the weakness of the structural system may be carried out if the observable crack pattern is compared with the building technique and with the construction phases. It is therefore possible to highlight the following crucial points:

1. The corners feature an intrinsic weakness which is mainly due to the construction technique; in other words, some discontinuities in the structure are due to the organization of the building site and construction in phases, portion after portion. These discontinuities are not necessarily dangerous, but may become a risk if subsequent repairs have modified the texture of the corners (Fig. 3). There are no signs of a subsidence of the soil.
2. Similar considerations apply to the compactness of the rammed earth masonry. In this case too, some longitudinal discontinuities are part of the nature of the structure (Fig. 2).
3. The brick facework of the western wall shows some fractures crossing its surface. It is necessary to understand their real origin. Generally speaking, the fracture pattern visible on the brick surface displays outside what actually is happening inside the wall. In other words we can observe, externally, discontinuities and weaknesses belonging to the earthen rammed filling. But we must not forget that the wall structure consists of a big mass of earth that constantly breathes and exchanges moisture with the external environment. Compared to this massive inner structure, the brick outer layers are a very thin skin that can only follow the movements led from the interior. In many situations (generally when they appear in the large inclined surfaces) the cracks are not necessarily the clue of a remarkable disconnection between the external brick facework and the rammed earth filling (Fig. 5).

   On the other hand, some corners (only a few ones) show a crack pattern that could be the external witness of a potential detachment of the brick facework. Should this be the case, the structure would be subject to dangerous movements. As raised earlier, such phenomenon can depend on the localized weakness of the earthen masonry of some corners (Fig. 6).
4. Many parts of rammed earth masonry at the bottom of the enclosure wall show an advanced state of corrosion and loss of material. A similar phenomenon also affects the lower parts of the brick facework of the western wall, very pronounced along the western facade (both on the internal and the external faces).

   In both these situations the cause can be attributed to the rising moisture from the soil (Fig. 7). In many parts of the perimeter of the Fortress the ground level raised a few feet, which certainly prevents the old drainage system from properly functioning. The entity of such phenomenon is not too worrying, but the decay process is ongoing.

Figure 5. Fractures in the brick facework (not worrying), not always expression of an important disconnection between the bricks and earth filling (Rovero & Tonietti).

Figure 6. Some corners show fractures that can be the clue of a potential local detachment of the brick facework (Rovero & Tonietti).

Figure 7. Advanced state of corrosion and loss of material at the bottom of the walls (Rovero & Tonietti).

A further consideration concerns the brick facework: it may be an obstacle to a correct transpiration especially where the brick layer is applied onto both surfaces of the wall and where the stone made basement is completely covered by the new level of the soil.

5. The drainage system on walkways is, in many parts, out of order. Where the old (yet simple and effective) system is present (ancient enclosure walls) it provides effective protection to the walls if continuous maintenance is ensured. The most worrying situation regards the walkway of the western wall, where no water drainage system seems to be currently present.

4 GUIDELINES FOR CONSOLIDATION AND CONSERVATION INTERVENTIONS

4.1 *Completion of the diagnostic phase*

Some aspects need to be investigated further:

a. a full understanding of the linkage between brick facework and inner earthen wall is of paramount importance if the structural vulnerability of the system has to be assessed.
b. the mechanical and geotechnical characteristics of the construction materials used for the walls, i.e.
   - the lime mortar used for the construction of the brick facework masonry,
   - the possible layer of earth and lime mix used between brick facework and earthen structure,
   - the mineralogical composition, granulometry and moisture content of the rammed earth wall. (Fratini et al. 2011, Rovero & Tonietti 2012)

4.2 *Consolidation of the fortress wall*

4.2.1 *Wall corners*

Almost all of the wall corners feature cracks whose origin has to be found in the construction techniques and also in past consolidation works. Since these cracks are unlikely due to foundations' sinking or to ongoing pushing forces, it would be sufficient to seal the cracks and enact the careful maintenance of the walls corners over time.

4.2.2 *Brick facework masonry*

As explained earlier, cracks of the brick facework are the visible expression of an inner movement of the earthen wall filling.

Basically, there are two types of cracks, those that are visible on large areas of the walls and those linked to the more critical discontinuities near the corners. For the formers, which are not in rapid evolution, light consolidation measures, aiming at reestablishing the original compactness of the masonry, would be adequate.

These light consolidation interventions could consist in (i) the re-pointing of the joints with mortar compatible with and similar to the ancient one in composition, or (ii) the resetting of the bricks position and masonry fabric through the insertion of new bricks in the most serious cases.

The situation at the wall corners is more critical. In this case, the detachment of the bricks facework may anticipate a movement inside, allowed by the conformation of the corner edge (which is less constrained and structurally weaker). It will be necessary to understand the gravity of the detachment by dismantling a small portion of the brick facework at the corner edge and later by reconstructing it. In the case of detachments with the danger of a collapse, a proposal of using transversal connecting elements is not to be rejected.

The fact that the most endangered brick faceworks are those parts of the Fortress characterized by a double layer of brick facework, suggest that such an exceedingly thick layer prevents the inner earthen filling from 'breathing normally', or evacuating normally any excess of moisture. A simple method to improve the situation would be the realization of small (and well hidden) ventilation holes at the bottom of the wall. This light intervention should be tested first on a limited area of the walls and in case of positive results, extended accordingly. In respect of the brick facework of the western wall, a further issue to be considered are the movements to which the wall is inevitably prone (swelling, shrinking of the rammed earthen core). Such movements are acceptable in the earthen walls with no or partial brick facework, where they are detectable through the presence of small 'natural' discontinuities. The phenomenon is much more visible where the bricks need to rip off in order to bear them. In this case, the cracks in the wall cannot be repaired by a simple localized intervention on the brickwork. A possible solution, the invasiveness of which has to be assessed, would consist in the creation of expansion joints in the brickwork.

4.2.3 *Enemy towers*
A peculiar case of instability concerns the Enemy Towers, especially the southern one. Here the upper wall displays a vertical discontinuity that reflects non homogeneous construction phases (Fig.8). In spite of the fact that a widening of such discontinuity appears not to be imminent, for the sake of safety it is recommended, to introduce wooden reinforcing ties in the floor of the upper level, in addition to localized repair of the brickwork, using the same technique described above.

4.3 *Protection against water infiltration*

4.3.1 *Water infiltration by capillarity*
In order to limit the consequences of infiltration by capillarity, it is needed a restoration of the original ground level all around the outer perimeter of the wall, so that the original drainage system (a layer of pebbles and stones at wall's base) be brought back to operational conditions.

4.3.2 *Drainage of flowing waters*
Regarding the upper parts of the wall, the simple drain holes existing in the enclosure walls have proved effective and are still playing their role (if accompanied by a regular maintenance of the plaster). In the western wall, where these drain holes in the battlements are missing, they could be created. In any case, the use of modern drainage methods based on the insertion of a downpipe at the interior of the earthen wall mass should strongly be avoided. A long history of maintenance of earthen structures has shown that systems that cannot be constantly inspected, repaired or provide for the replacement of their constituent parts are not appropriate. For this reason, alternatives to the creation of drain holes should consist in the creation of external eaves on the walkway (that should be newly created) and in external downpipes (made by appropriate materials).

Figure 8. A wooden belt is required at the floor of the upper level to prevent widening of the vertical discontinuity (Rovero & Tonietti).

NOTE

This paper collects part of the study that the authors have completed on the Jiayuguan Fortress in October 2011 within the *Gansu Cultural and Natural Heritage Protection and Development Project* at the request of the World Bank, Gansu PPMO, and HYDEA srl., aimed to address conservation design issues. The Authors wish to thank for this opportunity Mr. Alberto Brizio from HYDEA, Arch. Conservation Management Specialist Silvia Cravero, National Consultant Prof. Xuan Guo, finally the Head and staff of the local PIU in Jiayuguan and the personnel of the Scenic Area, together with Mr. Hugh Zhang, for their kind availability and hospitality.

REFERENCES

Baglioni, E. Rovero, L. & Tonietti, U. 2012. The Moroccan Drâa Valley earthen architecture: Pathology and intervention criteria. In: *RESTAPIA 2012 International Conference on Rammed Earth Conservation,* Valencia. CRC Press—Taylor & Francis Group.

Caesar G. J. 2001. *De bello gallico*. Bristol classical Press.

Fodde E. 2009. Traditional earthen building techniques in Central Asia. *International Journal of Architectural Heritage*.3.

Fratini, F. Pecchioni, E. Rovero, L. & Tonietti U. 2011. The earth in the architecture of the historical centre of Lamezia Terme (Italy): characterization for restoration. *Applied clay science* 53.

Gamrani, N. Chaham, K. R. Ibnoussina, M. Fratini, F. Rovero, L. Tonietti, U. Mansori, M. Daoudi, L. Favotto, C. & Youbi, N. 2011. The particular "rammed earth" of the Saadian sugar refinery of Chichaoua (XVIth century, Morocco): mineralogical, chemical and mechanical characteristics. *Environmental earth sciences*. 66 (1).

Houben, H & Guillaud, H. 1994. *Earthen Architecture: a comprehensive guide*. London. Bristol University Press.

Jaquin, P. Augarde, C. & Gerrard, C. 2008. Chronological description of the spatial development of rammed earth techniques. *International Journal of Architectural Heritage*, 2.

Lamberini, D. 2007. *Il Sanmarino: G. B. Belluzzi, architetto militare e trattatista del cinquecento*. Olschki, Firenze.

Magini, V. 2013. *Cuore di terra. L'anima della difesa nella storia del fortificare e il caso di S. Miniato al Monte (FI)*. Tesi di Laurea. Università di Firenze.

Rovero, L. & Tonietti, U. 2012. Structural behavior of earthen corbelled domes in the Aleppo's region. *Materials and structures* 45.

Vitruvius, P. M. *I dieci libri dell'Architettura.* Translated and commented by Daniele Barbaro. Roma. Bardi 2006.

Warren J. 1993. *Earthen Architecture: the conservation of bricks and earth structures. A hand book.* Paris. ICOMOS.

Earthen Architecture: Past, Present and Future – Mileto, Vegas, García Soriano & Cristini (Eds)
 ISBN 978-1-138-02711-4

Guidelines elaboration for a self-building manual with earth in Chiapas State, Mexico

K. Sánchez Torres
University of Nova Gorica, Slovenia

ABSTRACT: The use of earth as a construction material has been a traditional part of a great number of ethnic groups in the Republic of Mexico. In the State of Chiapas there are a significant number of earth-based constructions due to the significant percentage of indigenous groups concentrated in that area. Due to migration and the arrival of new industrial materials, the community has gradually abandoned its earth-based building tradition. This has caused the ebbing away of its architectural heritage and traditional knowhow. The publication of a self-building manual would be a possible solution. It should meet the needs of the people, minimize costs and curtail the intervention of specialized workers. It would help conserve the earthen architectural heritage of this area and aid the passing on of this tradition's methods to the next generation.

1 INTRODUCTION

Chiapas, the southernmost State of Mexico, is home to a wealth of earthen vernacular architecture. And exemplar of such earth-based construction is found within the indigenous community of Laguna Petej (Fig.1).

Furthermore, their architecture holds significant cultural and historical interest for study. Unfortunately it is disappearing little by little, evermore rapidly, owing to a number of factors; the inclusion of industrial materials, migration and defective do-it-yourself construction.

These are contextual changes within the population, causing significant demise in architectural heritage and those traditional skills that would otherwise sustain it (Fig. 2).

This study seeks to reveal the analysis undertaken of the earthen architecture of the indigenous Tzotzil population in Laguna Petej. This research is based on a technical and social study for housing and improving that community's way of life, integration and encouraging the population in the self-building processes. The following is an elaboration of their self-building manual, a much needed aid.

Figure 1. Vernacular adobe houses in Laguna Petej (Sánchez).

Figure 2. New house built with "modern" materials (Sánchez).

2 METHOD

The elaboration of the manual will be preceded by a preliminary study, which will take into account the following points:

- Cultural and social context of the Laguna Petej: interviews with villagers.
- Analysis of raw earth construction techniques: field studies on the current state of raw earth architecture in Laguna Petej.

3 THE OBJECTIVES OF A SELF-BUILDING MANUAL

- The main objectives of the manual is to help develop a participatory process with the inhabitants in all stages of the project, to evaluate and improve materials and utilize local and human resources.
- To satisfy housing needs through a sustainable construction model that uses natural elements that are easily obtained by all inhabitants.
- The reclamation and promotion of a culture based on traditional construction, educating the younger generation so these methods will thrive in the future.
- Informing the population through an educational campaign in order to rekindle interest and promulgate confidence in this type of construction. The objective here is to disseminate the ecological and cultural values inherent in such construction techniques.
- The optimization of natural resources.

4 CONTEXT

Laguna Petej is located in the municipality of San Juan Chamula, Chiapas. The distance between the town centre and Laguna Petej is about 2.9 km by road. The average height above sea-level is 2320 m. The population is indigenous and there are 647 speakers of the original indigenous language (non counting the youngest children) (Fig. 3).

The landscape of Laguna Petej is rich in vegetation; houses are often isolated and integrated with this natural context (Fig. 4).

The rapid globalization and the abandonment of traditional community practices is damaging the traditional building methods and introducing less than suitable industrial materials. It is evident that traditional materials have been mixed with new industrial materials in a haphazard way. Self-build construction is consequently losing it is traditional, tried and tested basis. This results in urban areas that are altogether less liveable. Such impoverished building methods no longer guarantee a healthy and comfortable environment for local inhabitants.

The mass migration towards the big cities in search of better living conditions is considered a determining factor in the neglect of traditional house-building. Proof of this is the custom of migrants sending money to their families of origin, who then use this new economic resource as incentive to transfigure the house. There many cases where returning migrants reconstruct their own houses by copying the standard styles and methods that they have witnessed in the city.

Figure 4. The landscape of Laguna Petej (Sánchez).

Figure 3. Indigenous woman of Laguna Petej (Sánchez).

The ideal self-build policy should actually reduce constructions costs, to preserve the old local buildings and their tradition and preserve the natural environment. It was on this basis that the interviews were conducted, the main objective being to understand the impact that earthen architecture has on the Laguna Petej population.

The interviews were carried out using two questionnaire sheets, distributed throughout the community. The first questionnaire posits eleven questions to elucidate personal details, unaffected by subsequent interview.

The representative sample is composed of 100 people in a population of 752 inhabitants; 30 women, 30 men, 20 children and 20 elderly people. The second questionnaire has 24 questions, an interview checklist. The interviewees were 50 men, all heads of family households. The aim of this interview was to discern household composition as well as common issues.

5 SUMMARY OF HOUSING NEEDS AS EXPRESSED BY THE POPULATION VIA INTERVIEW

- Separate spaces.
- Larger constructions (so as to enable private rooms).

- The transmission of tradition between the generations.
- Greater privacy.

The age of the interviewed people has had an influence on their answers. In fact, people of the same age gave similar answers. 60% of men agreed on the necessity of having separate areas in order to improve intimacy, on "having a kitchen separated from the bedroom", sleeping and cooking in the same room represents a problem, especially for young people who consequently may experience a lack of privacy, since there are houses where ten people sleep in the same room. It is necessary to note that all the interviewed generally agreed on the necessity of preserving the building tradition, even if with the characteristics of urban houses, or, as they are defined, "modern" houses. In daily life, the house represents a refuge for the population of Laguna Petej.

6 ANALYSIS OF THE RAW EARTH CONSTRUCTION TECHNIQUE *IN LAGUNA PETEJ*

In Laguna Petej the basic construction techniques are: adobe and wattle-and-daub. For this study the adobe technique was selected as it represents the most widely used (Census INEGI 2000). In the first phase of this research, the analysis focuses on the description of adobe construction. The second phase focuses on the critical stages during such construction.

7 THE GENERAL FEATURES OF ADOBE CONSTRUCTION TECHNIQUES IN LAGUNA PETEJ

The adobe is moulded from a mix of earth, straw and water until a plastic consistency is attained. The mixture is placed in a wooden frame (after basting with oil) tamped and levelled off by hand. The bricks are then turned-out of the mould to dry in the sun. After 5 days of drying, the adobe are ready to be stored. The mould is made from pinewood and measures 50 × 30 × 10 cm (Figs 5–7).

The excavation trench for walls is approximately 30 cm deep and 30 cm wide. Adobe house foundations aggregates are either brick or river pebbles in an earth and water mix.

The plinth wall is of stone, adobe houses are rarely constructed over a plinth (Fig. 8).

Walls are built above the foundation trench or onto the plinth directly. The maximum wall height is 2.80 m. The mortar is made with exactly the same pre-dried materials as the adobe and is troweled on to a thickness of 1 to 2 cm (Fig. 9).

The roof structure is gabled and generally constructed in pinewood. This comprises rafters with

Figure 5. Production of adobe block (Sánchez).

Figure 6. Production of adobe block (Sánchez).

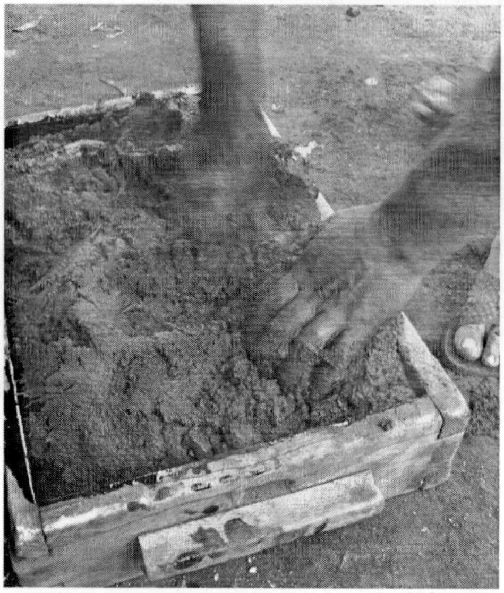

Figure 7. Production of adobe block (Sánchez).

Figure 8. Stone plinth (Sánchez).

Figure 9. Building of the adobe wall's (Sánchez).

smaller roof battens at spacing of 30 to 50 cm. The roof may be covered with sheet metal, tile or straw (Fig. 10).

The typical Tzotzil house has a single entrance door in the center of the façade, and rear door giving access to the garden. The house generally has two windows. Both doors and windows frames are constructed in pinewood. It is common for the door to be divided in two parts, stable-door fashion (Fig. 11). A closed lower section blocks entry to animals while an open upper section allows for natural light and ventilation (Fig. 12). The most common plaster employed, if used at all, is a lime plaster.

Figure 10. Sheet metal roof (Sánchez).

Figure 11. Door in pinewood (Sánchez).

Figure 12. Window in pinewood (Sánchez).

8 CRITICAL STAGES OF THE CONSTRUCTION TECHNIQUE

In the second phase of the study a number of crucial stages were identified and focused upon.

- The mould with handles to facilitate an easier release of the brick.
- Submerging the mould in water before filling it with the mixture and sprinkling the inside of the mould with fine sand and thus preventing sticking was found to be a better technique than basting the mould with oil (Fig. 13).
- The adobe mixture's ingredients have usually been measured approximatively and this imprecision has consequences for its quality.

Figure 13. Basting the mould with oil (Sánchez).

Figure 14. Introduction of concrete lintels (Sánchez).

- No field test was conducted to rate the quality of adobe.
- The lack of foundation or of insufficient depth can cause rising damp.
- In some cases the original structure undergoes alteration by the introduction of inappropriate industrial materials (Minke 2005). For example concrete lintels can produce cracks in walls as they are highly rigid with respect to an adobe wall.
- The lack of collar beam can reduce the torsion resistance.
- Poorly attached tiles can result in rain penetration.
- A roof overhang of just 20 cm is insufficient to shield walls from rain.
- Cement plaster lacks physical symbiosis with an adobe wall (Fig. 14).
- A lack of plaster can facilitate cracks and fissures.

9 STRUCTURE OF THE SELF-BUILDING MANUAL

The self-building manual will be divided in two parts. The first part describes the technical aspects for the production of adobe. The second part describes the technical aspects of adobe construction.

9.1 *First part*

The first step to deal with for the making of the self-construction manual is the selection of the soil used to make the adobes.

- Soil selection. Simple field test: Sedimentation test, dry strength test, adhesion test and a hand-washing test.
- Mixing
- Moulds made of: pine wood, measure: $40 \times 20 \times 10$
- Moulding
- Drying
- Strength test of adobe bricks

9.2 *Second part*

The part of the manual concerning the technical aspects begins with the excavation of the trench for the foundations.

- Foundations
- Plinth wall
- Walls
- Openings
- Roofs
- Plaster

10 CONCLUSION

The analysis of the existing earthen architecture in the village of LagunaPetej and the involvement of the population through the interviews were fundamental instruments for the identification of the inhabitants' needs and demands related to the issue of housing. In spite of the factors previously described, such as the inclusion of industrial materials, migration and defective do-it-yourself construction, it is possible to argue that, in this area; residential architecture is still clearly based on earthen construction.

The local population has shown a particular interest for the maintenance and continuity of this building tradition. However, at present, they are experiencing a moment of history in which modernity and transformations lead people, especially new generations, closer to new building technologies which are considered superior to the traditional ones.

The aim of the manual is to improve the traditional earthen building system by introducing a number of minor variations to the management of the living space and to the building technologies, without sacrificing the flair and simplicity found in the building tradition. The hope is to encourage and stimulate the continuous use of earthen construction by making it more "modern" and "interesting" for the population. The self-building manual, which is still a work in progress, hopes to realize the earthen architectural of Laguna Petej, which must be preserved, and possibily to serve as an inspiration for the indigenous populations of Mexico in order to create other such manuals.

NOTE

The interviews were conducted using a very clear language, almost employing a colloquial "slang"; in some cases, Jose Luis Patisthan (secretary of the directive board of the *Asosiaciòn del comité de educación)* partecipated as a Spanish-Tzotzil translator in order to make possible some of the interviews. An interview was conducted with the *adobero* (the worker responsible for the production of adobe) which has had the aim of understanding the level of knowledge of the technology employed create other such manuals.

REFERENCES

Gernot, M. 2005. *Manual de construcción en tierra. La tierra como material de construcción y su aplicaciòn en la arquitectura actual,* Editorial fin de siglo.

INEGI. 2000, *Censo general de poblaciòn y vivenda 2000.*

Lee Whiting, T. 1992. *Tipos de arquitectura vernacula de Chiapas: perspectiva histórica cultural.* Instituto Chiapaneco de cultura.

Lopez Morales, F. 1987. *Arquitectura vernacula en México.* Editorial Trillas.

Moya Rubio, V. 1982. *La vivienda indígena de México y del mundo*, México: UNAM.

Obregón Rodríguez, M. 2003. *Tzotziles. Comisión Nacional para el desarrollo de los Pueblos Indígenas.* México.

Pérez López, E. 1997. *Chamula: un pueblo Tzotzil Gobierno del Estado de Chiapas,* México. *CONECULTA-CELALI.*

Sànchez, K. 2014. *Thesis: Earthen architecture, tradition and modernity: self construction through building tradition and technological innovation opportunities in Chiapas, Mexico.* Mentor Mauro Bertagnin, University of Nova Gorica.

Earthen Architecture: Past, Present and Future – Mileto, Vegas, García Soriano & Cristini (Eds)
© 2015 Taylor & Francis Group, London, ISBN 978-1-138-02711-4

Clay boards and panels—an innovative earth building product for drywall construction

H. Schroeder
Dachverband Lehm e. V., Weimar, Germany

ABSTRACT: Industrially produced earth building materials are currently used in many areas of building. In the last 15 years, clay boards and panels have become an important part of the production of earth building materials in Germany. Clay boards and panels are for dry construction as an alternative to gypsum, lime or cement-based building boards. Several manufacturers in Germany currently offer a range of different clay board and panel products for different uses, for example as an ecological replacement for conventional plasterboard panels, as "masonry" panels for non-load bearing partitioning walls in interiors or as wall heating panels, either with embedded heating pipes or with air channels for hypocaustal wall heating systems. The German earth building association, the Dachverband Lehm e.V. (DVL), is currently preparing a draft of a national DIN norm for clay panels and boards. This paper provides an overview of the different kinds of clay panels currently available in Germany.

1 INTRODUCTION

Building boards are planar building elements characterized by load action perpendicular to their plane. They are used for a range of different applications in the building process, in particular for dry building construction.

Unlike most clay plasters, clay boards and panels are finished prefabricated dry building products and do not require water for their installation on site.

The primary ingredients of clay boards and panels are earth and clay as well as clay minerals as a binding additive, sand and different kinds of mostly organic fibrous aggregates. Clay boards and panels are not water resistant.

Clay boards and panels differ from earth blocks in size and thickness as well in their method of manufacture. Depending on their thickness, clay boards and panels are mounted on a supporting construction or can be used as freestanding elements or in non-structural/structural wall constructions.

Clay boards and panels contribute to creating a healthy indoor environment in several ways: they are able to absorb and dissipate airborne moisture, thereby naturally regulating the level of air humidity, and they have a good thermal capacity, absorbing heat from the room when it is hot and re-radiating heat back into the room when it cools down. In addition to the good sorption properties of earth building panels, their aesthetic qualities, such as their colour and haptic surface qualities, are also highly prized by consumers. As a consequence, a market for these products has emerged over the past decade.

2 DEFINITION

Generally, building boards and panels are planar building elements characterized by a main direction of load action: for thinner boards load action is typically *perpendicular* to their plane while for thicker slab-like panels load action is *in direction* of their plane. The Lehmbau Regeln [DVL 2009] defines clay panels as panel-shaped building materials that, depending on their thicknesses, can be laid like masonry or used as dry lining boards.

Currently, there is no standard for clay boards and panels and different manufacturers use different names for their products, such as earth building boards, clay boards, light clay boards, clay panels etc.

3 DIMENSIONS AND FORMATS

Clay boards and panels are available in a variety of non-standardized formats for different kinds of uses.

There is no definition for when an element should be classed as a panel or as an earth block. In Germany, earth blocks are produced in standardized dimensions as given in DIN 4172 "Modular Co-ordination in Building Construction" and

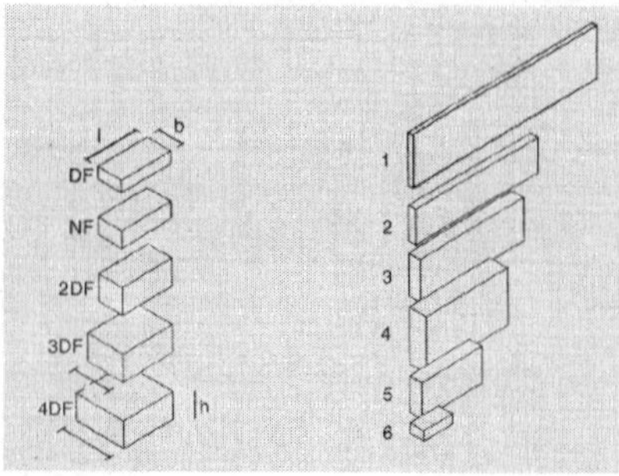

Earth block sizes

	L x B	x H [cm]
DF	24 x 11.5 x	5.2
NF	24 x 11.5 x	7.1
2DF	24 x 11.5 x	11.3
3DF	24 x 17.5 x	11.3
4DF	24 x 24.0 x	11.3

Clay board and panel sizes

	L x	B	x H [cm]
1	150 x	62.5 x	2.5
2	100 x	24.5 x	6.0
3	85 x	29.5 x	9.5
4	67 x	67.0 x	14.0
5	50 x	25.0 x	12.0 (10.0)
6	NF earth block for comparison		

Figure 1. Earth block sizes according to DIN V 105-1 and clay board and panel sizes, non-standardized [Schroeder 2013].

Table 1. Clay boards and panels, classes of thickness and corresponding kinds of use.

Class of thickness	Thickness [mm]	Kind of use [Schroeder 2013] [DVL 2004] [DVL 2010]
I	12,5–50	a) dry lining earth building boards, b) cladding of a dry stacked wall construction, c) dry lining with clay plaster boards, d) wall heating/ cooling elements, embedded pipes
II	51–125	a) partitioning walls, free-standing, b) inserted panels between ceiling beams, c) wall heating elements, hypocaustal system
III	250–300	a) load-bearing external/ interior wall elements

DIN 18945. Typical formats include *Thin Format* (DF) and multiples thereof (2DF, 3DF), as well as *Normal Format* (NF). Figure 1 provides an overview of different sizes of clay boards and panels and earth blocks.

4 APPLICATION

The different kinds of clay panels on offer can be classified into classes of different thickness and dry densities depending on their kind of use [Schroeder 2013] [DVL 2004] [DVL 2010]. Table 1 provides an overview of the thickness classes and their corresponding kinds of use.

Thinner boards of about 15 to 50 mm (class I) usually require a supporting construction, e.g. a framework of studs or a fully flat subsurface. Thicker panels (class II) can be used as self-supporting wall lining elements and do not need for a supporting construction. These class II panels are for non-structural use only. Finally, prefabricated rammed earth panels or slabs are offered for use as external/interior structural walls (class III).

The dry density of clay boards and panels can be divided into dry density classes according to DIN 18945, 18946 and 18947.

Clay boards and panels of low dry density classes contribute to thermal insulation of wall and ceiling construction depending on their class of thickness (usually classes I and II). Prefabricated rammed earth panels used for structural wall construction must have a high dry density class of at least class 2.2.

4.1 *Dry lining—earth building boards (class Ia)*

Lightweight earth building boards with a thickness of between 15 and 40 mm can be used for partitioning wall constructions e.g. a freestanding stud wall construction or a single-sided wall lining (Figure 2 [DVL 2004] [DVL 2010]). They can also be used for cladding suspended ceilings and the underside of pitched roofs.

The boards can be cut with a jigsaw or hand-held circular saw. For dry walling, panels are fixed with galvanised wood screws without washers. Washers are necessary (d = 15 mm) when mounting panels on ceilings or the underside of roofs inclines. For walls, 8 fixing points per m^2 are sufficient, for ceilings and roof inclines 12 fixing points per m^2, e.g. 2 or 3 screws for each batten of the subconstruction.

The final construction can be plastered over immediately. Scrim tape is laid over the joins and held in place with a slurry coat of thinned fine-finish plaster applied with a wide brush. The joint reinforcement tape serves to bridge panel joins and reduces susceptibility to cracking in the skim coat or plaster should the underlying construction be subject to movement.

Before applying the plaster coat, the entire area should be carefully wetted to prolong the workability of the thin layer of fine-finish plaster. A 2 to 3 mm thick layer of fine-finish plaster is then applied.

Figure 2. Dry lining with earth building boards, heat insulation with sheep wool.

4.2 *Cladding of a dry stacked wall construction (class Ib)*

Dry stacked wall construction can be used as wall infill or as an interior wall lining for increased thermal mass. They are constructed with earth blocks or green unfired earth bricks. While interior walls made of earth are usually masonry constructions, dry stacked wall construction do not require wet mortar and therefore have no drying time. This is especially well-suited to the speed and ease of timber frame construction, and avoids the introduction of moisture into the building site.

Dry stacked constructions can be used out as a wall lining applied to the interior of a planked timber frame construction or as a wall infill between the posts of a timber frame construction. The concentrated load of a stacked wall construction can be borne by the foundation when applied to the inner surface of external walls. Earth blocks and unfired green bricks can also be used as infill for interior walls provided the structure can support the load.

When using dry stacked construction as a wall lining, a wooden board or batten is first screwed to the floor. The batten should slope slightly towards the wall rather than the room, which can be achieved by inserting thin wedges beneath the batten.

The bricks or blocks are then stacked against the wall, each course offset so that the vertical joints are staggered. After approx. 50 cm, or three to four courses, the next wooden batten is inserted and screwed to the wall lining or the timber posts so that blocks or bricks are firmly clamped in place and cannot fall out. The wooden battens also serve as a subconstruction for fixing the wall cladding. The stacked wall construction is continued in the same manner up to the ceiling.

The stacked wall lining can be clad with dry lining boards such as earth building boards (Figure 3 [DVL 2004] [DVL 2010]). Because the front edge of the wooden battens lies slightly behind the plane of the surface of the earthen bricks or blocks, the earth building board is pressed against the bricks firmly when screwed to the batten. Where board joins are fixed with a backing strip or when working overhead, partially threaded screws should be used. These allow the boards to be tightly pressed against the underlying construction.

Dry lining plasterboards can be used as an alternative to the application of wet plaster. These 15 to 40 mm thick light earth boards can be applied to the surface of a wall or a ceiling with a clay-based bonding mass or adhesive (Figure 4 [DVL 2004] [DVL 2010]). This kind of surface is used where wet layers of plaster are excessively time-consuming to undertake or where longer drying times are problematic.

Clay plasterboards can also be used for repairing defects in plaster, for example caused by installation works. Because the panels are applied with water soluble mortar and can be removed again later without damaging the substructure, they can be used to cover up historically valuable surfaces that need protecting but cannot be left exposed. A further important area of use is for cladding

Figure 3. Dry stacked wall construction with cladding of earth building boards.

Figure 4. Dry lining with clay plaster boards.

composite wood panel materials in modern timber frame constructions.

After any necessary priming of the subsurface, the adhesive is applied with a notched adhesive spreader (5 mm) to the subsurface. Because the material dries relatively quickly the area of adhesive mortar applied should not be too large. The individual panels are then immediately applied and pressed on firmly. The boards are applied with staggered joins that is the joins should not meet in a cross-shape.

Clay plaster boards can be cut with a jigsaw or hand-held circular saw.

4.3 *Partitioning walls (class IIa)*

Earth panel elements are produced with the dimensions 500 × 250 × 100 mm and have tongue-and groove-edging. Using a notched base batten, accurate "masonry" walls can be quickly and simply erected (Figure 5 [Schroeder 2013] [DVL 2004] [DVL 2010]). The panel elements are "glued" together using a cellulose-based adhesive applied as a thin-bed or by immersing the edges in water to activate the adhesive forces of the clay minerals.

Junctions with other walls constructed in the same or different materials, and fixings to the base batten can be achieved using standard nails and screws.

4.4 *Inserted panels between ceiling beams (class IIb)*

Inserted ceilings are ceiling constructions in which panels are inserted between the ceiling beams. The resulting space between the beams is then filled with earth blocks, unfired bricks or a dry earthen loose-fill material. It is also possible to fill the space between the ceiling beams with preformed earthen hollow block elements (Figure 6 [DVL 2010]). No additional planking is required but the distance between the ceiling beams must be uniform. The earthen hollow block elements cannot sustain point loads.

Figure 5. Earth panel elements used for freestanding and non-structural partition wall.

The side-mounted battens (e.g. 40 × 60 mm) for the inserted hollow block elements must be securely fixed to the ceiling beam as they bear the full load of the ceiling fill material. The distance between the timber beams must be chosen to match the size of the elements. The hollow blocks can be slung directly between the side-mounted bearing battens and the ceiling can be covered immediately by a timber floor or other flooring surface.

4.5 *Wall heating/cooling elements (classes Id, IIc)*

An alternative to conventional heating is wall heating where the entire wall or part of the wall acts as a gentle heat radiator. Wall heating typically consists of loops of heating pipes made of copper, plastic sheathed aluminium or plastic which transport warm water as heating medium. The heating pipes can be fixed directly to the wall in a regular pattern and plastered over with an earth plaster or

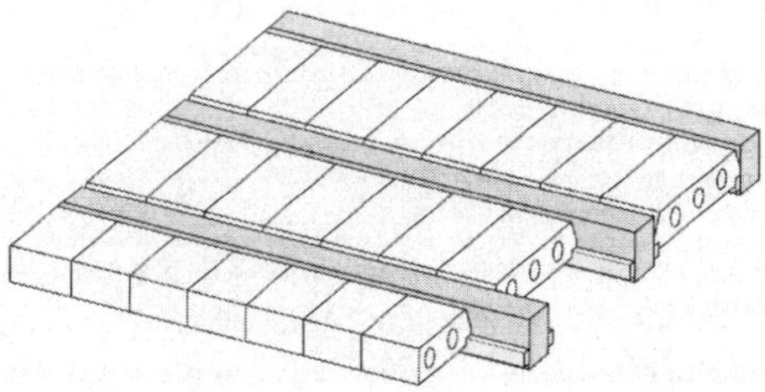

Figure 6. Hollow earth block elements slung between the ceiling beams.

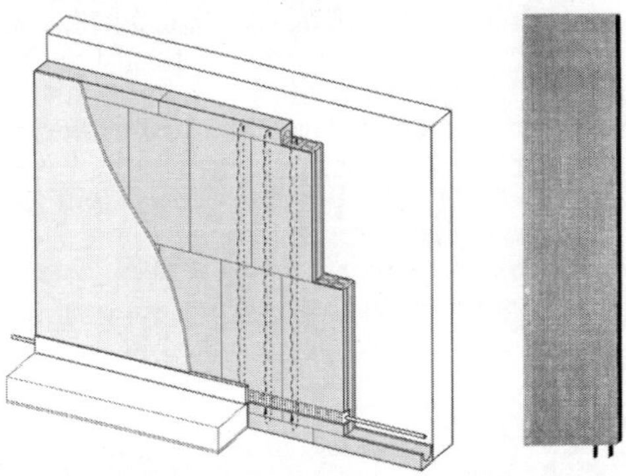

Figure 7. Wall heating/cooling with clay panels (7a) Hollow earth panels forming a (7b) Clay panels with embedded heating/cooling hypocaustal system (WEM).

embedded in preformed earthen slabs that are then mounted as a wall lining of an existing wall (class Id, Figure 7a).

Hypocaustal wall heating systems use air rather than water to warm the wall. A wall lining of hollow earth panels with U-blocks at top and bottom form a closed cavity in which air can circulate (class IIc, Figure 7b [WEM]). The air is heated using a tubular radiator or heat exchanger at skirting board level that is connected with to the heating pipework.

The latter method allows wall fixings to be mounted without fear of damaging pipework.

4.6 *Structural external/interior wall elements (class IIIa)*

Some manufacturers make prefabricated rammed earth elements for use as structural external or internal walls. Because of their use as structural wall elements, they typically have a thickness of 25 to 30 cm and can weigh up to 5 tons. As such, the rammed earth elements require a special system of transport to and on the site (Figure 8 [Claytec 2012]).

Figure 8. Prefabricated rammed earth elements, use as structural wall element.

These wall elements require a minimum compressive strength of 2.0 N/mm^2 (dry density 2.400 kg/m^3). The direction of load action is in the plane of the element.

5 OUTLOOK

Over the past decade, prefabricated earth building boards and panels have increasingly been used in many areas of building, in particular in dry construction in combination with other building materials. In Germany, a number of manufacturers offer earth building boards and panels for different uses. For earth building products that are industrially manufactured and made available for sale on the building materials market, the State Building Authority requires the development of national standards (DIN norms), which currently do not exist for this kind of product.

National DIN standards for earth blocks and earth mortars (DIN 18945, 18946 and 18947) were introduced in Germany in 2013. These national standards follow the general and basic requirements for the formulation of European regulations for building products defined in the European Regulation No. 305/2011 concerning harmonised conditions for the marketing of construction products in the EU (Construction Products Directive CPD).

The requirements of the CPD include:

- Definition of basic requirements to be met by the building,
- Terminology and area of application,
- Definition of essential features describing the performance of a building product,
- Quality control of the production process by control of the confirmation with standardized testing procedures,
- Uniform declaration of the building product.

According to the CPD standard, all industrially-produced building products must fulfil fitness-for-use requirements (performance) and verification thereof (CE-symbol for European, Ü-symbol for German standards).

Essential features are those features of a building product that relate to the basic requirements of a building. The performance of a building product describes how they perform with regard to their essential features. These are expressed in steps or classes or in a description and are declared by the product manufacturer. Basic requirements such as "mechanical stability" and "fire resistance" of the listing with relevant essential features have general priority.

Essential features must be defined for each building product separately with reference to its area of use. In the case of earth building boards and panels, this paper provides an overview of the main kinds of use (classes Table 1) covered by products that are currently available on the building materials market in Germany.

This classification needs to be discussed and where necessary expanded to include any additional basic requirements that are identified. A particular problem is the development of appropriate procedures for testing the physical and mechanical parameters at the level of national or approved standards.

A project proposal entitled "StandardBoard" addresses this problem and is currently being submitted to the German Ministry of Economy (BMWi). This project will be conducted by the Federal Institute of Research and Testing of Materials (BAM) and a German producer of earth building boards and panels (conluto) in cooperation with the German Association for Earth Building, the Dachverband Lehm e.V. (DVL). On the base of this research, the DVL will develop a draft for a DIN norm "Earth building boards and panels" for discussion by the "Lehmbau" DIN Technical Committee on Earth Construction.

REFERENCES

Claytec e.K. 2012 *Prefabricated rammed earth elements*—Technical information (Issue 11–2012).

Dachverband Lehm e. V. (DVL) (Ed.) 2004. *Building with Earth—Consumer Information*. Weimar: Dachverband Lehm e. V.

Dachverband Lehm e. V. (DVL) (Ed.) 2009. *Lehmbau Regeln*—Begriffe, Baustoffe, Bauteile. Wiesbaden: Vieweg + Teubner | GWV Fachverlage, 3rd revised edition.

Dachverband Lehm e. V. (DVL) (Ed.) 2010. *Specialist for Building with Earth—Course handbook*. Weimar: Dachverband Lehm e. V., unpublished.

Schroeder, H. 2013. *Lehmbau—Mit Lehm ökologisch planen und bauen*. Wiesbaden: Springer Vieweg Fachmedien, 2nd revised edition.

Wall heating WEM, flyer.

Earthen Architecture: Past, Present and Future – Mileto, Vegas, García Soriano & Cristini (Eds)
© 2015 Taylor & Francis Group, London, ISBN 978-1-138-02711-4

Earth as a construction material conserving Addiriyah of Saudi Arabia

M. Sharif Zami
Department of Architecture, King Fahd University of Petroleum & Minerals, Saudi Arabia

ABSTRACT: Addiriyah is one of the most prominent architectural heritages of Saudi Arabia amongst hundreds of rich heritage sites. It is located on the banks of the Wadi Hanifah elevated 700 m above sea level at Najd central plateau of the Arabian Peninsula. Today it is situated almost within the limits of the metropolitan Riyadh, the capital city of the Saudi Arabia. Fortunately, the architectural heritage of Addiriyah is taken care by the government of Saudi Arabia and they launched a large-scale program to restore this rich architectural heritage. This paper focuses on the construction technique and material used conserving the heritage of Addiriyah. The paper critically analyzes the ongoing earth construction of Addiriyah conservation program in the light of contemporary stabilized earth construction. To achieve the aim of this paper the author adopts several research methods, such as, desktop survey, case study, photography, and drawings.

1 INTRODUCTION

According to World Travel Web Site (2014) Addir'iya was an important station along the trade rout which linked the Arabian Peninsula from its west to its east in addition to its control of the pilgrimage road up to Makka. It was controlling a number of villages of Wadi Hanifa and its power extended up to Darmain on the western side of Twaiq Mountain and the Aba Al Kabash to the North Dir'iyah. Accoding to King (1998, 156) it was the most famous city in Wadi Hanifa during the twelfth and thirteenth century Hijri (17th and 18th century AD) and remained famous even after the armies of the Ottoman Empire caused the destruction of its environment and geographical location in the year of 1233 AH / 1818 AD.

The development of Addir'iyah Program includes historic villages and neighborhoods along the west to east of the valley including Qasiba, Al Zahra, Al Zowaihira, Al Bejairi, Al Mlaibeed, on the west bank the Al Turaif Quarter (ArRiyadh City Web Site 2014). The scope of the development program emphasizes Addir'iyah at the height of its prosperity during the reign of the first Saudi state. The project is implemented in partnership with the Ar-Riyadh Development Authority (ADA), the Saudi Commission for Tourism and Antiquities (SCTA) and the Addir'iyah Municipality. On 31st July, 2010 Al Turaif Quarter in Addir'iyah municipality was registered in the World Heritage List of UNESCO (World Travel Web Site 2014). Mud brick is used as primary building material conserving the old settlement buildings and palaces in Addiriyah. Reinforced concrete and load bearing walls is used in all the newly developed areas of Addiriyah, such as Tourist Centre. It is important note that these newly developed buildings are only plastered with mud to give a wrong image of mud building to the visitors. This paper aimed to highlight and analyses the scope of using contemporary stabilized earth construction in the conservation program and development of Addiriyah.

2 LOCATION AND HISTORICAL BACKGROUND OF ADDIR'IYAH

According to ArRyadh City Web Site (2014) Addir'iyah is a settlement located on Wadi Hanifah to the northwest of Arriyadh (Al-Rashid 1999, 220), covering an area of about 10 square kilometers. Figure 1 shows the location of Riyadh in Saudi Arabia. The historical precinct of Addir'iyah covers about 6 square kilometers of that area along the Wadi Hanifah, and comprises a group of settlements spread among date palm farms and vegetable gardens. As a whole, the Historical Precinct of Addir'iyah, within its historical boundaries, is one of the most important national heritage sites of Saudi Arabia as for many years since it was the capital of the first Saudi State (King 1998, 156). Wadi Hanifah is the most prominent feature of Addir'iyah. It is also an important geographic element and topographic natural formation within the Najd region. Its natural features, rich fertile soil and water combined with land formation suitable for settlement attracted many tribes throughout centuries (ArRiyadh City Web Site 2014).

Figure 1. Location of Riyadh in Saudi Arabia (Google image).

The First Saudi State rose at Addir'iyah very rapidly in the 18th century, inspired by the teachings of the great Sheikh Muhammad bin Abdulwahhab and under his guidance (King 1998, 157). He was the chief of the House of Saud, Muhammad bin Saud that led to the success of his Reform Movement (Salfi Dawa), and to the prominence for the first time of the House of Saud (Al-Rashid 1999, 224). The new state grew rapidly until, in the first years of the 19th century, it controlled all of central and eastern Arabia, as well as the holy cities of Makkah and Madinah, the Hijaz and Asir. Seen by the Ottoman Empire as a threat to its supremacy in the region, the Saudi State fell victim to successive invasions from Egypt. Addir'iyah itself fell to besieging Ottoman forces in 1818, and much of the city of Turayf was destroyed. In the following two centuries, Arriyadh was established as the capital of the kingdom (Eshteeaque et al 2008, 51). Addir'iyah declined and its building structures decayed.

The historic settlement of Addir'iyah represents an example of an urban development within the natural environment (King 1998, 158). The settlement's size was directly related to the area of farms in the Wadi that could sustain its inhabitants. The buildings responded to the need for shelter, privacy and security. They used local materials such as stone, clay and wood for construction and remained in harmony with the environment. They are now great examples of Najd vernacular architecture (Eshteeaque et al 2008, 55). Addir'iyah is a part of the great history of the region and consists of several historic and antiquities sites. It is composed of unique architectural buildings, old quarters, oasis, palm gardens, farms and agricultural areas.

3 CONSTRUCTION TECHNIQUES AND BUILDING MATERIALS USED IN ADDIR'IYAH

The major building material of Turayf was sun-dried mud brick. According to Facey (1997, 81) mud construction in Arabia traces back its ancestry to pre-Islamic times (King 1998, 12). The durability of mud walls can vary depending on the quality of the raw materials available, the recipe used by the builder, and the care that is taken preparing the mix. The mud mix used in the recent construction of traditional Najdi buildings, such as the houses built when Turayf was briefly re-occupied in the mid-20th century, seems to be of an inferior quality compared to that used in the early palace construction (Facey 1997, 82).

According to King (1998, 159) the essential ingredient of mud bricks is clay, found in natural deposits in the Najd, usually in the presence of ground water. Clay is itself an adhesive and a thick mass of clay will crack and break up when it dries up unless mixes with other materials. Therefore in order to make good mud bricks clay was mixed with fine soil taken from silt deposits in Wadi banks, chopped wheat or barley straw and water (Facey 1997, 82). After mixing, sometimes by trampling, the mix was then poured into a brick-shaped wooden frame on the ground. The mud bricks were then left in the sun in rows to dry. Alternatively, this mix was used directly for constructing a wall making continuous layers on top of each other. The mix might be left for several hours to "ferment". It was considered that, the longer the mix was left to ferment the better the quality of the building material became. If it was supposed to be used for a mud plaster finish on the completed wall it was left for up to twenty-four hours. The mud bricks used constructing palaces at Diriyah in the late 18th and early 19th century was considered fine quality compared to the mud bricks commonly used in the recent past. The mud is rather creamy, with very little gravel, suggesting that only a minimum of fine wadi silt was used to temper the clay. There is hardly any straw in the mix. These observations suggest that the builders wanted to maximize the proportion of clay, for the sake of hardness and durability, and added just enough silt and straw to ensure that the bricks remained crack-free.

As mentioned earlier the mud bricks were used on top of the stone foundations. At the plinth level wall was very thick, a meter or more, especially if the building was a tall one. For example, the wall thickness of ground floor of Salwa Palace was equal to three mud bricks width, two mud bricks width on the first floor, tapering to one mud brick width on the upper floors. This tapering construction was typical of all Najdi mud walls, and can be clearly seen in the ruins of Turayf palaces (Facey 1997, 83). The buildings were mud plastered once they were nearing completion, and the decorative triangular ventilation openings (Fig. 2) were created.

Timber was used to make structural components of the buildings such as ceilings and lintels. The Tamarisk wood and palm trunks were commonly used. Tamarisk wood is light, strong and structurally good for its tensile quality. It does not split easily, and branches and trunks grow to a sufficient length for the use as rafters. Rafters were laid close together to span a small room, or in larger room to span across to a central column-supported beam (Fig. 3). Palm leaves were then laid over the rafters, and a thick layer of rubble and mud was laid on top of that to form the floor slab or the roof.

Very occasionally, instead of palm leaves and rubble, flat stone slabs were used for roofing. This can be seen in small rooms of the Salwa Palaces and the Palace of Thunayyan. Tamarisk or palm planks were also used to make doors in Diriyyah and elsewhere in Najd. They were colorfully decorated, though lintels and cross-beams were also often decorated with patterns of scorched dots and some color.

Stone was also used extensively in the construction of foundations, sometimes rising to a few feet above ground (Facey 1997, 80; King 1998, 11).

Most of the superstructures of the building are constructed with mud bricks. Stone courses above the ground level (Fig. 4) and mud brick courses in the superstructure all were covered with mud plaster, therefore, different construction techniques and materials are not visible. Dressed stone blocks were used extensively at Dirlyyah in the construction of the lower courses of the great palaces, such as the Salwa and Thunayyan; such fine masonry work is not found elsewhere in Najd. Stone was also extensively used in Diriyyah in the construction of columns. Limestone blocks were shaped to cylinders (Fig. 5) which were then laid one on another and the finished columns were covered with mud plaster.

Another common use of stone was in the construction of keel arches (Fig. 6) which were commonly used in mosque arcade (Facey 1997) and

Figure 2. Triangular ventilation openings were found in many of the palaces at Addiriyah (Facey 1997, 80).

Figure 4. Dressed limestone courses were used in the ground level of the palaces (Facey 1997, 81).

Figure 3. Column, beam and rafter construction details of 20th century houses in Turayf (Facey 1997, 83).

Figure 5. Fallen cylindrical drums of stone colums (Facey 1997, 96).

Figure 6. Keel Arches in mosque (Facey 1997, 81).

Figure 7. Herringbone pattern of stone used in wall construction, western Turayf (Facey 1997, 81).

triangular openings in the palaces and houses. According to Facey (1997, 80) at Dir'iyyah, stone was often also used to construct entire walls; in the western half of the site, where the common people of Turayf are thought to have lived, this can be seen clearly today. Rough cut, flat blocks of the plentiful local limestone are laid on edge in mud mortar, sometimes laid in a herringbone pattern (Fig. 7).

4 RENOVATION AND DEVELOPMENT PROGRAM OF HISTORIC ADDIR'IYAH

A Royal decree was issued on 17/6/1419H approving the development program of historic Addir'iyah and an executive committee was established under the chairmanship of HRH Prince Salman bin Abdulaziz with its members from various concerned organizations (ArRiyadh City Web Site 2014). High Commission for the Development of Arriyadh was responsible to supervise and execute this program. They carried out specialized studies on historic Addir'iyah, and prepared a final executive plan in cooperation with the General Commission for Tourism and Antiquities for the restoration and rehabilitation of Addir'iyah. According to the plan Addir'iyah was supposed to be restored and renovated into its original shape. The development program of Addir'iyah composed of large number of cultural, architectural, social, economic projects and upon the execution of these projects Addir'iyah planned to become a model cultural tourism hub of the region (ArRyadh City Web Site 2014). According to Facey (1997, 84) in 1974 the Department of Antiquities and Museums became interested in preserving the remains of Dir`iyah, and UNESCO sponsored a study of the site which was carried out by the Egyptian Antiquities Organization and in 1978 a second study had been carried out. The aim of both studies was to recommend structures for preservation and restoration, and propose methods for doing so. The Department of Antiquities has faced this challenge by selecting only certain buildings for restoration. The approach recommended in the UNESCO reports. Inside Turayf, since the 1970s, it has concentrated on restoring first the Palace of Nasr, and then, in the late 1980s, the Palace of Sa'd.

The restoration of sun-dried mud buildings involves difficult decisions. Such buildings naturally suffer from heavy rain, and require a constant maintenance. Up-to-date there is no additive found which can be added in mud mix to get 100% satisfactory result in terms of increased durability. The decision to use only original materials means that a continuing program of repair of restored buildings has to be undertaken. A further problem with the restoration of a mud building was that it was almost impossible to tell which part was original, and which one was re-built or renovated. Mud walls tend to collapse completely into shapeless forms which tell little about their original appearance except for their alignment. They have to be re-built as new, therefore, and inevitably "restored" structures take on the appearance of brand new buildings. In the process, the atmosphere of age is lost, however much care is taken, as it must be, to use original materials Facey 1997, 85). With time, the decision may be taken to re-build other structures inside Turayf, such as the Salwa Palaces, the Bait al-Mal, the Turayf Mosque and the Baths. However, at present it is seen as a priority to consolidate many of the remains as they are, to prevent further deterioration. The Department faces its task with the determination to preserve this historic site for future generations of Saudi Arabians. No visitor, whether Saudi or foreign, can fail to be impressed by the work that has been carried out. In this new phase of its existence, as the great symbol of Saudi Arabian nationhood, Dirlyyah's survival is assured.

However, the recent development and method of construction work at the visitor's center of Addiriyah raised a lot of criticism amongst the earth construction researchers. The entire visitor's center is constructed out of reinforced concrete and load bearing wall (Fig. 8) and the exterior of the walls are unceremoniously plastered with mud (Fig. 9). It is important to note that the mud plasters were washed away with the rain (Fig. 10) and the original brick structure was exposed and visible (Fig. 11).

Figure 8. The reinforced concrete construction work of the visitor's center of Addiriyah (Author 2013).

Figure 9. The exterior wall of the visitor's center of Addiriyah plastered with mud (Author 2013).

Figure 10. Mud plaster washed away by the rain from the exterior walls of the visitor's center of Addiriyah (Author 2013).

Figure 11. The original load bearing wall structure of the visitor's center of Addiriyah was exposed (Author 2013).

Figure 12. The restoration work of fortification wall of Addiriyah (Facey 1997, 85).

The department of Antiquities put a lot of effort restoring fortification wall (Fig. 12) of Addiriyah in the 1980s. Locally available Limestone and mud mortar was used for the reconstruction of this wall. The wall still stands as a witness of appropriate building material and construction technique. The construction work of the Visitor's Center or any other newly developed building could be carried out using stabilized earth construction method which would be way durable compared to traditional mud brick construction.

5 CONCLUSIONS

This paper presented the case of Addiriyah to the international audience and enhanced the need of using appropriate building material and construction method for conserving architectural heritage in the Kingdom of Saudi Arabia. The conservation work of the old settlement of Addiriyah is using un-stabilized mud bricks but stone could be used as the major building material considering unstabilized mud brick is way less durable compared to the stone. Furthermore the newly developed and constructed zones such as Visitor's Centre could use contemporary stabilized earth construction technologies instead of faking its original structural system. One must recognize and appreciate the importance of conserving its architectural heritage. The Saudi government needs to save its past architectural heritage in the best and appropriate manner and build a future that respects pride in its own history and culture. Besides, the government alone cannot be responsible for urban conservation but must fully mobilize the public and private sector or public, private partnerships. In addition, international organizations can be engaged with the local organizations to work on the best and appropriate construction techniques and building materials for conservation work.

ACKNOWLEDGEMENTS

The author is grateful to King Fahd University of Petroleum and Minerals (KFUPM), Dhahran, Saudi Arabia for their support to write and present this paper to this International Conference.

REFERENCES

ArRiyadh City Web Site 2014. *High Commission for the Development of Arriyadh*. http://www.arriyadh.com/Eng/ADA/Left/DevProj/getdocument.aspx?f=/openshare/Eng/ADA/Left/DevProj/AddiriyaEn093.doc_cvt.htm (Accessed 01.03.2014).

Al-Rashid, Saad A. 1999. *An introduction to Saudi Arabian Antiquities*. Riyadh: Deputy Ministry of Antiquities and Museums, Ministry of Education, Kingdom of Saudi Arabia.

Facey, William 1997. *Diriyyah and the first Saudi state*. London: Mark Petre. Stacey International.

Google image 2014. http://joshuaproject.net/profiles/maps/sa-map.gif. (Accessed 25.03.2014).

Ishteeaque, Ellahi M. & Al-Said, Fahad A. 2008. *The Native Architecture of Saudi Arabia, Architecture and Identity*. Riyadh: Riyadh Municipality.

King, Geoffrey 1998. *The Traditional Architecture of Saudi Arabia*. London: I. B. Tauris & Co Ltd.

World Travel Web Site 2014. *Historic Al-Diriya*. http://world-travel7.blogspot.com/2013_07_01_archive.html (Accessed 10.02.2014).

Earthen Architecture: Past, Present and Future – Mileto, Vegas, García Soriano & Cristini (Eds)
© 2015 Taylor & Francis Group, London, ISBN 978-1-138-02711-4

Architecture and landscape in Sharq al-Andalus. The conservation of an Islamic *qarya*

A. Soler Estrela
Universitat Jaume I, Castellón, Spain

R. Soler Verdú
Universidad Politécnica de Valencia, Valencia, Spain

E. Paredes Vañó
Technical Architect, Alcoi, Alicante, Spain

ABSTRACT: Most of the rural settlements (*qarya*) of Sharq al-Andalus have unfortunately disappeared. It is only possible to find some barely recognisable ruins. La Cairola (Vall d'Ebo, Alicante, Spain) is an exception, situated on a rocky plateau on the edge of farmland, overlooking the valley, with a view of other settlements and significant points of the Andalusian cultural landscape. Currently, despite its ruined state, it is a material document of exceptional value because of its authenticity. It accurately reflects vernacular building techniques and even makes it possible to reconstruct the spaces for this unique architecture, dating back to the late 10th century, the early Islamic period. This paper deals with a proposed intervention to solve serious stability and conservation issues, respecting the original architectural design, materials, construction systems and techniques such as rammed earth. The project clearly aims to recall the Andalusian architecture inserted in a cultural landscape.

1 INTRODUCTION

The valleys of the county of Marina Alta (Alicante, Spain) are special places of outstanding natural and historical value. They were profusely inhabited in the Mudejar and Morisco (Moorish) periods, between the 13th and 17th centuries. The importance of this cultural landscape has to be valued as the area that possibly best preserves the authentic dispersed Andalusian settlements throughout the Hispanic world. Besides, the remaining architecture is in great danger of extinction, something which would entail the loss of the memory of more than seven centuries of Muslim culture.

During this long period, the way in which the population settled in small villages or farmsteads cannot be ignored. Each of these places gathered just a few families, distributed along the valleys in a hilly landscape characterised by agricultural terraces built with stone walls. Some of these places were left uninhabited and, although they are very badly ruined and in danger of disappearing, they are still useful for observing the features of the hamlets of the time. They have been called "abandoned Morisco villages", given that the expulsion in 1609 of the Moriscos who inhabited these lands was a point of cultural separation and abandonment. Some of these places have not been re-inhabited but have been re-used as sheds for sheep (Christie el al. 2004). However, others were temporarily repopulated by Christian settlers.

The case of La Cairola is one of those which retain greater authenticity, as a dispersed settlement marking the territory with its presence in these valleys. It is representative of the long period from the Islamic period until the expulsion of the Moors.

Since the Venice Charter of 1964, it has been accepted that the concept of a historic monument

Figure 1. View of the group 1 (A. Soler).

embraces not only the individual architectural work but also the urban or rural setting in which the evidence of a particular civilisation, a significant development or a historic event is found. This applies not only to great works of art but also to more modest works from the past which have acquired cultural significance with the passing of time.

A restoration project was drafted (Soler 2011), which proposed to retrieve the architectural spaces, both in their historical and landscape dimension. It is based on direct observation of the current state and the reading of the architecture preserved. All the documentation necessary for the definition of the pathology was compiled beforehand, with special attention being paid to the graphic documentation that allows metrics and dimensional characteristics to be defined, also including specific studies such as the readings of constructive stratum, states of degradation, construction techniques and materials.

1.1 *Historical value*

Accounts of the existence of La Cairola "qarya" go back to the 10th or 11th centuries, according to ceramics found. It was possibly cited in the documents as Ebu in 1245 and 1279, as a single settlement in the valley. There is documentary evidence that it had twenty one houses in 1361, twenty in 1391, eleven in 1571, and just seven houses in 1602. Depopulated in 1609 (Torró, Ivars 1990), it is an example of extraordinary antique architecture dating back to the early Islamic period, and is therefore of special historical value.

Although unfortunately this information is barely relevant to an understanding of the architecture of the site, the research of distinguished experts (Bazzana 1992, Bazzana et al. 1988) provides unanimously accepted theoretical reference models that help to interpret the relationship between these sites, the fortresses and the territory.

1.2 *Architectural value*

The state of ruin is due to an abandonment spanning centuries, which has caused the buildings to lose roofs, floors and lintels. Nonetheless, it conserves most of the walls, still standing as an extraordinarily valuable document because of its authenticity. They accurately reflect the construction techniques and even make it possible to reconstruct the original spaces.

This is especially important due to the fact that they are one of the few that remain today. The structure has survived several earthquakes and abandonment during many centuries. Unfortunately other examples have disappeared and in these sites it is only possible to find some barely recognisable walls or even worse, only the name in a map. Therefore, the value of the architecture preserved is highly exceptional as a testimony of disappeared architecture that deserves to be rescued from oblivion.

2 DESCRIPTION

La Cairola is located at the entrance to the valley, from Alcalà fortress, on a hillside facing northeast, near the plain of the valley, which it overlooks. The land has a steep slope, which on the opposite side drops steeply to the Girona River.

The buildings that remain are located along the road, the construction of which has destroyed part of a house.

At present, the historical settlement can be divided into two existing groups of buildings.

Group 1: This is located on very hilly terrain, half hidden by vegetation and in ruinous condition. Wooden beams and roofs have almost completely disappeared, but their outlines can still be made out. The current state can only be understood as a result of multiple interventions before its abandonment, which is why this description is made with difficulty. The northwest area has two L-shaped rooms connected to a patio. A very long

Figure 2. The Cairola. Group 1 and group 2 locations (A. Soler).

wall extends the group to the other side. In these two rooms between walls, the main one is longitudinal in the direction of the maximum slope of the land. Traces of the missing beams can be observed, and the slope of the pitched roof can be deduced. The southeast part is actually closed for private use, so it is not accessible. From the outside it can be described as a single floor with a pitched roof, but at the bottom, two walls forming a corner remain with a height of up to 3 floors.

The observation of the walls makes it possible to perceive different techniques: tapia and masonry. The analysis of phases confirms greater antiquity for the tapia walls. The group provides extensive data for understanding the construction technique of tapia, making it possible to establish the dimensions and details of the forms used. The use of three wooden provisional stays per board can be appreciated. Executed tapia lengths are close to 1.65 m. Traces of wooden beams show that their diameter was 15 cm, and that they were positioned at a distance of 60 cm. As regards typology, the buildings are two storeys high with pitched roofs. Inner widths are 2.60 and 2.85 m. The situation of the openings is known in the longer side of the room, but their exact form is unknown.

Group 2: Unfortunately, this group is in ruins and has been destroyed by the current road. Nonetheless important features can still be observed. One of them is the former existence of two storeys, with an intermediate floor that has disappeared. The entrance door width was 1.40 m. The way of access to the first floor is unknown, but a window can be observed.

At the end of the patio, currently divided by other walls, there is another room, not accessible due to abundant vegetation. Most of the group is constructed using the tapia technique, so that its study provides interesting data on this method of construction. The data collected, in accordance with Group 1 indicates executed tapia lengths are close to 1.70 m.

Figure 3. Detail of traces of disappeared beams (A. Soler).

Figure 4. Current state, group 2. Analysis of wall techniques (A. Soler).

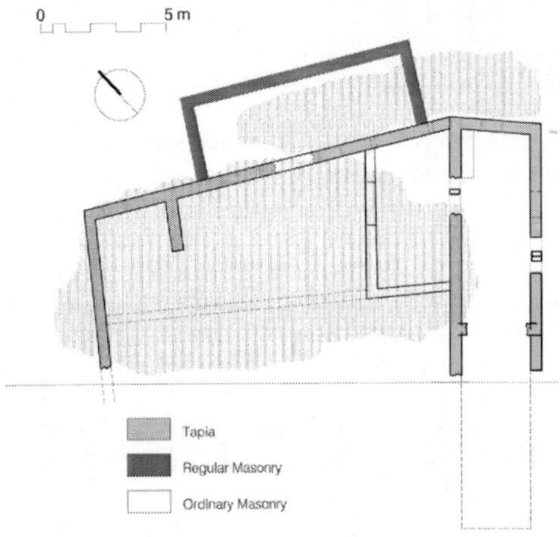

Figure 5. Current state, group 2 (A. Soler).

Near Group 2, across the road, numerous masonry and tile fragments seem to indicate the former existence of another building, called Group 3. To gain further knowledge, more archaeological work is required.

3 PROPOSED INTERVENTION

Taking into account the current state of this representative architectural heritage, a restoration Project has been drafted, aiming to simultaneously safeguard the material remains and their value. This architectural intervention goes beyond the strictly technical dimension and delves into the heritage values of the monument.

Heritage restoration must recognise historical, symbolic, compositional, aesthetic, constructive, metric aspects... all values of material culture that are also intangible. Restoration is essential in enabling a personal reading of the monument, as a text, as a work of art, and as architecture.

The aim is to recover the historical dimension of the rural settlement, remembering the houses which have disappeared, with vegetation recreating the immediate environment, connecting with the landscape and making visual contact with other places.

3.1 *General criteria for intervention*

The proposed intervention should be considered primarily as a safeguard, respecting the original architectural design, materials and construction systems. A primary objective is to facilitate public access, establishing an itinerary to guarantee the safety of visitors and facilitate observation and respect for the monument. Another main objective is to halt the deterioration process through the consolidation and stabilisation of the walls.

The project follows the recommendations for restoration of ancient buildings (ICOMOS 2003), taking into account the specific characteristics of the case. The current state is the result of different interventions over centuries. The restoration would not try to recover the initial state of a particular phase, but rather to assess the interest of historical stages. Replacements of missing parts must be integrated harmoniously into the whole, while still being distinguishable from the original and ensuring that restoration does not falsify the historic evidence.

The process of restoration is based on respect for the original architecture, and on the information existing regarding construction processes. One value of this architectural heritage is as a unique product of the building technology of its time. The intervention should respect the tapia technique, and leave evidence that can be recognised. The restoration must be preceded and followed by an archaeological and historical study of the monument.

3.2 *Architecture and landscape*

The characteristics of these Andalusian settlements, spread out through the region, make the relationship between architecture and landscape evident. The concept of cultural landscape, as the combined work of nature and humankind, expressing a long-lasting and intimate relationship between peoples and their natural environment, fits perfectly (Mitchell et al. 2009).

The territory of Ebo valley is an extremely valuable example of a cultural landscape with interesting historical stratifications that have been studied (Muñoz, Soler 2006).

Figure 6. Proposed intervention: Itinerary and viewpoints (A. Soler).

According to the project objectives, it is necessary to define an itinerary for visitors, paying special attention to the route itself and the perceived sensations (Careri 2002). The path begins at an existing parking area beside the road (close to a cave that is open to the public, La Cova del Rull), from where it is possible to walk to a picnic area, under the shade of tall pines. Continuing along a dirt road, between the stone walls of the fields of olive trees, visitors reach Group 1 of The Cairola. From this point on, the priority is to provide a safe walk through the location, taking into account that the terrain is very uneven and rocky.

The building must be freed of the vegetation that invades it, recreating a perimetral path. As a corrective measure for the adverse visual impact of the road, a narrow strip of vegetation is maintained to conceal the road from the visitor's point of observation.

The amount of vegetation removed should be kept to a minimum, just the indispensable to respect the architectural environment and the path. At the same time, the environmental study will enhance the autochthonous flora and research on species that were predominant in the Muslim period, with a view to possible recovery.

Considering La Cairola as a historical site, a major aim is to understand the place and its relationship with other historical sites. In some sites, platforms or viewpoints are provided as observation points of the landscape, enjoying scenic views. These points are designed to direct the vision towards cultural or natural landmarks.

3.3 *Restoration of the tapia*

Stone tapia walls are a very specific type of tapia, to be found, within the Islamic culture, in Mediterranean mountains where stone is abundant. The technique shares all the elements of formwork, but cannot be considered to be rammed earth. Instead of the characteristic process of tamping, it follows

Figure 7. The stone tapias (A. Soler).

a specific procedure for filling, positioning rough stones and compacting. The proposed intervention is focused on the restoration of the tapia walls, which are in an advanced state of degradation. There are predominant building systems that still remain, but are affected by important pathologies such as:

- Loss of the upper crown.
- Absence of the top connection of the roof.
- Cracks and breaks.
- Loss or damage to the outer crust

The actions aimed at correcting the above anomalies are reflected in the plans and adhere to the following general methodological guidelines:

- Intervention should place special emphasis on respecting the original technique of the tapia. In the event of reconstruction, the maximum height will take into account the existing remains and dimensional modulation of original tapia constructions.
- Reconstruction of missing parts must be integrated harmoniously into the whole, while remaining distinguishable from the original
- The aim is to preserve all the imprints and patina, healing wounds and only replacing specific points for conceptual or technical reasons.

3.4 *Indoor spaces*

The project has a clear intention to recover the architecture of Andalusian culture. The current state of ruin prompted some questions on what the original construction was like.

Proposed intervention as the first phase is focused on Group 2. It is possible to restore a sequence of different types of rooms, corresponding to indoor spaces or outdoor patios. Despite having lost the intermediate floor and the roof, the existence of the rectangular two-storey unit is highly relevant, and it is still possible to determine the position and characteristics of the elements. The intervention aims to recover these spaces by reconstructing the floor and roof. This not only contributes to the stability of the walls, but also recovers a space that is considered exceptional, given that there are very few Moorish rural houses which still conserve two storeys.

The house outline can still be read from the ruin. It refers to an ideal type with rooms on both sides of the patio, hence the interest in recovering the room at the bottom of the patio.

Analyses of the current state suggest that it was a one-storey construction with a pitched roof. A minimum intervention to recover the interpretation of the space is the replacement of the wooden beams but not the whole roof, given the limitations posed by its current state of ruin.

The project aims to implement this solution, but the process is conditioned by the information obtained at the beginning of the work, which requires the completion of archaeological tasks to verify the hypothesis on the building and provide specific data for intervention.

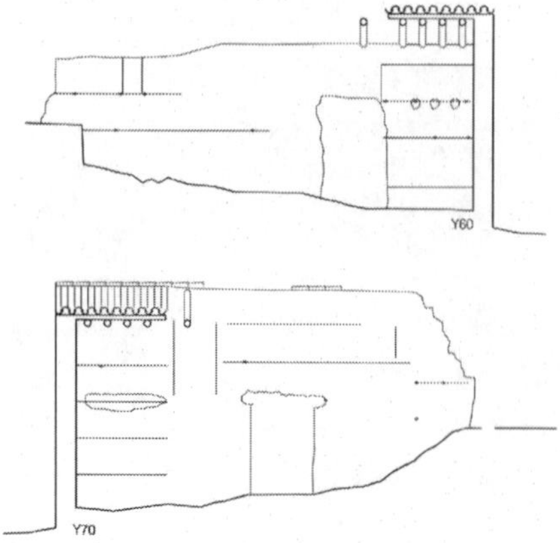

Figure 8. Current state. Indoor spaces, group 2 (A. Soler).

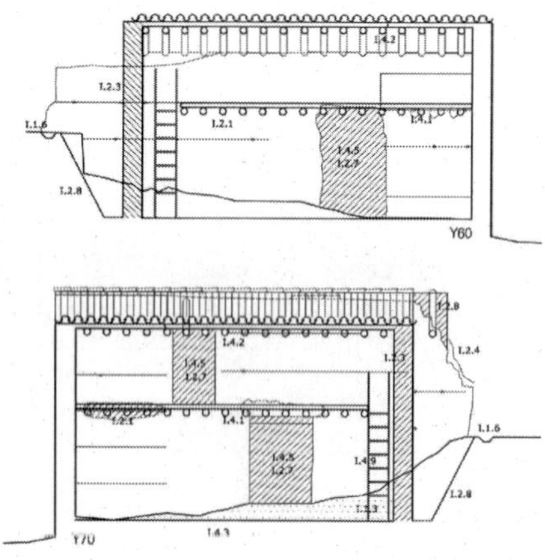

Figure 9. Proposed intervention. Indoor spaces, group 2 (A. Soler).

The location was inhabited for over 6 centuries (11th to 17th century). Rather than trying to define its exact state at a specific historical moment, the intention is to allow the evocation of a type of habitat with common features that match its historical and cultural background, and geographical environment, climate, materials, suggesting uses of spaces but allowing different interpretations.

4 CONCLUSIONS

The current state of the houses must be interpreted from a diachronic point of view. The ancient village has been affected by superimpositions over various centuries and subsequently been abandoned and ruined. Architectural spaces and typology have been determined. The architecture can be analysed at various levels, from landscape to the domestic sphere, in a succession of scales running from the smallest to the largest grouping. It begins with the smallest unit of constructed room detected and continues by juxtaposing these units to form houses, which together form villages. These, in turn, were distributed in the region in a particular way, configuring a humanised landscape.

In the period studied, architecture shows a series of invariable features, such as the use of building techniques like the tapia, within a Muslim-Andalusi cultural concept that extended into the Mudejar and Morisco periods (17th century).

Following a long period of neglect, the project proposes the beginning of the recovery of an architectural heritage in danger of disappearing.

The aim is to transmit to the monument visitors the values of the site's history. The recovery of architecture considering its heritage category must be understood as part of a cultural movement of a community to keep its identity.

REFERENCES

Bazzana, A. 1992. *Maisons d'Al-Andalus. Habitat medieval et structures du peuplement dans l'Espagne orientale*. Madrid: Casa de Velázquez.

Bazzana, A., Cressier, P & Guichard, P. 1988. *Les chateaux ruraux d'al-Andalus. Histoire et archeology des husn du sud-est de l'Espagne*. Madrid: Casa de Velázquez.

Careri, F. 2002. *Walkscapes. Walking as an aesthetic practice.* Barcelona: Gustavo Gili.

Christie, N., Beavitt, P., Gisbert, J.A., Seguí, J. & Gil, M.V. 2004. *Ethnography and Archaeology in Upland Mediterranean Spain. Manolo's World: Peopling the Recent Past in the Serra de L'Altmirant*. University of Leicester.

ICOMOS. 2003. Principles for the analysis, conservation and structural restoration of architectural heritage. http://www.international.icomos.org/charters/structures_e.pdf.

Mitchell, N., Rössler, M., & Tricaud P-M. 2009. *World Heritage Cultural Landscapes, A Handbook for Conservation and Management.* UNESCO.

Muñoz, A, Soler, R. & Soler, A. 2006. *Proyecto URBA-COST, corredor del Serpis.* Valencia: Generalitat Valenciana.

Soler, A. & Soler, R. 2011. Estudio arquitectónico, salvaguarda y puesta en valor del despoblado morisco de La Cairola, Vall d'Ebo, Alacant: Ajuntament de la Vall d'Ebo. Unpublished.

Soler, A., Soler, R. & Ortega Pérez, J.R. 2012. Restoration of the stone tapia of hsin al-Qala (Castell d'Alcalà de Gallinera). Alicante. Spain. In Mileto, Vegas, Cristini (eds.) *Rammed Earth Conservation*. London, pp. 413–418.

Torro, J. & Ivars, J. 1990. La vivienda rural mudéjar y morisca en el sur del País Valenciano. En *La casa hispano musulmana. Aportaciones de la arqueología*. Granada.

Earthen Architecture: Past, Present and Future – Mileto, Vegas, García Soriano & Cristini (Eds)
© 2015 Taylor & Francis Group, London, ISBN 978-1-138-02711-4

Study of the Pre-Hispanic Road located in the PUCP Campus, Lima, Peru

J. Vargas-Neumann, F. Jonnard, S. Gil Zacarias & J. Montoya Robles
Centro Tierra, Pontificia Universidad Católica del Perú, Lima, Perú

ABSTRACT: The Pre-Hispanic Road located at the Campus of Pontificia Universidad Católica del Perú (PUCP) is one of the few remains of the Qhapaq Ñan road system that crosses the city of Lima. Its 467 meters of two parallel walls that have withstood severe earthquakes can provide valuable information on the techniques used by the Ichsma (1100–1472 A.D.) and Incas (1472–1532 A.D.) in the construction. Centro Tierra-INTE, a PUCP research team and UNESCO Chair in Earthen Architecture, is conducting a study to gain an accurate knowledge of the material and the construction technique and to propose guidelines for the conservation of this type of earthen structures so recurrent in archeological sites of the Peruvian Central Coast. The interdisciplinary study includes a bibliographic, historical, archeological compilation, soil mechanic tests, structures laboratories tests, and electronic microscopy sweep assays. Conclusions with a conservation methodology for Ichsma and Inca archaeology sites are presented.

1 INTRODUCTION

The Maranga Complex is one of the most important archaeological sites in the Rimac valley. The complex covered 150 hectares (370 acres) in its heyday (Canziani, 1987). Its archaeological sites are part of what is nowadays the *Parque de las Leyendas* zoo (52 *huacas*/burial grounds), Pontificia Universidad Católica del Perú, Universidad de San Marcos, and archaeological sites including Mateo Salado and Huaca La Luz, among many others.

This complex dates back to the Early Intermediate Period (Lima Culture 100–600 A.D.), and was occupied until the Late Inca Horizon (1472–1532 AD) (Canziani, 2012). The main construction material was earth, used in applications ranging from the mortar for setting stones and small and large adobes to earthen walls and plasterwork.

The site reveals the large variety and combination of building methods using earth and stone used in the Rimac valley—many of them with modifications and extensions made by various sequential cultures—and particularly the convergence of influences from the Lima, Ichsma and Inca cultures in the Maranga Complex, which flourished during the latter two cultures.

A part of this complex, the Ichsma-Inca Road (see location in Figure 1), linked important sites and stretched to the Pachacamac Sanctuary in the Lurin valley. It is one of the few remains of the Qhapaq Ñan road system across the Rimac valley, and ends at an outcrop located between the Huaca Tres Palos and the Huaca La Cruz, major sites of the Maranga Complex. (L. Carrión, personal communication, March 7, 2014).

This project focuses on the walls bordering this road along a stretch of 467 meters (1532 feet) within the Pontificia Universidad Católica del Perú (PUCP) campus.

Since the beginning of 2013, studies and interventions to this stretch of the road commissioned by PUCP have been underway. Masonry with new adobes has been used for consolidation and reconstruction of the original walls.

As a result of this intervention, the Centro Tierra INTE, a PUCP research group specializing in earthen constructions, set out to conduct an in-depth study of this type of structures, given their importance in the Rimac valley, and proceeded to request approval from the Vice President of the PUCP for this project, which began in September 2013.

2 PRECEDENTS

In the archaeological sites located in the Rimac River valley, there are several precedents of wall repairs using adobe masonry to reconstruct missing portions, especially in the eroded lower wall segments, in order to regain their stability. Such repairs included emergency work, for example along the main road of the Cajamarquilla archaeological site and, later, intervention work such as in the Mateo Salado archaeological complex (Figure 2) and Huaca San Miguel in the Parque de las Leyendas, among others.

On the other hand, a large variety of vertical, sloping walls, with either constant or trapezoid sections, exist in Maranga. These have been

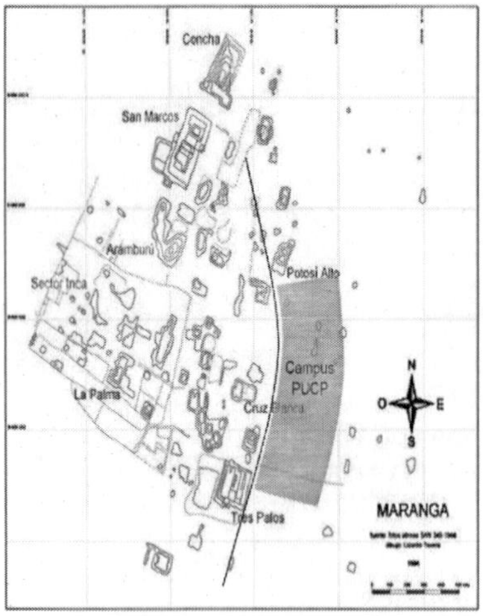

Figure 1. Ground plan of the Maranga Complex. Location of the Maranga Complex *huacas* and the Pre-Hispanic Road in the southern section of the PUCP campus. Credit: L.Tavera. Modified by F. Jonnard.

Figure 2. Consolidation with new adobe masonry. Pyramid A of Mateo Salado. Photo: Mirna Soto.

graphically identified to a certain degree for classification purposes.

These pre-Hispanic high solid earthen walls are commonly called *tapiales* (rammed-earth walls). However, definitions for these structures are still relatively unclear and ambiguous. In this paper, we conclude that they are not rammed-earth walls.

As to the mechanical characteristics of the walls, they are described in generic terms (hard, weak, filled with loose soil or soil mixed with riverbed stones), but no specific studies have been conducted to arrive at conclusions as to their restoration. There are no precedents of numerical results from lab tests that may serve as comparisons in order to classify the walls or to assess their structural behavior.

As to their construction, the first reference available to us was provided by Spanish Father Bernabé Cobo: *"The Indians in ancient times did not make these mud walls as we do, with slightly moistened loose soil, but with well-kneaded and soft mud, just as we make adobes. Their walls were very straight and smooth because, instead of using wooden formwork on each side, they used cloth and fiber mesh, and subsequently plastered them with the same mud. We have come to determine their shape and construction because of the large number of ancient mud walls found throughout the Lima valley that are made in this manner."* (Cobo 1964: 241).

This description reveals the differences between the *tapial* as the term is understood today (rammed-earth), and these walls mentioned above.

In general, in Spanish, rammed earth fences are called *tapias* (a word derived from Arabic). These walls were and are made by compacting moist earth with wooden rams inside forms originally made with wooden planks, in 100 to 150 mm (4" to 6") layers and approximately 0.60 (23") courses, dried for a period of time before making the top courses. Therefore, any wall made in a way similar to this Mozarabic technique is called tapia, without making any construction distinctions.

This study of the road's wall materials, as well as the porosity, strength and other characteristics of its composition, propounds the need for a more accurate definition of the method used, given the great importance of this information for any recommendation on repairs or restoration of these walls.

3 OBJECTIVES AND METHODOLOGY

The purpose of this research is to study the walls bordering the Pre-Hispanic Road, their geometric characteristics, the mechanics of its materials, their stability and strength characteristics, and a definition of the construction method used.

Likewise, the study seeks to promote sharing and generating new knowledge about pre-Hispanic construction methods and, as a result, propose guidelines for conservation of these types of earthen structures that are prevalent in the Peruvian central coast (Maranga, Mateo Salado, Cajamarquilla, Puruchuco, Huaycán de Pariachi and others).

Given the historical, archaeological and technical components of this study, it has been conducted in two stages:

The first stage was a brief bibliographic, historical and archaeological compilation to provide a reference framework. For this purpose, Inés del Águila, a researcher at the PUCP's Riva Agüero Institute, and the archaeologist Lucénida Carrión, director of Parque de las Leyendas, were interviewed. Both researchers are considered leading experts on pre-Hispanic cultures in the Rimac valley and, particularly, on the Maranga Complex, where they have carried out heritage maintenance, restoration and conservation work for many years.

The technical stage covers studies conducted at the PUCP Soil Mechanics and Structures labs (granulometry, Atterberg limits, volumetric contraction, compressive strength, indirect traction, consolidation and other field tests). Additionally, the assessment of degree of compaction, porosity, particle size and nature, and chemical composition was complemented by electron microscope scanning.

This task was carried out in sections to allow for subsequent diagnosis based on sections and cases leading to a set of recommendations for intervention. As this is an archaeological site, studies were conducted on portions dislodged from wall surfaces buried along the sides of the road. The 19 samples studied were provided under supervision by the team of archaeologists, verifying their location and orientation with respect to the wall.

Figure 3. South earthen wall of the road. The red arrow indicates the fallen riverbed stones. Photo: J.Vargas N.

4 RESULTS

The Pre-Hispanic Road was built with earth and stone, using various methods from two different cultures (Ichsma and Inca) and from different stages of these cultures. There are also vestiges of four centuries of Hispanic interventions.

At a late stage of the Ichsma culture, pyramids were built using an organized filler method. Cells were formed by orthogonal walls made of riverbed stones with mud mortar, filled unsystematically with earth and stone. Perimeter walls were sloping and made of poured earth in layers from 0.15 to 0.20 meters (6" to 8") high. Walls can be made of two or three sloping layers as necessary to make the pyramid's platform. This structure can be seen in the Huaca Mateo Salado. Figure 3 shows the described structural elements and concepts used in the same manner as in the Road. (Vargas J. 2009). In both constructions, the top sections are completed with solely earthen walls, possibly of Inca construction, sometimes supported by riverbed stone layers, simulating Ichsma earth and stone methods.

During Colonial or Republican times, stretches of walls with parallel faces may have been built using *tapial* methods in order to complete stretches missing as a result of earthquake damage. We must not forget that, at the least, the Pre-Hispanic Road has survived the 1687 and 1746 earthquakes, which practically leveled Lima.

At a macroscopic study, the road samples collected show various compactness rates. Some are very hard and solid, while others completely disintegrate when handled. Some include a large amount of riverbed stones, while others are made of very fine soil. On limited occasions, fibers and small ceramic fragments were found.

The results of the lab assays and field tests show, firstly, a lack of uniformity due to the difficulty in maintaining a source of consistent soil and consistent quality along the road's extensive length and also its probable construction in successive stretches.

It is believed that the walls were built using top soil from the Rimac River deposits and river stones from its alluvial fan, which is very vast. The top layers usually have a larger content of wind-borne fine sands, while the lower layers contain clay, an essential component for good earthen construction.

Plasticity limits, as determined for each sample, as well as from granulometric analysis, show that the soil is a fine particle (more than 50% of the material passes through a 200 mesh screen), slightly plastic (with a < 50 liquid limit) clayey-silt.

On one hand, observation by scanning electron microscope confirms this high proportion of fine sands and silts, resulting in no tensile strength and low compression.

On the other hand, most samples from the assayed sections show very little grain to grain contact, leading to the assumption that little or no compacting was carried out. Clays in the global matrix are very low and the degree of porosity reaches almost 50% in the more fragile samples. This confirms that no major compacting was used in the construction method, leading to believe that poured earth methods were used using soft and deformable molds, and the presence of imprints from 0.15 to 0.20 meters (6" to 8") high, as observed in the Maranga Complex. Lastly, the presence of montmorillonite was determined. The percentage of montmorillonite, a highly hygroscopic clay, in relation to the total material analyzed was very low.

This fact stresses the importance of geographical conditions in the wall's conservation, since the ever-present high degree of moisture in Lima is an important factor in sustaining these structures.

To test the level of material strength, compression testing was done in carved/cut probes with samples extracted from the Road. Resulting strengths ranged from 0.15 to 0.43 MPa (1.56 to 4.4 kgf/cm^2). Current Peruvian standards specify a minimum strength of 1.2 MPa in compressed assayed cubes. (Figure 4).

Assays in carved/cut probes were also done on the new adobes used in the intervention work, with results of 1.8 MPa, much higher than the original road materials. It was concluded that the original construction material has very low strength, compared to the adobe strength, which is important information for restoration or intervention efforts.

It is important to note that at present there are no serious studies on the mechanical characteristics of the materials and methods used by builders in pre-Hispanic cultures. However, current consolidation methods, such as the adobe masonry method being used in the Cajamarquilla archaeological site, are interventions made without deep understanding of the immense Peruvian heritage in earth and stone, as is the case of the Pre-Hispanic Road.

The compression strength of adobe pile masonry using mud mortar, according to the Peruvian Standard NTE E.080 Adobe, is 50% of the compression strength of adobe blocks. The compression strength used in the intervention conducted in the PUCP campus is 0.9 MPa, that is, approximately 4 times more than the average strength of the Pre-Hispanic Road walls.

The elasticity modulus of fragile materials, though not linear, has an initial portion that is almost linear, which is usually considered linear based on the modulus of elasticity tangential to the curve of unit efforts-deformations in compression assays. The deformability is thereby considered linear and proportional to the strength.

A simple example: an original wall with a trapezoidal cross section that has lost half of its volume is replaced by an adobe masonry wall in the same shape. In an earthquake, the two half-walls, due to their differences in deformability and strength, will vibrate at different basic frequencies and will knock against each other, damaging or causing the collapse of the one with lower strength—which we know in this case to be the original heritage wall.

The consolidation required in the Road must use materials and methods similar to those used originally, that is, low walls (maximum 0.20 m [8"] high) molded in layers made of poured mud. The material must be as similar as possible to the original in granulometry, deformability and strength.

A method must also be developed for placing the last top layer of the new material, to be applied one month after laying and drying. We propose using a mortar of the same material, braced by thin wood slabs of sufficient size to ensure total filling of the empty space left after the drying process.

Figure 4. Compression assays on the original material of the road.

5 CONCLUSIONS

Our findings have allowed us to establish a technique that would resemble the poured earth method more closely than the *tapial* method. However, an important component of the second part of our research is to actually experiment with such technique. This experience, based on the studied bibliography and the test results, will define a construction method with even greater certainty, since certain field criteria are difficult to be determined on the basis of the results obtained so far.

It is necessary to call as much attention as possible to the problems that may arise from interventions using adobe adjacent to a much less resistant material in the walls, considering that original walls are not rammed-earth.

Dissemination of this information and specific studies need to be systematized and must be supported, since the components in the earth change according to the context. This will allow for developing restoration techniques that are appropriate for each case.

REFERENCES

Canziani J. 1987. Análisis del complejo urbano Maranga Chayavilca. *Gaceta Arqueológica Andina n.14,* pp. 10–17, Lima.

Canziani J. 2012. *Ciudad y territorio en los Andes. Contribuciones a la historia del urbanismo Prehispánico.* Fondo Editorial de la PUCP. Lima.

Carrion L. 2014. Personal Communication (authors). Lima.

Cobo B. 1964. *Crónicas.*

Del Águila I. 2014. Personal Communication (authors). Lima.

Vargas J. 2009. Conservación del Complejo arqueológico Mateo Salado. *APT's Annual Conference.* Los Angeles.

Earthen Architecture: Past, Present and Future – Mileto, Vegas, García Soriano & Cristini (Eds)
© 2015 Taylor & Francis Group, London, ISBN 978-1-138-02711-4

Rammed earth and prestige in Rincón's region, the Abbey House in Castielfabib, Spain

À. Vázquez-Esparza
Universitat Politècnica de València, Valencia, Spain

ABSTRACT: Rammed earth is generally considered a humble building technique, which lacks character and prestige. However, in the Rincón de Ademuz, the vernacular architecture is built almost always resorting to gypsum, wood and masonry walls, whereas rammed earth rarely appears. The reason for this is the shortage of this specific material in the area, which appears only in the valley of Turia's River, especially in Ademuz, Torrealta and Torrebaja. That is why the use of the mixed rammed earth with gypsum, *brencas* and *rafas* is so relevant. It is sometimes reinforced on the outside face with masonry, as a rural version of the more common brick-reinforced rammed earth which is used in very specific cases such as the House of the Garcés de Marcilla in Torrealta and the Abbey House in Castielfabib. Within the specificities of the vernacular architecture of this area, both examples demonstrate that rammed earth became in some cases a feature of prestige or social significance.

1 INTRODUCTION

The region of El Rincón is located in the Northwest of the Valencian Country, in an isolated enclave surrounded by the regions of Aragon and Castilla—La Mancha. In spite of the geographical and socio-cultural relationship with Aragon, the region has been part of Valencia since the beginning of the 13th century. During this period the towns of Castielfabib and Ademuz were conquered to the Muslims first by Pedro II, and again later by Jaume I who incorporated them into the Kingdom of Valencia.

The isolation caused by the topography of the land, the lack of infrastructure as well as the great depopulation suffered since the 1960s, has encouraged much of their tangible and intangible heritage. In this way, a large amount of pre-industrial, residential vernacular architecture, as well as civil or religious buildings have been preserved without significant changes.

2 VERNACULAR ARCHITECTURE IN EL RINCÓN, THE GYPSUM

El Rincón landscape, rugged and varied, has naturally developed a vernacular architecture stylebased on the use of four materials; gypsum, stone, reeds and wood. The most significant element is gypsum, obtained after firing the gypsum stone, an abundant material found on the slopes of the region. It is used practically in all the constructive elements, "as the main binder to make walls, pillars, frameworks, roofs, built-in-furniture, etc. "(Mileto & Vegas 2008, p.174). The frequent use of this material is its high resistance and versatility of use, both indoors and outdoors. Moreover the quick setting time of gypsum enables fast construction, thus optimizing the use of formworks and execution times. This fact justifies the denomination of gypsum in El Rincón, as the "total source" in the vernacular architecture of the region. Other materials such as lime that require a higher temperature and longer production time, require a large amount of firewood, and are therefore scarcely used.

The constructions in the region have a similar structure in terms of use within the different existing typologies. No vernacular residential building has more than 4 levels, 3 being the most common number. The ground floor was kept for the animals, the corral, the house itself stood on the first floor, and on the top floor the *cambra*, a place for grain storage, food and other belongings. The existence of access to different levels is very common, especially on the ground and first floor, because of the topography of the region. Other types of structures such as wineries, straw lofts, etc. tends to have fewer levels: between 1 and 2, although it depends on the conditions of the site.

Overall, all of them share the same constructive features, in general, with different variations. Special mention is due to the few dry stone constructions like field cabins, etc., in areas where this material is plentiful and always with a temporary function or agricultural support in general.

The exception is found in the higher areas of the region as Arroyo Cerezo or Puebla de San Miguel, where the stone is frequently used.

At the structural level, the buildings are not built through load-bearing walls from the foundation to the roof. Load-bearing walls are used only on the ground floor, built with masonry for several reasons: to isolate the rest of building's dampness by capillarity, to receive and distribute the structural efforts satisfactorily, and finally to provide a greater resistance to these elements in front of the stakes of animals since on the ground floors in residential architecture the corral was located. On these walls, a vertical pillar structure formed by buttresses of gypsum mixed with masonry (a kind of cyclopean concrete) connected through logs of wood starts. The horizontal structure is formed by beams and joists of wood, creating ceilings using gypsum formwork in mass. Hence, "the structure in Domino... of Le Corbusier, already existed long ago, in the vernacular architecture of el Rincón de Ademuz" (Vegas et al. 2001, p. 47).

The enclosures employees in divisions, façades, interior walls, and other interior elements such as *vaseras, cantareras* or fireplace hoods, obey the same building technique. This technique is based on the placement of vertical stone slabs, taken with mortar of gypsum, with or without the help of formwork. The absence of this element shows the vertical slabs on both sides, however it was very common the use of the formwork in one side, applying a layer of gypsum mortar on it and placing immediately the slabs to the wall (Fig. 1). In cases of external enclosures, these elements were built confined between the mentioned gypsum buttresses.

It should be noted that gypsum varies its properties according to the calcination temperature. When it gets up to 200°C it contains a high amount of anhydrite and sets slowly. However when it reaches up to 900°C the gypsum obtained does not contain any anhydrite, so it is very resistant and sets quickly. For this reason, the gypsum resulting from the traditional calcination kiln is called multi-phase, because it has components fired at different temperatures and with different properties. Moreover considering that the presence of non-fired gypsum stone parts increases the setting time of the material, the gypsum used in vernacular architecture is the result of a heterogeneous combination of components according to the quality of the calcinations process (Villanueva, 2004).

Figure 1. Gypsum formwork texture. External enclosure (Vázquez-Esparza, À).

3 RAMMED EARTH IN EL RINCÓN

In Spanish, *tapia* (rammed earth) and *tapial* (rammed earth formwork) is popularly used interchangeably, to designate the walls of rammed earth. However, it can also be included within this building technique, those walls erected "with formwork as a box or mold, where a wide range of materials such as masonry with concrete of lime or concrete mixed with crushed stone of stone can be poured" (Font & Hidalgo 2009, 59). Using this definition as a reference, we find examples of formwork walls (*tapiales*) of different types, from the gypsum walls in the vernacular architecture described, as well as many other forms of wall that will be associated within the context that concerns us.

With regards to rammed earth, "despite the apparent existence of earth suitable for immediate use in the erection of buildings, the construction with raw earth does not abounds in the Rincón de Ademuz" (Vegas et al. 2001, 34), therefore their presence is linked to those areas where scarce stone and gypsum is found, the Ebrón and Turia's Valley. Either way, in the cases of the known rammed earth examples in the area, the use of gypsum will often be present in one way or another, as we will explain.

3.1 *Popular architecture*

In almost all the cases of rammed earth in El Rincón vernacular architecture, the role played by this technique was to build enclosures within a preexisting structure of gypsum buttresses, usually not as the load-bearing element itself. These enclosures can be accompanied by gypsum joints between *modules* or *brencas* of gypsum as a reinforcement and protection addition (Mileto et al. 2012).

3.2 *Unique architecture*

The most significant architecture of the region is associated with one built with snecked masonry, combined with blocks of stone or masonry, as it is shown on the existing palaces or mansions.

The same happens when it comes to the towers, walls, or churches; although it doesn't need to be like this.

Those examples built in rammed earth, have either disappeared due to their fragility or lack of appreciation by the population, or sometimes there are hidden under layers of coating. Some cases of rammed earth construction can be found in the Hermitage of the Virgen de la Huerta in Ademuz, or the northern and east facades in the Castielfabib church on 4th floor, both built in the 13th century with lime concrete mixed with masonry. In the case of Castiefabib, these cloths are located in the most hidden part of the building, while the rest of facades were made in masonry. Civil buildings such as the House of the Villa in Castielfabib, 16th century (Lara 2007, 77), have a top floor built with rammed earth hidden behind a thick coating of gypsum, which has come to light after recent renovation works. We find remarkable examples of rammed earth in several walls in the towns of Torrebaja and Torrealta, built with earth and small joints of gypsum (*brencas*) by way of large ashlars, or some rests of the dependencies of the former convent of San Guillermo in Castielfabib. In terms of defensive architecture, the presence of lime concrete walls with masonry is very common in the west walls of the Castielfabib Castle, as well as on the inside part of the "Torreta", also in Castielfabib, unlike the outer wall formed by ashlars and masonry of limestone and tosca stone (Fig.4). Another example is the western wall of "el Carrerón" in the access to the fortified church, measuring more than 2 m thick, is part of the current complex original building, dating prior to the Christian conquest. In any case, the most symbolic rammed earth building, in the region is Torrealta Watchtower, a tower of the 14th century originally exempt (Eslava 2012). It retains its original structure including the machicolations of wood, and it is the only example existing in the region built with the classical technique of the brick-reinforced rammed earth (Mileto et al. 2012). Annexed to this tower, is the Garcés de Marcilla House, which together with the Abbey House of Castielfabib, constitute as the only two examples of stone-reinforced rammed earth with *brencas* (Fig. 2).

4 ONE-REINFORCED RAMMED EARTH WITH BRENCAS

As previously described, in the area of rammed earth constructions, the emergence of gypsum joints is frequent between the different modules of rammed earth, known as *brenca,* when this joint is thicker and has a wavy shape (Mileto et al. 2012) like a half moon or gutter. It is common that this half moon shape, the *brenca*, may appear also in the corners of the building, creating an embedded buttress in the wall but executed at the same time as the rest of the construction. This mixed solution between the traditional buttresses of gypsum and the load-bearing wall, receives the name of *rafa*. The presence of *brencas* and *rafas*, creates a wall much more durable and expensive to build due to the higher workload required in gypsum geometry. This geometrical composition as well as the colour contrast and material, creates a very attractive and distinctive image comparing with the rest of the buildings of the place. This matter combined with the cost, makes the few buildings that follow this pattern be linked to local noble families. Rammed earth with gypsum *brencas* is used especially in the regions of the South of Aragon, without proof of examples in the Valencian Country according to Fermí Font (2014). However we can say that there are significant examples in the regions of El Rincón or the Serranía, as we will describe below.

Figure 2. Watchtower and Garcés de Marcilla House in Torrealta (Vázquez-Esparza, À).

The rammed earth wall with *brencas*, in some special cases, is enhanced with the presence of masonry embedded in the rammed earth modules. This is a vernacular modification of the brick-reinforced rammed earth where apart from the *brencas* and gypsum *rafas*, the bricks are replaced by masonry. This reinterpretation is logical if we consider that in the traditional construction of the place, the use of brick was virtually nonexistent until the middle of the 20th century, when it was used just in the eaves of roofs. This typology of rammed earth is probably unique throughout Valencian country, and a contribution from the particularities of El Rincón with its Aragonese influences and the skillful handling of gypsum, to a constructive tradition entrenched in other areas of the territory such as the brick-reinforced rammed earth.

4.1 *Two unique cases, the Garcés de Marcilla House in Torrealta and the Abbey House of Castielfabib*

The only two cases of stone-reinforced rammed earth with *brencas*, are the Garcés de Marcilla House in Torrealta, and the Abbey House of Castielfabib. The Garcés de Marcilla family was of Navarrese origin and held great power within the Teruel nobility. They were the occupants of the Watchtower in Torrealta and the adjoining House. The tower is the oldest building in the town. It was built in the 16th century during the times of the Ruiz de Lihori, and it gave name to the population.

The role of this impressive building, was more than a noble residence: it used to be a shelter within the plain of the Turia's River. Attached to the Tower, is the Garcés de Marcilla House, a large mansion attached to the north facade of the Watchtower.

It was projected as a new residence with palatial aspirations for the noble family. Despite having undergone major renovations, it still retains the original access through an ashlars arch in the main façade, which presides over the square of the place, and where the formerly the Royal Road of Aragón was formerly situated (Eslava, 2012).

Today, beside the disfigurement that the building has suffered due to renivations, the different used wall modules can still be seen on the facade, as well as the reinforcement of masonry and *brencas*. However this structure appears only on the first floor of the building, where it is possible to see the existing *rafas* as reinforcement in the corners, as well as in the centre of the façade.

On the ground floor, walls are made of masonry, as usual, and upstairs the enclosures are apparently built with walls of slabs and gypsum, which may be due to a later enlargement, or most likely, due to the coexistence of both building techniques (Fig. 3). The current Abbey House in Castielfabib is a building which possesses documentary evidence of its existence in the 17th century, according to the historian Raúl Eslava Blasco and the building technique used on it. It is very probable that it belongs to the same period as the House of the Garcés de Marcilla in Torrealta. The Abbey House was built by a noble family. Perhaps the links between the Garcés and the Espejo family from Castielfabib, who shared both the patronage of the chapel of Santa Catalina Mártir in the Church of Castielfabib (Eslava, 2012), can explain the similarities of these two buildings. However this building, like many other Abbey houses of the region, was subsequently transferred to the Catholic Church, at which moment in which its use became the Abbey House of Castielfabib.

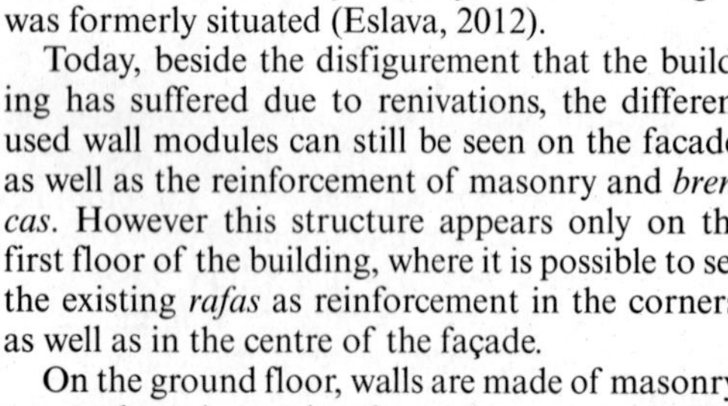

Figure 3. Garcés de Marcilla, stone-reinforced rammed earth with brencas (Vázquez-Esparza, À).

5 CASTIELFABIB ABBEY HOUSE

5.1 *Description*

The Abbey House, popularly known as the "Casa el Cura" (the Priest's House), is a building which sits at the foot of the town's castle, where a tower of the fortified enclosure of Castielfabib was placed. Its location, under the castle, on a rocky promontory to the east, allows a great visual control of the entire Ebrón's Valley, and the rest of the population. This fact justifies the settlement of a House originally linked to the nobility, in the same way that the Garcés was built next to the Watchtower in Torrealta. The housing consists of three dislocated units, one of greater importance, with three floors, and two smaller units with two floors. Of these two units, which were built later, the one in the northeast is located on a lower altitude than the main unit. Throughout this unit we can accede to the building, and it has various rooms that articulate the access to the rest of the volumes, even to the ancient latrine of the house situated in a cantilever which was the first toilet in the town. In the unit located in the northwest, we find a bread oven and a large *solanar* (traditional inner big balcony), as well as other private rooms. The third unit placed in the southern side, constitutes as the main piece, in terms of its representativeness within the urban area. It is located on the rock of three floors, and a three-slope roof. It is an important landmark in a village composed by dwellings of reduced dimensions in general, with a pitched roof between party walls. Inside, we can find several rooms and alcoves, except on the top floor where we find an open space under the structure of the roof, the traditional *cambra*. This last unit will be the one which we will focus our analysis on with regards to the stone-reinforced rammed earth with *brencas* (Fig. 4).

5.2 *Building system*

The constructive system of the Abbey House, specifically the main volume, is formed by load-

Figure 4. The main volume of the Abbey House, west, south and east façade. (Vázquez-Esparza, À).

bearing walls in all the structure, and there are two very distinct parts: the ground floor and the upper floors. The ground floor consists of a load-bearing masonry wall, with ashlars in the corners. As typical in the area, the ashlars are made by *tosca* stone, due to its easiness of carving, while the rest of the wall consists of limestone masonry. It should be noted that in the west facade, a change of the stone type and bond is perceived at the base, which corresponds to the ancient tower of the defensive enclosure whereon this building was placed.

The rest of the house is composed by walls of stone-reinforced rammed earth with gypsum *brencas* as we have already discussed. The existing modulation is shaped by five rows of rammed earth reinforced with *brencas*, and each row has a height of approximately 80–85 cm. This modulation does not correspond to the heights of the ceilings, so the division between the first and second floor is located in the middle of the third row. The *brencas* made by gypsum mixed with masonry, constitute a horizontal reinforcement as a crossed buttress between *rafas*, and generally with an alternating rhythm from the upper and lower rows.The modulation of this elements ranges between 40 cm to 60 cm between peaks, although very often this measure is irregular and depends on the distance to save (Fig. 5). The *rafas*, are placed in the corners or in the span jambs such as balconies, and they have been executed at the same time as the *brencas*, like on the rest of the wall. The part of the wall built with rammed earth strictly speaking, is located apparently filling the gap between the *rafas* and *brencas*, and it is made of lime mortar with a lime/sand proportion of 1:2.5 (Vegas et al., 2009, 104–105), which seems to vary between the different facades as it is shown in their physical appearance and degree of deterioration. In each one of the five existing modules, we find three rows of irregular masonry of reinforcement, except for the upper row, where it is possible to see just two, or even only one on the east façade. Both the masonry reinforcement and the *brencas*, and in all probability the *rafas*, do not cross the thickness of the wall, penetrating just a quarter of the wall on both sides, the core of the wall is only plain rammed earth. The internal divisions and external walls of the rest of the House, apart from the access façade volumes, are built using gypsum formworks, as the vernacular tradition shows.

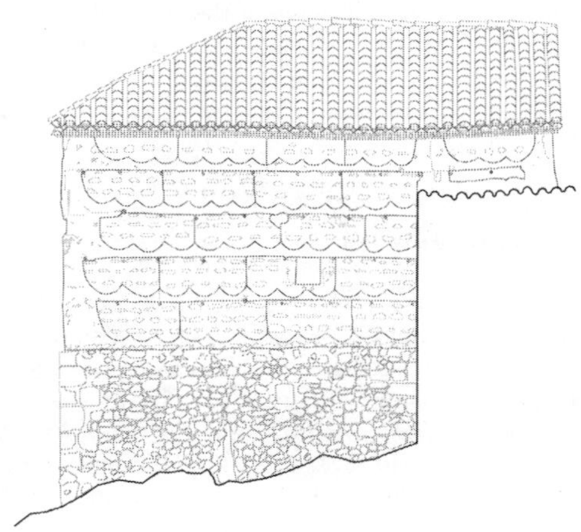

Figure 5. The *brenca's* modulation in the east façade (Vázquez-Esparza, À).

5.3 *Current state*

The Abbey House has undergone various modifications, as it is possible to perceive by reading their facades, but the original modulation and composition is still maintained. Major changes can be seen in the spans, where the vertical spans appear on the first floor. These spans appear flanked with *rafas*, subsequently reduced to windows with a parapet; a parapet built with the same technique and modulation as in the rest of the façade. This fact can tell us two different things: that this change was carried out during the building work, or maybe that when this modification took place they had enough sensitivity to use exactly the same architectural language. Later on, blinded spans can be appreciated, done with less care regarding the overall appearance of the building, as well as flaws of gypsum around some windows, especially in the west façade.

In general terms, this building is badly damaged by both the weather agents and the latest renovations carried out in the middle of the 20th century. As the images show, the south-west façade is the most damaged, most likely due to the great changes of temperature throughout the day as well as the freeze-thaw cycles, which may occur in an hygroscopic material such as earth, since gypsum (although deteriorated) has withstood the passage of time better. The south façade is in a better state of conservation and the east façade does not present any signs of deterioration, most likely because of a higher proportion of lime in the mass, as well as the shelter that the castle provides to the building in terms of weathering. On the other hand, the aforementioned latrine and the *solanar* are virtually destroyed, and in a progressive ruin state. Although the main pathology that the building presents is structural, and the origin is on the roof, where a rehabilitation of the 1960's, improved the watertightness of the element but did not solve the efforts that the roof structure creates on the corners; for this reason the building is opening itself up like a box made of cardboard.

6 CONCLUSIONS

To conclude, we can determine that the use of the rammed earth in El Rincón has coexisted with the vernacular architecture of gypsum, especially in those areas where there is scarcity of stone or deposits of gypsum, above all in the Valley of the Turia's River. However this architecture of earth was used as an enclosure, between buttresses with small thickness and gypsum boards. When it comes to singular architecture, such as churches or defensive structures, the use of lime formworks as a load-bearing wall is what has been more common, instead of rammed earth.

For this reason, the existence of two very specific cases such as the House of Garcés de Marcilla in Torrealta, and especially the Abbey House of Castielfabib, where the rammed earth is used in the load-bearing walls combined with large *brencas* of gypsum and *rafas*, suggests that this way of building was considered as a sign of prestige, since the use of masonry or stonework on the noble residences construction was common.

In Torrealta, the proximity of the earth, justifies its use, but in Castielfabib, where the stone abounds, the moving of huge amounts of earth to the top of the hill in order to build the walls, together with the careful design of the gypsum *brencas* and *rafas*, indicates that the use of this material by the noble family was not random, but a sign of prestige.

On the other hand, the masonry reinforcement, combined with the gypsum work, is a contribution of the region's particularities with its Aragonese influences and skillful handling of gypsum, to the constructive tradition entrenched in other areas of the territory as the brick-reinforced rammed earth is. Hence a reinforced rammed earth where the brick is replaced by the stone, creating a vernacular stone-reinforced rammed earth.

REFERENCES

Eslava Blasco, R. 2012. El señorío de Torrealta y su iglesia parroquial. *Ababol. Revista del Instituto Cultural y de Estudios del Rincón de Ademuz,* nº 8(68), pp. 4–14.

Font, F., Hidalgo, P. 2009. *Arquitecturas de tapia.* Castelló: Colegio Oficial de Aparejadores y Arquitectos Técnicos de Castellón.

Lara Ortega, S. 2007. *Las seis grandes lonjas de la Corona de Aragón.* Valencia: General de Ediciones de Arquitectura.

Mileto, C. & Vegas, F. 2008. *Homo faber. Arquitectura preindustrial del Rincón de Ademuz.* Casas Bajas: Mancomunidad de municipios del Rincón de Ademuz.

Mileto, C., Vegas López-Manzanares, F., García Soriano, L. & Mestre Sabater, M. 2012. Rammed earth construction in El Rincón de Ademuz (Spain). In Mileto, C., Vegas, F. & Cristini, V. (ed.), *Rammed Earth Conservation,* (pp. 181–186). London: 2012 Taylor & Francis Group.

Vegas, F., Mileto, C. & Cristini, V. 2009. Reinforcement of rammed earth constructions with gypsum in Aragon Area, Spain. In Achenza, M., Correia, M., Guillaud, H. (coord.), *Mediterra 2009, 1st Mediterranean Conference on Earth Architecture,* (pp. 99–108). Gorizia: Edicom.

Vegas, F., Mileto, C. & Zuccolotto, M. 2001. *Memoria construida. Arquitectura tradicional del Rincón de Ademuz.* Valencia: ADIRA.

Villanueva, L. 2004. Evolución Histórica de la construcción con yeso. *Informes de la Construcción,* Vol. 56, nº 493, pp. 5–11. Madrid: Csic.

Earthen Architecture: Past, Present and Future – Mileto, Vegas, García Soriano & Cristini (Eds)
© 2015 Taylor & Francis Group, London, ISBN 978-1-138-02711-4

Contributions of traditional façades to the thermal comfort

J. Vercher, F. Cubel, C. Lerma & Á. Mas
Department of Architectural Constructions, Universitat Politècnica de València, Valencia, Spain

E. Gil
Department of Continuum Media Mechanics and Theory of Structures, Universitat Politècnica de València, Valencia, Spain

ABSTRACT: Spanish vernacular architecture has been based essentially on stone, brick or rammed earth bearing walls. Enclosures could solve hygrothermal comfort using enough thickness. However, nowadays we have an extensive knowledge about physical phenomena and the façade's behavior against them, which has allowed us to reduce considerably the thickness of the enclosures. The vernacular architecture had thicker walls, but they were also efficient, and especially more sustainable. A theoretical study comparing the thermal performance of traditional enclosures, with different materials and thicknesses, are analyzed. Finally, a real case of an historic building has been studied by comparing the actual measurements taken in situ with the realized simulations, in the case of the corner section of The Seminary School of Corpus Christi of Valencia (Spain). This paper evaluates the solution for indoor climate of buildings with an enclosure composed of stone, rammed earth and brick.

1 INTRODUCTION

Heat transfer in buildings is produced by conduction, convection or radiation. The heat flux is a vector parameter (Kakaç 1994) analyzed by computer calculation using numerical simulations.

This study deals mainly with the transmission of heat by convection, between the outdoor and indoor air, and by conduction between different materials.

Similar studies have also been conducted with recent façades. Heat flow is analyzed through different materials (Vercher, 2013).

1.1 *Thermal impedance of the enclosure*

Impedance is the heat transfer of an enclosure in a dynamic way. Specifically, it is the ratio between the wave amplitude of external temperature and the wave amplitude of the heat flux.

Buildings façades should not only be considered as isolating systems. They also filter and flatten temperature waves. The temperature outside the building may fluctuate due to wind and the clouds that hide the sun.

The enclosure acts as a filter, so that these variations of external temperature are reduced in the interior. The façades with high thermal inertia do the same with daily temperature waves or with heat and cold waves. In summer, these enclosures prevent that the building increases substantially the temperature when the outer temperature is very high. In winter, these walls accumulate heat during daylight hours and emit this energy inwardly throughout the day (White 2011).

We must remember that the building façades suffer the action of sunlight periodically, so the thermal waves have a period of 24 hours.

Damping and phase shift of the waves produced by enclosure walls depend on the characteristics of the construction materials.

The sun's energy can be considered uniform. However, the energy density incident on the walls depends on the orientation of the façades, the season and weather conditions (Grinzato 1997).

1.2 *Material characterization*

The properties of the different materials involve a different response to temperature conditions, humidity, etc... In order to model a traditional enclosure is necessary to know the density of the material used (in our case, stone, brick and rammed earth), the specific heat, the thermal conductivity and the film coefficient to account for the phenomenon of convection which occurs along the surface of the enclosure.

1.3 *Application of thermography*

Sometimes it is difficult to perform laboratory tests in a historic building. Until now infrared thermography has been used to provide information of surface temperature of the materials, but a better analysis can provide more data to our research. Measuring and quantifying the parameters offered by this non-destructive technique is possible to identify different materials of the same enclosure or check differences on the same material (Lerma, 2012).

2 METHODOLOGY

The research is divided into two parts, an initial one theoretical and a second one with application to a real building.

The theoretical study compares the thermal performance and comfort inside buildings, based on different criteria. Various materials have been assessed, such as rammed earth, stone and brick. Furthermore, several wall thicknesses are analyzed, with dimensions common in the traditional building (20, 40, 60 and 80 cm).

The simulation of the thermal behavior allows calculating the thermal conditions of the enclosure for any thickness and material, simply by changing the parameters.

A history of real temperatures has been simulated with multiple steps to perform the calculations, covering a total interval of 60 hours. The graphs show the first 36 hours, as the solution rapidly converges because temperature cycles are 24 hours. And finally, a real case of a historic building has been studied.

2.1 *Material properties*

In order to get an approximate response to reality is very important simulating the actual properties of the different materials. The simulation considers the exterior temperature variations and the material properties.

The materials listed in Table 1 are the most common in traditional building. Brick, rammed earth and limestone have been used in the theoretical calculations and the simulation of real case, to compose the enclosures.

Concrete, wood, and sandy soil have been used to simulate the real case, where a more complete model is shown, with a stone base that starts from the ground and a wood pitched roofs.

2.2 *Simulation of the enclosures*

Simulations and calculations performed in this work have been carried out with ANSYS, finite element program used in engineering and architecture able to study multiple variables simultaneously (Ansys, 2013). Heat has been considered as an energy flux for the calculation of the thermal comfort.

Table 1. Properties of the used materials.

Material	Density (D) Kg/m³	Specific heat (Cp) J/Kg°K	Thermal conductivity (λ) W/m°K
Ram. earth	2000	900	0.85
Brick	1800	921	0.81
Limestone	2600	920	2.10
Concrete	2400	805	1.63
Wood	700	1340	0.14
Sandy soil	1400	795	0.42

Table 2. Film coefficients used in the calculations.

Flux W/(m² k)					
Horizontal		Vertical			
EXT	INT	Upward		Downward	
		EXT	INT	EXT	INT
0.06	0.11	0.05	0.09	0.05	0.17

2.3 *Boundary conditions*

Temperature conditions are dynamic on the outside of the building. It has been considered a variation of 20°C over 12 hours in 24-hour cycles. This generates an external temperature wave that also generates a temperature wave on the inner surface of the building. This temperature wave has been analyzed.

The film coefficient is another variable in this calculation. This coefficient is used to quantify the influence of air, the enclosure surface and the fluid which occurs when the heat transfer is produced by convection. The values used in this research are listed in Table 2.

3 RESULTS

3.1 *Theoretical study*

In a first stage, a study of theoretical cases has been realized to determine the influence of the material and thickness of the enclosure.

Figure 1 compares in summary the behavior of the walls from the thermal point of view. This figure represents the temperature distribution throughout the section of a enclosure. Results have been obtained after a half-cycle of 12 hours. Four thicknesses and three different materials have been analyzed.

The thermal behavior of a façade depends greatly on the properties of the materials that compose it. Figure 2 shows the evolution of temperatures in the inner surface within the enclosure depending on the time, with the three tested materials. These cases correspond with the lower row of Figure 1.

Figure 3 shows the evolution of the temperature in the inner surface according the different studied thicknesses (from 0.2 to 0.8 m), for the same material (limestone). These cases correspond with the first column specimens of Figure 1.

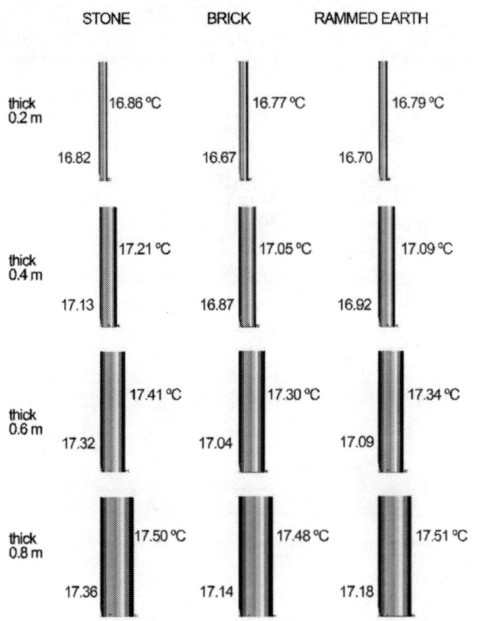

Figure 1. Comparison of enclosures with different materials and thicknesses (Authors).

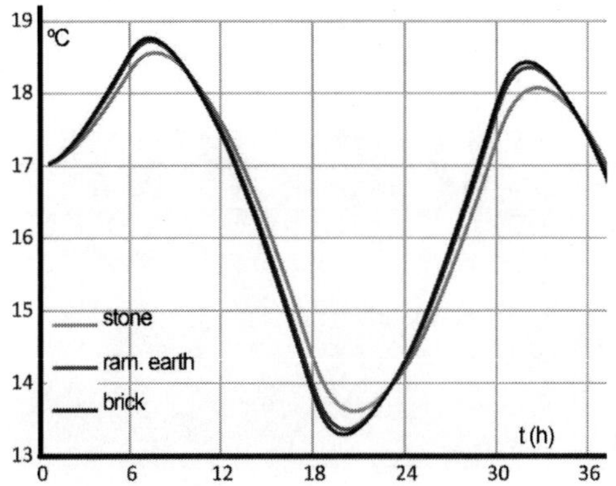

Figure 2. Thermal behaviour of the three enclosures (0.8 m thick): limestone, rammed earth and brick. Figure property of the authors.

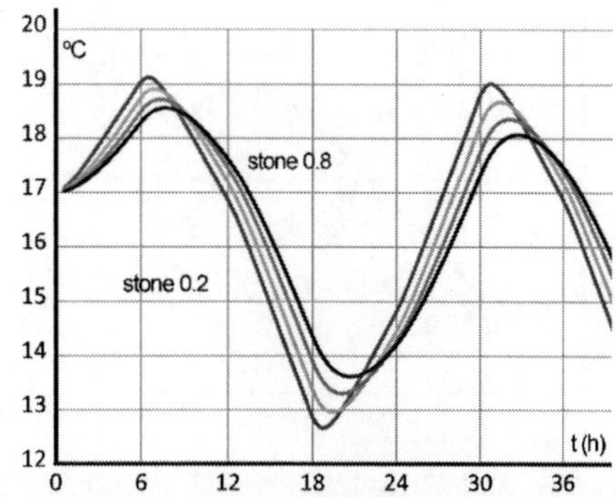

Figure 3. Evolution of temperature in a limestone enclosure for different thicknesses (Authors).

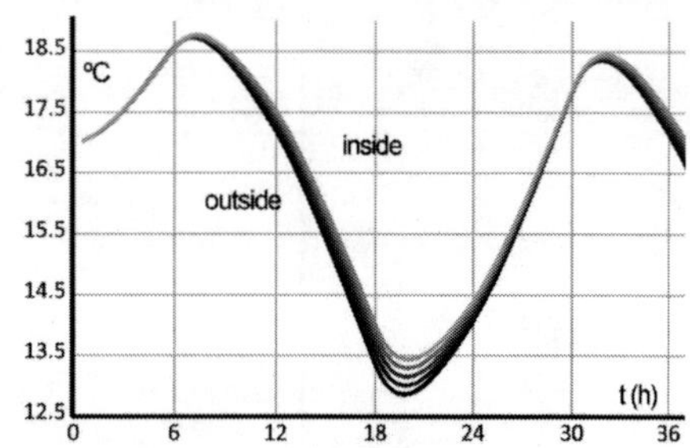

Figure 4. Temperature at different points in a brick enclosure (0.8 m thick) (Authors).

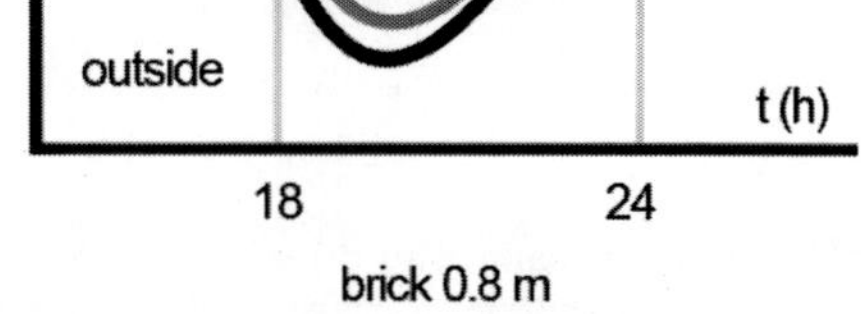

Figure 5. Detail of Figure 4 (Authors).

In the theoretical studied cases, the initial temperature of the enclosure is known and the same for all points. The temperature distribution throughout the thickness of the enclosure also has been evaluated. Changes of temperature are lower at the inner side of the enclosure and greater at the outer side, as shown in Figure 4 and in detail in Figure 5.

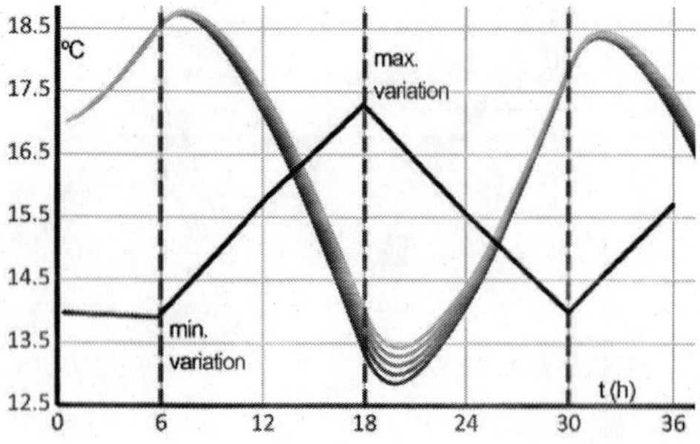

Figure 6. Variations in temperature between the outer and the inner surface of a brick enclosure (Authors).

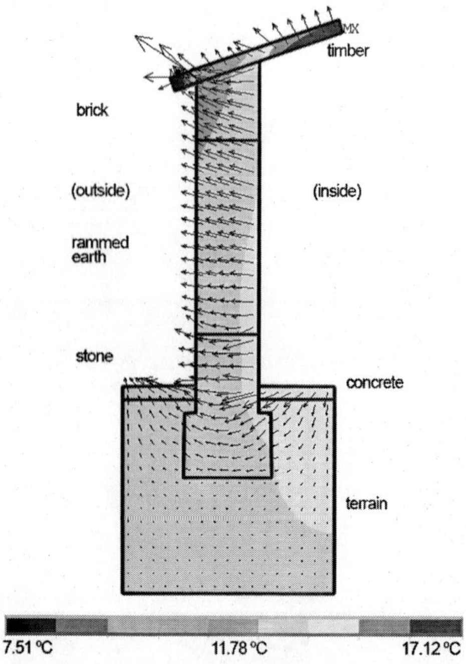

Figure 7. Temperature distribution and heat flux (2D) (Authors).

The minimal changes in temperature between the outer and the inner surface of the enclosure are produced at 6 hours. The maximum temperature variations are produced 12 hours later, as shown in Figure 6. In this graph, the brick is used as material.

3.2 *Case study*

In the second part of this paper, the case of the Seminary School of Corpus Christi in Valencia (Spain) has been studied. This studio has been performed in 2D and 3D.

In Figure 7 a portion of the section of the building is shown. It consists of a rammed earth wall supported by a stone base that starts from the ground. At the top, the wall changes its construction, having an area with brick. And the building is topped by wooden roof.

In this figure, the thermal flux is represented by arrows. The larger arrows indicate that the flux has a higher intensity, where the temperature is lost from the inside outwards.

By simulating it is possible to know the temperature at each point depending on the analyzed parameters (thickness, temperature, materials...).

Figure 8 shows the model in three dimensions from two points of view. A corner has been modeled with a length of 3 meters on each side.

To verify the results obtained by the simulation, these are compared with actual thermography taken in the building. Figure 9 shows a thermal image of the Seminary School of Corpus Christi of Valencia. The façade is composed of three materials from the ground to the top: a stone base at the bottom, a rammed earth wall (below) and brick (above) and a wooden roof.

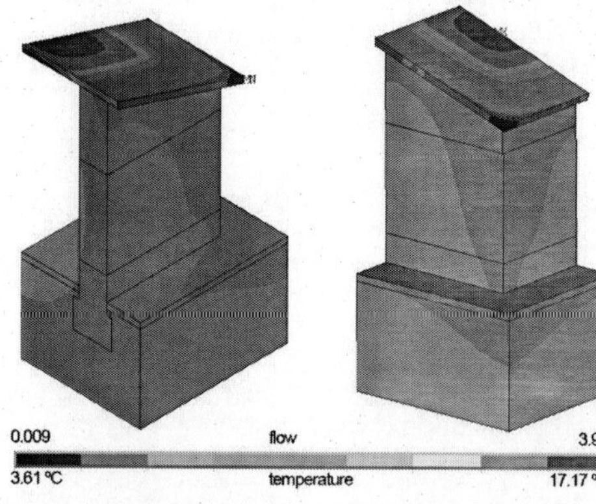

Figure 8. Temperature distribution (3D) (Authors).

Figure 9. Thermography of the Seminary School of Corpus Christi (Authors).

The correspondence of the temperature distribution in the corner between the simulation and reality can be observed in these figures.

4 CONCLUSIONS

This research approaches the analysis of thermal comfort in buildings that have enclosures that continue the building tradition. The usual materials in traditional buildings (limestone, brick, rammed earth) has been studied with different thicknesses (0.2, 0.4, 0.6 and 0.8 m).

The minimal changes in temperature between the outer and the inner surface of the enclosure are produced at 6 hours and the maximum temperature variations are produced 12 hours later. Every 12 hours this phenomenon occurs (Figure 6). It has been observed and quantified temperature variations in the outer and inner surfaces of an enclosure. This work reflects that while on the inner surface the temperature is stabilized, on the exterior surface the temperature varies widely and in a longer way.

Other parameters being equal, the stone is the traditional material with a higher thermal comfort (Fig. 2).

By increasing the wall thickness increases the thermal inertia, and temperature variations are smaller and softer. With reduced thickness, the differences are larger and more abrupt. Figure 1 shows how the wall reaches a higher temperature by increasing its thickness. A higher thermal inertia also reflects a higher gap in the indoor wave temperature (Fig. 3).

By analyzing the temperature through the thickness of an enclosure with 24-hour cycles, it is found that the inner surface is less cooled, and also is slower cooled and faster heated (Fig. 4).

When the temperature is lower on the outside, the temperature difference between the outer and the inner surface of the wall is maximal. When the temperature rises outside, this temperature difference is smaller (Fig. 6). The study of a real case shows that the temperature is not lost through the ground, because it has a high thermal inertia. The heat loss through the enclosure is uniform in all three building materials. In this case the maximal loss of temperature is located at the meeting of the façade with the roof (Fig. 7).

REFERENCES

Ansys. 2013. *Ansys 15.0 Help Manual*. Ansys Inc. Ansys Academic Research V15.0. USA.

Avdelidis, N.P. 2004. Moropoulou, A. Applications of infrared thermography for the investigation of historic structures. *Journal of Cultural Heritage* 5 pp. 119–127.

Blanco Álvarez, F. 2011. Personal website: http://www.etsimo.uniovi.es/usr/fblanco/.

González Cruz, E. M. 2003. *Selección de materiales en la concepción arquitectónica bioclimática*. Instituto de investigaciones de la Facultad de Arquitectura y Diseño, Universidad de Zulia, Venezuela.

Grinzato, E., Vavilov, V., Kauppinen, T. 1997. Quantitative infrared thermography in buildings. *Energy and Buildings* 29, pp. 1–9.

Kakaç, S., Yener, Y. 1994. *Heat transfer*. USA: CRC Press.

Lerma, C. 2012. *Análisis arquitectónico y constructivo del Real Colegio de Corpus Christi de Valencia*. PhD Thesis. Valencia: Universitat Politècnica de València.

Lerma, J. L., Mileto, C., Vegas, F., Cabrelles, M. 2007. Visible and thermal IR documentation of a masonry brickwork Building. *XXI International CIPA Symposium*, Athens, Greece.

Vercher, J., Lerma, C., Vidal, M., Gil, E. Analysis of Energy Efficiency in Construction Solutions at the Façade-slab Connection. *Advanced Materials Research*, vol 787, pp. 731–735.

Earthen Architecture: Past, Present and Future – Mileto, Vegas, García Soriano & Cristini (Eds)
© 2015 Taylor & Francis Group, London, ISBN 978-1-138-02711-4

Author index